T0238533

Communications in Computer and Information Science 761

Commenced Publication in 2007
Founding and Former Series Editors:
Alfredo Cuzzocrea, Orhun Kara, Dominik Ślęzak, and Xiaokang Yang

More information about this series at http://www.springer.com/series/7899

Minrui Fei · Shiwei Ma · Xin Li
Xin Sun · Li Jia · Zhou Su (Eds.)

Advanced Computational Methods in Life System Modeling and Simulation

International Conference on Life System Modeling
and Simulation, LSMS 2017
and International Conference on Intelligent Computing
for Sustainable Energy and Environment, ICSEE 2017
Nanjing, China, September 22–24, 2017
Proceedings, Part I

 Springer

Editors
Minrui Fei
Shanghai University
Shanghai
China

Shiwei Ma
Shanghai University
Shanghai
China

Xin Li
Shanghai University
Shanghai
China

Xin Sun
Shanghai University
Shanghai
China

Li Jia
Shanghai University
Shanghai
China

Zhou Su
Shanghai University
Shanghai
China

ISSN 1865-0929   ISSN 1865-0937   (electronic)
Communications in Computer and Information Science
ISBN 978-981-10-6369-5   ISBN 978-981-10-6370-1   (eBook)
DOI 10.1007/978-981-10-6370-1

Library of Congress Control Number: 2017951411

Printed on acid-free paper

This Springer imprint is published by Springer Nature
The registered company is Springer Nature Singapore Pte Ltd.
The registered company address is: 152 Beach Road, #21-01/04 Gateway East, Singapore 189721, Singapore

Preface

This book constitutes the proceedings of the 2017 International Conference on Life System Modeling and Simulation (LSMS 2017) and the 2017 International Conference on Intelligent Computing for Sustainable Energy and Environment (ICSEE 2017), which were held during September 22–24, in Nanjing, China. These two international conference series aim to bring together international researchers and practitioners in the fields of advanced methods for life system modeling and simulation as well as advanced intelligent computing theory and methodologies and engineering applications for sustainable energy and environment. The two conferences held this year were built on the success of previous LSMS and ICSEE conferences held in Shanghai and Wuxi, respectively. The success of the LSMS and ICSEE conference series were also based on several large-scale RCUK/NSFC funded UK–China collaborative projects on sustainable energy and environment, as well as a recent government funded project on the establishment of the UK-China University Consortium in Engineering Education and Research, with an initial focus on sustainable energy and intelligent manufacturing.

At LSMS 2017 and ICSEE 2017, technical exchanges within the research community took the form of keynote speeches, panel discussions, as well as oral and poster presentations. In particular, two workshops, namely, the Workshop on Smart Grid and Electric Vehicles and the Workshop on Communication and Control for Distributed Networked Systems, were held in parallel with LSMS 2017 and ICSEE 2017, focusing on the two recent hot topics on green and sustainable energy systems and electric vehicles and distributed networked systems for the Internet of Things.

The LSMS 2017 and ICSEE 2017 conferences received over 625 submissions from 14 countries and regions. All papers went through a rigorous peer review procedure and each paper received at least three review reports. Based on the review reports, the Program Committee finally selected 208 high-quality papers for presentation at LSMS 2017 and ICSEE 2017. These papers cover 22 topics, and are included in three volumes of CCIS proceedings published by Springer. This volume of CCIS includes 59 papers covering 7 relevant topics.

Located at the heartland of the wealthy lower Yangtze River region in China and being the capital of several dynasties, kingdoms, and republican governments dating back to the 3rd century, Nanjing has long been a major center of culture, education, research, politics, economy, transport networks, and tourism. In addition to academic exchanges, participants were treated to a series of social events, including receptions and networking sessions, which served to build new connections, foster friendships, and forge collaborations. The organizers of LSMS 2017 and ICSEE 2017 would like to acknowledge the enormous contribution of the Advisory Committee, who provided guidance and advice, the Program Committee and the numerous referees for their efforts in reviewing and soliciting the papers, and the Publication Committee for their editorial work. We would also like to thank the editorial team from Springer for their support and guidance. Particular thanks are of course due to all the authors, as without their high-quality submissions and presentations the conferences would not have been successful.

Finally, we would like to express our gratitude to our sponsors and organizers, listed on the following pages.

September 2017

Bo Hu Li
Sarah Spurgeon
Mitsuo Umezu
Minrui Fei
Kang Li
Dong Yue
Qinglong Han
Shiwei Ma
Luonan Chen
Sean McLoone

Organization

Sponsors

China Simulation Federation (CSF), China
Chinese Association for Artificial Intelligence (CAAI), China
IEEE Systems, Man & Cybernetics Society Technical Committee on Systems Biology, USA
IEEE CC Ireland Chapter, Ireland

Technical Support Organization

National Natural Science Foundation of China (NSFC), China

Organizers

Shanghai University, China
Queen's University Belfast, UK
Nanjing University of Posts and Telecommunications, China
Southeast University, China
Life System Modeling and Simulation Technical Committee of CSF, China
Embedded Instrument and System Technical Committee of China Instrument and Control Society, China
Intelligent Control and Intelligent Management Technical Committee of CAAI, China

Co-sponsors

Shanghai Association for System Simulation, China
Shanghai Association of Automation, China
Shanghai Instrument and Control Society, China
Jiangsu Association of Automation, China

Co-organizers

Swinburne University of Technology, Australia
Queensland University of Technology, Australia
Tsinghua University, China
Harbin Institute of Technology, China
China State Grid Electric Power Research Institute, China
Chongqing University, China
University of Essex, UK
Cranfield University, UK
Peking University, China

Nantong University, China
Shanghai Dianji University, China
Jiangsu Engineering Laboratory of Big Data Analysis and Control for Active
    Distribution Network, China
Shanghai Key Laboratory of Power Station Automation Technology, China

Honorary Chairs

Li, Bo Hu, China
Spurgeon, Sarah, UK
Umezu, Mitsuo, Japan

Advisory Committee Members

Bai, Erwei, USA
Ge, Shuzhi, Singapore
He, Haibo, USA
Hu, Huosheng, UK
Huang, Biao, Canada
Hussain, Amir, UK
Liu, Derong, USA
Mi, Chris, USA

Nikolopoulos,
    Dimitrios S., UK
Pardalos, Panos M., USA
Pedrycz, Witold, Canada
Polycarpou, Marios M.,
    Cyprus
Qin, Joe, HK
Scott, Stan, UK

Tan, KC, Singapore
Tassou, Savvas, UK
Thompson, Stephen, UK
Wang, Jun, HK
Wang, Zidong, UK
Wu, Qinghua, China
Xue, Yusheng, China
Zhang, Lin, China

General Chairs

Fei, Minrui, China
Li, Kang, UK
Yue, Dong, China

International Program Committee

Chairs

Chen, Luonan, Japan
Han, Qinglong, Australia
Ma, Shiwei, China
McLoone, Sean, UK

Local Chairs

Chiu, Min-Sen, Singapore
Cui, Shumei, China
Deng, Mingcong, Japan
Ding, Yongsheng, China
Ding, Zhengtao, UK
Fang, Qing, Japan

Fridman, Emilia, Israel
Gao, Furong, HK
Gu, Xingsheng, China
Guerrero, Josep M.,
    Demark
Gupta, Madan M., Canada

Hunger, Axel, Germany
Lam, Hak-Keung, UK
Liu, Wanquan, Australia
Luk, Patrick, UK
Maione, Guido, Italy
Park, Jessie, Korea

Peng, Chen, China
Su, Zhou, China
Tian, Yuchu, Australia
Xu, Peter, New Zealand

Yang, Taicheng, UK
Yu, Wen, Mexico
Zeng, Xiaojun, UK
Zhang, Huaguang, China

Zhang, Jianhua, China
Zhang, Wenjun, Canada
Zhao, Dongbin, China

Members

Andreasson, Stefan, UK
Adamatzky, Andy, UK
Altrock, Philipp, USA
Asirvadam, Vijay S.,
   Malaysia
Baig, Hasan, UK
Baker, Lucy, UK
Barry, John, UK
Best, Robert, UK
Bu, Xiongzhu, China
Cao, Jun, UK
Cao, Yi, UK
Chang, Xiaoming, China
Chen, Jing, China
Chen, Ling, China
Chen, Qigong, China
Chen, Rongbao, China
Chen, Wenhua, UK
Cotton, Matthew, UK
Deng, Jing, UK
Deng, Li, China
Deng, Shuai, China
Deng, Song, China
Deng, Weihua, China
Ding, Yate, UK
Ding, Zhigang, China
Du, Dajun, China
Du, Xiangyang, China
Ellis, Geraint, UK
Fang, Dongfeng, USA
Feng, Dongqing, China
Feng, Zhiguo, China
Foley, Aoife, UK
Fu, Jingqi, China
Gao, Shouwei, China
Gu, Dongbin, UK
Gu, Juping, China
Gu, Zhou, China
Guo, Lingzhong, UK

Han, Bo, China
Han, Xuezheng, China
Heiland, Jan, Germany
Hong, Xia, UK
Hou, Weiyan, China
Hu, Liangjian, China
Hu, Qingxi, China
Hu, Sideng, China
Huang, Sunan, Singapore
Huang, Wenjun, China
Hwang, Tan Teng,
   Malaysia
Jia, Dongyao, UK
Jiang, Lin, UK
Jiang, Ming, China
Jiang, Ping, China
Jiang, Yucheng, China
Kuo, Youngwook, UK
Laverty, David, UK
Li, Chuanfeng, China
Li, Chuanjiang, China
Li, Dewei, China
Li, Donghai, China
Li, Guozheng, China
Li, Jingzhao, China
Li, Ning, China
Li, Tao, China
Li, Tongtao, China
Li, Weixing, China
Li, Xin, China
Li, Xinghua, China
Li, Yunze, China
Li, Zhengping, China
Lin, Zhihao, China
Lino, Paolo, Italy
Liu, Chao, France
Liu, Guoqiang, China
Liu, Mandan, China
Liu, Shirong, China

Liu, Shujun, China
Liu, Tingzhang, China
Liu, Xianzhong, China
Liu, Yang, China
Liu, Yunhuai, China
Liu, Zhen, China
Ljubo, Vlacic, Australia
Lu, Ning, Canada
Luan, Tom, Australia
Luo, Jianfei, China
Ma, Hongjun, China
McAfee, Marion, Ireland
Menary, Gary, UK
Meng, Xianhai, UK
Menhas, Muhammad
   Ilyas, Pakistan
Menzies, Gillian, UK
Naeem, Wasif, UK
Nie, Shengdong, China
Niu, Yuguang, China
Nyugen, Bao Kha, UK
Ouyang, Mingsan, China
Oyinlola, Muyiwa, UK
Pan, Hui, China
Pan, Ying, China
Phan, Anh, UK
Qadrdan, Meysam, UK
Qian, Hua, China
Qu, Yanbin, China
Raszewski, Slawomir, UK
Ren, Wei, China
Rivotti, Pedro, UK
Rong, Qiguo, China
Shao, Chenxi, China
Shi, Yuntao, China
Smyth, Beatrice, UK
Song, Shiji, China
Song, Yang, China
Su, Hongye, China

Sun, Guangming, China
Sun, Xin, China
Sun, Zhiqiang, China
Tang, Xiaoqing, UK
Teng, Fei, UK
Teng, Huaqiang, China
Trung, Dong, UK
Tu, Xiaowei, China
Vlacic, Ljubo, UK
Wang, Gang, China
Wang, Jianzhong, China
Wang, Jihong, UK
Wang, Ling, China
Wang, Mingshun, China
Wang, Shuangxin, China
Wang, Songyan, China
Wang, Yaonan, China
Wei, Kaixia, China
Wei, Lisheng, China
Wei, Mingshan, China
Wen, Guihua, China
Wu, Jianguo, China
Wu, Jianzhong, UK

Wu, Lingyun, China
Wu, Zhongcheng, China
Xie, Hui, China
Xu, Sheng, China
Xu, Wei, China
Xu, Xiandong, UK
Yan, Huaicheng, China
Yan, Jin, UK
Yang, Aolei, China
Yang, Kan, USA
Yang, Shuanghua, UK
Yang, Wankou, China
Yang, Wenqiang, China
Yang, Zhile, UK
Yang, Zhixin, Macau
Ye, Dan, China
You, Keyou, China
Yu, Ansheng, China
Yu, Dingli, UK
Yu, Hongnian, UK
Yu, Xin, China
Yuan, Jin, China
Yuan, Jingqi, China

Yue, Hong, UK
Zeng, Xiaojun, UK
Zhang, Dengfeng, China
Zhang, Hongguang, China
Zhang, Jian, China
Zhang, Jingjing, UK
Zhang, Lidong, China
Zhang, Long, UK
Zhang, Qianfan, China
Zhang, Xiaolei, UK
Zhang, Yunong, China
Zhao, Dongya, China
Zhao, jun, China
Zhao, Wanqing, UK
Zhao, Xiaodong, UK
Zhao, Xingang, China
Zheng, Xiaojun, UK
Zhou, Huiyu, UK
Zhou, Wenju, China
Zhou, Yu, China
Zhu, Yunpu, China
Zong, Yi, Demark
Zuo, Kaizhong, China

Organization Committee

Chairs

Li, Xin, China
Wu, Yunjie, China
Naeem, Wasif, UK
Zhang, Tengfei, China
Cao, Xianghui, China

Members

Chen, Ling, China
Deng, Li, China
Du, Dajun, China
Jia, Li, China
Song, Yang, China
Sun, Xin, China
Xu, Xiandong, China
Yang, Aolei, China
Yang, Banghua, China
Zheng, Min, China
Zhou, Peng, China

Special Session Chairs

Wang, Ling, China
Meng, Fanlin, UK

Publication Chairs

Zhou, Huiyu, UK
Niu, Qun, China

Publicity Chairs

Jia, Li, China
Yang, Erfu, UK

Registration Chairs

Song, Yang, China
Deng, Li, China

Secretary-General

Sun, Xin, China
Wu, Songsong, China
Yang, Zhile, UK

Contents

Medical Apparatus and Clinical Applications

Bionics Control Methods, Algorithms and Apparatus

Modeling and Simulation of Life Systems

Data Driven Analysis

Image and Video Processing

Biomedical Signal Processing

Research of Rectal Pressure Signal Preprocessing Based on Improved FastICA Algorithm

Peng Zan[1], Yankai Liu[1], Suqin Zhang[2], Chundong Zhang[1], Hua Wang[1], and Zhiyuan Gao[1(✉)]

[1] Shanghai Key Laboratory of Power Station Automation Technology, School of Mechatronics Engineering and Automation, Shanghai University, Shanghai, China
gaozhiyuan86@shu.edu.cn
[2] Naval Aeronautical University Qingdao Campus, Qingdao, China

Abstract. In view of some shortcomings of the existing rectal function diagnosis method, we propose that use the artificial anal sphincter system to collect the human rectal pressure signal, and then achieve the diagnosis of human rectal status through the rectal function diagnosis model. Since the collected signal is not pure rectal pressure signal, the single-dimensional pressure signal is extended to a multidimensional time series by phase space reconstruction. And then preprocessing of the reconstructed signal is carried out by the improved fifteenth order Newton iteration Fast ICA algorithm. The improved algorithm is simulated and the better separation effect is realized, proving the feasibility of the algorithm.

Keywords: Rectal pressure signal · FastICA · Signal preprocessing

1 Introduction

Fecal incontinence is that the anal sphincter loses control capacity of fecal and gas discharge, which is a common clinical disease. There are a variety of treatments for fecal incontinence, including conservative treatment represented by controlling diet, surgical treatment represented by colostomy, and other treatments between them [1, 2]. But they all have some deficiencies. So bionic artificial anal sphincter system that do not change the body's traditional defecation model came into being, which will greatly reduce the pain of patients with fecal incontinence [3].

At present, the diagnosis of rectal function are: digital rectal examination, proctoscopy, rectal pressure measurement, X-ray examination, CT, MRT and other methods [4]. However, there are more or less drawbacks. This article designs a new type of rectal function diagnosis model, which is characterized by the use of artificial anal sphincter system implanted in human body for intestinal pressure information in a real-time, sustained and convenient way. The one-dimensional signal is reconstructed into multidimensional by phase space reconstruction [5], and then the preprocessing of the signal is realized by using the improved FastICA. Finally, a series of further signal

© Springer Nature Singapore Pte Ltd. 2017
M. Fei et al. (Eds.): LSMS/ICSEE 2017, Part I, CCIS 761, pp. 3–12, 2017.
DOI: 10.1007/978-981-10-6370-1_1

processing of the rectal pressure signal after pretreatment is used to realize the rectal function diagnosis. In addition, the effectiveness of the method is proved by simulation experiments.

2   The Principle of Rectal Pressure Signal Preprocessing

The rectal function diagnostic model relies on the platform of the artificial anal sphincter system and obtains rectal pressure information through the sensor located in the sensing bag around the rectum continuously. Since the signals we receive exist some Interference signals (breathing, muscle movement, noise, etc.) and various interference signals have independent sources. Therefore, this paper uses FastICA algorithm for rectal pressure signal pretreatment. For characteristic of the algorithm that the initial value is more sensitive, we adopt an improved fifteenth order convergence of the iterative algorithm [6]. FastICA algorithm requires multidimensional time series, but the signal measured is one-dimensional pressure signal. Thus, phase space reconstruction technology is adopted to rebuild the one-dimensional pressure signal to multi-dimensional. Rectal pressure signal preconditioning will lay the foundation for the realization of rectal function diagnostic model. The principle of rectal pressure signal preprocess is shown in Fig. 1.

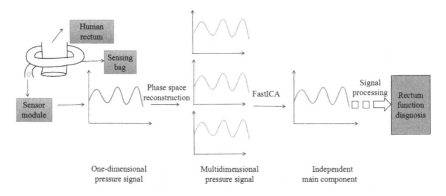

Fig. 1. The principle of rectal pressure signal preprocessing

3   Algorithm Analysis of Rectal Pressure Signal Preprocessing

3.1   Phase Space Reconstruction

Phase space reconstruction is a method proposed by Takens to construct the phase space structure of the original system by one-dimensional time series. The delay coordinate method is commonly used. The selection of time delay τ and embedding dimension m is very important. In this paper, the embedded dimension and time delay automatic algorithm is used to reconstruct the one-dimensional pressure sequence [7]. The principle is as follows:

(1) Set the pressure sequence as $X = \{xi(t)\}$, $i = 1,2...N$. Let $m = m_0$ and τ varies from small to large. Construct X as vectors $\{y_i\}$, $i = 1,2...M$, $y_i = (x_i, x_{i+1}...x_{i+}(m-1)\tau)$, $M = N - (m-1)\tau$. Then Calculate the average amount of displacement:

$$s(\tau) = \frac{1}{M}\sum_{i=1}^{M}\sqrt{\sum_{j=1}^{m-1}(y_{i+j\tau} - y_i)^2} \tag{1}$$

Next it will be derivative of τ. When the value is close to 0, the value of $s(\tau)$ reaches the saturation.

Substitute the calculated τ into $\Gamma_{-\text{test}}$ to calculate the corresponding best m. Set $z = f(x_1, x_2...x_m) + \gamma$. Give a value of m, and then reconstruct vectors space of X.

$$\varepsilon_i = \{x(i), x((i+1)\tau),...x((i+m-1)\tau)\} \tag{2}$$

(2) Set $z_i = x((i+m)\tau)$, $i = 1,2..M$. Create M groups of input, output vector pairs. Find the neighbor vectors of p in the vector space. Calculate

$$\begin{cases} dx(h) = \frac{1}{p}\sum_{h=1}^{P}\frac{1}{M}\sum_{i=1}^{M}|\varepsilon(N(i,p)) - \varepsilon(i)|^2 \\ dz(h) = \frac{1}{p}\sum_{h=1}^{P}\frac{1}{2M}\sum_{i=1}^{M}[z(N(i,p)) - z(i)]^2 \end{cases} \tag{3}$$

Use linear interpolation $dz = Adx + \Gamma$ to estimate the approximate value of γ, Γ. Increase m, and then repeat (1). Calculate the corresponding the minimum value of the approximation of τ, γ, where m, τ is the optimal value.

3.2 FastICA Algorithm

FastICA algorithm is presented by Hyvärinen et al., University of Helsinki, Finland [8]. In this paper, an improved FastICA algorithm based on the largest negative entropy is used. For the shortcoming of basic Newton iteration Fast ICA algorithm that is more sensitive to the initial value, the improved algorithm takes the modified form of the Fifth - order Newton iteration with better convergence rate [6].

Before carrying out the FastICA, data needs to be removed mean and whited.

$$x = x' - E\{x'\} \tag{4}$$

$$E\{yy^T\} = I \tag{5}$$

The objective function of Fast ICA algorithm based on negative entropy is:

$$J(W) = \left[E\{G(W^T Z)\} - E\{G(V)\} \right]^2 \tag{6}$$

The algorithm estimates an independent component of the source signal by maximizing the objective function. And the maximum value of J (W) is obtained at the extreme point of $E\{G(W^T Z)\}$. According to the Lagrangian condition, the extreme value of $E\{G(W^T Z)\}$ can be obtained by solving the following condition under the constraint condition of $E\{(W^T Z)^2\} = \| W \|^2 = 1$:

$$E\{Zg(W^T Z)\} + \beta W = 0 \tag{7}$$

$\beta = E\{W_0^T Zg(W_0^T Z)\}$, W_0 is the initial value of W, and g (.) is the derivative of G (.).

The improved Newton iteration method is as follows:

$$\begin{cases} x_{n+1}^* = x_n - \dfrac{f(x_n)}{f'(x_n)} \\ z_n = x_n - \dfrac{2f(x_n)}{f'(x_{n+1}^*) + f'(x_n)} \\ y_n = z_n - \dfrac{f(z_n)}{f'(x_{n+1}^*)} \\ x_{n+1} = y_n - \dfrac{f(y_n)f'(y_n)}{f'^2(y_n) - f''(y_n)f(y_n)/2} \end{cases} \tag{8}$$

The improved iterative scheme is at least 15th order convergent, and the iterative efficiency is much larger than the Newton iteration method. Use the modified Newton iteration method to solve Eq. (7):

$$W_{k+1} = W_1 - \frac{\left[E\{Zg(W_1^T Z)\} + \beta W_1 \right] \left[E\{g'(W_1^T Z)\} + \beta \right]}{\left[E\{g'(W_1^T Z)\} + \beta \right]^2 - \frac{\left[E\{g''(W_1^T Z)\} \right] \left[E\{Zg(W_1^T Z)\} + \beta W_1 \right]}{2}} \tag{9}$$

Among them,

$$\begin{cases} W_1 = W_2 - \dfrac{E\{Zg(W_2^T Z)\} + \beta W_2}{E\{g'(W_3^T Z)\} + \beta} \\ W_2 = W - \dfrac{2[E\{Zg(W^T Z)\} + \beta W]}{[E\{g'(W_3^T Z)\} + \beta] + [E\{g'(W^T Z)\} + \beta]} \\ W_3 = W - \dfrac{E\{Zg(W^T Z)\} + \beta W}{E\{g'(W^T Z)\} + \beta} \end{cases} \tag{10}$$

4   Experimental and Simulation Analysis

In order to validate the improved FastICA method in this paper, simulation experiments are carried out. First, four common non-Gaussian signals are generated: sine wave, square wave, sawtooth wave, and a random noise. They are expressed as: sine wave

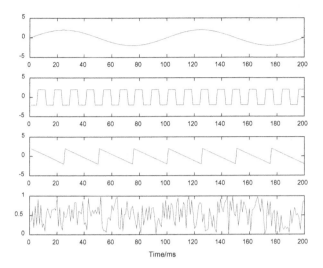

Fig. 2. Source signal waveform

signal $s_1 = 2 * \sin (0.02 * pi * n)$; square wave signal $s_2 = 2 * square (100 * t, 50)$; sawtooth signal $(1, -1, n)$; $s_3 = 2 * [a, a, a, a, a, a, a, a]$; random noise is $s_4 = rand (1, N)$. The waveform of the source signal is shown in Fig. 2.

The sinusoidal signal is viewed as a valid signal, and the others are viewed as three kinds of interference signals. Mix the four signals into one-dimensional signals, as shown in Fig. 3. The mixing matrix is A = [0.7847 0.5379 0.0529 0.0891].

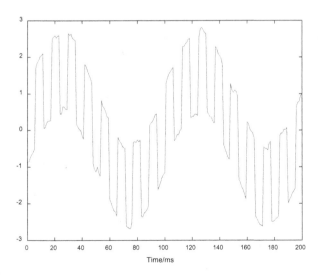

Fig. 3. One-dimensional mixed signal

The obtained one-dimensional mixed signal is phase-reconstructed to obtain the reconstructed 4-D signal as shown in Fig. 4.

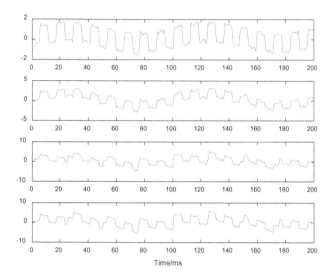

Fig. 4. Phase space reconstructed signal

Use the improved FastICA algorithm described above to obtain the Separation signal as shown in Fig. 5. As can be seen from the figure, the output of the classification of the waveform roughly unchanged, proving a better separation effect and achieving the desired purpose. The output order of the waveform of the components after the remolding changed and the amplitude and phase of each component changed contrast to the source signal. This is determined by the inherent characteristics of the ICA algorithm [6]. But we generally pay more attention to the waveform of the signal and the amplitude of the change doesn't affect our extraction and analysis.

The reconstructed signal is processed by using the basic FastICA algorithm to obtain the separation result as shown in Fig. 6. From Figs. 5 and 6, we can see that the two algorithms can separate the various components, and the separation effect is similar. Then, each algorithm is run five times under the condition that the initial vector is indefinite. The number of iterations and the running time of the two algorithms are compared and the results are shown in Table 1. It can be seen from the table that the improved FastICA algorithm has less iterations. The iterative stability is better and the running time has been reduced. It is proved that the improved algorithm is insensitive to the initial value and the effect is better.

And then the real human rectal pressure signal is used for the validation of the improved the FastICA algorithm. The data used in this paper is from the pressure data collected in tester whose gut is healthy by biological parameters telemetry capsule invented by Shanghai Jiao tong University [9]. Select the fragment of one of the signal sequences as shown in Fig. 7 as the mixed signal to be separated.

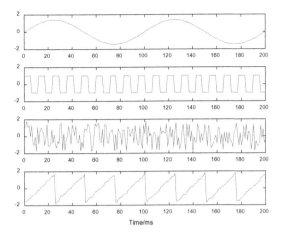

Fig. 5. Improved FastICA separation results

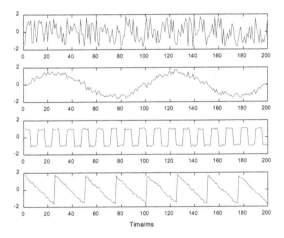

Fig. 6. Basic FastICA separation results

Table 1. Comparison of the number of iterations and run time of two algorithms

Number	Improved FastICA		Basic FastICA	
	Number of iterations	Run time	Number of iterations	Run time
1	152	0.817	714	1.646
2	134	0.834	384	0.957
3	167	0.822	217	0.867
4	179	0.834	206	0.805
5	157	0.818	482	0.943
Average value	157.8	0.825	400.6	1.0436

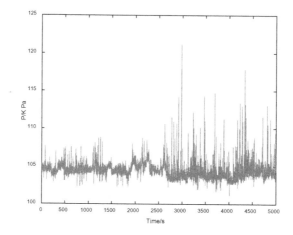

Fig. 7. Rectal mixed signal

According to the phase space reconstruction technique proposed above, the multidimensional pressure sequence reconstruction is carried out. Experiments show that $m = 3$, $\tau = 5$ are the optimal values, and the effect of reconstruction is the best. The reconstructed observation signal is shown in Fig. 8. The reconstructed multidimensional pressure sequence is separated by FastICA, and the results of the separation are shown in Fig. 9. It can be seen from the figure that the improved FastICA algorithm achieves the separation of the signal. The first component has the main trend and characteristics of the corresponding parent signal and it should be the pressure component caused by rectal contraction. The other two components should be abdominal muscle contraction, respiratory movement. The improved FastICA algorithm achieves the preconditioning of rectal pressure signals, laying the foundation for further online diagnosis of rectal function by further analysis and processing of pressure signals.

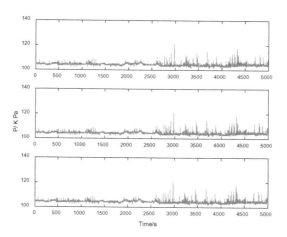

Fig. 8. The reconstructed observation signal

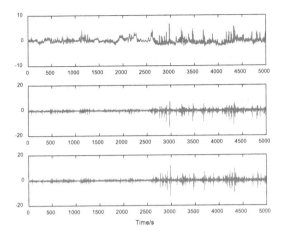

Fig. 9. Rectal pressure signal separation results

5 Conclusion

In order to achieve on-line, continuous detection of rectal function, using artificial anal sphincter system to collect the patient's rectal pressure signal for analysis is proposed. The adaptive pressure signal is reconstructed into multidimensional by phase space reconstruction technique, and then the improved FastICA algorithm is used to pre-process the signal. Finally, the simulations of the simulated data and real data are analyzed. The results show that the modified method can achieve the purpose of pretreatment, laying the foundation for the diagnosis of rectal function.

Acknowledgments. This work was supported by National Natural Science Foundation of China (No. 31570998).

References

1. Rezvan, A., Jakuswaldman, S., Abbas, M.A., et al.: Review of the diagnosis, management and treatment of fecal incontinence. Female Pelvic Med. Reconstr. Surg. **21**(1), 8–17 (2014)
2. Bochenska, K., Boller, A.M.: Fecal incontinence: epidemiology, impact, and treatment. Clin. Colon Rectal Surg. **29**(3), 264 (2016)
3. Fattorini, E., Brusa, T., Gingert, C., et al.: Artificial muscle devices: innovations and prospects for fecal incontinence treatment. Ann. Biomed. Eng. **44**(5), 1355–1369 (2016)
4. Saldana, R.N., Kaiser, A.M.: Fecal incontinence - challenges and solutions. World J. Gastroenterol. **23**(1), 11–24 (2017)
5. Song, J., Meng, D., Wang, Y.: Analysis of chaotic behavior based on phase space reconstruction methods. In: Sixth International Symposium on Computational Intelligence and Design, vol. 2, pp. 414–417. IEEE (2013)
6. Luo, W.J., Yuan, L.F., He, Y.G.: Improved fastICA algorithm based on fifteen-order Newton iteration. Comput. Eng. Appl. **52**(20), 108–113 (2016)

7. Otani, M., Jones, A.J.: Automated Embedding and the Creep Phenomenon in Chaotic Time Series (2001)
8. Huang, X.: An improved FastICA algorithm for blind signal separation and its application. In: International Conference on Image Analysis and Signal Processing, vol. 8556, pp. 1–4. IEEE (2013)
9. Li, Q.R., Yan, G.Z.: A non-invasive capsule and clinic experimentation for deteeting the physiological parameters of gastrointestinal tract. Beijing Biomed. Eng. 27(3), 281–285 (2008)

A Noncontact Measurement of Cardiac Pulse Based on PhotoPlethysmoGraphy

Xiaohua Wu, Xin Li[✉], Yulin Xu, and Lang Zhang

School of Mechatronics Engineering and Automation, Shanghai University,
Shanghai 200072, China
su_xinli@aliyun.com

Abstract. Heart rate measurement is important for monitoring people's phys-iological and body state. In this paper, a heart rate measurement methodology based on PhotoPlethysmoGraphy (PPG) signal is proposed. Human face positions are detected and tracked in real time by using facial color videos taken from cameras by non-contact shooting. Signals containing pulse components are extracted from images of the forehead skin area for the purpose of calculating blood volume pulse waves via wavelet filtering. Hence, heart rates are calculated after energy spectrum analysis using Fourier transform. The method realizes non-contact measurement, which avoids potential discomfort caused by direct skin contact, and has the advantages of simple operation and low costs. The result indicates that it is sensible to apply this method to daily family heart rate moni-toring and remote medical monitoring equipment.

Keywords: PPG · Face detection and tracking · Heart rate · Noncontact

1 Introduction

Family medical services in today more and more disciplines more and more. The remote monitoring of vital signs includes not only high-precision diagnostic equipment, but also simple diagnostic equipment, and for everyone to facilitate. One of the most common examinations in health care monitoring is cardiac pulse measurement. There are many different ways to measure heart rate contact measurements, where the gold standard is ECG. However, the heart of records generated by the potential need for proper appli-cation of the electrodes, which may be too complicated and annoying in a home envi-ronment. Other methods of measuring cardiac pulse include thermal imaging [1], Doppler optical [2] and ultrasound [3] or piezoelectric measurements. However, these methods are too expensive and can not be used for home care, so finding low-cost, non-contact heart rate measurement is the key to solving the problem.

Photoplethysmography is a new type of photoelectric detection method, first proposed by Schmitt et al. in 2000, and applied to the detection of skin blood flow and related fluctuations in the phenomenon, to study the skin surface wound healing

---

This work is supported by science and technology commission of shanghai (15411953500).

M. Fei et al. (Eds.): LSMS/ICSEE 2017, Part I, CCIS 761, pp. 13–22, 2017.
DOI: 10.1007/978-981-10-6370-1_2

Happening. According to Poh et al. [4], Harvard and MIT Joint Research Group, 2010, a method of measuring the heart rate and respiration rate based on independent component analysis (ICA) was proposed. The motion artifacts caused by motion disturbances seriously affect the accuracy of the measurement results [5]. The In this paper, an improved measurement algorithm is proposed, and the accuracy of the experiment is evaluated by experiment.

The paper is organized as follows: Sect. 2 talks about the related work for the topics discussed in this paper. Section 3 describes the results of the experiments done and compared with the ground truth. Section 4 provides a conclusion and outlook on future works.

2   Noncontact Cardiac Pulse Measurement

2.1   Photoplethysmography (PPG)

Photoplethysmography (PPG) is an optical measurement technique that can be used to detect blood volume changes in the micro vascular bed of tissue. It has widespread clinical application, for example in pulse oximeters, vascular diagnostics and digital beat-to-beat blood pressure measurement systems. The pulsatile component of the PPG waveform is often called the "AC" component and usually has its fundamental frequency, typically around 1 Hz, depending on heart rate (Fig. 1). This AC component is superimposed onto a large quasi-DC component that relates to the tissues and to the average blood volume. This DC component varies slowly due to respiration, vasomotor activity and vasoconstrictor waves. With suitable electronic filtering and amplification both the AC and DC can be extracted for subsequent pulse wave analysis.

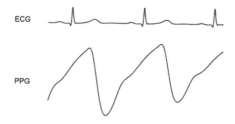

Fig. 1. The pulsatile (AC) component of the PPG signal and corresponding electrocardiogram (ECG). The AC component is actually superimposed on a much larger quasi-DC component that relates to the tissues and to the average blood volume within the sample. It represents the increased light attenuation associated with the increase in microvascular blood volume with each heartbeat.

2.2   Measurement Procedures

An ordinary camera is used to capture face image information in real time with the help of the face detection algorithm designed to detect the human face. Besides, the human face is tracked via the tracking algorithm. After the acquisition of signals from the forehead area of the face image, the ROI image of each frame is divided into three primary colors to generate *R*, *G* and *B* three channel images. Take the gray space average value of each channel as the signal value of the frame image. Consequently, here come the three original signals *R (t)*, *G (t)* and *B (t)*, which will change into three independent source signals after blind signal separation. Then the correlation analysis of the three independent source signals is carried out, and the signal containing the maximum blood volume change is selected as the signal to be analyzed. Lastly, the signal is analyzed via frequency analysis to obtain the heart rate.

Measurement will only need household PCs and ordinary consumer class cameras with subjects to be located in front of the camera at *0.5* m, facing the camera without significant movements. The measurement devices are shown in Fig. 2.

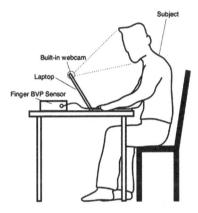

Fig. 2. Experimental setup

First, set the camera resolution pixel as *640 × 480*, frame rate of *30* frames/s and the image color space as RGB. After getting *450* frames of face video from shooting *15* s, process the face video, extracting source signals which contain pulse components. Then wavelet filter for blood volume pulse waves. Finally, analyze heart rate by energy spectrum analysis. The algorithm flow chart is shown in Fig. 3, which mainly includes the following parts: face detection and tracking in Images; choosing the Region of Interest; source signal extraction; signal filtering based on Wavelet Transform; Heart rate calculation based on energy spectrum analysis.

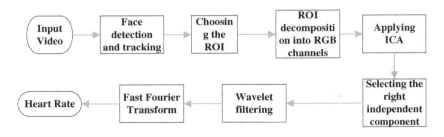

Fig. 3. The algorithm flowchart

2.3   Face Detection and Tracking

One of the main challenges in video-based non-contact heart rate measurement systems is the presence of motion artifacts [8], which has great influence on the accuracy of the whole system and cannot be completely eliminated. In the process of collecting videos, even if the subjects are asked to remain still, some tiny jitter is unavoidable, which will produce great interference on the extraction of weak PPG signals. Therefore, this paper selects the Adaboost algorithm to detect faces, and then combines the KLT feature point tracking algorithm to experiment with face tracking.

The specific algorithm is as follows: Capture video frames, to shorten the processing time to ensure tracking effect, detect human face every *300* frames based on Adaboost to return the location of the face in the image, positioning target tracking area; KLT feature points are used to track the feature points of the target area to achieve face tracking and the rest of the video frame only needs to locate the target tracking region in the next image according to the feature points of the target tracking region in the video image. In order to avoid the loss of tracking, the feature points are updated every *300* frames, and the number of tracking points is set as less than 4 thus the Adaboost tracking algorithm is used to update the feature points of the target tracking region. This iteration to achieve face tracking can effectively shorten the calculation time. The Specific face detection and tracking algorithm flow chart is shown in Fig. 4.

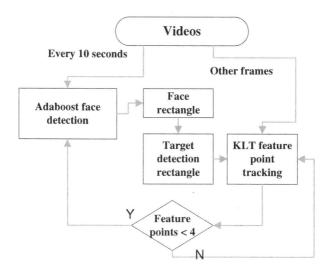

Fig. 4. Face detection and tracking algorithm flow chart

Suppose the coordinate of face detection in video images is *(x, y, w, h)*, wherein *x* represents the face in the image horizontal coordinate position, *y* the face position in the image vertical, *w* face frame width and *h* face frame height. According to face distribution proportion, adjust the target area coordinates to *(x', y', w', h')*, which is shown in Fig. 5(a) while the relation between *(x', y', w', h')* position and face tracking *(x, y, w, h)* is $x' = x + 0.2 \times w$, $y' = y + 0.25 \times h$, $w' = 0.58$ w, $h' = 0.58$ h. The rectangular box calibration area shown in Fig. 5(b) is the new target tracking area.

Fig. 5. (a) Face tracking area (b) New target tracking area

In the new target tracking area, the region with rich features such as the eyebrows, mouth and nose is selected which can exclude backgrounds, face edges, hairs and other tracking feature points of interference. Therefore, it can avoid the late ROI tracking deviation from the face area, to achieve a better ROI tracking and positioning, thus inhibit the movement of artifacts.

2.4   Choosing the Region of Interest (ROI)

Previous HR measurement methods use the Viola-Jones face detector [9] of Opencv [10] to detect faces. It only finds coarse face locations as rectangles, which is not precise enough for HR measurement task since non-face pixels at corners of rectangles are always included. The case becomes even worse when the face rotates. To this end, we first apply Adaboost face detector to detect the face rectangle on the first frame of the input video, Then select a rectangle as the region of interest in the forehead according to the method shown in Fig. 6(b).

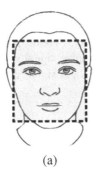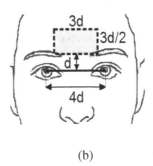

(a)   (b)

Fig. 6. Definition of ROI's for two approaches utilized in the paper, (a) the whole face as a ROI, (b) selected part of the forehead as a ROI

Two rules are followed for defining the ROI: the first one is to exclude the eye region since blinking may interfere with the estimated HR frequency; the second one is to indent the ROI boundary from the face boundary, otherwise non-face pixels from background might be included during the tracking process.

2.5 Recovery of BVP from Video Recording

In the current commonly used method [6], the source signal is separated by color separation of the image, and a color channel was selected for spatial pixel averaging. ICA [7] is a technique used for isolating independent signals from a set of vectors that consist of a linear combination of these signals [11, 12]. The difference between ICA and other methods is that ICA searches for statistically independent and non-Gaussian signals [13]. ICA is applied to many applications in various areas.

When taking a facial video, RGB (red, green and blue) sensors will capture a blend of the reflected PPG signal along with other noise originating artifacts. Each color sensor will record a blend of the initial source signals with a little change in their weight as a result of the differences in the hemoglobin absorptivity in the visible and near-infrared spectral range. In our method, the detected signals from the RGB sensors are denoted as $y_1(t)$, $y_2(t)$ and $y_3(t)$ which represent the amplitudes of the saved signals at time t. Also, it supposes that $x_1(t)$, $x_2(t)$ and $x_3(t)$ are the three fundamental source signals that have been linearly combined to generate $y_1(t)$, $y_2(t)$ and $y_3(t)$. Hence, Eq. (1) describes the aforementioned relationship between captured and source signals.

$$y(t) = \mathbf{A}x(t) \tag{1}$$

Where $y(t)$ is vector $[y_1(t), y_2(t), y_3(t)]^T$, $x(t)$ is vector $[x_1(t), x_2(t), x_3(t)]^T$ and A is a square matrix (called mixture matrix) of size 3×3 that include the mixture coefficients a_{ij}. Therefore, in order to model the relationship between $y(t)$ and $x(t)$, we need to determine a separation matrix W that is estimates the inverse of the mixture matrix A. Once W is determined, the source signals (i.e. $x(t)$) can be estimated. Some of the independent random variables is more Gaussian comparing to the original variables, that is why the non-Gaussian of all sources should be maximized by W to reveal the independent sources (Fig. 7).

The three independent source signals obtained from ICA are disordered. Since the green signal best reflects heart beats, the independent source signal is correlated with the green channel signal and the most relevant independent signal is selected as the effective signal U(t).

2.6 Signal Filtering Based on Wavelet Transform

In order to filter the baseline drift and high frequency noise in the source signal, wavelet transform and wavelet inverse transform are used to filter the passband frequency of 0.75–2.0 Hz (corresponding to heart rate of 45–120 times/min).

Wavelet transform (WT), a time-frequency analysis method, is similar to Fourier transform. However, unlike the Fourier transform, the CWT can detect the abrupt changes in the signal due to the variable sampling step, so it is more suitable for the analysis of biomedical signals. Wavelet transform can be divided into Continuous

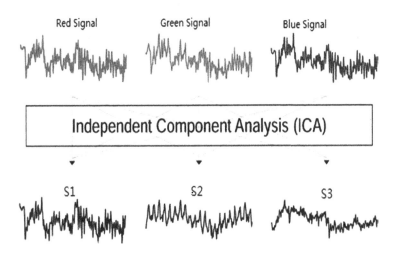

Fig. 7. After the raw signals are detrended and normalized, ICA is applied to separate three independent sources.

Wavelet Transform (CWT) and Discrete Wavelet Transform (DWT). Since the DWT discretizes the scale factor and the translation factor in the CWT, some frequency information may be ignored. Therefore, the system chooses CWT to convolution the source signal $u(t)$ and $\psi_{\tau,s}$ according to the Eq. (2). Where $\psi_{\tau,s}$ is a wavelet basis function, derived from the mother wavelet translation, s is the scale factor that is the expansion factor, and τ is the translation factor. The scale factor is related to the frequency, the larger the scale factor is, the longer the sampling step is whereas the lower the frequency resolution is. At present, the commonly used wavelet basis functions include *sym* wavelet, *dB* wavelet, *Haar* wavelet, *Morlet* wavelet and so on. The selection of wavelet basis function should be considered from two aspects: general principles and specific objects. The general principles are: (1) orthogonality; (2) compact support; (3) symmetry; (4) smoothness. But it is very difficult to fully satisfy these principles, as compact support contradicts with smoothness and the compactness of orthogonality makes it impossible to achieve symmetry, so it is reasonable and meaningful to find a way to properly balance these principles. Choices are made depending on processing signals. *Morlet* wavelet is validated for PPG signal analysis [20].

$$
\begin{cases}
CWT_G^\psi(\tau, s) = \int_{-\infty}^{+\infty} u(t)\psi_{\tau,s}(t)dt \\
\psi_{\tau,s}(t) = \dfrac{1}{\sqrt{|s|}}\psi\left(\dfrac{t-\tau}{s}\right)
\end{cases}
\tag{2}
$$

In the Eq. (2), the inverse wavelet transform is used to inverse transform the signal in the scale range of 0.75–2 Hz, and the reconstructed signal is the pulse wave BVP, as shown in Fig. 8.

$$\begin{cases} \text{BVP} = \dfrac{1}{C_\psi} \displaystyle\int_0^{+\infty} \int_{-\infty}^{+\infty} \text{CWT}_G^\psi(\tau, s)\dfrac{1}{\sqrt{|s|}}\psi\left(\dfrac{t-\tau}{s}\right)d\tau ds \\[4mm] C\psi = \displaystyle\int_0^\infty \dfrac{|\widehat{\psi}(\zeta)|^2}{|\zeta|}d\zeta < \infty \end{cases} \tag{3}$$

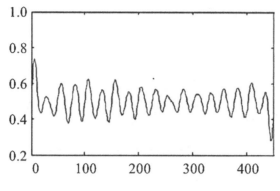

Fig. 8. BVP wave

2.7 Heart Rate Calculation Based on Energy Spectrum Analysis

The frequency domain is the main interest; therefore, to convert the time domain signal to the frequency, the Discrete Fourier Transform (DFT) is usually used. For less processing time the FFT is applied. The FFT's complexity is much lower than DFT. After performing the FFT, the highest peak detected in the interest band (which is the HR band between *45* and *120*). The detected peak is then converted to the right frequency in the FFT vector. In the Fig. 9, *HR = 1.203 × 60 ≈ 72* beats/min.

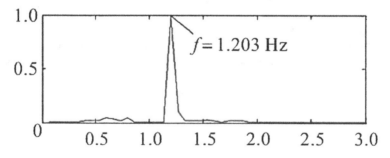

Fig. 9. Energy spectrum of BVP

3   Experiments

In this experiment, 20 subjects (15 males and 5 females) were recruited. The selected subjects are between 20 and 30 years old. They were tested in two groups under the uniform distribution of light. In the first group, the 20 subjects' heart rates were detected in a resting state, and the subjects were measured by the finger BVP sensor at the same time. In the second group, the 20 subjects' heart rates were detected after 50 deep squats, and the subjects were measured by the finger BVP sensor at the same time. The data obtained are shown in Table 1.

Table 1. Accuracy experiment

The first group				The second group			
Sub. (Resting state)	Our method	Finger BVP sensor	Error	Sub. (After exercise)	Our method	Finger BVP sensor	Error
1	66	63	3	1	113	116	3
2	54	59	5	2	94	99	5
3	81	77	4	3	110	108	2
4	53	52	1	4	100	98	2
5	50	52	2	5	95	99	4
6	74	74	0	6	104	100	4
7	60	61	1	7	101	100	1
8	71	70	1	8	110	113	3
9	56	58	2	9	100	99	1
10	58	58	0	10	112	110	2
11	54	55	1	11	90	93	3
12	50	53	3	12	99	98	1
13	62	62	0	13	103	102	1
14	68	67	1	14	105	105	0
15	59	62	2	15	98	97	1
16	72	71	1	16	103	100	3
17	65	66	1	17	92	97	5
18	80	75	5	18	105	103	2
19	71	69	2	19	101	102	1
20	63	63	0	20	99	98	1

As can be seen from Table 1, the maximum error of the measured value of this method is about 5 beats/min; the average error is about 2 beats/min. The average correlation coefficient between the measurement results by our method and by the ECG measurements was 0.982, and the result meets the requirements of the People's Republic of China's pharmaceutical industry standards (error ≤ 5 beats/min). So this method can be used in home medical system.

4   Conclusions

This method intends to achieve accurate non-contact heart rate measurement with the help of ordinary color camera under natural light conditions. It is a good choice when traditional measurement methods are not suited to some subjects. Besides, in remote medical inquiry system, this method can monitor heart rates whenever necessary. Hence, it has a bright prospect in the future family medical service. However, in the case of too short video length and poor lighting conditions, this method may not measure heart rates accurately. Therefore, in order to improve the robustness of heart rate measurement, further study is needed to solve these problems.

References

1. Garbey, M., et al.: Contact-free measurement of cardiac pulse based on the analysis of thermal imagery. IEEE Trans. Biomed. Eng. **54**(8), 1418–1426 (2007)
2. Shastri, D., et al.: Imaging facial signs of neurophysiological responses. IEEE Trans. Biomed. Eng. **56**(2), 477–484 (2009)
3. Holdsworth, D.W., et al.: Characterization of common carotid artery blood-flow waveforms in normal human subjects. Physiol. Meas. **20**(3), 219 (1999)
4. Poh, M.Z., Mcduff, D.J., Picard, R.W.: Advancements in noncontact, multiparameter physiological measurements using a webcam. IEEE Trans. Biomed. Eng. **58**(1), 7 (2011)
5. Lewandowska, M., et al.: Measuring pulse rate with a webcam - a non-contact method for evaluating cardiac activity. In: Proceedings Federated Conference on Computer Science and Information Systems - FedCSIS 2011, Szczecin, Poland, 18–21 September 2011, DBLP, pp. 405–410 (2011)
6. Tsouri, G.R., et al.: Constrained independent component analysis approach to nonobtrusive pulse rate measurements. J. Biomed. Optics **17**(7), 077011 (2012)
7. Allen, J.: Photoplethysmography and its application in clinical physiological measurement. Physiol. Meas. **28**(3), R1–R39 (2007)
8. Viola, P., Jones, M.: Rapid object detection using a boosted cascade of simple features. In: Proceedings of the 2001 IEEE Computer Society Conference on Computer Vision and Pattern Recognition, CVPR 2001, vol. 1, pp. I-511–I-518. IEEE (2001)
9. Bradski, G.: The Opencv library. Doct. Dobbs J. **25**(11), 384–386 (2000)
10. Addison, P.S., Watson, J.N.: Rapid Communication: a novel time frequency-based 3D Lissajous figure method and its application to the determination of oxygen saturation from the photoplethysmogram. Meas. Sci. Technol. **15**(11), L15 (2004)
11. Comon, P.: Independent component analysis, a new concept? Sig. Process. **36**(3), 287–314 (1994)
12. Stone, J.V.: Independent component analysis. Trends Cogn. Sci. **6**(1), 529 (2005)
13. Kwon, S., Kim, H., Park, K.S.: Validation of heart rate extraction using video imaging on a built-in camera system of a smartphone. In: International Conference of the IEEE Engineering in Medicine & Biology Society PubMed, pp. 2174–2177 (2012)

Classification of MMG Signal Based on EMD

Lulu Cheng[1], Jiejing Wang[1], Chuanjiang Li[1(✉)], Xiaojie Zhan[1],
Chongming Zhang[1], Ziming Qi[2], and Ziqiang Zhang[1]

[1] Mechanical and Electrical Engineering, College of Information,
Shanghai Normal University, Shanghai 201418, China
Licj@shnu.edu.cn
[2] Otago Polytechnic, Dunedin, Otago, New Zealand

Abstract. Mechanomyography (MMG) signal is the sound from the surface of a muscle when the muscle is contracted. The traditional filtering algorithms for the processing of MMG signal would make most useful signal filtered when they are used to remove noise. According to MMG signal's characteristics, a new signal filtering method is presented in this paper based on combining empirical mode decomposition with digital filter, which has a better performance on MMG signal filtering processing in experimental analysis. With extracting the energy feature of wavelet packet coefficient as the feature of classifier, the BP neural network classifier gets a better classification results. The average classification results showed that the best performance for recognizing hand gestures with the energy feature of wavelet packet coefficient features was achieved by BP neural network with the accuracy of 86.41%. This work was accomplished by introducing the new signal filtering method for the recognition of different hand gestures; And suggesting basing on combining empirical mode decomposition with digital filter as a new filtering method in MG-based hand gesture classification.

Keywords: MMG · Empirical mode decomposition · Wavelet packet coefficient · BP neural network

1 Introduction

Prosthetic research focuses on the pretreatment of physiologic signal processing, classifier algorithm design, and prosthetic hand control, especially how to use the user's own signal flexibly and effectively to control an upper limb prosthesis [1, 4]. Prosthetics based on MMG signals have a very broad prospect in medical rehabilitation for the disabled, therefore its application on the medical industry is a development trend in artificial intelligence.

The vast majority of MMG signal ability focuses on the frequency range of 8–70 Hz. In this study, we investigated the potential of a multi-function muscle-interface that is controlled by MMG signals from forearm muscles.

Aiming at the characteristic of non-linear, non-stationary, and multi-forms of the MMG signal, the EMD method is used to preprocess the collected signal, and combined with the BP neural network, to research the hand gesture of classification. The EMD method is used to decompose the MMG signal containing noise, and extract

© Springer Nature Singapore Pte Ltd. 2017
M. Fei et al. (Eds.): LSMS/ICSEE 2017, Part I, CCIS 761, pp. 23–34, 2017.
DOI: 10.1007/978-981-10-6370-1_3

the effective IMF component of the signal to reconstruction. The reconstructed signal filtered with a Chebyshev band-pass filter can obtain the effective MMG signal. Then, the effective MMG signal is decomposed by a wavelet packet to get the wavelet packet energy feature that is used as the input of the BP neural network that is established to classify the hand gesture.

2 Experiments and MMG Signal Acquisition

A convenience sample of 5 healthy individuals (4 males and 1 female), 23.5 ± 4 years of age, provide a written consent to participate in the study.

In this study, signals in the X, Y, and Z axes are detected with two three-axis Motion Tracker. A custom terminal box is built to amplify the accelerometer signals and interface the accelerometers with a terminal block. Vibrations of known amplitude are applied to each accelerometer-amplifier assembly via a mechanical shake to ensure uniform gain across all MMG sensors.

Participants sit on a chair fitted with a custom arm-rest and keep body's static state as far as possible. Two MMG sensors, manufactured according to the method of Silva J. et al. [2], are affixed to the participants' dominant forearm over extrinsic hand muscles. The two typical muscles, monitored by the sensors, are the extensor carpi radialis longus and extensor digitorum communis. Each sensor is individually affixed with a small Velcro strap.

A professional collecting software is used to start data acquisition and visually cue the participants to perform the following eight hand motions: natural state (Motion 1); hand open (Motion 2); hand close (Motion 3); wrist flexion (Motion 4); wrist extension (Motion 5); gesture "Two" (Motion 6); gesture "Five" (Motion 7); and gesture "Eight" (Motion 8). The 8 different motions are as shown Fig. 1:

Fig. 1. The eight different motion modes

The acceleration signals generated at the two muscle sites from the terminal block channeled through an analog signal conditioning input module, sampled at a rate of 200 Hz, and the digitized signals were stored on the controller's hard drive. Participants perform 80 repetitions of each of the eight motions in a pre-defined order. This process execute 200 times in total. Each motion is comprised of the full range of motion from the resting position to the target position, followed by 5 s of the hand being held in the target position. Two typical arm positions(extensor carpi radialis longus and extensor digitorum communis) commonly use in daily life activities are considered in this study for evaluating the possible effects of arm position variation on motion classification performance, which is shown on Fig. 2.

Fig. 2. Tracking sensor placement in the experiment

The three-axis Motion Tracker could collect availably three signals containing the calibrated data output of the accelerations, rate of turn and magnetic field in X, Y and Z axes. The data of the Z axis contains a majority of the effective MMG data and the other two directions' data should be thrown out. The original acceleration signal of the motion 1 collected at the extensor carpi radialis longus is shown Fig. 3.

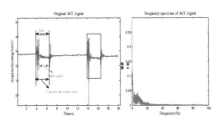

Fig. 3. Typical original signal from a forearm muscle

The acceleration signal in Z axes contains many interfering signals. In order to extract the effective MMG signal, the acceleration signals in Z axes must be made a pre-processing as shown in the following paragraphs.

3   Data Filtering Pre-processing

This paper filters the noisy signal with the Chebyshev digital filter with the method of empirical mode decomposition (EMD) combined. The Chebyshev digital filter mainly filters the direct current signal. The EMD method is used to remove the overlapping interference signals.

IMF is an intermediate product of EMD, a pre-processing algorithm of Hilbert–Huang transforms (HHTs). There are two steps involved in HHT. The first step involves the EMD to extract IMF. The second step is the Hilbert transform of the decomposed IMF to obtain a time-frequency distribution.

The EMD decomposes any signal, into a set of IMFs, each of which has a distinct time scale. It results that $x(t)$ is expressed in formula (1), as follows:

$$x(t) = \sum_{j=1}^{C} IMF_j(t) + r_C(t) \tag{1}$$

Every IMF component all are stationary signal, and directly perform Hilbert transform. The Hilbert transformation form is expressed in formula (2) as follows:

$$\overline{c_i}(t) = H[c_i(t)] = \frac{1}{\pi} \int_{-\infty}^{+\infty} \frac{c_i(\tau)}{t - \tau} d\tau, (i = 1, \ldots, n) \tag{2}$$

$c_i(t)$ and $\overline{c_i}(t)$ can constitute a complex signal $z_i(t)$ in formula (3) as follows:

$$z_i(t) = c_i(t) + j\overline{c_i}(t) = a_i(t)e^{y_i(t)}, (i = 1, \ldots, n) \tag{3}$$

It can get instantaneous envelope $a_i(t)$, instantaneous phase $\theta_i(t)$ and instant frequency $\omega_i(t)$ in formulas (4), (5) and (6) as follows:

$$a_i(t) = \sqrt{c_i(t)^2 + \overline{c_i}(t)^2} (i = 1 \cdots n) \tag{4}$$

$$\theta_i(t) = arctan\left\{\frac{Im[z_i(t)]}{Re[z_i(t)]}\right\} = arctan\left\{\frac{C_i(t)}{\overline{C_l}(t)}\right\} \tag{5}$$

$$\omega_i(t) = \frac{d\theta_i(t)}{dt} = \frac{\overline{C_l}(t) * C_i(t) - C_i'(t) * \overline{C_l}(t)}{C_i(t)^2 + \overline{C_l}(t)^2} \tag{6}$$

Where $i = 1, \ldots, n$.

The low IMF scales mainly are the high-frequency components of signal, while the high IMF scales are the low-frequency components of signal. Thus, an EMD-based band-pass filter is developed to be use of the partial reconstruction of the selected IMF scale, which is given in formula (7) as follows:

$$EMD_K = \sum_{i=k}^{n+1} IMF_j(t) \tag{7}$$

The effective MMG signal is decomposed to ten IMFs and every IMF is shown in the Figs. 4, 5, 6 and 7:

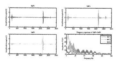

Fig. 4. The signal and frequency spectrum of IMF1- IMF3

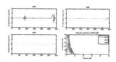

Fig. 5. The signal and frequency spectrum of IMF1- IMF3

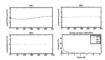

Fig. 6. The signal and frequency spectrum of IMF7- IMF9

Fig. 7. The signal and frequency spectrum of IMF10- IMF12

The EMD method as a filter is essentially a partial reconstruction process of relevant modes. These modes are selected based on a given criterion that identifies the modes carrying information relevant to the main structures of the input signal [10–12].

In this paper, firstly, the EMD method is selected to filter the some interference signal overlapping the effective MMG signal in frequency spectrum. Each group data from every Tracking sensor can be decomposed to eleven IMFs. Compared with the frequency spectrum of different IFMs and the frequency band of effective MMG signal, the IMF1-IMF7 and IMF9 are selected to reconstruct the filtered signal that can represent major information of the effective MMG signal, which is mainly to solve an open question about frequency aliasing.

Then, the Chebyshev band-pass filter of Matlab library is used for filtering the low and high frequency noise. According to the effective frequency range of MMG signal, the "stop band cutoff frequency" is settled at 1 Hz to 85 Hz. Combined with Silva Jand Natasha Alves's research [2, 3], this paper sets that the "bandpass cutoff frequency" range from 8 Hz to 70 Hz. The "minimum stop band reduction" equal to 40 dB and the "maximum passband reduction" equal to 1 dB.

Compared with the Figs. 8 and 9, it could be obviously obtain that the EMD filter makes the low frequency signal significant reduction in the amplitude of 8 Hz and the signal of other frequency also reduce a bit. The low frequency interference signal is not filtered by the classic digital filter and only filtered by the EMD method.

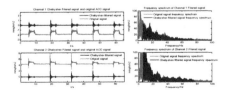

Fig. 8. Two channel MMG signal filtering using Chebyshev digital filter

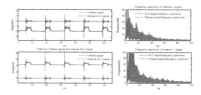

Fig. 9. Two channel MMG signal filtering using the EMD combined with Chebyshev digital filter

Although the EMD method filter remove small effective MMG signal, the feature of different motion's MMG signal becomes more obvious and it has a propitious result to classification

4 Feature Extraction

Feature extraction plays a vital role in any hand gesture recognition system based on accelerometer. Instead of dividing only the approximation spaces, wavelet packet bases present both approximation and detail spaces in a binary tree by recursive splitting of vector spaces. Any node of a binary tree is labeled by its depth j and number p of nodes (packets). Each node (j,p) corresponds to a space W_j^p, which admits an orthonormal basis $\Psi_j^p(n-2t)_{n\in Z}$. The two wavelet packet orthogonal bases at the children nodes are defined by recursive relations in formulas (8) and (9) as follows

$$\varphi_j^{2p-1}(t) = \sum_{n=-\infty}^{\infty} h[n]\varphi_{j-1}^p(n-2t) \tag{8}$$

$$\varphi_j^{2p}(t) = \sum_{n=-\infty}^{\infty} g[n]\varphi_{j-1}^p(n-2t) \tag{9}$$

Two orthogonal spaces W_j^{2p} and W_j^{2p-1} can be defined as closure spaces of time-varying signals $x_j^{2p}(k)$ and $x_j^{2p-1}(k)$ respectively. Meanwhile, they can also be represented as:

$$W_j^p = W_j^{2p} \oplus W_j^{2p-1} \tag{10}$$

This recursive splitting defines a binary tree of wavelet packet spaces where each parent node is divided into two orthogonal subspaces. When the domain signal $x(t)$ satisfies two-scale relations in formulas (11) and (12) as follows:

$$x_{j+1}^{2p}(t) = \sqrt{2}\sum h(n)x_{j+1}^{2p}(2t-n) \tag{11}$$

$$x_{j+1}^{2p}(t) = \sqrt{2} \sum g(n) x_{j+1}^{2p}(2t - n) \tag{12}$$

The expanding coefficients $h[n]$ and $g[n]$ can be expressed in the frequency domain as in formula (13) as follows:

$$H\{\cdot\} = \sum_{n=-\infty}^{\infty} h(n - 2r), G\{\cdot\} = \sum_{n=-\infty}^{\infty} g(n - 2r) \tag{13}$$

Let $x_j^p(k)$ be the pth packet on the jth resolution; hence, the wavelet packet transform can be computed by the following recursive algorithm as in formulas (14), (15), (16) and (17) as follows:

$$x_0^1(k) = x(t) \tag{14}$$

$$x_j^{2p-1}(k) = Hx_{j-1}^p(k) \tag{15}$$

$$x_j^{2p}(k) = Gx_{j-1}^p(k) \tag{16}$$

$$x_j^p(k) = x_j^{2p-1}(k) + x_j^{2p}(k) \tag{17}$$

Where $k = 1, 2, \ldots, 2^{J-P}, P = 1, 2, \ldots, J$ and $J = log_2 N$.

The paper uses 6th wavelet packet to decompose the effective MMG signal and each one senor signal can get twenty-five wavelet trees with different frequency. The order of wavelet packet is determined by empirical value. The 6 order is optimal result and it is determined with a mount of experiments using different order of wavelet packet to extract features to test the classification accuracy. Wavelet coefficient energy can be counted by extracting the coefficients of the wavelet packet frequency bands. S_{6j} represents the fifth layer wavelet of the jth node of packet signal $x_i^j(k)$ coefficient. The jth node of wavelet coefficient energy can be represented as in formula (18) as follows:

$$E_N^j = \sum_{i=1}^{N-1} (S_{6j})^2, (i = 0, 1, \ldots, N - 1) \tag{18}$$

Where N represents the wavelet packet decomposition layer, j represents the nodes of the wavelet packet.

The feature vector M that it is determined by the BP neural network structure as follows in formula (19), which is consisted of the vector E_{ab}^j. The vector E is consisted of the all 25 nodes of the wavelet packet energy value of two Tracking sensors, is as follows in formula (20):

$$E_{ab}^j = [E_{11}^0, \ldots, E_{11}^{24}, E_{21}^0, \ldots, E_{21}^{24}] \tag{19}$$

$$M_{bn} = [E_{11}E_{12}\ldots E_{1n}E_{21}\ldots E_{2n}\ldots E_{81}E_{82}\ldots E_{8n}] \tag{20}$$

Where $a = 1, 2$, represent Tracking sensor's number, $b = 1, 2, 3, \ldots 8$, represent the type of gesture movement, n represent the number of sample.

5 Classifier Processing

BP neural network is essentially a sample set of the input and output that is transformed into a nonlinear optimization problem. It is a learning algorithm through the gradient algorithm to solve the question of the weight.

This paper adopts standard three-layer BP neural network and uses self-adapting gradient descent algorithm to improve network performance with the optimal parameters for the best recognition rate for Finger motion. The standard three-layer BP neural network model is shown in Fig. 10:

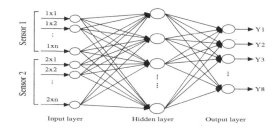

Fig. 10. BP neural network structure mode

The hidden layer node: The number of neurons in the hidden layer is usually more than the number of the input factors and hidden layer neurons of 7–15 are appropriate, so the number of hidden neurons is 12.

Output layer node: BP neural network number of hand gesture is the number of output neurons, so the output layer of the network of each neuron corresponds to a class. The eight motion modes respectively corresponding to the target output vector. According to the above method, the node number is: 50, 12 and 8.

Learning rate: the value is 0.08 according to many experimental results.

6 Pattern Recognition Analysis and Results

In this chapter, the experiment selects the first 100 groups of data to train the sample to acquire the proper weights and thresholds and the remaining 100 groups' data to test the classification accuracy of the finger motion.

In the experiment, the two groups' data pass pretreatment, and are extracted the training feature sets and test feature sets. Then, BP neural network is used to train characteristics training set and sets the maximum training accuracy of the training network at 10^{-4}. BP neural network achieves the expected accuracy requirements after 8000 iterations training.

In the study, experiment will use two methods to confirm the effectiveness of finger motion's recognition, which is different filter method for the purpose of individual-validation and cross-validation. Detailed experimental procedure is described below.

Firstly, in the experiment, the two groups' data pass Chebyshev filter, and the features of wavelet packet coefficient energy separately are extracted to form the training set and test set. Then, the training sets are imported to train the three layers of BP neural network and the optimal parameter matrix is acquired. The average false recognition rate of each group of samples and the false recognition rate of each action was calculated. The results are shown in Fig. 11:

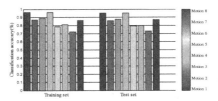

Fig. 11. The classification recognition accuracy of hand gestures with the Chebyshev digital filter.

Secondly, the two groups' data pass the Chebyshev band-pass filter combined with the EMD method, and the features of wavelet packet coefficient energy separately are extracted to form the training set and test set. Figure 12 shows the recognition precision of hand gesture recognition.

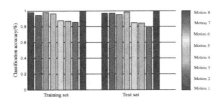

Fig. 12. The classification recognition accuracy of hand gestures with the EMD filtering method combined with Chebyshev digital filter

Finally, we use the cross-validating method to validate the advantage of EMD method in the hand gesture recognition. In this paper, cross-validating method is used for testing the classification accuracy. This method is explained that the prostheses is only designed for special disabled, so the validation sets must come from entirely same subjects in consideration to the prostheses characteristic. Research should be sure that the result of classification accuracy is not influenced by the collecting order of training sets. In order to avoid the over-fitting of BP neural network, the remaining 100 groups' data randomly is divided into five groups. then, among four groups' data are used for training set and the remaining one group data is used for test set until the every group

Table 1. The recognition accuracy of hand gestures cross-validation method.

Example	Subject 1	Subject 2	Subject 3	Subject 4	Subject 5
Test 1	87.78%	86.89%	89.07%	85.75%	82.29%
Test 2	89.03%	88.21%	90.64%	86.06%	81.77%
Test 3	88.69%	89.42%	88.88%	88.30%	79.84%
Test 4	86.34%	87.09%	90.05%	86.66%	81.63%
Test 5	90.01%	89.44%	86.39%	87.69%	82.76%
Average	88.31%	87.08%	88.74%	86.34%	81.52%

data all is used for test set. The average recognition rate of 5 times examples are counted and the results is shown in the Table 1 (The Subject 1, Subject 2, Subject 3 and Subject 4 is male with strong body and muscle and the subject 5 is a health female. The result in Table 1 is the average classification.):

The Table 1 can get the conclusion that the difference of muscle leads to that the MMG signals collected by Motion Tracker have apparent difference, which directly influence the classification accuracy. The recognition accuracy of cross-validating is close to the recognition accuracy of test set using the EMD method.

The experiment results show that the maximum error of eight kinds of gesture recognition using BP neural network algorithm for EMD signal is about 13.59%, and the average accuracy of eight kinds of gesture motion pattern recognition rate is 86.41% in the condition of individual-validation. But the accuracy of the cross-validation decreases in a degree, because the different subject's muscle is different and leads to that the MMG signal collected has a bit diversity.

7   Conclusion

In the analyzing of the MMG signal with non-stationary and nonlinear characteristic, this study uses the method of combining the EMD with the classical digital filter to remove the noise and the better results are obtained. Compared with the traditional time domain and frequency domain analysis method, the EMD combined with the band pass filter can effectively filter out the noise signal that is same with the MMG signal frequency spectrum. The study show that this method can effectively identify the pattern classification of eight kinds of action and the average recognition rate reached 86.41% using 50 features.

Compared with Natasha Alves's mean accuracy of $93 \pm 0.9\%$ using 15 features for eight classes of forearm muscle activity using LDA in 2009 and 2010 [1], Yanjuan Geng's the average 90% classification error with two-channel ACC-MMG signals about sevsen classes of forearm muscle activity using LDA in 2012 [4] and Lou'l Al-Shrouf's 75% accuracy to classify five hand-motions with five ACC sensors using STFT-SVM in 2014 [6], a certain practical value can be shown in this study, which only use two-channel ACC-MMG signals to classify eight forearm muscle activities.

Acknowledgments. 1. Supported by research project of Science and Technology Commission of Shanghai Municipality (Project Number: 16070502900) 2. Supported by the Program of Shanghai Normal University (A-7001-15-001005)

References

1. Alves, N., Chau, T.: Classification of the mechanomyogram its potential as a multifunction access pathway. In: Annual International Conference of the IEEE Engineering in Medicine and Biology Society, pp. 2951–2954 (2009)
2. Silva, J., Heim, W., Chau, T.: MMG-based classification of muscle activity for prosthesis control. In: Proceeding of IEEE Conference on Engineering Medical Biology Society science, vol. 2, pp. 968–971 (2004)
3. Alves, N., Sejdi, E., Sahota, B., Chau, T.: The effect of accelerometer location on the classification of single-site forearm mechanomyograms. BioMed. Eng. Online **9**(1), 1–14 (2010)
4. Geng, Y., Zhou, P., Li, G.: Toward attenuating the impact of arm positions on electromyography pattern-recognition based motion classification in trans radial amputees. J Neuro. Eng. Rehabil. **9**, 74 (2012)
5. Khan, A.M., Siddiqi, M.H., Lee, S.-W.: Exploratory data analysis of acceleration signals to select light-weight and accurate features for real-time activity recognit ion on smartphones. Sensors **13**(10), 13099–13122 (2013)
6. Al-Shrouf, L., Saadawia, M.S., Söffker, D.: Improved process monitoring and supervision based on a reliable multi-stage feature-based pattern recognition technique. Inf. Sci. **259**, 282–294 (2014)
7. Daoud, H.-G., Ragai, H.-F.: Mechanomyogram signal detection and decomposition. conceptualisation and research. Int. J. Healthcare Technol. Manag. **13**(1–3), 32–44 (2012)
8. Xie, H.-B., Zheng, Y.-P., Guo, J.-Y.: Classification of the mechanomyogram signal using a wavelet packet transform and singular value decomposition for multifunction prosthesis control. Physiol. Meas. **30**(5), 441–457 (2009)
9. Chang, K.-M., Liu, S.-H.: Gaussian noise filtering from ECG by wiener filter and ensemble empirical mode decomposition. J. Signal Process. Syst. **64**(2), 249–264 (2011)
10. Komaty, A., Boudraa, A.-O., Augier, B., Dare-Emzivat, D.: EMD-based filtering using similarity measure between probability density functions of IMFs. Instrum. Meas. IEEE Trans. **63**(1), 27–34 (2014)
11. Yang, Z., Yu, Z., Xie, C., Huang, Y.: Application of Hilbert-Huang transform to acoustic emission signal for burn feature extraction in surface grinding process. Measurement **47**, 14–21 (2014)
12. Guo, K., Zhang, X., Li, H., Meng, G.: Application of EMD method to friction signal processing. Mech. Syst. Signal Process. **22**(1), 248–259 (2008)
13. Li, M., Wu, X., Liu, X.: An improved emd method for time–frequency feature extraction of telemetry vibration signal based on multi-scale median filtering. Circuits Syst. Signal Process. **34**(3), 815–830 (2015)
14. Rilling, G., Flandrin, P., Goncalves, P.: On empirical mode decomposition and its algorithms. In: IEEE - EURASIP Workshop on Nonlinear Signal and Image Processing. Grado(I), pp. 8– 11 (2003)

15. Liu, Y., Li, Y., Lin, H., Ma, H.: An amplitude-preserved time-frequency peak filtering based on empirical mode decomposition for seismic random noise reduction. Geosci. Remote Sens. Lett. IEEE **11**(5), 896–900 (2014)
16. Chatlani, N., Soraghan, J.J.: EMD-Based Filtering (EMDF) of low-frequency noise for speech enhancement. IEEE Trans. Audio Speech Lang. Process. **20**(4), 1158–1166 (2012)

Adaptive KF-SVM Classification
for Single Trial EEG in BCI

Banghua Yang[1(✉)], Chengcheng Fan[1], Jie Jia[2(✉)], Shugeng Chen[2],
and Jianguo Wang[1]

[1] Key Laboratory of Power Station Automation Technology,
Department of Automation, College of Mechatronics Engineering
and Automation, Shanghai University,
No. 149, Yanchang Road, Shanghai 200072, China
yangbanghua@shu.edu.cn
[2] Department of Rehabilitation Medicine, Huashan Hospital, Fudan University,
No. 12, Wulumuqi Middle Road, Shanghai 200040, China

Abstract. Single trial electroencephalogram classification is indispensable in
online brain–computer interfaces (BCIs) A classification method called adaptive
Kernel Fisher Support Vector Machine (KF-SVM) is designed and applied to
single trial EEG classification in BCIs. The adaptive KF-SVM algorithm
combines adaptive idea, SVM and within-class scatter inspired from kernel
fisher. Firstly, the within-class scatter matrix of a feature vector is calculated.
And to construct a new kernel, this scatter is incorporated into the kernel
function of SVM. Ultimately, the recognition result is calculated by the SVM
whose kernel has been changed. The proposed algorithm simultaneously maximizes the discrimination between classes and also considers the within-class
dissimilarities, which avoids some disadvantages of traditional SVM. In addition, the within-class scatter matrix of adaptive KF-SVM is updated trial by trail,
which enhances the online adaptation of BCIs. Based on the EEG data recorded
from seven subjects, the new approach achieved higher classification accuracies
than the standard SVM, KF-SVM and adaptive linear classifier. The proposed
scheme achieves the average performance improvement of 5.8%,5.2% and 3.7%
respectively compared to other three schemes.

Keywords: Brain computer interface (BCI) · Support vector machine (SVM) ·
Adaptive classification · Kernel fisher · Within-class scatter

1 Introduction

Brain–computer interfaces (BCIs) is a direct communication pathway between brain
and external device which is independent from muscle pathway [1]. The inherent
nonstationarities existed in the sampled EEG data makes a principal problem in electroencephalogram based brain computer interfaces [2]. These nonstationarities are
caused by many factors such as variations of the concentration and excitation level,
fluctuations in the involved subjects' mental task, the impedance variations or positions
movement of the electrodes, affection of feedback, fatigue, and swallowing and
blinking artifacts [3–6]. In addition, the characteristics of EEG signals may vary

© Springer Nature Singapore Pte Ltd. 2017
M. Fei et al. (Eds.): LSMS/ICSEE 2017, Part I, CCIS 761, pp. 35–45, 2017.
DOI: 10.1007/978-981-10-6370-1_4

significantly from person to person [7]. In order to track non-stationary EEG, reduce the subject training, some adaptive algorithms are investigated extensively, from its feature extraction to its classification [8–12]. This present work mainly emphasizes on the analysis of motor imagery electroencephalogram signal in BCI for the ideal of adaptive classification. As a kind of spontaneous EEG signal, Motor imagery signal is commonly used in BCIs because of the reason that it is much more natural compared with evoked EEG signal for example as P300 and VEP (Visual Evoked Potential).

Some typical classification methods include adaptive linear discriminant analysis (LDA), adaptive support vector machine (SVM) The adaptive linear discriminant analysis is investigated for left and right hand motor imagery classification. By using Kalman filter, the adaptive LDA parameters are constantly updated trial by trial, which get better results than the non-adaptive LDA classification. It is a simple and efficient method, but it cannot avoid shortcomings of linear classifiers. An adaptive SVM classification for BCIs is proposed in [10], which attains much higher classification accuracy than the non-adaptive SVM. However, the classification performance of standard SVMs is restricted by many factors such as data noise, unbalanced data points, complexity of data points and so on. When the classification examples are difficult data classification for standard SVMs, it is impossible to obtain optimal classification results [11]. To classify imagery EEG data in BCIs, this article describes a novel and adaptive method called adaptive kernel fisher SVM (KF-SVM), which combines adaptive idea, SVM, kernel fisher inspiring within-class scatter.

Due to the powerful classification ability, the SVM has become a major method to make electroencephalogram(EEG) based BCI classification, thus to overwhelm other classifiers in many other applications [13, 14]. Nevertheless, the SVM algorithm only considers the discrimination between classes, but neglects within-class scatters information. As an algorithm improvement, the proposed algorithm not only maximizes the discrimination between classes but also considers the within-class dissimilarities inspired from kernel fisher simultaneously. Firstly, the within-class scatter matrix of a feature vector is calculated. And after that this scatter is incorporated into the kernel function of SVM to reconstruct a new kernel. Finally, the recognition result is calculated by SVM whose kernel has been changed. At the same time, the adaptability of the proposed algorithm is improved by updating the within-class scatter simultaneously and continuously. The proposed method was tested on dataset collected from seven subjects. Its performance is compared to other classification algorithms including SVM, KF-SVM and adaptive LDA. The highlighted point in this paper is the employment of common spatial patterns (CSP) method for feature extraction of the three classifiers.

2   Methods

2.1   Dataset

Dataset was experimented from laboratory. A 16-channel electrode cap is used for EEG signal recording. The authors didn't use too many channels considering that fewer channels were more practical for online application of BCIs. The EEG biological amplifier was developed by Tsinghua University research group with its high quality of precision. The EEG signals were transformed by a 24-bit A/D converter and collected with the sampling frequency of 100 Hz through acquisition software. In the process of the experiment, each of seven subjects was asked to complete one session containing 60 trials. For each trial, a 4 s left or right hand motor imagination task is included. For each subject, the total accepted sessions are eight and corresponding eight datasets for each subject were acquired. These subjects had no experience of the BCI experiment. The aim of selecting them was to check if the proposed algorithm had better generalization capability for naïve BCI users. The datasets were filtered between 8 Hz and 30 Hz by band-pass filter (the usual range for motor imagery EEG data).

2.2 Research Scheme

A flowchart of the adaptive KF-SVM is illustrated in Fig. 1. In part 1, the trials from session 1 are extracted features by the CSP. For motor imagery features recognition, the PP-SVM via adding within-class scatter is used firstly and it takes five-fold cross validation secondly, so for the testing data, the average recognition accuracies are calculated across these five folds. In part 2, the parameters of adaptive KF-SVM via adding within-class scatter are initialized by the training data of all the trials in session 1. After completing initialization, the obtained trials from session 2 to session 8 are used to evaluate the performance of the classifier. At the simultaneous time, within class scatter of the proposed adaptive algorithm is updated trial by trial continuously.

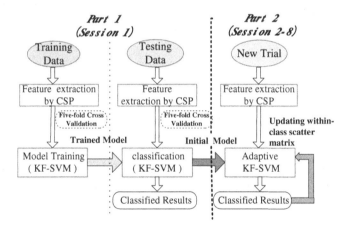

Fig. 1. Flowchart of adaptive KF-SVM.

2.3 SVM Introduction

The purpose of SVM algorithm is to search the optimal hyperplane to separate the two classes of samples [11, 12]. The SVM has good generalization ability and its optimization problem is defined as:

$$\min(1/2)w^T I w + \gamma(1/2) \sum_{i=1}^{n+m} e_i^2 \tag{1}$$

where the parameter I represent the identity matrix, the regularisation term γ remains positive, $n + m$ reflects the training samples number, $\{e_i\}_{i=1}^{n+m}$ represents the error vector, $y_i \in \{-1, 1\}$ represents the label for sampling, $\phi(x)$ is a mapping function, the vector W is weighting vector and scalar $b \in R$ is bias vector. According to formula (1), the Lagrange multipliers are used to solve the optimisation problem:

$$\begin{bmatrix} 0 & -Y^T \\ Y & zz^T + \gamma^{-1}I \end{bmatrix} \begin{bmatrix} b \\ a \end{bmatrix} = \begin{bmatrix} 0 \\ I \end{bmatrix} \tag{2}$$

where $(zz^T)_{ij} = y_i y_j k(x_i, x_j)$, α reflects the dual variable vector, and this function $k(x_i, x_j) = \phi(x_i) \cdot \phi(x_j)$ is called a kernel function.

From formula (2), the non-zero parameter a and parameter b can be obtained. So the decision function could be shown as the following equations:

$$f(x) = \sum_{i=1}^{n+m} a_i y_i k(x_i, x) + b \tag{3}$$

According to formula (3), the classification surface of SVM depends on two classes' boundary samples and misclassified samples. In other words, the SVM depends on samples that makes a non-zero and ignores samples within the boundaries. This may cause deviation when facing strong noise interference or an uneven data distribution. The output modality of the SVM classifier is expressed as the following equation:

$$z(x) = \begin{cases} 1 & f(x) > 0 \\ -1 & f(x) < 0 \end{cases} \tag{4}$$

2.4   KF-SVM

Inspiring from Kernel fisher, the SVM could make an integration with within-class scatter.. Fisher discriminant analysis can make input data relationship conversion from non-linear into linear form. Fisher's linear discriminant is found by maximising $J(w)$ equation:

$$J(w) = w^T M w / w^T N w, \quad N = \sum_{j=1,2} k_j (I - 1_{l_j}) k_j^T, \quad k_j = k(x_{n+m}, x^j) \tag{5}$$

Where the parameter N represent the within-class scatter, k_j represents not only the matrix of $(n + m)*n$ or $(n + m)*m$ but also the kernel function matrix of class j, m represents the first class's sample number, n is the second class's sample number, the parameter I represents the unit matrix, 1_{l_j} represents the matrix with all the inner elements are n-1 or m-1, w represents the transform vector, and M denotes the distance between classes.

On practical data, the Kernel fisher has got excellent results. Its classification error rate could be as low as or even lower compared with the SVM. Considering that SVM algorithm ignores within-class scatter and meanwhile to improve the adaption of a classifier, so the current work proposes adaptive KF-SVM classification method through the way to add within-class scatter inspired from kernel fisher. Combing with the kernel fisher, the SVM can be optimized which could be described as the following formula

$$\min(1/2)w^T(\lambda N + I)w + \gamma(1/2)\sum_{i=1}^{n+m} e_i^2 \tag{6}$$

$$s.t. \ y_i[(w \cdot \phi(x)) + b] = 1 - e_i \ e_i \geq 0, i = 1, 2, 3 \ldots, n + m$$

In a method similar to the SVM, according upper Eq. (6), the optimization matter could be resolved by implementing Eq. (7), with $(zz^T)_{ij} = y_i y_j k^*(x_i, x_j)$. The kernel function has been changed after incorporating within-class scatter to the kernel function. Now, $k^*(x_i, x_j) = \varphi(x_i) * \Sigma^{-1} * \varphi(x_j)^T$, where $\Sigma = \lambda N + I$. To obtain $k^*(x_i, x_j)$, on the basis of the Mercer condition, the afterwards deduction could be achieved: suppose input space $x = x_1, x_2, \ldots, x_n$ and k(x, y) is a symmetric function, for all the involved samples, the matrix can be given by: $k = (k(x_i, x_j))(i, j = 0, 1, \ldots, n)$, which is apparently a symmetric matrix. Definitely, there exists an orthogonal matrix to form $P^T k P = \Lambda$, where Λ represents a diagonal matrix which consists of eigenvalue λ_i, then the eigenvector of λ_i is $v_t = (v_{t1}, v_{t2}, \ldots, v_{tn})^T$, where n represents the size of the sample. The input space could be mapped as the following equation:

$$\phi : x_i \rightarrow (\sqrt{\lambda_1} v_{1i}; \sqrt{\lambda_2} v_{2i}; \ldots, \sqrt{\lambda_n} v_{ni}) \in R^n (i = 1, 2, \ldots, n)$$

Where $\phi_i(x_j) = \sqrt{\lambda_i} v_{ij}$. Then:

$$\phi(x_i) = (\phi_1(x_i), \phi_2(x_i), \ldots, \phi_n(x_i))$$

$$<\phi(x_i), \phi(x_j)> \ = \sum_{t=1}^{n} \lambda_t v_{ti} v_{tj} = k(x_i, x_j)$$

So $k^*(x_i, x_j) = \varphi(x_i) * \Sigma^{-1} * \varphi(x_j)^T$ is obtained and new a and b values are acquired from formula (7). On the basis of this new function with its new a and b values, f(x)function can be expressed as the following equation:

$$f(x) = \sum_{i=1}^{n+m} a_i y_i k^*(x_i, x) + b \tag{7}$$

The decision function f(x) in Eq. (7) varies the traditional SVM algorithm in formula (3). The final class label is determined by whiten the spatial coefficient matrix S and transformation matrix P jointly. So the KF-SVM is formed.

2.5   Comparison Between SVM and KF-SVM

To further explain the difference between SVM and KF-SVM, optimization objective, constraint function and formula are given in Table 1.

Table 1. Comparison between SVM and KF-SVM

	SVM	KF-SVM
Optimization objective, Constraint Function	$\min\ (1/2)w^T I w + \gamma(1/2)\sum_{i=1}^{n+m} e_i^2$ $s.t.\ y_i[(w \cdot \phi(x)) + b] = 1 - e_i$ $e_i \geq 0, i = 1, 2, \ldots n + m$	$\min\ (1/2)w^T(\lambda N + I)w + \gamma(1/2)\sum_{i=1}^{n+m} e_i^2$ $s.t.\ y_i[(w \cdot \phi(x)) + b] = 1 - e_i$ $e_i \geq 0, i = 1, 2, \ldots, n + m$
Formula and steps	Step1: $k(x_i, x_j) = \phi(x_i) \cdot \phi(x_j)$ Step2: $f(x) = \sum_{i=1}^{n+m} a_i y_i k(x_i, x) + b$ Step3: $z(x) = \begin{cases} 1 & f(x) > 0 \\ -1 & f(x) < 0 \end{cases}$	Step1: $k^*(x_i, x_j) = \phi(x_i) * \Sigma^{-1} * \phi(x_j)^T$ Step2: $f(x) = \sum_{i=1}^{n+m} a_i y_i k * (x_i, x) + b$ Step3: $z(x) = \begin{cases} 1 & f(x) > 0 \\ -1 & f(x) < 0 \end{cases}$

2.6   Adaptive KF-SVM

A whole scheme is described in section "research scheme" and we have known that the scheme is divided into two parts including part 1 and part 2. Detailed flow of adaptive KF-SVM to part 1 and part 2 is shown in Fig. 2.

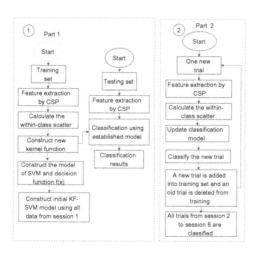

Fig. 2. Detailed flow of adaptive KF-SVMP

The realization steps involved in proposed adaptive algorithm are given below:
Initialization Process (part 1)

Step 1: The feature vector of training data are extracted by the CSP.

Step 2: The within-class scatter N of feature vectors are calculated by formula (5).

Step 3: The new kernel function is constructed by $k^*(x_i, x_j) = \phi(x_i) * \Sigma^{-1} * \phi(x_j)^T$.

Step 4: The decision function $f(x)$ and final classification result are obtained according formula (2) and CSP whiten transformation matrix and CSP spatial coefficient matrix respectively. An initial model is formed.

Step 5: The established initial model is used to classify testing data.

Step 6: The five-fold cross validation is implemented according above steps and afterwards the average recognition rate for the testing data is calculated across these five folds.

Step 7: The adaptive KF-SVM is initialized by implementing the session 1 trials as training data.

Validation Process (part 2)

Step 1: The feature vector of a new single trial i from next sessions is extracted by CSP.

Step 2: The initialized model obtained in training process is used to classify the feature vector and the corresponding values of the decision function $f(x(i))$ and class label can be obtained.

Step 3: The new trial is added into training dataset and at the same time the oldest trial is deleted from the training dataset in order to keep training number constant.

Step 4: The within-class scatter is updated by using new training dataset. New training dataset are trained by the KF-SVM to get a new model, which will replace the old model.

Step 5: When next trial comes, the above four steps are repeated again until all data from session 2 to session 8 are finished. The classification results of all data from these seven sessions are implemented to estimate the performance of the algorithm.

3 Discussions and Results

3.1 Classifier

For the purpose to check the performance of the upper proposed algorithm, the four bellowing classification methods were used in the experiments.

1. SVM: as described above, it uses the following RBF kernel function:

$$K(x, y) = \exp\left(-\frac{||x - y||^2}{\sigma^2}\right) \tag{8}$$

2. KF-SVM: the within-class scatter based on kernel fisher is added into the SVM, so the optimization object, kernel function and decision function of SVM are changed. The model of KF-SVM is established based on these changed parameters. The detailed process can be seen in above Section.
3. Adaptive LDA: the LDA is used as a classifier. The classification result from latest trial will serve as the label of the trial. Meanwhile, the latest trial is added into training set and the oldest trial in the training set will be deleted. By this manner, the classification model is trained and updated trial by trial.
4. Adaptive KF-SVM: the adaptive idea is added into the KF-SVM and the classification model is continuously updated trial by trial.

3.2 Parameters Selection in Proposed Algorithm

In the analysis of the proposed algorithm, two indispensable parameters are required to make optimal selection, that is to say, the parameter of kernel function σ and the parameter of within-class scatter λ. Here, the best σ and λ are selected from the sets $\sigma \in \{1, 2, 3, \ldots, 20\}$ and $\lambda \in \{0, 1, 2, 3. \ldots \ldots, 100\}$ respectively. Five-fold cross-validation was performed using multigroup values of σ and λ on the training data and those resulting in the minimum error are chosen.

3.3 Experiment Results

For each trial, the 2.1 s to 3.1 s time duration is selected as the signal processing period from the total 4 s for each imagination based trial. So every trial's time length was 1 s. A 4-dimensional feature vector $F = \{f_1, f_2, f_3, f_4,\}$ is calculated to every trial. Finally, a 60×4 (60 trials, four dimension of every trial) feature vector from session 1 is used for the initialization of classification model. After initialization, the trials from session 2 to session 8 are classified to make estimation of the different classifiers' performances. Table 2 lists accuracies for seven subjects (Sub1–Sub7) with different classifiers.

3.4 Discussions

The row in blue shade in Table 2 shows that the average classification accuracy of the adaptive KF-SVM is 5.8%, 5.2% and 3.7 higher than those obtained with other three methods (SVM, KF-SVM, Adaptive LDA) in part 2. Meanwhile, the row in grey shade indicates that the adaptive KF-SVM has a slight higher classification accuracy over SVM and adaptive LDA during session 1. It should be noted that the adaptive KF-SVM has the same classification accuracy and model with the KF-SVM because they have the same initialization procedure during session 1. The main difference between adaptive KF-SVM and KF-SVM lies in part 2 (from session 2 to session 8). In order to more clearly describe differences among these three classification methods, all classification results are plotted in Fig. 3. From this figure, we can see that classification accuracies varied significantly from person to person, in which subject Sub7 obtains the best performance with proposed method over other subjects. Classification performance of SVM and KF-SVM classifiers declines during part 2 in relation to part 1, implying that the nonstationarity of EEG affects the classification performance.

Table 2. Classification accuracies with different classifiers

	Subject	Classification accuracy (%)			
		SVM	KF-SVM ($\lambda = 21$, $\sigma = 5$)	Adaptive LDA	Adaptive KF-SVM ($\lambda = 21$, $\sigma = 5$)
Part 1 (Session 1, Initialization)	Sub1	75	77	74.3	77
	Sub2	71	63.33	65.6	63.33
	Sub3	87.5	89.5	85.1	89.5
	Sub4	78.5	86.25	80.8	86.25
	Sub5	75	75	73.2	75
	Sub6	77.5	79	73.8	79
	Sub7	87.5	85	78.5	85
	Mean	78.9	79.3	75.9	79.3
Part 2 (From Session 2 to Session 8)	Sub1	75	75	76.9	85
	Sub2	70	62.5	69.1	72.5
	Sub3	87.5	87.5	86.1	87.5
	Sub4	77.5	80	83.1	85
	Sub5	72.5	75	76.2	77.5
	Sub6	70	75	73.9	77.5
	Sub7	81.5	83	83.5	89.5
	Mean	76.3	76.9	78.4	82.1

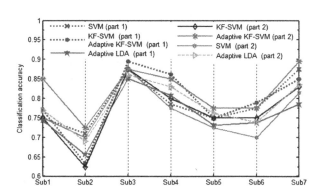

Fig. 3. Comparison of classification results among different subjects and methods

However, adaptive KF-SVM and LDA presents minimal reduction or even some increase, which implies that adaptive idea is helpful to analyze varied EEG. Meanwhile, adaptive KF-SVM achieves the best performance and has better adaptation over other three algorithms.

4   Conclusion

This paper presents a new adaptive KF-SVM classification method combining the kernel fisher, adaptive idea, and the SVM. It takes advantage of the properties of the kernel fisher and overcomes some defects inherent to the SVM. Meanwhile, the within-class scatter in the adaptive KF-SVM is continuously updated trial by trial, which could improve adaptation of the new classifier. The upper proposed method is verified by comparing it with other three algorithms. The results show that the upper proposed method could obtain satisfying recognition accuracy. It may be practical for online application in BCIs. The next-step research should take aim at the verification of the algorithm on bigger data.

References

1. Wolpaw, J.R., Birbaumer, N., Heetderks, W.J., McFarland, D.J., Peckham, P.H., Schalk, G., et al.: Brain–computer interface technology: a review of the first international meeting. IEEE Trans. Rehabil. Eng. **8**, 164–173 (2000)
2. Arvaneh, M., Guan, C.T., Ang, K.K., Quek, C.: Optimizing spatial filters by minimizing within-class dissimilarities in electroencephalogram-based brain–computer Interface. IEEE Trans. Neural Netw. Learn. Syst. **24**(4), 610–619 (2013)
3. Liu, G.Q., Zhang, D.G., Meng, J.J., Huang, G., Zhu, X.Y.: Unsupervised adaptation of electroencephalogram signal processing based on fuzzy C-means algorithm. Int. J. Adapt. Control Signal Process. **26**, 482–495 (2012)
4. Wang, Y.-T., Nakanishi, M., Wang, Y., Wei, C.-S., Cheng, C.-K., Jung, T.-P.: An online brain-computer interface based on SSVEPs measured from non-hair-bearing areas. IEEE Trans. Neural Syst. Rehabil. Eng. **25**(1), 11–18 (2017)
5. Mondini, V., Mangia, A.L., Cappello, A.: EEG-Based BCI system using adaptive features extraction and classification procedures. Comput. Intell. Neurosci. **2016**, 1–14 (2016)
6. Yan, W., Ge, Y.B.: A novel method for motor imagery EEG adaptive classification based biomimetic pattern recognition. Neurocomputing **116**, 280–290 (2013)
7. Seo, D., et al.: Wireless recording in the peripheral nervous system with ultrasonic neural dust neuron neuroresource wireless recording in the peripheral nervous system with ultrasonic neural dust. Neuron **91**, 1–11 (2016)
8. Yang, H., Sakhavi, S., Kai, K.A., Guan, C.: On the use of convolutional neural networks and augmented CSP features for multi-class motor imagery of EEG signals classification. In: 2015 IEEE 37th Annual International Conference on Engineering in Medicine and Biology Society (EMBC), pp. 2620–2623 (2015)
9. Hsu, W.Y.: EEG-based motor imagery classification using enhanced active segment selection and adaptive classifier. Comput. Biol. Med. **41**, 633–639 (2011)
10. Wu, D.: Online and Offline Domain Adaptation for Reducing BCI Calibration Effort. IEEE Transactions on human-machine Systems, pp. 1–14 (2016)
11. Ye, Q.L., Zhao, C.X., Ye, N.: A New SVM classification approach via minimum within-class variance. J. Comput. Inf. Syst. **6**(1), 39–45 (2011)
12. Spüler, M., Rosenstiel, W., Bogdan, Martin: Adaptive SVM-based classification increases performance of a MEG-based brain-computer interface (BCI). In: Villa, A.E.P., Duch, W., Érdi, P., Masulli, F., Palm, G. (eds.) ICANN 2012. LNCS, vol. 7552, pp. 669–676. Springer, Heidelberg (2012). doi:10.1007/978-3-642-33269-2_84

13. Kaper, M., Meinicke, P., Grossekathoefer, U., Lingner, T., Ritter, H.: BCI competition 2003-data set IIb: support vector machines for the P300 speller paradigm. IEEE Trans. Biomed. Eng. **51**, 1073–1076 (2004)
14. Hema Rajini, N., Bhavani, R.: Automatic classification of computed tomography brain images using ANN, k-NN and SVM. AI Soc. **29**, 97–102 (2014)

Research on Non-frontal Face Detection Method Based on Skin Color and Region Segmentation

Haonan Wang[(✉)] and Tianfei Shen

Department of Automation, School of Mechatronic Engineering and Automation,
Shanghai University, Shanghai 200040, China
w478586630@gmail.com

Abstract. The detection of face region can be divided into two kinds: frontal and non-frontal faces. This thesis focuses on the detection of human face region in non-frontal cases. A method of separating face and neck region is presented to extract the non-frontal face in the image. Facial features are usually used in frontal face detection, such as eyes, mouth and etc. With complete facial features, the frontal face can be easier to detected with high accuracy now. However, the research on non-frontal face detection is just beginning. Since the non frontal face image can not provide complete facial features information, it is necessary to develop a new method. Skin color is the most prominent facial feature in the non-frontal cases. It is found that the skin color has better clustering capability in YCbCr color space. According to the skin color characteristics and illumination conditions in the YCbCr color space, the Gaussian model and the Otsu method are used to segment the skin color to extract the non-frontal face region in the images. But the segmented skin color area often contains the neck region. In this paper, the contour line of the chin is fitted by illumination intensity and position information, remove the neck area and get a face region without the neck. Simulation results show the effectiveness of the proposed method for the detection of non-frontal face region.

Keywords: Non-frontal face detection · Gaussian model · Edge detection · Region growing

1 Introduction

Face detection is a kind of computer technology which can detect the position and size of faces in any digital images. Traditional face detection is aimed at frontal faces, while non frontal face is based on side faces. Traditional face detection usually uses facial features to detect face, however facial features may not be captured or only partially captured in non frontal case. Instead of facial features, skin color model will be used in this thesis. As the most prominent feature of human face, skin color is more and more widely used in face detection system. YCbCr is a kind of color space which contains illumination information. Skin color has good clustering in this color space. The distribution of Cb and Cr components of skin color is almost the same in YCbCr color space which is consistent with two-dimensional Gauss distribution. An improved Gaussian skin color model based on Otsu method has better skin color segmentation effect.

© Springer Nature Singapore Pte Ltd. 2017
M. Fei et al. (Eds.): LSMS/ICSEE 2017, Part I, CCIS 761, pp. 46–52, 2017.
DOI: 10.1007/978-981-10-6370-1_5

In non-frontal cases, the segmented skin area is made up of the face region and the neck region. Typically, the face region is located above the neck region. Edge detection and eight connected domain method are used to find a suitable fitting curve between the face region and the neck region. A distinct feature in neck region is the shadow of the chin. Fitting the curve of the chin through the brightness changes in the neck area. Region growth is used to remove the are below the chin.

2   Skin Color Segmentation

2.1   YCbCr Color Space

YCbCr is a kind of color space which separating color and brightness. YCbCr can be transformed from the RGB color space [2, 9], where Y represents the brightness information, Cb and Cr on behalf of the color information. The transform matrix is as follows:

$$
\begin{bmatrix} Y \\ Cb \\ Cr \end{bmatrix} = \begin{bmatrix} 16 \\ 128 \\ 128 \end{bmatrix} + \begin{bmatrix} 65.481 & 128.553 & 24.996 \\ -37.797 & -374.207 & 112 \\ 112 & -93.786 & -18.214 \end{bmatrix} \begin{bmatrix} R \\ G \\ B \end{bmatrix} \tag{1}
$$

Figure 1 shows the aggregation of skin color in YCbCr color space [3]. The blue dots represent all collected pixels in the image, and the red dots represent the skin color pixels.

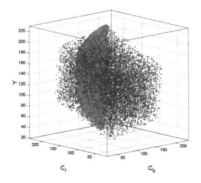

Fig. 1. The aggregation of skin color in YCbCr color space (Color figure online)

2.2   Otsu (Maximum Inter-class Variance Method)

The maximum inter-class variance method was proposed by Otsu [11] in 1978. This is a stable and commonly used algorithm which based on the principle of least squares method. In Gaussian skin color model, the selected segmentation threshold should make the difference between the average gray level of the whole image and the target area largest, the same to the non target area. Otsu can be used to adjust the segmentation threshold adaptively.

The specific algorithm of Otsu is as follows:

Gray level range is $[0, L-1]$, the number of pixels of gray level i is n_i, then the total number of pixels is $N = \sum_{i=0}^{L-1} n_i$, the probability of each gray level is $p_i = \frac{n_i}{N}$, and $\sum_{I=0}^{L-1} p_i = 1$. Using threshold T to divide the pixels into c_0 and c_1, pixels in $[0, T-1]$ belong to c_0 and $[T, L-1]$ belong to c_1. Then the probability of regional c_0 and c_1 are:

$$p_0 = \sum_{i=0}^{L-1} p_i$$
$$p_1 = \sum_{i=T}^{L-1} p_i = 1 - p_0 \tag{2}$$

The average gray level of area C1 and C2 are:

$$\mu = \frac{1}{P_0}\sum_{i=0}^{L-1} ip_i = \frac{\mu(T)}{p_0}$$
$$\mu_1 = \frac{1}{P_0}\sum_{i=T}^{L-1} ip_i = \frac{\mu - \mu(T)}{1 - p_0} \tag{3}$$
$$\mu = \sum_{i=0}^{L-1} ip_i = \sum_{i=0}^{L-1} ip_i + \sum_{i=T}^{L-1} ip_i = p_0\mu_0 + p_1\mu_1$$

(μ is the average gray level of whole image)

The total variance of the two regions are:

$$\sigma_B^2 = p_0(\mu_0 - \mu)^2 + p_1(\mu_1 - \mu)^2 = p_0p_1(\mu_0 - \mu)^2 \tag{4}$$

2.3 Improved Gaussian Skin Color Model

In the YCbCr color space, the distribution of color components Cb and Cr of different skin color tends to be consistent, which is similar to the two-dimensional Gaussian distribution N(M, C) (Fig. 2).

By calculating the probability of each pixel, a skin color similarity face image is formed. Finally, according to the segmentation threshold of Otsu, the skin color segmentation area is obtained. The probability of the pixels is calculated by the Gaussian probability density function:

$$P(Cb, Cr) = \exp\left[-0.5(\mathbf{x} - \mathbf{M})^T\mathbf{C}^{-1}(\mathbf{x} - \mathbf{M})\right] \tag{5}$$

X is the Cr and Cb value of sample pixels in YCbCr color space, $\mathbf{x} = [Cb.Cr]^T$, M is the sample mean value of Cr and Cb, M = E(x), C is the variance matrix of skin

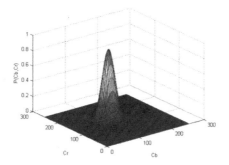

Fig. 2. Two dimensional Gaussian distribution of skin color in YCbCr color space

color similarity model, C = E((x − M)(x − M)T). Through a large number of simulation experiments, it is concluded that the mean M and variance C values are:

$$C = \begin{bmatrix} 160.130 & 12.1430 \\ 12.1430 & 299.457 \end{bmatrix}$$

$$M = (148.5599, 117.4361)$$

(6)

Gaussian skin color model [8] transfer the RGB image into gray image. The gray value of pixels is represented by the possibility which pixels belonging to the skin area. Then, the gray image can be further converted to binary image by threshold. The general method of binarization is to set the fixed threshold value, but the effect is not ideal. In this thesis, we will use Otsu to automatically adjust the threshold of skin segmentation in Gaussian model (Fig. 3).

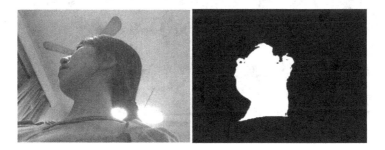

Fig. 3. Improved Gaussian skin color model skin segmentation

3 Edge Detection and Eight Connected Domain Method

In order to realize the approximate separation of neck and face, the edge detection and eight connected domain method are used to find the suitable segmentation line [10]. Canny operator is used to detect the facial contour after skin color segmentation, and then the eight domain method is used to find the coordinates of all pixels. The eight connected domain method can detect and extract all the closed curve among all the edges in binary image. We choose the eight connected domain as the standard algorithm. As we can see, there will be only one edge curve in the skin color segmentation image. This algorithm can collect all pixels of the skin color contour (Fig. 4).

Fig. 4. Skin color contour extraction

In the experiment, because the position of the camera is fixed, the position of the mouth in the face contour is about 1/3 of the whole contour height. After the approximate segmentation, we can guarantee that the upper part dose not contain the neck, the lower is the part of the chin and neck area (Fig. 5).

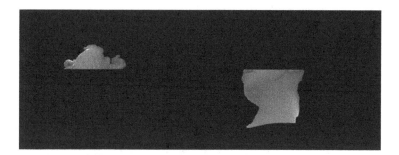

Fig. 5. Separation of face region and neck region

4 Chin Fitting

A large number of experiments show that there is a linear relationship between the average brightness of the neck and the chin. Converting the neck part to YCbCr color space to extract the brightness of the pixels [1, 7]. According to the average brightness value of the neck part, dynamically adjust the threshold of the chin shadow to find the edge of the chin (Fig. 6).

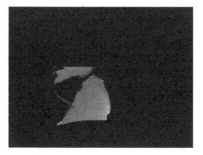

Fig. 6. Neck part chin fitting

5   Region Growing Algorithm for Neck Culling

Region growing is the process of integrating pixels or sub regions into larger regions based on a previously defined. The basic idea is to start with a set of growth points which can be either a single pixel or a small area. Combining the adjacent pixels or regions which are similar to the growth point to form a new growth point. Repeating this process until it cannot grow. The similarity judgment of growing points and adjacent regions can be gray level, texture, color and other image information. We choose gray level to judge the growth point because there will be a binary image of face.

Using eight field method to find the face contour and neck contour. In order to culling the neck area, the center coordinates of the face contour are calculated as the starting point for the region growth (Fig. 7).

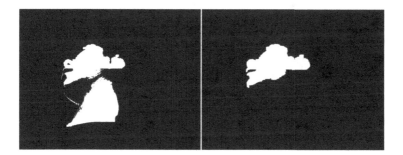

Fig. 7. Regional growth culling neck part

6   Conclusion and Further Work

The experimental results show that this method has a good effect on non-frontal face detection which can fitting the contour of the chin and culling the neck part. However, the experimental results are obtained under ideal illumination conditions. When the position and strength of the light source changed, the separation effect may be poor. Adapting to more complex conditions will be the focus of future research.

References

1. Finlayson, G.D., Hordley, S.D., Drew, M.S.: Removing shadows from images. In: Heyden, A., Sparr, G., Nielsen, M., Johansen, P. (eds.) ECCV 2002. LNCS, vol. 2353, pp. 823–836. Springer, Heidelberg (2002). doi:10.1007/3-540-47979-1_55
2. Teng, Q., Shen, T., Yang, J.: Research on face detection system based on multi-skin color models. Electron. Measur. Technol. **38**(9), 47–51 (2015)
3. Teng, Q., Yang, J., Fang, Y.: Research on face detection system under multiple head gesture. Ind. Control Comput. **29**(1), 91–95 (2016)
4. Tsitsoulis, A., Bourbakis, N.: A methodology for detecting faces from different views. In: IEEE 24th International Conference on Tools with Artificial Intelligence (ICTAI), vol. 1, pp. 238–245. IEEE (2012)
5. Jain, V., Patel, D.: A GPU based implementation of robust face detection system. Proc. Comput. Sci. **87**, 156–163 (2016)
6. Orozco, J., Martinez, B., Pantic, M.: Empirical analysis of cascade deformable models for multi-view face detection. Image Vis. Comput. **42**, 47–61 (2015)
7. Hua-nan, Z., Quan, F., Mei, Y., Miao-Qi, L.: Shadow detection and removal of blade on YCbCr color space. Comput. Syst. Appl. **24**(11), 262–265 (2015)
8. Zhou, L., Gu, L.: The detection of face and chin based on Gaussian skin color model. J. Xi'an Polytech. Univ. **29**(6), 751–755 (2015)
9. Jin, X., Chang, Q.: RGB to YCbCr color space transform based on FPGA. Mod. Electron. Tech. **18**, 73–75 (2009)
10. Hong-ke, X., Yan-yan, Q., Hui-ru, C.: An improved algorithm for edge detection based on Canny. Infrared Technol. **36**(3), 210–214 (2014)
11. Qi, L., Zhang, B., Wang, Z.: Application of the OTSU method in image processing. Radio Eng. **36**(7), 25–26 (2006)

Modelling and Control Design for Membrane Potential Conduction Along Nerve Fibre Using B-spline Neural Network

Qichun Zhang$^{(\boxtimes)}$ and Francisco Sepulveda

University of Essex, Colchester CO45FT, UK
{qichun.zhang,f.sepulveda}@essex.ac.uk

Abstract. Based on B-spline neural network, the analysis of membrane potential conduction has been presented for peripheral nerve fibres whereby the effects of the interactions between axons have been taken into account. In particular, the modelling problem is investigated firstly with the vector-valued weight transformation and parameter identification. Using the presented model, the control design is proposed to reproduce the membrane potential along nerve fibres. The algorithm procedure and interaction characterization for coupled axons are given while the numerical simulation illustrates the effectiveness of the presented algorithm.

Keywords: Membrane potential conduction · B-spline neural network · Neural interaction · Stimulation design · Computational model

1 Introduction

As a significant research topic, the mechanism description of the membrane potential conduction of nerve axons has been investigated since the famous Hodgkin-Huxley model [1] was presented in 1952, e.g., [2–4]. Moreover, the conduction of membrane potentials along nerve fibres can be further analysed combining this model and the cable equation. All of these results are very helpful for medical applications such as modelling peripheral fibres towards sensory rehabilitation using next generation prosthesis [5,6].

However, there are two main problems that cannot be ignored. Firstly, the solution of the cable equation based model is very difficult to obtain, even the numerical solution, due to the fact that it is a partial differential equation. Secondly, the neural interaction between the coupled axons have to be taken into account because this interaction widely exists in vivo, e.g., in the form of neuron-to-neuron ephaptic interactions [7]. This means that the membrane potential produces the associated electrical field [4] which would affect other membrane potentials of coupled axons in a physical sense.

Motivated by the modelling of stochastic distribution systems [8], the PDE description of the nerve signal conduction can be restated by dynamic weights

© Springer Nature Singapore Pte Ltd. 2017
M. Fei et al. (Eds.): LSMS/ICSEE 2017, Part I, CCIS 761, pp. 53–62, 2017.
DOI: 10.1007/978-981-10-6370-1_6

using an NN-based transformation. Due to the fact that NNs can in principle approximate any nonlinear function, then the coupled membrane potential of nerve fibres at any location can be any general function in terms of time which means that the interaction has been considered. Along the nerve fibres, the conduction of nerve signal with various locations can be described by the dynamics of the transformed weights. Based on the data and on system identification methods, the dynamic function can be estimated by the linear format, and the coefficient matrices can be identified. Following this approach, we can predict the membrane potential at any location and avoid solving PDEs. Similarly, the stimulation signal of the nerve fibre can also be rewritten by weights and the relationship between the membrane potential and stimulation can be further formulated as an input vector. Once the model is obtained, the membrane potential at any location can be obtained with initial weights. Moreover, the stimulation signal can be designed at any location to control the shape of the membrane potential. In addition, a novel approach is presented based on this model where the interaction can be further characterized. Furthermore, motivated by the mutual coupling factor [6], the neural interaction can also characterized by the model with the identified coefficient matrices. As a summary, the contributions of this paper include as follows: (1) a novel NN-based approach for modelling peripheral nerve fibres; (2) the control design for reshaping the membrane potential along the nerve fibres; (3) a novel interaction characterization method.

2   Modelling

2.1   Measurement and Dynamics of Membrane Potential

Suppose that the membrane potentials can be measured individually for each axon. Then, the nerve signal would be collected along the nerve fibres at various locations. Without loss of generality, the measurement and dynamics of the membrane potential are shown in Fig. 1 where the stimulation is applied to the nerve fibres at the same location of the membrane potential collection.

2.2   Weight-Based Transformation via Neural Network

Basically, the NN can be used to approximate any nonlinear function such as the membrane potential at location sampling instance which is a nonlinear function of time. Due to the sample structure, the B-spline NN has been chosen in this paper. Therefore, the membrane potentials for n coupled axons can be restated as follows:

$$V_{j,k}(t) = \sum_{i=1}^{m} w_{j,k,i} B_i + e_{j,k}, j = 1, \ldots, n \tag{1}$$

where j and k denote the indexes of location sampling and individual axon, respectively. w and B denote the weight and B-spline basis function while m is the pre-specified number of basis functions. e stands for the approximation

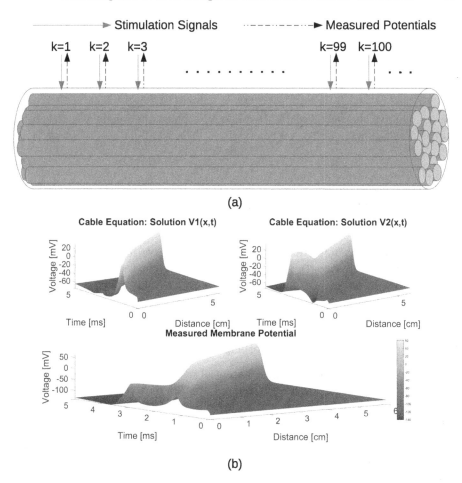

Fig. 1. The problem description of this paper: (a) the signals are collected along the nerve fibre with multiple coupled axons where the index k denotes the location sampling; (b) the measured signal is shown in a 3D mesh while the distance is discretized by k and the membrane potential is not with the typical shape due to the neural interaction.

error and it is ignorable if value of m is sufficiently large and the weighs can be trained by data sets [9].

In other words, the membrane potential at location k can be represented by the m-dimensional weight vector, which can be further denoted by the following vector.

$$W_{j,k} = [w_{j,k,1}, w_{j,k,2}, \ldots, w_{j,k,m}]^T, j = 1, \ldots, n \tag{2}$$

Similarly, the stimulation signals can also be restated following the NN-based transformation. Thus, we have

$$I_{j,k}(t) = \sum_{i=1}^{m} s_{j,k,i} B_i + e_{j,k}, j = 1, \ldots, n \tag{3}$$

where s denotes the weight of the B-spline NN. Then, the weight vector can be expressed as follows.

$$S_{j,k} = [s_{j,k,1}, s_{j,k,2}, \ldots, s_{j,k,m}]^T, j = 1, \ldots, n \tag{4}$$

Notice that the number of basis functions for these two transformation can be different which implies that the dimensions of $W_{j,k}$ and $S_{j,k}$ can be unmatched. However, in this paper, we select the same dimension for mathematical simplicity purposes.

As a result, the complete neural signal of the nerve fibre with interaction can be represented by the compact vector while n weight vectors are taken into account.

$$W_k = \left[W_{1,k}^T, W_{2,k}^T, \ldots, W_{n,k}^T \right]^T, S_k = \left[S_{1,k}^T, S_{2,k}^T, \ldots, S_{n,k}^T \right]^T \tag{5}$$

2.3   Weight-Based Dynamic Description

Next, as we know that the membrane potential would be slightly changed by the conduction which means the weight vectors are different at different locations, the dynamics of the conduction can be described by the dynamic of the weight vectors which can be expressed as a general nonlinear function.

$$W_{j,k+1} = f(W_1, W_2, \ldots, W_k), j = 1, \ldots, n \tag{6}$$

For the more general situation, the stimulation signals should also be taken into account along with the investigated nerve fibre. Thus, the function should be further formulated as follows.

$$W_{j,k+1} = \bar{f}(W_1, W_2, \ldots, W_k, S_1, S_2, \ldots, S_k), j = 1, \ldots, n \tag{7}$$

It shows that the weights of the individual axon would be affected by other axons due to the effect of the inter-neural interaction.

To obtain the relationship between the weight vectors, the linearisation operation can be considered to simplify the identification due to the fact that the structure of the general nonlinear function is unknown. Then the dynamics can be suitably described by the following recursive format.

$$W_{k+1} = A_k W_k + G_k S_{k+1} \tag{8}$$

where A_k and B_k are sampling-varying coefficient matrices with appropriate dimension. Meanwhile, Eq. (1) can be rewritten by

$$V_{j,k}(t) = C W_{j,k}, C = [B_1, B_2, \ldots, B_m] \tag{9}$$

Combining Eqs. (8) and (9), the complete B-spline NN based model is obtained as a representative of the nerve signal conduction.

Next, the system identification method should be used here to estimate the matrices A_k and G_k. Once the signals $V_{j,k}$, $V_{j,k-1}$ and $I_{j,k-1}$ are collected, the coefficient matrices A_k and G_k can be obtained since matrix C is known. In particular, the simplified form can be used to reduced the complexity if the weight vector can be collected for each location sampling. Then, we can have

$$w_{k,i} = \sum_{j=1}^{m} a_{ij} w_{k-1,j} + \sum_{j=1}^{m} g_{ij} s_{k-1,j}, i = 1, \ldots, m \tag{10}$$

where a_{ij} and b_{ij} denote the elements of matrices A_k and G_k.

Rearranging Eq. (10), it results in the standard form for parameter identification.

$$w_{k,i} = \theta_{k,i}^T \Phi_k, i = 1, \ldots, m \tag{11}$$

$$\theta_{k,i} = [a_{i1}, \ldots, a_{im}, g_{i1}, \ldots, g_{im}]^T, \Phi_k = \left[W_{k-1}^T, S_{k-1}^T \right]^T \tag{12}$$

Then, the recursive least square (RLS) algorithm [10] can be used to estimate the parameter vector θ as follows.

$$\theta_{k,i}(l+1) = \theta_{k,i}(l) + \frac{P(l-1)\Phi_k \varepsilon(l)}{1 + \Phi_k^T P(l-1)\Phi_k} \tag{13}$$

$$\varepsilon(l) = w_{k,i} - \theta_{k,i}^T(l)\Phi_k \tag{14}$$

$$P(l) = \left(\bar{I} - \frac{P(l-1)\Phi_k}{1 + \Phi_k^T P(l-1)\Phi_k} \right) P(l-1) \tag{15}$$

where l is the index of the identification iteration.

Once the parameter vector $\theta_{k,i}$ estimation is completed, the modelling of the simplified peripheral nerve fibre with neural interaction is done.

3   Control Design

Stimulation signal design is a significant challenge for many neural medical applications, such as sensory rehabilitation via next generation prostheses. In particular, the response of the nerve fibres should be reproduced by electrical stimulation with pre-specified location.

Based upon the presented model, the stimulation design can be transformed as the weights tracking control problem. Suppose that we have the reference weights for the location samplings which is denoted by $W_{ref,k}$.

To achieve the near-perfect tracking, the integrator should be introduced into the stimulation design which is defined as follows.

$$\varepsilon_{k+1} - \varepsilon_k = W_{ref,k} - W_k \tag{16}$$

Thus, the model with the tracking integral can be formulated by

$$\begin{bmatrix} \varepsilon_{k+1} \\ W_{k+1} \end{bmatrix} = \bar{A}_k \begin{bmatrix} \varepsilon_k \\ W_k \end{bmatrix} + \bar{G}_k S_k + \begin{bmatrix} W_{ref,k} \\ 0 \end{bmatrix} \tag{17}$$

where

$$\bar{A}_k = \begin{bmatrix} I & -I \\ 0 & A_k \end{bmatrix}, \bar{G}_k = \begin{bmatrix} 0 \\ G_k \end{bmatrix} \tag{18}$$

Based on the control theory of linear system [11], the stimulation signal can be designed as a closed-loop input with a linear gain matrix:

$$S_k = K_k \begin{bmatrix} \varepsilon_k \\ W_k \end{bmatrix} \tag{19}$$

where K_k should be selected to hold the inequality $\bar{A}_k + \bar{B}_k K_k$ is Hurwitz. Also, K_k can be further optimized following performance criteria such as the minimum input energy or LQR design [12].

In practice, we can use one electrode as the stimuli at one single location sampling, which means that the model can be divided into two parts: The stimulation signal leads to a membrane potential response as a static function. Then, the stimulated membrane potential can be considered as the initial value and recursively update as an autonomous system. Also notice that S_k is the compact vector of the weights for individual stimulation.

4   Simulation

Following the presented algorithm, the membrane potential can be transformed to weight vector since the B-spline NN can be used with training. In this section, we consider two coupled axons. The membrane potential of axon 1 at $k = 50$ is shown by Fig. 2 where all the simulation data is in the 3D mesh in Fig. 1.

Based on the neural networks, the training performance is given by Fig. 3, which indicates that the performance with mean-square criteria is reduced by the number increase of the basis functions.

Notice that the computational loads would be drastically increased if the number of basis functions is chosen to be large. Without loss of generality, the simple numerical example is given below where we only pre-specified 4 basis functions. Although the accuracy of the weight transformation is weak, the performance and procedure of the algorithm can be demonstrated.

The stimulus is set up as an intracellular current density of $0.1 \, \mathrm{mA/cm^2}$ and the duration is $0.2 \, \mathrm{ms}$, which is similar to the settings in [3]. When this stimulus is applied to one axon, the responses of other axons would achieve the action potential due to the interaction. Moreover, the measured membrane potentials at different location samples are expressed by the following table via weights (Table 1).

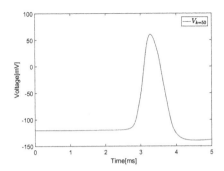

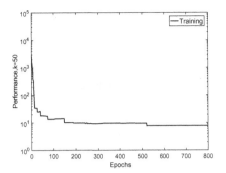

Fig. 2. The membrane potential of axon 1 at location sampling $k = 50$

Fig. 3. The training performance of the neural networks for membrane potential $V_{k=50}$

Table 1. The weight transformation of membrane potential and stimulation

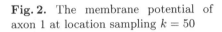

Weight transformation			
Sampling index	Weight vector for axon 1	Weight vector for axon 2	Stimulation weight for axon 1
$k = 1$	1.425 3.295 0.545 2.615	3.325 5.000 2.785 4.995	0 0.005 0 0
$k = 2$	1.460 3.295 0.505 2.615	3.330 5.000 2.780 4.995	0
$k = 3$	1.505 3.300 0.535 2.630	3.335 5.000 2.770 4.995	0
$k = 5$	1.600 3.350 0.650 2.710	3.355 5.000 2.740 4.995	0
$k = 10$	1.840 3.510 1.015 3.085	3.360 5.000 2.750 4.995	0
$k = 20$	2.245 3.805 1.570 3.710	3.050 4.815 2.580 5.000	0
$k = 50$	3.270 5.000 2.625 4.995	3.270 5.000 2.625 4.995	0
$k = 70$	3.880 5.000 3.935 4.995	3.880 5.000 3.935 4.995	0

From the adjustment of the weights, the conduction of the nerve signal is very clear. Then, the dynamic of these weights can be estimated following the presented algorithm. Notice that we assume $W_{k=0} = 0$ then the matrix G_1 can be calculated directly based on the static relationship between the membrane potential and stimulation with $k = 1$. Thus we have $G_{21} = G_{22} = G_{12} = 0$ and

$$G_{11} = \begin{bmatrix} 0 & 755 & 0 & 0 \\ 0 & 178 & 0 & 0 \\ 0 & 1000 & 0 & 0 \\ 0 & 609 & 0 & 0 \end{bmatrix}$$

Based upon the RLS algorithm, the estimation error curves of the membrane potential weights are shown by Figs. 4 and 5 while all the curves are convergent and the errors are very small. Furthermore, the coefficient matrix A is obtained at $k = 3$ and $k = 6$.

$$A_3 = \begin{bmatrix} 0.0969 & -0.0250 & -0.1896 & -0.0250 & 0.9118 & 0.0456 & -0.0413 & 0.1733 \\ 0.1218 & 0.1851 & 0.1035 & 0.1849 & 0.0456 & 0.1228 & 0.0422 & 0.0994 \\ -0.0361 & -0.0224 & -0.0482 & -0.0223 & -0.0413 & 0.0422 & 0.9653 & 0.1592 \\ 0.0959 & 0.1225 & 0.0351 & 0.1223 & 0.1733 & 0.0994 & 0.1592 & 0.1201 \\ 0.1297 & 0.1854 & 0.0972 & 0.1852 & 0.0969 & 0.1218 & -0.0361 & 0.0959 \\ 0.1854 & 0.2977 & 0.1894 & 0.2974 & -0.0250 & 0.1851 & -0.0224 & 0.1225 \\ 0.0972 & 0.1894 & 0.1555 & 0.1892 & -0.1896 & 0.1035 & -0.0482 & 0.0351 \\ 0.1852 & 0.2974 & 0.1892 & 0.2971 & -0.0250 & 0.1849 & -0.0223 & 0.1223 \end{bmatrix}$$

$$A_6 = \begin{bmatrix} 0.0126 & -0.0049 & 0.0052 & -0.0049 & 0.9998 & -0.0016 & 0.0007 & -0.0010 \\ 0.0864 & -0.0332 & 0.0353 & -0.0332 & -0.0016 & 0.9889 & 0.0050 & -0.0066 \\ -0.0386 & 0.0149 & -0.0158 & 0.0148 & 0.0007 & 0.0050 & 0.9978 & 0.0030 \\ 0.0516 & -0.0199 & 0.0211 & -0.0198 & -0.0010 & -0.0066 & 0.0030 & 0.9960 \\ 0.3281 & 0.2584 & -0.2743 & 0.2581 & 0.0126 & 0.0864 & -0.0386 & 0.0516 \\ 0.2584 & 0.4011 & 0.1055 & 0.4007 & -0.0049 & -0.0332 & 0.0149 & -0.0199 \\ -0.2743 & 0.1055 & 0.8880 & 0.1054 & 0.0052 & 0.0353 & -0.0158 & 0.0211 \\ 0.2581 & 0.4007 & 0.1054 & 0.4003 & -0.0049 & -0.0332 & 0.0148 & -0.0198 \end{bmatrix}$$

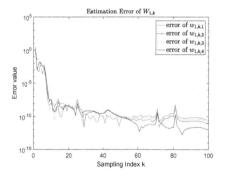
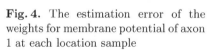

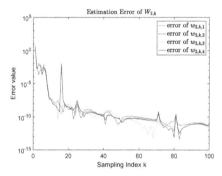

Fig. 4. The estimation error of the weights for membrane potential of axon 1 at each location sample

Fig. 5. The estimation error of the weights for membrane potential of axon 2 at each location sample

Therefore, the complete model can be formulated by $W_{k+1} = A_k W_k + G_k S_k$. Due to the fact that the membrane potentials are strongly nonlinear, the estimated coefficient matrix A should be considered as time-varying or interval-based time varying matrix, which would improve the accuracy of the model.

Following the presented control design, near-perfect tracking can be achieve based on the linear control system design. The effectiveness of this control design has been demonstrated in previous studies such as [13]; we therefore omit the control simulation here.

5   Further Discussion on Interaction Characterization

The interaction between the coupled axons in nerve fibres cannot be ignored, as discussed in Sect. 1. Following the existing analysis in [6], this interaction can be characterized by the mutual coupling factor matrix based on the Hodgkin-Huxley model. Following a similar idea, the norm-based mutual coupling factors are developed as a characterization function in terms of location sampling index k.

Based on the presented model with weight vector, the interaction has already been included in the transformed weight vector. In order to characterise the interaction, we only need to investigate the characterisation of the weight vector dynamics. Thus, we divide the coefficient matrices A_k and G_k into sub-matrices, as follows:

$$A_k = \begin{bmatrix} A_{1,k} & A_{12,k} & \cdots & A_{1n,k} \\ A_{21,k} & A_{2,k} & \cdots & A_{2n,k} \\ \vdots & \vdots & \ddots & \vdots \\ A_{n1,k} & A_{n2,k} & \cdots & A_{n,k} \end{bmatrix}, G_k = [G_{1,k}, \ldots, G_{n,k}] \tag{20}$$

Since the interaction is independent of the stimulation, only A_k should be considered to characterise the interaction. Based on the mutual coupling factor, the generalized mutual coupling factor is described by the following norm function:

$$\gamma_{ij,k} = \|A_{ij,k}\|, i, j = 1, \ldots, n, i \neq j \tag{21}$$

In this case, the norm of the associate sub-matrix can be considered as a gain which can be used to describe the coupling effects from other axons. Furthermore, the interaction should be characterised by the following matrix:

$$\Lambda_k = \begin{bmatrix} 0 & \gamma_{12,k} & \cdots & \gamma_{1n,k} \\ \gamma_{21,k} & 0 & \cdots & \gamma_{2n,k} \\ \vdots & \vdots & \ddots & \vdots \\ \gamma_{n1,k} & \gamma_{n2,k} & \cdots & 0 \end{bmatrix} \tag{22}$$

Notice that the generalized mutual coupling factor matrix is not symmetric, which means that the neural interaction is an axon-to-axon asymmetric mutual influence.

6   Conclusion

The modelling problem has been investigated for complex nonlinear dynamics of nerve signal conduction along coupled axons. Different from the existing PDE approach, a B-spline neural network based model has been presented whereby the membrane potentials at any location along the axon are restated by the weight vector of the neural network. Furthermore, the dynamics of the weight vector can be estimated by parameter identification methods. In addition, the membrane potential can be controlled by stimulation design using the presented model with linear control theory, and the interaction characterisation problem is also addresser by the generalized mutual coupling factor matrix. As a summary, the contributions of this paper include: (1) a novel NN-based approach for modelling peripheral nerve fibres; (2) the control design for reshaping the membrane potential along the nerve fibres; (3) a novel interaction characterisation method.

Acknowledgements. This work is supported by UK's EPSRC-funded SenseBack project with grant number EP/M025977/1.

References

1. Hodgkin, A.L., Huxley, A.F.: A quantitative description of membrane current and its application to conduction and excitation in nerve. J. Physiol. **117**(4), 500 (1952)
2. Frankenhaeuser, B., Huxley, A.: The action potential in the myelinated nerve fibre of Xenopus laevis as computed on the basis of voltage clamp data. J. Physiol. **171**(2), 302 (1964)
3. Rattay, F.: The basic mechanism for the electrical stimulation of the nervous system. Neuroscience **89**(2), 335–346 (1999)
4. Joucla, S., Yvert, B.: Improved focalization of electrical microstimulation using microelectrode arrays: a modeling study. PloS one **4**(3), e4828 (2009)
5. Kolbl, F., Juan, M.C., Sepulveda, F.: Impact of the angle of implantation of transverse intrafascicular multichannel electrodes on axon activation. In: 2016 IEEE Biomedical Circuits and Systems Conference (BioCAS), pp. 484–487, October 2016
6. Zhang, Q., Sepulveda, F.: A statistical description of pairwise interaction between the nerve fibres. In: 2017 8th International IEEE/EMBS Conference on Neural Engineering (NER). IEEE (2017, in press)
7. Buzsáki, G., Anastassiou, C.A., Koch, C.: The origin of extracellular fields and currents – EEG, ECoG, LFP and spikes. Nat. Rev. Neurosci. **13**(6), 407–420 (2012)
8. Wang, H.: Bounded Dynamic Stochastic Systems: Modelling and Control. Advances in Industrial Control. Springer, London (2000)
9. Billings, S.A., Zheng, G.L.: Radial basis function network configuration using genetic algorithms. Neural Netw. **8**(6), 877–890 (1995)
10. Marple, S.L.: Digital Spectral Analysis: With Applications, vol. 5. Prentice-Hall, Englewood Cliffs (1987)
11. Kwakernaak, H., Sivan, R.: Linear Optimal Control Systems, vol. 1. Wiley-Interscience, New York (1972)
12. Kirk, D.E.: Optimal Control Theory: An Introduction. Courier Corporation, Chelmsford (2012)
13. Zhang, Q., Wang, Z., Wang, H.: Parametric covariance assignment using a reduced-order closed-form covariance model. Syst. Sci. Control Eng. **4**(1), 78–86 (2016)

Study of Perfusion Kinetics in Human Brain Tumor Using Leaky Tracer Kinetic Model of DCE-MRI Data and CFD

A. Bhandari[1], A. Bansal[2], A. Singh[3,4], and N. Sinha[1(✉)]

[1] Department of Mechanical Engineering,
Indian Institute of Technology, Kanpur 208016, India
nsinha@iitk.ac.in
[2] Department of Mechanical and Industrial Engineering,
Indian Institute of Technology, Roorkee 247677, India
[3] Centre for Biomedical Engineering, Indian Institute of Technology,
New Delhi 110016, India
[4] Department of Biomedical Engineering,
All India Institute of Medical Sciences, New Delhi 110016, India

Abstract. A computational fluid dynamics (CFD) model based on realistic voxelized representation of human brain tumor vasculature is presented. The model utilizes dynamic contrast enhanced magnetic resonance imaging (DCE-MRI) data to account for heterogeneous porosity and permeability of contrast agent inside the tumor. Patient specific arterial input function (AIF) is employed in this study. Owing to higher accuracy of Leaky Tracer Kinetic Model (LTKM) in shorter duration human imaging data, the model is employed to determine perfusion parameters and compared with General Tracer Kinetic Model (GTKM). The developed CFD model is used to simulate and predict transport, distribution and retention of contrast agent in different parts of human tissue at different times. In future, a patient specific model can be developed to forecast the deposition of drugs and nanoparticles and tune the parameters for thermal ablation of tumors.

Keywords: Voxelized model · Human brain tumor · AIF · LTKM · GTKM · DCE-MRI · CFD

1 Introduction

Cancer, a deadly disease occurs as a consequence of abnormal cell growth. It is the leading cause of human's death in developed as well as developing countries [1]. Most of the human cancers are solid tumors (approximately 85%) [2]. These tumors depend on other normal tissues for their nutritional material and thus grow in size. Chemotherapy and hyperthermia are widely used for cancer treatment. However, physicochemical properties of drug and biological properties of tumors put many limitations in all treatment strategies. Limited penetration of drug to tumor cells and difficulty in targeting sufficient amount of heat only to tumor tissue are the primary reasons of failure of chemotherapy and hyperthermic treatment respectively. On the other hand,

© Springer Nature Singapore Pte Ltd. 2017
M. Fei et al. (Eds.): LSMS/ICSEE 2017, Part I, CCIS 761, pp. 63–73, 2017.
DOI: 10.1007/978-981-10-6370-1_7

biological properties of tumors such as irregular vasculature, impaired lymphatic system and hypoxic conditions also lead to failure of these treatments [3]. Therefore, there is an urgent need to have proper knowledge of transport barriers that the drug or a drug carrying nanoparticle encounters when administered systematically inside a human's body. To this end, mathematical models are an excellent tool in investigating these transport barriers.

In order to study the transport barriers, Baxter and Jain developed a theoretical model to analyze a uniform as well as non-uniform perfused tumor. They demonstrated the effect of interstitial fluid pressure (IFP) and necrotic core on drug delivery [4, 5]. Soltani and Chen developed a homogenous tumor model and concluded that IFP becomes less than the effective pressure below the critical tumor radius, which makes the chemotherapeutic drug transport to tumor site easier [6]. Wang *et al.* used two different drug delivery modes to simulate the delivery of carmustine to brain tumors and concluded that polymeric drug delivery is better [7]. Pishko *et al.* modelled the drug delivery through the tumor tissue by using magnetic resonance imaging (MRI) technique and demonstrated the effects of heterogeneous vasculature and porosity [8]. Later, Magdoom *et al.* came out with a voxelized approach to model the drug delivery process that helped in reducing the computational time and increased accuracy [9]. Zhan *et al.* modeled the delivery of thermo-sensitive liposomes to solid tumors and concluded that thermo-sensitive liposome delivery leads to higher intra-cellular concentration of drug, enhancing the therapeutic effect of the drug [10].

All the above mentioned models have unfolded new insights in understanding the transport process that can be extended to study the transport of nanoparticles for thermal ablation of tumors. However, the assumption of spherical shape and homo-geneous vasculature are far from reality. Studies incorporating heterogeneous vascu-lature of tumors too had mostly focused on animal models with Simple Tofts Model and global arterial input function (AIF), which is time dependent concentration of tracer in blood plasma.

The main objective of the present study is to model the interstitial fluid flow parameters (pressure and velocity) and tracer transport in realistic human brain tumors with help of DCE-MRI data. Heterogeneous vasculature of tumor and spatially varying permeability and porosity have been taken into account. In addition, local or patient specific AIF has been measured and used for accurate determination of the perfusion parameters. General Tracer Kinetic Model (GTKM) [11] also called Extended Tofts Model has been employed for calculation of permeability and porosity maps. This model has been used because blood volume fraction in case of human tumors is quite large and this is taken care of by including intravascular term in this model [12]. In addition to the GTKM model, Leaky Tracer Kinetic Model (LTKM) [13] has also been used for generating perfusion parameters. CFD results (IFP, Interstitial fluid velocity (IFV) and tracer concentration) have been obtained by using perfusion parameters from both the models and are compared with experimental DCE-MR results. To the best of authors' knowledge, this is the first study related to CFD analysis of realistic human brain tumors based on DCE-MRI data and LTKM.

2   Data and Methodology

To obtain the permeability and porosity maps of tumor, a compartment model is fitted to the DCE-MRI data. The GTKM model assumes that tissue is divided into two compartments: plasma space and extravascular extracellular space (EES), also called interstitial space. It assumes bidirectional exchange of contrast agent from plasma space to EES. Contrast agent from plasma compartment permeates into EES. This model basically calculates the three important perfusion parameters namely rate transfer constant from plasma space to EES (K^{trans}), fractional plasma volume (v_p) and fractional EES volume or porosity (v_e). On the other hand, LTKM is a three compartment model in which three compartments are plasma, permeable and leakage compartments. This model basically assumes that EES is composed of two compartments permeable space and leakage space. In permeable space bidirectional exchange of contrast agent takes place where as in leakage space only unidirectional exchange of contrast agent occurs i.e. contrast does not flow back to vasculature. Figure 1 gives a clear picture of these two models depicting only bidirectional exchange in GTKM and unidirectional as well as bidirectional exchange in LTKM.

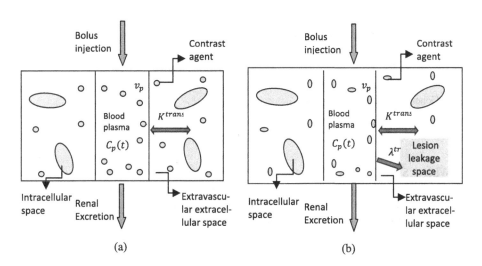

(a)   (b)

Fig. 1. Schematic of both models (a) GTKM (b) LTKM

2.1   MR Imaging and 3D Porous Media Computational Model

DCE-MR imaging was performed on a 3.0T Ingenia MRI scanner (Philips Healthcare, The Netherlands). Written consent from each patient was obtained before MRI study. Imaging was performed using a fast field echo (T1-FFE) sequence (TR/TE = 4.38 ms/2.3 ms, flip angle = 10°, field of view (FOV) = 240 × 240 mm², slice thickness = 6 mm, matrix size = 256 × 256). A dose of 0.1 mmol/kg body weight of Gd-BOPTA (Gadobenate Dimeglumine) (Multihance, Bracco, Italy) was administered intravenously with the help of a power injector. A total of 384 images at 32 time points for 12 slices were acquired with a temporal resolution approximately of 4 s for each time point. From DCE-MRI images, contrast concentration on each voxel was calculated from signal intensity with the help of SPGR/FFE signal equation. Pre contrast T1 (T10) was estimated using 3 fast spin echo (FSE) image (T1 - weighted, T2 - weighted, and proton density - weighted) [14]. The longitudinal and transverse relaxivity (R_1 and R_2) of contrast agent in body were taken as 6.3 mmol^{-1}s^{-1}L and 17.5 mmol^{-1}s^{-1}L respectively [15]. Local AIF was estimated using a method described by Singh et al. [16]. These concentration values obtained from Eq. (1) are used to fit Eqs. (2) and (3) i.e. two compartment model and three compartment model respectively. Equations (1, 2 and 3) are listed in Table 1. The second and third equations listed in Table 1 are fitted to get perfusion parameter maps (K^{trans}, v_e, v_p) by GTKM and ($K^{trans}, v_e, v_p, \lambda^{tr}$) by LTKM at each voxel respectively. The perfusion parameter maps got from both the models were imported in 3D porous media model made in OpenFOAM. Porous media model used in this study consists of fluid flow and tracer transport equations and has been described in our previous study [17].

Table 1. Equations used for analysis of MR images

Name of Equation	Equation
1. SPGR/FFE Signal Equation	$\frac{S(t)}{S(0)} = k_0 \exp(-TER_2 C(t)) \frac{1-\exp\left(-TR(T_{10}^{-1} + R_1 C(t))\right)}{1-\cos(\theta)\exp\left(-TR(T_{10}^{-1} + R_1 C(t))\right)}$
	Where $k_0 = \frac{1-\cos(\theta)\exp\left(-TRT_{10}^{-1}\right)}{1-\exp\left(-TRT_{10}^{-1}\right)}$
2. General tracer kinetic model	$C_t = v_p C_p(t) + K^{trans} \int_0^t C_p(\tau) e^{\frac{K^{trans}}{v_e}(\tau-t)} d\tau$
3. Leaky tracer kinetic model	$C_t = v_p C_p(t) + K^{trans} \int_0^t C_p(\tau) e^{\frac{K^{trans}}{v_e}(\tau-t)} d\tau + \lambda^{tr} \int_0^t C_p(\tau) d\tau$

Where, S (0) is the signal intensity when no contrast agent is given, S (t) is the signal intensity at a particular time point, **TE** is the echo time (msec), **TR** is the repetition time (msec), θ is the flip angle, C_t is the total tissue contrast agent concentration (mmol/Lt), $C_p(t)$ is the time dependent concentration of contrast agent in blood plasma (mmol/Lt) and λ^{tr} is the rate transfer constant between plasma and leakage compartment.

For reduction in the computational time, only the tumor part and the surrounding brain normal tissue were modelled. A rectangular volume of size $40 \times 36 \times 72$ mm^3 enclosing the tumor and the normal tissue was created and meshed using the Open-FOAM software. The mesh element size in the rectangular volume was same as that of voxel size in MRI slice ($0.9375 \times 0.9375 \times 6$ mm^3). The values of tracer kinetic parameters obtained from both the models were entered at each voxel inside the OpenFOAM CFD model. The SIMPLE (semi implicit method for pressure linked equations) algorithm [18] was used and standard interpolation schemes used by OpenFOAM were used to discretize the equations. Zero fluid pressure boundary conditions were applied at all the boundaries. Zero gradient boundary conditions were applied for interstitial fluid velocity and concentration of contrast agent. Initial condition for contrast agent transport was set to zero ($C_t = 0$). Grid independence test was done to see the effect of change in mesh size on the simulated tracer concentration. Increasing the number of mesh elements to four times the original value resulted in less than 3% change in tracer concentration.

3   Results and Discussion

Pre-contrast and post-contrast images of brain of one slice are shown in Fig. 2(a), (b) respectively. Local or patient specific AIF used in this study is shown in (Fig. 2(c)).

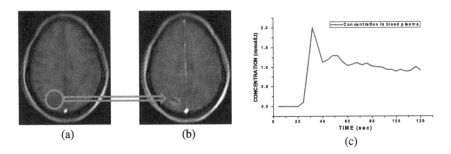

(a)   (b)

(c)

Fig. 2. MR images of brain (a) Pre contrast (b) Post contrast (c) AIF of the patient

Permeability and porosity maps of a particular slice (slice 10) obtained by fitting DCE-MRI data to both the models (GTKM and LTKM) are shown in Fig. 3. As can be observed, perfusion parameters obtained from both the models have significant difference. K^{trans} and porosity maps derived from LTKM were found to be more heterogeneous as compared to those got from GTKM. Figure 4 shows the contour plots of IFP and IFV, showing no significant difference in the values obtained from both the models. Figure 4(a) shows higher IFP inside the tumor with value equal to 1530 Pa, which rapidly decreased at the tumor boundary. IFV contour plot (Fig. 4(b)) was completely reverse of IFP with higher values of 0.04 μm/s at the tumor periphery. Simulated values of IFP and IFV were validated with the experimental values previously measured in the literature for human brain tumors [19, 20]. The higher and uniform value of IFP within the tumor is responsible for negligible convective transport of tracer within the tumor interstitium. Transport of tracer within tumor interstitium takes place mainly by diffusion. Convective transport of tracer is only significant at the periphery of tumor. This is due to the reason of steep pressure gradient and higher IFV at the tumor periphery, which helps in outward convection of tracer from the tumor.

Next, the interstitial tracer concentration was simulated using perfusion parameters of both the models. The tracer concentration simulation was carried out for two minutes since experimental data was available only for two minutes.

The comparison with experimental data was done at 14th time point (56 s) and 28th time point (112 s). Figure 5 shows contour plots of the tracer concentration simulated by using perfusion parameters of both the models (GTKM and LTKM) and

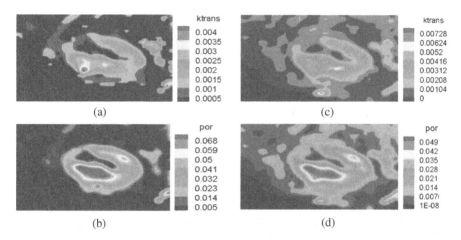

Fig. 3. Contour plots of perfusion parameters by GTKM (a) Permeability (K^{trans} (sec^{-1})) maps (b) Porosity (v_e) maps and LTKM (c) Permeability (K^{trans} (sec^{-1})) maps (d) Porosity (v_e) maps.

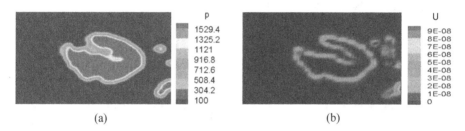

Fig. 4. Contour plots of (a) Interstitial fluid pressure (IFP) and (b) Interstitial fluid velocity (IFV).

experimental data at both the time points. As seen from contour plots tracer concentration obtained from LTKM perfusion parameters was more heterogeneous and more close to experimental results qualitatively. To make quantitative comparison, line plots were plotted along the horizontal and vertical bisector of the slice for both time points. Line plots of tracer concentration give an accurate picture of how closely the simulated results overlap with experimental ones. Figure 6 shows line plots of tracer concentration (both experimental and simulated) at horizontal and vertical bisector for 14th and 28th time points.

It is clear from line plots that simulated tracer concentration from LTKM obtained perfusion parameters was much close and correlates well with experimental concentration as compared to GTKM obtained perfusion parameters. Similar analysis was done on three more tumor data sets. Similar results were obtained, with simulated concentration from LTKM obtained perfusion parameters being more close to experiments. Tracer concentration peaks at tumor site in approximately 90–100 s after the administration intravenously, and then contrast begins to wash out from tumor site.

However, the wash out rate found in this tumor can't be extrapolated to other tumors as it highly depends on the type and characteristics such as size, grade and volume of tumors. It can be concluded that the developed computational model accurately captures the tracer concentration in tissues qualitatively as well as quantitatively. Further, the LTKM model gives more accurate perfusion parameters for short duration MRI data, as used in this study, which further help in predicting more accurate tracer concentration by CFD. For short duration MRI data, the GTKM model does not give correct estimation of volume fraction of extravascular extracellular space (EES) or porosity [13]. For correct estimation of porosity, data acquisition is suggested to be long enough for concentration of contrast agent to become stabilized (approximately 15–20 min) [21]. This is not always possible in case of clinical data due to many issues such as long scan time and blurring of image caused due to motion of patient over long time leading to inaccurate analysis. Also porosity and K^{trans} values in the GTKM model keep on varying with time for contrast enhancing tissues, which is not the case with LTKM. By changing temporal resolution of DCE-MR scans GTKM gives varying estimates of perfusion parameters whereas with LTKM, the perfusion parameters remain constant [13]. Thus, for shorter duration of MRI data, LTKM is preferable and provides better estimate of perfusion parameters. To further confirm the accuracy of results, a statistical analysis was performed. Root mean square (RMS) error and Pearson product moment correlation coefficient (PPMCC) were calculated along the values at horizontal and vertical bisector for both the time points as shown in Table 2.

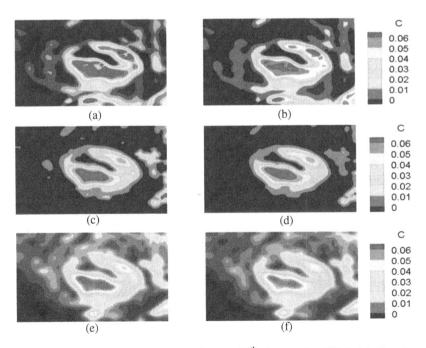

Fig. 5. Contour plots of tracer concentration at 14th time point (56 s) (a) Experimental (c) Simulated with GTKM perfusion parameters (e) Simulated with LTKM perfusion parameters and at 28th time point (112 s) (b) Experimental (d) Simulated with GTKM perfusion parameters (f) Simulated with LTKM perfusion parameters. Units of concentration are mmol/Lt.

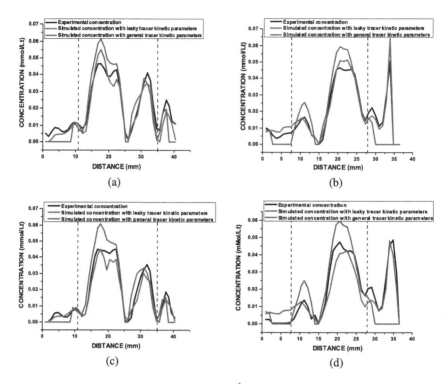

Fig. 6. Line plots of tracer concentration at 14th time point (56 s) (a) Horizontal Bisector (b) Vertical Bisector and at 28th time point (112 s) (c) Horizontal Bisector (d) Vertical Bisector

Table 2. Statistical analysis comparing Experimental and Simulated results

Variable	Model	Quantity	RMS error	PPMCC
C_t	Tracer concentration simulated with GTKM perfusion parameters	t = 14th time point		
		Horizontal Bisector	0.0079 mmol/Lt	0.91055
		Vertical Bisector	0.0155 mmol/Lt	0.74357
		t = 28th time point		
		Horizontal Bisector	0.0071 mmol/Lt	0.91944
		Vertical Bisector	0.0143 mmol/Lt	0.70202

(*continued*)

Table 2. (*continued*)

Variable	Model	Quantity	RMS error	PPMCC
	Tracer concentration simulated with LTKM perfusion parameters	$t = 14^{th}$ time point		
		Horizontal Bisector	0.0044 mmol/Lt	0.97273
		Vertical Bisector	0.0036 mmol/Lt	0.97731
		$t = 28^{th}$ time point		
		Horizontal Bisector	0.0041 mmol/Lt	0.98563
		Vertical Bisector	0.0060 mmol/Lt	0.95278

RMS error decreased by 44% and 76% at 14th time point and by 38% and 58% at 28th time point along horizontal and vertical bisector respectively when tracer concentration was simulated using LTKM parameters. Tracer concentration simulated with help of LTKM parameters always show a good correlation (>0.9) at both time points and for both bisectors as compared to those simulated with GTKM parameters.

4   Conclusions

A computational model based on DCE-MRI data was developed to study transport of contrast agent in realistic heterogeneous human brain tumor. Two different models (GTKM and LTKM) were used to obtain perfusion parameters. Simulated contrast agent concentration obtained by LTKM perfusion parameters showed better agreement with the experimental MRI data as compared to those obtained by GTKM perfusion parameters. Also, simulated IFP and IFV values correlated well with experimentally measured human brain tumor values reported in literature. The developed model is patient specific and can be used to select the most suitable chemotherapeutic drug for a specific patient before starting the treatment. Moreover, the developed CFD model in future can be used to predict the deposition of nano-particle encapsulated drugs. Once the deposition of nano particles in tumor area is known the porous media transport model used in this study can be coupled with heat transfer equations to study the effect of hyperthermic treatment in tumor microenvironment [22].

Acknowledgements. The authors would like to thank Dr. R.K. Gupta for providing clinical data, Prof. R.K.S. Rathore and Dr. Prativa Sahoo for technical support in DCE-MRI data analysis. This research was supported by grants from IIT Kanpur and Science and Engineering Research Board (grant number: YSS/2014/000092).

References

1. Siegel, R., Naishadham, D., Jernal, A.: Cancer statistics, CA cancer. J. Clin. **63**, 11–30 (2013)
2. Jain, R.K.: Normalization of tumor vasculature: an emerging concept in antiangiogenic therapy. Science **307**, 58–62 (2005)
3. Etryk, A.A., Giustini, A.J., Gottesman, R.E., Kaufman, P.A., Jack Hoopes, P.: Magnetic nanoparticle hyperthermia enhancement of cisplatin chemotherapy cancer treatment. Int. J. Hypertherm. **29**, 845–851 (2013)
4. Baxter, L.T., Jain, R.K.: Transport of fluid and macromolecules in tumors. I. Role of interstitial pressure and convection. Microvasc. Res. **37**(1), 77–104 (1989)
5. Baxter, L.T., Jain, R.K.: Transport of fluid and macromolecules in tumors. II. Role of heterogeneous perfusion and lymphatics. Microvasc. Res. **40**(1), 246–263 (1990)
6. Soltani, M., Chen, P.: Numerical modeling of fluid flow in solid tumors. PLoS ONE **6**(6), e20344 (2011)
7. Wang, C.H., Li, J., Teo, C.S., Lee, T.: The delivery of BCNU to brain tumors. J. Control Release **61**, 21–41 (1999)
8. Pishko, G.L., Astary, G.W., Mareci, T.H., Sarntinoranont, M.: Sensitivity analysis of an image based solid tumor computational model with heterogeneous vasculature and porosity. Ann. Biomed. Eng. **39**(9), 2360–2373 (2011)
9. Magdoom, K.N., Pishko, G.L., Kim, J.H., Sarntinoranont, M.: Evaluation of a voxelized model based on DCE-MRI for tracer transport in tumor. J. Biomech. Eng. **134**, 091004 (2012)
10. Zhan, W., Xu, X.Y.: A mathematical model for thermo-sensitive liposomal delivery of doxorubicin to solid tumour. J. Drug Del., Article ID 172529 (2013)
11. Tofts, P.S., Parker, G.J.M., DCE-MRI: acquisition and analysis techniques. In: Clinical Perfusion MRI: Techniques and Applications, pp. 58–74 (2013)
12. Parker, G.J.M., Buckley, D.L.: Tracer kinetic modelling for T1-weighted DCE-MRI. In: Jackson, A., Buckley, D.L., Parker, G.J.M. (eds.) Dynamic Contrast-Enhanced Magnetic Resonance Imaging in Oncology. Medical Radiology (Diagnostic Imaging), pp. 81–92. Springer, Heidelberg (2005). 10.1007/3-540-26420-5_6
13. Sahoo, P., Rathore, R.K.S., Awasthi, R., Roy, B., Verma, S., Rathore, D., Behari, S., Husain, M., Husain, N., Pandey, C.M., Mohakud, S., Gupta, R.K.: Sub compartmentalization of extracellular extravascular space (EES) into permeability and leaky space with local arterial input function (AIF) results in improved discrimination between high- and low-grade glioma using dynamic contrast-enhanced (DCE) MRI. J. Magn. Reson. Imaging **38**, 677–688 (2013)
14. Singh, A., Haris, M., Purwar, A., Sharma, M., Husain, N., Rathore, R.K.S., Gupta, R.K.: Quantification of physiological and hemodynamic indices using T1 DCE-MRI in intracranial mass lesions. J. Magn. Reson. Imaging **26**, 871–880 (2007)
15. Pintaske, J., Martirosian, P., Graf, H., Erb, G., Lodemann, K.P., Claussen, C.D., Schick, F.: Relaxivity of gadopentetate dimeglumine (magnevist), gadobutrol (gadovist), and gadobenate dimeglumine (multihance) in human blood plasma at 0.2, 1.5 and 3 tesla. Investigat. Radiol. **41**(3), 213–221 (2006)
16. Singh, A., Rathore, R.K.S., Haris, M., Verma, S.K., Husain, N., Gupta, R.K.: Improved bolus arrival time and arterial input function estimation for tracer kinetic analysis in DCE-MRI. J. Magn. Reson. Imaging **29**, 166–176 (2009)
17. Bhandari, A., Bansal, A., Singh, A., Sinha, N.: Perfusion kinetics in human brain tumor with DCE-MRI derived model and CFD analysis. J. Biomech. **59**, 80–89 (2017)

18. Anderson, D.A., Tannehill, J.C., Pletcher, R.H.: Computational Fluid Mechanics and Heat Transfer, pp. 671–674. Hemisphere, New York (1984)
19. Jain, R.K.: Delivery of molecular and cellular medicine to solid tumors. Adv. Drug Deliv. Rev. **46**, 149–168 (2001)
20. Guttman, R., Leunig, M., Feyh, J., Goetz, A.E., Messmer, K., Kastenbauer, E.: Interstital hypertension in head and neck tumors in patients: correlation with tumor size. Cancer Res. **52**, 1993–1995 (1992)
21. Donaldson, S.B., West, C.M., Davidson, S.E., Carrington, B.M., Hutchison, G., Jones, A.P., Sourbron, S.P., Buckley, D.L.: A comparison of tracer kinetic models for T1 weighted dynamic contrast enhanced MRI: applications in carcinoma of the cervix. Magn. Reson. Med. **63**, 691–700 (2010)
22. Nabil, M., Decuzzi, P., Zunino, P.: Modeling mass and heat transfer in nano-based cancer hyperthermia. R. Soc. Open Sci. **2**, 150447 (2015)

Computational Methods in Organism Modeling

Modelling and Analysis of the Cerebrospinal Fluid Flow in the Spinal Cord

Xiaode Liu[1], Danmei Luo[1], Panpan Hu[2], Miao Yu[2(✉)], and Qiguo Rong[1(✉)]

[1] Department of Mechanics and Engineering Science, College of Engineering,
Peking University, Beijing, China
qrong@pku.edu.cn
[2] Orthopaedic Department, Peking University, Third Hospital,
Beijing 100191, People's Republic of China
miltonyu@126.com

Abstract. The cerebrospinal fluid (CSF) flow in the spinal cord is important in maintaining the stability of the central nervous system. However, the interaction between CSF and spinal cord is not well understood. A three-dimensional (3D) simplified finite element model (FEM) of a sheep CSF and spinal cord segment was developed, verified using clinical experimental data, and used to investigate the effect of deformations and stress distributions on spinal cord in normal physiological conditions. The commercial software ANSYS Workbench was adopted to simulate the unidirectional CSF flow along the coaxial tube, which considered the bi-directional fluid-solid coupling. It was demonstrated that CSF had a slight impact on the spinal cord, which was transmitted to the white and gray matter through the pia mater. The pia mater protected the normal physiological function of the white and gray matter while the spinal dura mater ensured the regular rate and pressure of CSF. It was also showed that the CSF flow in the spinal cord was laminar. This model might help us to better understand the mechanism of interaction between CSF and spinal cord and provide a baseline for mechanical comparisons in spinal cord injury.

Keywords: Cerebrospinal fluid · Finite element analysis · Spinal cord · Fluid-Solid coupling

1 Introduction

The CSF was contained within and surrounded the ventricular system and spinal cord, serving to provide spinal cord with nutrients and maintain the homeostasis of metabolism of central neural system. The volume, estimated to be about 150 ml in adults, was distributed between 125 ml in cranial and spinal subarachnoid spaces and 25 ml in the ventricles, but with marked interindividual variations.

In recent years, with the development of imaging and finite element techniques, the intracranial circulation of CSF as well as its physiology and pathology has been well-studied. However, the physiological and pathological processes of CSF flow in the spinal cord are still in its infancy. Mardal et al. [1] had calculated the flow resistance in the

© Springer Nature Singapore Pte Ltd. 2017
M. Fei et al. (Eds.): LSMS/ICSEE 2017, Part I, CCIS 761, pp. 77–85, 2017.
DOI: 10.1007/978-981-10-6370-1_8

cervical spinal canal in a group of subjects with and without the Chiari malformation. The result showed resistance tended to be lower in Chiari I patients than in healthy volunteers. Cirovic [2] considered the effect of finite thickness of the tube walls and calculated four roots of the characteristic equation correspond to four modes of wave propagation. When the thickness of the spinal cord was reduced below its normal value, the first mode became dominant in terms of the movement of the CSF, and its speed dropped significantly.

Based on previous works, many researchers developed 3D FEMs, and more attention was paid to the mechanism of spinal cord injury. Greaves et al. [3] reconstructed a human cervical spine and spinal cord segment using a 3D FEM, investigating the difference in cord strain distributions during various column injury patterns. Li et al. [4] created a 3D FEM of the cervical spinal cord enlargement and simulated a hyperextension injury of the cervical cord. The simulation showed high localized stress at the anterior and posterior horn in the gray matter, which probably accounted for the predominance of the hand weakness in patients with central cord injury. Maikos et al. [5] simulated the impactor weight-drop experimental model of traumatic spinal cord injury based on the rat. The model predicted stress and strain patterns that matched patterns of primary injury, and also indicated that separate descriptions of the material and failure properties of gray and white matter were important. All the above study proved the finite element method could improve the understanding of the biomechanical behavior of the spinal cord.

However, the FEM analysis was mostly focused on exploring the mechanism of spinal cord injury, and less involved in the effect of CSF flow on the spinal cord. In our research, a 3D simplified FEM of a sheep CSF and spinal cord segment was developed. The influence of CSF flow on the displacement and stress of each part of the spinal cord was investigated. It was of great significance in theoretical analysis for the further study of abnormal CSF flow in the spinal cord.

2   Materials and Methods

2.1   Model Description

Based on the Magnetic Resonance Imaging (MRI) scanning data of T8–T12 segment of sheep thoracic vertebra, the commercial software Mimics was used to reconstruct the 3D solid contour of the CSF and spinal cord. In the process of data acquisition, the posture of the sheep was procumbent. A simplified model of CSF and spinal cord with the length of 50 mm was established, which contained gray matter, white matter, pia mater spinalis, CSF and spinal dura mater (Fig. 1).

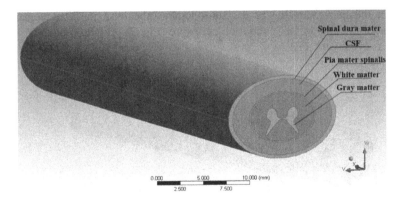

Fig. 1. Simplified geometric model of CSF and spinal cord.

The gray matter, white matter, pia mater spinalis and spinal dura mater were defined as homogeneous, isotropic, linear elastic material in the relevant literatures, as was shown in Table 1. The CSF was assumed as an incompressible viscous fluid.

Table 1. Material properties of the spinal cord and CSF [6–10].

Material	Density (tonne/mm^3)	Young's Modulus (MPa)	Viscosity Coefficient (tonne/(mm*s))	Poisson ratio
CSF	9.98e-10	\	1e-9	\
Gray matter	1.04e-9	0.656	\	0.499
White matter	1.04e-9	0.277	\	0.499
Pia mater	1.13e-9	11.5	\	0.470
Dura mater	1.13e-9	142	\	0.450

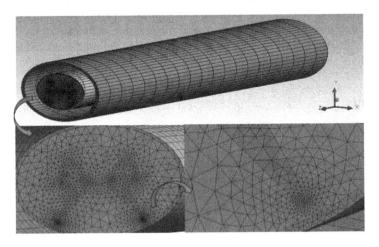

Fig. 2. The mesh of solid-domain.

As for the mesh partition of the solid-domain, the local mesh refinement was performed at the area of large variation curvature (Fig. 2). The fluid boundary layers were set as five layers (Fig. 3).

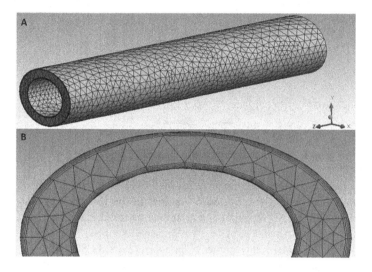

Fig. 3. The mesh of fluid-domain: (A) total; (B) local boundary layers.

2.2 Boundary Conditions and Solution

In the simulation, the analysis of bi-directional fluid-solid coupling was completed in the CFX module in ANSYS Workbench, and the ANSYS took a part of role in post processing.

The CSF flowed in a channel consisted of the dura mater and pia mater wall along one direction of the coaxial tube. Based on the clinical experimental data, the velocity of the flow entrance was set as 50 mm/s, and the pressure at the end of the flow field was 50 Pa. We used the MFX-ANSYS and CFX to solve the solid-field and fluid-field, respectively.

3 Results

3.1 Solid-Domain

In our study, the displacement of spinal dura mater was much small and negligible. The pattern of the spinal cord of each part was similar, which displayed large displacements at the center and small ones at both ends (Fig. 4).

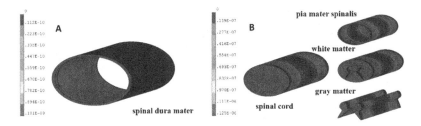

Fig. 4. The displacement [m]: (A) spinal dura mater and (B) spinal cord and its parts.

The first principal stress at the entrance of pia mater spinalis was larger, calculated to be about 80 to 260 Pa. It was showed that the entrance was the pulling state while the middle and posterior were compression (Fig. 5A). Stress distributions of gray matter and white matter were similar, both of which were negative. The result indicated it was a triaxial compression state. There was an increased tendency from the inlet to the outlet (Fig. 5B, C).

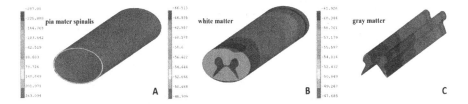

Fig. 5. The first principal stress [Pa] of the spinal cord of each parts.

The distribution of the third principal stress was shown in Fig. 6. The stress at the entrance of the pia mater spinalis was positive, which showed it was a triaxial tensile state (Fig. 6A).

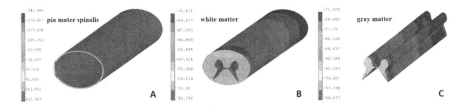

Fig. 6. The third principal stress [Pa] of spinal cord and its parts.

The von Mises stress distribution of the pia mater was with marked variations. The value was smaller (2.92–33.3 Pa) at the middle compared to the both ends (about 130 Pa). Stresses of the white and gray matter increased from the middle to the ends, but the difference was small. It was also noticed that the stress of the pia mater was much higher than that of the white and gray matter at the middle section (Fig. 7D). There were stress concentrations at the anterior and posterior horn in gray matter (Fig. 7E).

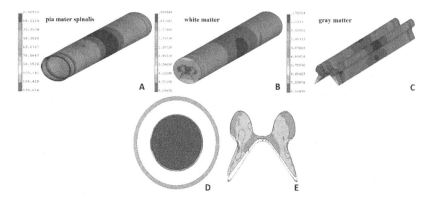

Fig. 7. The von Mises stress [Pa]: (A, B, C) spinal cord and its parts; (D) middle section; (E) the gray matter of middle section.

3.2 Fluid-Domain

The pressure of CSF flow field from the inlet to the outlet showed a gradient descent. The result of streamline indicated that the velocity of the left and right sides was larger than that of the upper and the lower (Fig. 8B).

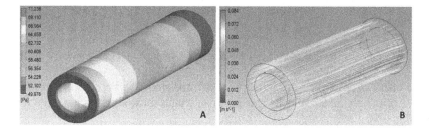

Fig. 8. (A) The flow field pressure [Pa] and (B) the streamline.

4   Discussion

In this study, we developed a 3D simplified FEM of the CSF and spinal cord model in the sheep based on the MRI scanning data of T8–T12 segment of thoracic vertebra. The model was based on the procumbent posture of the sheep. It was used to investigate the mechanical interaction between the CSF flow and the spinal cord.

4.1 Solid-Domain

In the natural state, CSF flow had a very small influence on the spinal dura mater (Fig. 4A). The spinal dura was made of dense connective tissues, which increased the stiffness of the spinal cord and enhanced its shape recovery after removal of the compression. It firmly covered the spinal cord and had a high elastic modulus. Therefore, it provided a constraint on the spinal cord surface. The spinal dura sac was a dynamic structure, readily changing its capacity in response to prevailing pressure gradients across its walls. By its bladder-like ability to alter its capacity, the spinal dura sac provided the 'elasticity' of the covering of the central nervous system [11]. These characteristics ensured the stability of CSF flow in the spinal cord. The result also revealed that the possibility of lesions was slight under the effect of CSF, which could maintain the normal flow rate and channel.

It could be inferred that influences of CSF flow on spinal cord parenchyma (gray matter, white matter and pia mater spinalis) were greater than that of the dura mater according to the material properties and computational results (Figs. 4 and 7(A, B, C)). The spinal cord parenchyma was produced tiny displacement at the effect of shear stress along the CSF flow direction. Deformations of the parenchyma were larger in the longitudinal direction than that of in the transverse direction. On the basis of the numerical analysis and the published properties for spinal cord tissues [5], we could conclude that the effect was little.

Stresses of white matter and gray matter were about 2 Pa and 3 Pa, respectively. The difference was quite slight, which could be neglected. The value of the pia mater was about 35 times and 16 times that of the white matter and gray matter at the same position, respectively, varying from 18 Pa to 48 Pa. These showed that the order of stress effects of CSF flow on spinal cord was: pia mater > gray matter ≈ white matter. The pia mater protected the gray and white matter to avoid large compressive stress due to CSF flow. Accordingly, their deformation energy were slight. There was little risk of destruction in the normal physiological condition.

Previous studies [3, 4] had demonstrated that functional areas of the spinal cord were concentrated in the gray and white matter. In order to analyze possible damage patterns of the spinal cord in abnormal flow of CSF, the first and third principal stresses of the spinal cord were calculated. The pia mater might be pulled or compressed, while the white and gray matter were throughout in triaxial compression state (Fig. 5B, C). It revealed that the white and gray matter mainly bear pressure in the natural state. When the pressure changed, structures would be altered and diseases might subsequently appear. It could also be inferred that the damage form was compression failure. Owing to the uncertainty of bearing tension or pressure of the pia mater (Fig. 6A), the stress or strain might be difficult to figure out when the shape of CSF channel altered due to pathological or surgical complications.

There were high localized stress at the anterior and posterior horn in the gray matter (Fig. 7E). The area with large variation curvature was sensitive, and damages might firstly occur in these places.

4.2 Fluid-Domain

In the fluid-domain, the pressure had a trend of gradient decrease, which showed the CSF flow was stable. The speed of CSF flow was not constant, which was related with the curvature (Fig. 8B).

In order to investigate the CSF flow pattern in spinal cord, we chose the velocity distribution of flow field in the middle section (Fig. 9). It could be found that maximum speed occurred at the center and the minimum lied in the boundary. Velocities closed to the boundary were approximately zero, which satisfied the non-slip condition. The fluid moved along a straight line parallelled to the axis of the tube (Fig. 8B). So, the present findings revealed that CSF flow in the spinal cord was laminar.

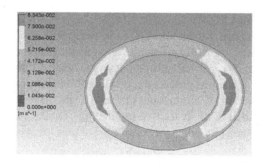

Fig. 9. The velocity distribution of flow field in the middle section.

5   Conclusion

We conducted a 3D FEM of the CSF and spinal cord segment based on the anatomy data of the sheep spinal cord taken from MRI. The model considered the unidirectional flow of CSF along the coaxial tube, and predicted displacement and stress patterns. It concluded that: (1) the spinal dura mater ensured the regular rate and pressure of CSF flow; (2) the pia mater played a role in protecting the natural state of the gray and white matter; (3) CSF flow in the spinal cord was laminar.

We have shown that it is possible to combine anatomy, MRI scans with CFD simulations to reconstruct the CSF flow in the spinal cord, and our findings may be important in understanding the mechanism of interaction between CSF and spinal cord, which can ultimately be used to predict, and develop means to prevent spinal cord injury in humans.

In future computer simulations, due to the spine was movable in reality, changes in stress and strain of the spinal cord under different postures will be clarified in more detail. The impact of gravity would also be considered in the ANSYS workbench simulation process. These should help elucidate the interaction between CSF and spinal cord in human body.

Acknowledgments. The study was financially supported by the National Science Foundation of China (Grant Ref. No. 81541122).

References

1. Mardal, K.A., Rutkowska, G., Linge, S., et al.: Estimation of CSF flow resistance in the upper cervical spine. Neuroradiol. J. **26**(1), 106–110 (2013)
2. Cirovic, S.: A coaxial tube model of the cerebrospinal fluid pulse propagation in the spinal column. J. Biomech. Eng. **131**(2), 021008 (2009)
3. Greaves, C.Y., Gadala, M.S., Oxland, T.R.: A three-dimensional finite element model of the cervical spine with spinal cord: an investigation of three injury mechanisms. Ann. Biomed. Eng. **36**(3), 396–405 (2008)
4. Li, X.F., Dai, L.Y.: Three-dimensional finite element model of the cervical spinal cord: preliminary results of injury mechanism analysis. Spine **34**(11), 1140–1147 (2009)
5. Maikos, J.T., Qian, Z., Metaxas, D., et al.: Finite element analysis of spinal cord injury in the rat. J. Neurotrauma **25**(7), 795–816 (2008)
6. Lakin, W.D., Stevens, S.A., Tranmer, B.I., Penar, P.L.: A whole-body mathematical model for intracranialpressure dynamics. J. Math. Biol. **46**(4), 347 (2003)
7. Zhong, Z.C., Wei, S.H., Wang, J.P., et al.: Finite element analysis of the lumbar spine with a new cage using a topology optimization method. Med. Eng. Phys. **28**(1), 90–98 (2006)
8. Denoziere, G., Ku, D.N.: Biomechanical comparison between fusion of two vertebrae and implantation of an artificial inter-vertebral disc. J. Biomech. **39**(4), 766–775 (2006)
9. Ozawa, H., Matsumoto, T., Ohashi, T., et al.: Mechanical properties and function of the spinal pia mater. J. Neurosurg. Spine **1**(1), 122–127 (2004)
10. Zarzur, E.: Mechanical properties of the human lumbar duramater. Arq Neuropsiquiatr. **54**(3), 455–460 (1996)
11. Xu, C., Yan, Y.-B., Wu, Z.-X., et al.: 3D finite element analysis on injury mechanism of spinal cord compression. Med. Biomech. **4**, 306–312 (2014)

Fracture Prediction for a Customized Mandibular Reconstruction Plate with Finite Element Method

Danmei Luo[1], Xiangliang Xu[2], Chuanbin Guo[2], and Qiguo Rong[1(✉)]

[1] Department of Mechanics and Engineering Science, College of Engineering, Peking University, Beijing 100871, China
qrong@pku.edu.cn
[2] Department of Oral and Maxillofacial Surgery, Peking University School and Hospital of Stomatology, Beijing 100081, China

Abstract. The use of customized reconstruction plate is an effective method for reconstruction of mandibular continuity defects. Plate fracture is one of the most common postoperative complications. The aim of this study was to investigate the biomechanical behavior of the customized reconstruction plate by finite element method. The geometry model was created from computed tomography (CT) data of a patient. The muscle forces for the defected mandible under two common static biting tasks were estimated by a numerical optimization strategy with the objective function of minimization of overall muscle force. The simulation results revealed that changing bite from molar region to incisor region increased the maximum stress in the plate. The position of stress concentration, the upper-inner edge of the plate near ramus-end, was in agreement with that of fracture, which indicated that stress concentration regions were critical regions for fracture failure.

Keywords: Finite element analysis · Muscle forces · Reconstruction plate · Fracture

1 Introduction

Mandible defects without timely reconstruction can lead to disturbed mastication, impairment of speech, and facial deformity, which seriously affect patients' quality of life. The objective of mandibular reconstruction is to restore both the shape and the function of the mandible.

The use of titanium reconstruction plate with or without bone grafting has become a popular choice for repairing mandibular defects. Standard titanium reconstruction plates are commonly used in surgery. Currently, with the development of computer-aided design techniques and additive manufacturing techniques, customized titanium reconstruction plates with specific geometrical shapes can be fabricated and are gradually used in oral maxillofacial surgery. Compared with standard reconstruction plates, customized plates offer many benefits, including better bone surface adaption and superior facial recovery. Nevertheless, there are some serious long-term complications after plate implantation surgery, such as plate exposure, plate fracture and screws loosening.

© Springer Nature Singapore Pte Ltd. 2017
M. Fei et al. (Eds.): LSMS/ICSEE 2017, Part I, CCIS 761, pp. 86–94, 2017.
DOI: 10.1007/978-981-10-6370-1_9

According to the previous literature, incidence of plate fracture ranges from 2.9% to 10.7% [1]. Plate fracture mostly occurs in less than 6–9 months after surgery. It is more frequent in patients with a bone resection including the mandibular angle, most commonly occurring near the anterior region of the mandibular angle [2]. By material analysis, it was observed that the origin of cracks was stress concentration regions in the plate and metal fatigue was caused by frequently masticatory function. Thus, it becomes important to validate the biomechanical behavior of the reconstruction plate.

The objective of this study was to estimate reasonable muscle forces of the defected mandible with an optimization algorithm and to investigate the biomechanical behavior of the reconstruction plate under masticatory loads with finite element method. The position of the plate fracture and that of the stress concentration were also compared.

2   Materials and Methods

A 45-year-old female patient with ameloblastoma was chosen for this study. She got extensive tumor resection and immediate mandibular reconstruction surgery with a customized titanium plate. The customized titanium reconstruction plate extending from the chin to the mandibular angle was anchored to the residual mandibular bone with seventeen 2.7-mm-diameter bicortical osteosynthesis screws. Unfortunately, plate fracture occurred in several months after surgery.

2.1   Generation of the Geometric Model

The anatomical geometry of the mandible with defects was reconstructed from postoperative CT data of the patient by using MIMICS (Version 10.01, Materialise, Inc.). By employing threshold segmentation method, regions of bone and titanium plate were separated and reconstructed separately. Screws were modeled as 2.7-mm-diameter cylindrical pins in Geomagic Design X (Version 2016. 0. 1, 3D systems, Inc.). After position aligning, boundary smoothing and volume forming, the geometric model of the assembly of plate/mandible/screw system was obtained in Geomagic Studio (Version 12.0, Geomagic, Inc.).

2.2   Computation of Muscle Forces

The two most common chewing tasks, i.e. incisal clenching (INC) and left molar clenching (LMOL) were simulated in this study. Due to the segmental mandibular resection, a reduced biting force of 300 N in the vertical direction was chosen. The extensive mandiblular resection (including the mandibular angle) meant that right masseter muscles were destroyed. To determine reasonable muscle and joint forces of the defected mandible, an optimization strategy with the objective function of minimization of overall muscle force was carried out [3]. Seven pairs of major masticatory muscles, namely the superficial masseter, the deep masseter, the anterior temporalis, the middle temporalis, the posterior temporalis, the lateral pterygoid and the medial pterygoid, with the absence of right masseter muscles were included in this study. The joint load was set as a single component with a constrained direction. The direction of joint force was considered to be normal to the joint surface at the articular contact, which was measured in the geometric model.

The resultant bite force, muscle and joint forces must fulfill the following six equilibrium conditions:

$$\sum_{i=1}^{12} \mathbf{M}_i + \mathbf{J}_r + \mathbf{J}_l + \mathbf{B} = \mathbf{0} \tag{1}$$

$$\sum_{i=1}^{12} (\mathbf{r}_i \times \mathbf{M}_i) + (\mathbf{r}_{jr} \times \mathbf{J}_r) + (\mathbf{r}_{jl} \times \mathbf{J}_l) + (\mathbf{r}_b \times \mathbf{B}) = \mathbf{0} \tag{2}$$

where $\mathbf{M}_i, \mathbf{J}_r, \mathbf{J}_l$ and $\mathbf{B}$ represent muscle force, right joint force, left joint force and vertical bite force, respectively. The magnitude of vertical bite force was set to be 300 N in incisor region and left molar region successively, for simulating incisor clenching and left molar clenching.

The magnitude of muscle and joint forces must fulfill the following constraint conditions:

$$0 \leq |\mathbf{J}_r| \text{ and } 0 \leq |\mathbf{J}_l| \tag{3}$$

$$M_i = |\mathbf{M}_i| \text{ and } 0 \leq M_i \leq m_i \tag{4}$$

where m_i represents maximum tensile force of each muscle.

During optimum processing, muscle forces were determined when the following objective function achieved. Afterward, these calculated muscle forces were applied in finite element analysis.

$$f = \min[\sum_{i=1}^{12} M_i] \tag{5}$$

2.3 Finite Element Analysis

The geometric model of the assembly of plate/mandible/screw system was imported into ANSYS (Version 14.0, ANSYS, Inc.) for mechanical analysis. The finite element model was shown in Fig. 1. The model was meshed with linear tetrahedrons elements. The mesh consists of 799,327 elements and 151,045 nodes, which was dense enough to ensure calculation accuracy. A bond-type connection was applied to the interface of locking screws and the reconstruction plate. Material properties involved in the model were all considered isotropic, homogeneous and linear elastic (Table 1).

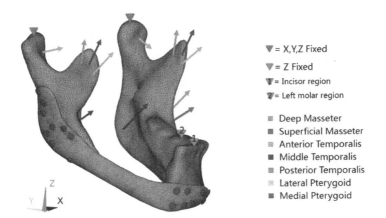

Fig. 1. Finite element model of the assembly of the mandible and the reconstruction plate. Boundary conditions and loads are also briefly illustrated.

Table 1. Material properties of different parts in the finite element model.

Type of material	Young's modulus [MPa]	Poisson's ratio
Mandible (cortical bone)	8700	0.28
Ti6Al4 V (plate and screw)	105,000	0.30

Incisor clenching and left molar clenching were simulated by applying corresponding muscle forces to the mandible. The top surfaces of two condyles were fully restrained to prevent the rigid-body displacement of the mandible. Displacement in vertical direction of corresponding occlusal contacts was constrained too. Values of muscle forces were obtained from calculated results.

3   Results

Calculated muscle forces and joint forces are shown in Table 2. Joint force can only be transmitted by compression and its direction was normal to the joint surface at the articular contact. On the basis of the condyle morphology, the measured joint force vector pointed back and downward in the saggital plane with an angle of 26° away from vertical. Under both chewing tasks, the superficial masseter, the anterior temporalis and the medial pterygoid were the most heavily loaded muscles, and the superficial masseter was proportionally larger than the medial pterygoid. Deep masseter and posterior temporalis were not active during any biting task. It seems that the superficial masseter was always activated in preference to the deep masseter. During incisor clenching, due to the absence of right masseter muscles, the right and the left muscle groups were not recruited equally. During left molar clenching, muscles on the balancing side except the anterior temporalis were totally suppressed. It was observed that the joint force at the left side was larger than that at the right side under both loads. In addition, changing bite from incisor region to left molar region largely reduced joint loads.

Table 2. Calculated muscle force, joint forces and corresponding directions.

	Max. force (N)	Calculated muscle force (N)				Direction of force		
		INC		LMOL		Cos-x	Cos-y	Cos-z
		Right	Left	Right	Left			
Superficial Masseter	190.4	–	190.4	–	129.1	−0.21	−0.42	0.88
Deep masseter	81.6	–	0	–	0	−0.55	0.36	0.76
Medial pterygoid	174.8	123.4	174.8	0	93.9	0.49	−0.37	0.79
Lateral pterygoid	66.9	0	25.2	0	0	0.63	−0.76	−0.17
Anterior temporalis	158.0	158.0	158.0	89.1	158.0	−0.15	−0.04	0.99
Middle temporalis	95.6	9.5	16.1	0	39.0	−0.22	0.50	0.84
Posterior temporalis	75.6	0	0	0	0	−0.21	0.86	0.47
Joint force	–	141.8	340.3	42.8	141.0	0	0.44	−0.90

The displacement distribution in the reconstructed mandible under incisor clenching and left molar clenching was analyzed (Fig. 2). For better display, the deformation was magnified by three times in Fig. 2. Unsymmetrical displacement distribution was observed on the mandible under both biting tasks. Specifically, the displacement on the defected side of the mandible was larger than that on the opposite side. The maximum displacement in the plate was 0.83 mm under incisor clenching, decreasing to 0.25 mm under left molar clenching. Under incisor clenching, the plate slightly moved towards the left side. The largest displacement occurred close to the resection site on the right ramus end. When the biting position moved to left molar, displacements on both sides of the mandible decreased and the largest displacement was observed in the overhanging part of the plate.

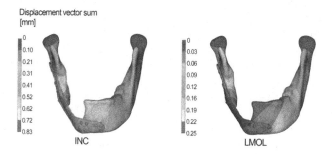

Fig. 2. Total displacement [mm] in the mandible and the plate during INC and LMOL. The undeformed model is meshed with black lines and the deformed model displays in color.

Figure 3 shows the von Mises stress distribution in the reconstruction plate during incisor clenching and left molar clenching. Under both clenching tasks, the patterns of stress distribution in the plate were quite similar: the overhanging part of the plate was more critically loaded than two fixed ends and the maximum von Mises stress in the plate was both observed on the inner-upper edge near the ramus end-incisor clenching: 260.55 MPa and left molar clenching: 92.91 MPa. Another high stress concentration region in the plate was also observed on the inner-lower edge near the chin end with a stress of approximately 233 MPa. The screw holes near the resection sites exhibited higher stress. Figure 4 shows the fractured plate reconstructed geometrically from the CT data after plate failure, in which the fracture of the plate is highlighted by a red circle. The position of the plate failure and that of the largest von Mises stress in the plate calculated by FEM agreed well.

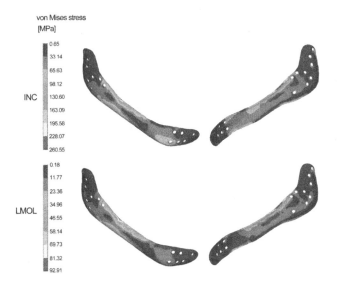

Fig. 3. von Mises stress [MPa] in the reconstruction plate during INC and LMOL.

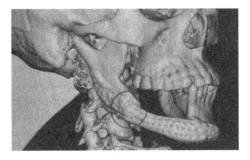

Fig. 4. The fractured plate reconstructed geometrically from the CT data after plate failure. (Color figure online)

The maximum and minimum principal strain distributions in the mandible were shown in Fig. 5. In general, the screw holes near the resection sites on the chin and ramus ends exhibited higher strain. The upper ramus-end hole was considered the most critical area as it always demonstrated the highest tension and compressive strain under both loading tasks.

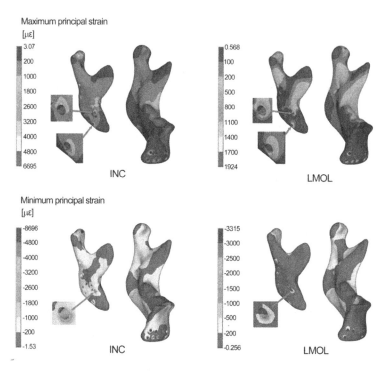

Fig. 5. Maximum and minimum principal strain [με] in the mandible during INC and LMOL.

4   Discussion

The female patient who had an ameloblastoma on her right mandible underwent segmental bone resection surgery. Then a customized titanium plate was inserted to the defect site. Plate fracture occurred in several months after surgery. To study the failure of the reconstruction plate, a numerical simulation using finite element method under different static chewing loadings was performed to calculate von Mises stress distribution in the plate and to determine the position of high stress concentration.

A reconstruction plate for repairing a mandible with continuity defects must reconstruct the bone shape anatomically. The external shape of the plate directly affects the recovery of facial appearance. Stock plates must be bended to adjust the plate shape to the mandible morphology for suitable placement, while customized plates with specific shape don't require intraoperative bending and offer better fitting.

In addition, the reconstruction plate must withstand the repeated forces created during mastication as well. In our simulations, the maximum von Mises stress in the plate was 260.55 MPa under incisor clenching, decreasing to 92.91 MPa under left molar clenching. These values are lower than the material failure limits of Ti6Al4 V (yield strength: 800 MPa and fatigue strength: 600 MPa) [4]. The maximum von Mises stress in the plate was observed on the inner-upper edge near the ramus end under both biting loads. Previously published literature found that the origin of cracks was stress concentration regions in the plate and stress concentration during repeated mastication was the origin of cracks, and it finally resulted in fatigue fracture failure. Although the calculated peak stress in the plate was below the failure limit of Ti6Al4 V, plate failure occurred. This might be explained by the existence of manufacturing defects in the plate which reduce its strength. The position of the plate fracture (Fig. 4) was in agreement with that of peak stress in the plate (Fig. 3). The result supports the reported findings that plate failure tends to occur at the stress concentration region. To prevent plate fracture, it is necessary to redesign the morphology of the plate for avoiding stress concentration. Therefore, smoothing the inner side of the plate and increasing heights and widths in the critical region may be helpful.

After segmental mandible resection surgery, the occlusal force was reduced in some degree. Muscle forces also changed due to bone resection and the remove of relevant tissue. There is still lack of information about muscle forces in patients with extensive mandibular defect. Previous work suggested that objective function-based numerical models provided a method to study muscle and joint forces for the individual. In this study, although the calculated muscle forces are exceedingly difficult to validate, it is useful to compare the plate biomechanical behavior under loads that it may be subjected to.

5   Conclusions

In conclusion, the finite element simulative results indicated that changing bite from molar region to incisor region increased both the maximum stress in the reconstruction plate and the maximum strain in the bone. It is suggested that the patient with lateral mandibular defects should reduce incisor clenching frequency. Additionally, stress concentration regions in the plate during mastication were likely regions of fracture failure. These critical regions should be strengthened in height and thickness for avoiding stress concentration.

Acknowledgements. The study was financially supported by the National Key Research and Development Program of China (Grant Ref. No. 2016YFB1101503).

References

1. Martola, M., Lindqvist, C., Hanninen, H., Al-Sukhun, J.: Fracture of titanium plates used for mandibular reconstruction following ablative tumor surgery. J. Biomed. Mater. Res., Part B **80**, 345–352 (2007)
2. Katakura, A., Shibahara, T., Noma, H., Yoshinari, M.: Material analysis of AO plate fracture cases. J. Oral Maxil. Surg. **62**, 348–352 (2004)
3. Trainor, P.G.S., Mclachlan, K.R., Mccall, W.D.: Modelling of forces in the human masticatory system with optimization of the angulations of the joint loads. J. Biomech. **28**, 829 (1995)
4. Niinomi, M.: Mechanical properties of biomedical titanium alloys. Mater. Sci. Eng., A **243**, 231–236 (1998)

Three-Dimensional Pathological Analysis of Cerebral Aneurysm Initiation

Xinning Wang[1(✉)], Kenta Suto[1], Takanobu Yagi[1,2], Koichi Kawamura[1], and Mitsuo Umezu[1]

[1] Center for Advanced Biomedical Science (TWIns), Waseda University, Wakamatsu-cho, Shinjuku-ku, Tokyo 162-8480, Japan
wangxinning1202@fuji.waseda.jp, kamokenkenpa@gmail.com, takanobu_yagi@aoni.waseda.jp, nao1223@cna.ne.jp, umezu@waseda.jp
[2] EBM Corporation, 4-16-15-508, Ohmoriminami, Ohta-ku, Tokyo 143-0013, Japan

Abstract. Cerebral aneurysm is known to initiate at the cerebral artery bifurcation. The pathological mechanism of cerebral aneurysm awaits further understanding especially on its initiation. This study sought to elucidate the three-dimensional structure of cerebral vascular bifurcations with and without aneurysms using human cadavers. The two cases had aneurysmal initiations out of total 7 cases. The studied structure was intimal hyperplasia, tunica media and internal elastic lamina, which were recognized by elastica masson staining. The results showed that the non-existence of tunica media and internal elastic lamina was found in the lesion without aneurysm. The non-existence of intimal hyperplasia was only found in the lesion with aneurysm. These data suggest that the formation of intimal hyperplasia may be related with the initiation of aneurysm. We regarded the boundary of existence arteriosclerosis as the position for new arteriosclerosis occurs and thought the direction of new arteriosclerosis grows would influence whether the cerebral aneurysm initiates or not.

Keywords: Cerebral aneurysm · Initiation · Intimal hyperplasia · Tunica media · Internal elastic lamina

1 Background and Purpose

The rupture of cerebral aneurysm is a major cause of subarachnoid hemorrhage. Cerebral arterial bifurcations are known to be its common sites. The aneurysm initiation is generally believed to occur as the mechanical strength is weakened. Although there are numerous studies of the mechanism of cerebral aneurysm, full understanding of initiation, growth and rupture is still unknown.

Currently, there is no treatment that can cure cerebral aneurysm completely, and no preventative therapy of cerebral aneurysm is developed. The rupture of cerebral aneurysm causes 2/3 of the patients death or after effect [1] so that it is impossible to rehabilitate them in society. Therefore, it is necessary to develop a new treatment to prevent the initiation of cerebral aneurysm to reduce the number of patients suffering

© Springer Nature Singapore Pte Ltd. 2017
M. Fei et al. (Eds.): LSMS/ICSEE 2017, Part I, CCIS 761, pp. 95–103, 2017.
DOI: 10.1007/978-981-10-6370-1_10

from the cerebral aneurysm. For reaching this purpose, full understanding of initiation of cerebral aneurysm is required.

The human cadavers and animal models were used to study the initiation of cerebral aneurysm traditionally. The human cadavers were first used for pathological analysis to study about the initiation of cerebral aneurysm in the early years. The data showed the non-existence of tunica media [2], internal elastic lamina [3] and the existence of intimal hyperplasia [4] was recognized as a phenomenon of aneurysm initiation. Actually, the non-existence of the tunica media and the internal elastic lamina was recognized as the factors of aneurysm initiation still nowadays [5]. However, the reason why this phenomenon was caused is still unknown. Later, hemodynamics of cerebral arteries was thought to influence the initiation of cerebral aneurysm, and animal experiments were occurred for this kind of study. As a result, hemodynamics seems to have something to do with initiation of cerebral aneurysm [6, 7], but still the cause of effect relationship between hemodynamics and cerebral aneurysm did not clear. Considering the differences of body condition between animals and human beings. It is still necessary to use human cadavers for this study. In order to approach to elucidating the mechanism of initiation of cerebral aneurysm, we developed a new idea of studying human cadavers. Both of the human cerebral arteries with an aneurysm and with no aneurysm should be elucidated in 3D so that the temporal and spatial pathological changes of the cadavers might be clear. Therefore, the purpose of this study is to elucidate the vascular structure of human bifurcations before and after initiation in 3D pathologically.

2 Method

2.1 Sections Cutting

The pathological analysis method used in this research was cutting subcontinuous tissue sections and occurring pathology staining. Then, the optical microscope was used to observe and extract the pathological characteristics of the tissue. The details of analysis objects were shown in Table 1.

Table 1. Detailed information of the analysis objects.

ID	Age/sex	Location	With an aneurysm or not
Case 1	Unavailable	BA	With an aneurysm
Case 2	Unavailable	BA	With an aneurysm
Case 3	74/M	Lt IC-PC	With no aneurysm
Case 4	74/M	Rt IC-PC	With no aneurysm
Case 5	74/M	BA	With no aneurysm
Case 6	78/F	BA	With no aneurysm
Case 7	33/M	BA	With no aneurysm

2.2   Observing and Analyzing

In order to analyze accurately and on basis, we used qualitative analysis and quantification method. Explaining in detail, as shown in Fig. 1A, we took the part of cerebral arteries which the lumen appeared as the analysis area, and took the Sects. 300 μm intervals as measurement objects. Then, as shown in Fig. 1B, 3 mm from the apexes of bifurcation of cerebral arteries regarded as the localization that the aneurysms initiate easily was dealt with examining areas. The apex of the bifurcation was set to zero point, left direction was set to minus distance, right direction was set to plus distance. Measurement was occurred from zero point to both directions. The measure points were 11 points in total and were set to zero point and the points 600 μm intervals from zero point to each direction. Finally, in order to elucidate the distribution of tunica media and intimal hyperplasia, it is necessary to measure the thickness of tunica media and intimal hyperplasia. However, as shown in Fig. 1C1, depending on the situation, internal elastic lamina might be meandering so that we had to add imaginary lines for measuring. Therefore, as shown in Fig. 1C2, while measuring the thickness of tunica media or intimal hyperplasia, we defined the length of the vertical direction of internal elastic lamina as the thickness. In this research, we defined non-existence of the intimal hyperplasia or the tunica media, while the thickness was 0 μm.

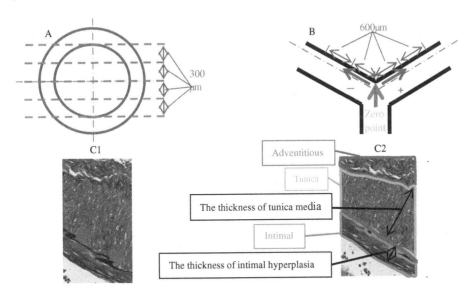

Fig. 1. Image of the cross section of the vessel lumen (A). The tissue sections were expected to be took as 300 μm intervals for measuring. Also, image of the longitudinal section of the vessel lumen (B). Each section was measured the pathological characteristics 600 μm intervals from zero point to each direction. Because of the internal elastic lamina meandering (C1), imaginary lines was added for the measurement of the thickness of intimal hyperplasia or tunica media (C2).

3   Results

Results of the pathological characteristics of cerebral arteries with an aneurysm and cerebral arteries with no aneurysm were showed below in details.

3.1   Cerebral Arteries with An Aneurysm

The subcontinuous partial images located on the bifurcation of tissue sections of the typical example were shown as Fig. 2A. As a result, the non-existence of internal elastic lamina, tunica media and intimal hyperplasia was found in aneurysmal parts. Here indicates the intimal hyperplasia by yellow.

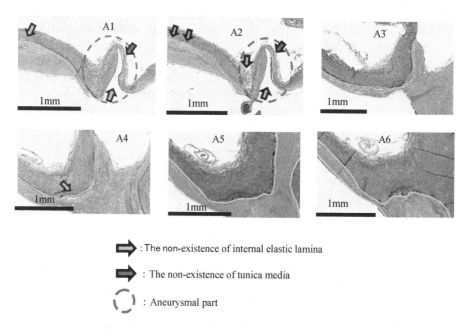

➡ : The non-existence of internal elastic lamina

➡ : The non-existence of tunica media

◯ : Aneurysmal part

Fig. 2. The subcontinuous partial images of the typical case with an cerebral aneurysm (A). The existed only in two of the all partial images (A1) (A2). The non-existence of intimal hyperplasia, tunica media and intimal hyperplasia was found only in the aneurysmal part (A1) (A2). (Color figure online)

3.2   Cerebral Arteries with No Aneurysm

The subcontinuous partial images located on the bifurcation of tissue sections of the typical example were shown as Fig. 3A. As a result, the non-existence of internal elastic lamina and the non-existence of tunica media was found in the non-aneurysmal part. However, the non-existence of intimal hyperplasia were not found in the non-aneurysmal part. Here indicates the intimal hyperplasia by yellow.

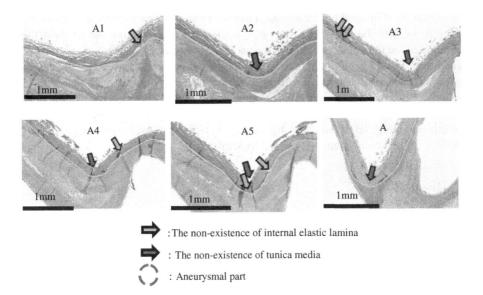

➡ : The non-existence of internal elastic lamina

➡ : The non-existence of tunica media

◌ : Aneurysmal part

Fig. 3. The subcontinuous partial images of the typical case with no cerebral aneurysm (A). The non-existence of internal elastic lamina was found in some of the non-aneurysmal part (A1) (A3) (A4) (A5). Also, the non-existence of tunica media was found in some of the non-aneurysmal part (A2) (A3) (A4) (A5) (A6). However, the non-existence of intimal hyperplasia was not found in the non-aneurysmal part. (Color figure online)

3.3 Summary of the Results

We quantified these results and made a summary of the results of all of these cases. Both of aneurysm cases and non-aneurysm cases had the pathological characteristic of the

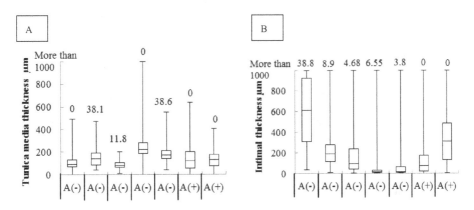

Fig. 4. The thickness of tunica media of all the cases (A) showed the non-existence of tunica media occurred in both of the aneurysm cases and the non-aneurysm cases. On the other hand, the thickness of intimal hyperplasia of cases (B) showed the non-existence of intimal hyperplasia occurred only in the aneurysm cases

non-existence of internal elastic lamina. Here we expressed the aneurysm cases as the mark "A(+)" and the non-aneurysm cases as the mark "A(−)". In order to show the minimum of the thickness, the numbers were wrote over the error bars of each sections. In addition, while the minimum showed "0", it meant non-existence. Also, the pathological characteristic of tunica media was shown in Fig. 4A. It indicated that both of aneurysm cases and non-aneurysm cases had the pathological characteristic of the non-existence of tunica media. Finally, the pathological characteristics of intimal hyperplasia were shown in Fig. 4B. It indicated that only aneurysm cases had the pathological characteristic of the non-existence of intimal hyperplasia.

4   Discussion

4.1   Discussion on the Relationship Between Aneurysm Initiation and the Distribution of Internal Elastic Lamina or Tunica Media

It is said that the non-existence of internal elastic lamina, the non-existence of tunica media and the existence of intimal hyperplasia are the reason why cerebral aneurysm initiate through previous studies [2, 3]. However, as shown in Fig. 5, compared the aneurysm cases with the non-aneurysm cases, the phenomenon of the non-existence of internal elastic lamina and the non-existence of tunica media were found in both of the cases. Therefore, we thought that the non-existence of internal elastic lamina or the non-existence of tunica media may be some kinds of phenomenon but not be the original factor of the cerebral aneurysm initiation.

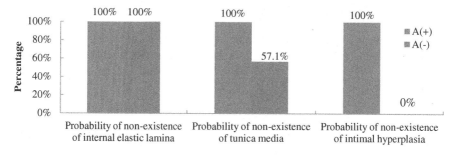

Fig. 5. The condition of pathological characteristics of all the cases showed the non-existence of internal elastic lamina and tunica media was possibly found in both of aneurysm cases and non-aneurysm cases non-existence of intimal hyperplasia was possibly found only in aneurysm cases. This indicated that the formation of intimal hyperplasia may be concerned with aneurysm initiation.

4.2   Discussion on the Relationship Between Aneurysm Initiation and the Distribution of Intimal Hyperplasia

The non-existence of intimal hyperplasia was found only in the aneurysm cases. Therefore, we thought that it is possible that the formation of intimal hyperplasia may be concerned with the cerebral aneurysm initiation. In order to surmise the relationship between the distribution of intimal hyperplasia and mechanism of initiation of cerebral aneurysm, we used the heatmap to show the thickness of intimal hyperplasia in detail. As shown in Fig. 6, all the two cases with an aneurysm have the common characteristic that the aneurysm parts (the parts of where the intimal hyperplasia loss) located at the boundary of where the thickness of intimal hyperplasia changed distinctively (so called the bottom of intimal hyperplasia).

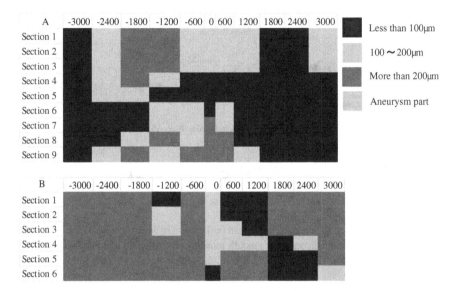

Fig. 6. The distribution of thickness of intimal hyperplasia of case 1 with an aneurysm (A) and case 2 with an aneurysm (B). The common characteristic of the aneurysmal parts (the parts of where the intimal hyperplasia loss) located at the boundary of where the thickness of intimal hyperplasia changed distinctively.

4.3 Discussion on the Mechanism of Aneurysm Initiation Based on the Results

Previous studies showed that some factors such as hemodynamics may cause the migration of cells [8] and the intimal hyperplasia occurs because of the migration of smooth muscle cells, and the condition of endothelial cells decides whether the smooth muscle cells migrate or not [9]. Here we thought the boundary of where the thickness of intimal hyperplasia changed distinctively as the location that the intimal hyperplasia would occur in near future. As shown in Fig. 7, based on these evidence, we thought that the smooth muscle cells of boundary migrated to the location that arteriosclerosis occurred because the topical migrated factors only existed at where the intimal hyperplasia occurred. As a result, the vessel wall of boundary of where the thickness of intimal hyperplasia changed distinctively became thinner and thinner and finally a cerebral aneurysm initiated there. On the other hand, we thought that the smooth muscle cells of boundary of where the thickness of intimal hyperplasia changed distinctively migrated to the location no matter whether arterioscleriosis occurred or not because the migrated factors existed at all the locations. As a result, the vessel wall of boundary of where the thickness of intimal hyperplasia changed distinctively did not become thin because of the neogenetic intimal hyperplasia and the cerebral aneurysm initiation did not occur.

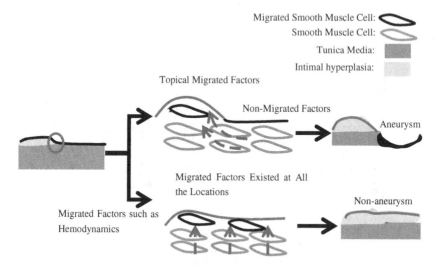

Fig. 7. The mechanism of the cerebral aneurysm initiation. The location where migrated factors distributed may influence the direction of smooth muscle cells migrated and decided whether an aneurysm initiates or not.

5   Summary

5.1   Finding of This Research

We studied the pathological characteristics of cerebral vascular bifurcations with and without aneurysm for elucidating the cerebral aneurysm initiation. Here we would like to summarize the findings of this research.

(1) The non-existence of internal elastic lamina and the non-existence of tunica media were found in both cases, whereas the non-existence of intimal hyperplasia was found only in the case with an aneurysm.
(2) Not the non-existence of internal elastic lamina or tunica media but the non-existence of intimal hyperplasia may be the reason of initiating cerebral aneurysm.

5.2   Contribution of This Research

These results make a new view of point for elucidating cerebral aneurysm initiation, and might be useful for developing a new kind of treatment for preventing the initiation of cerebral aneurysm. We also expect this research to be useful for the development of a new system of prediction of the cerebral aneurysm initiation.

Acknowledgements. This work was mainly supported by a research project at Waseda Research Institute for Science and Engineering, Project No.: #13L02, Title: Biomedical Engineering Research for Advanced Medical Treatment Using Nonclinical Study.

References

1. Massachusetts Medical Society: Unruptured intracranial aneurysm–risk of rupture and risks of surgical intervention. New Engl. J. Med. **339**(24), 1725–1733 (1998). The International Study of Unruptured Intracranial Aneurysms Investigators
2. Forbus, W.D.: On the orgin of military aneurysms of the superficial cerebral arteries. Bull. Johns Hopkins Hosp. **47**, 239 (1930)
3. Glynn, L.E.: Medial defects in the circle of Willis and their relation to aneurysm formation. J. Pathol. Bacteriol. **51**, 213–222 (1940)
4. Walker, A.E., Allegre, G.W.: The pathology and pathogenesis of cerebral aneurysms. J. Neuropathol. Exp. Neurol. **13**, 248–259 (1954)
5. Tulamo, R., Frosen, J., Heenesniemi, J., Niemela, M.: Inflammatory changes in the aneurysm wall: a review. J. NeuroIntervent. Surg. **2**, 120–130 (2010)
6. Meng, H., Wang, Z., Hoi, Y., Gao, L., Metaxa, E., Swartz, D.D., Kolega, J.: Complex hemodynamics at the apex of an arterial bifurcation induces vascular remodeling resembling cerebral aneurysm initiation. Stroke **38**, 1924–1931 (2007)
7. Metaxa, E., Tremmel, M., Natarajan, S.K., Xiang, J., Paluch, R.A., Mandelbaum, M., Siddiqui, A.H., Kolega, J., Mocco, J., Meng, H.: Characterization of critical hemodynamics contributing to aneurysmal remodeling at the basilar terminus in a rabbit model. Stroke **41**, 1774–1782 (2010)
8. Chatzizsis, Y.S., Coskun, A.U., Jonas, M., Edelman, E.R., Feldman, C.L., Stone, P.H.: Role of endothelial shear stress in the natural history of coronary atherosclerosis and vascular remodeling. J. Am. Coll. Cardiol. **49**, 2379–2393 (2007)
9. Rudijanto, A.: The role of vascular smooth muscle cells on the pathogenesis of atherosclerosis. Acta Med. Indones. **39**, 86–93 (2007)

Technology of Cortical Bone Trajectory on the Influence of Stability in Fixation of Burst Fracture of Thoracolumbar Spine: A Finite Element Analysis

Jianping Wang[1], Juping Gu[1(✉)], Jian Zhao[2], Xinsong Zhang[1],
Liang Hua[1], and Chunfeng Zhou[3]

[1] College of Electrical Engineering, Nantong University,
Nantong 226019, China
gu.jp@ntu.edu.cn
[2] Department of Orthopaedics, Changzheng Hospital,
Second Military Medical University, Shanghai 200003, China
[3] Department of Orthopaedics, Rich Hospital of Nantong,
Nantong 226019, China

Abstract. Objective: To study the biomechanical stability of a new screw-setting technique, we used cortical bone trajectory (CBT) in injury vertebra relative to the traditional pedicle screw-setting technique.

Methods: We used thoracolumbar spine CT data of a healthy adult male volunteer and engineering data of internal fixation system of spine to simulate intact state, burst fracture state and combination of three kinds of internal fixation state of the spine: (1) 4 pedicle screws cross segment and 2 rods (P4); (2) 4 pedicle screws, 2 CBT screws at injured vertebrae and 2 rods (P4C2); (3) 6 pedicle screws and 2 rods (P6). Then we compared differences of the stability of the corresponding fixed system and stress distribution of fixation models of three groups above.

Results: The total deformation of all nodes of the fracture spine model of P4C2 was less than the fracture spine model node group of P4 and larger than the fracture spine model node group of P6 during normal weight status, rotation (right), bending forward, stretch and lateral bending(right) state. The equivalent stress of all nodes of internal fixation system of P4C2 was smaller than the fixation model node group of P4 and bigger than the fixation model node group of P6 during normal weight status, rotation(right), bending forward, stretch and lateral bending(right) state.

Conclusion: CBT technology for injured vertebra fixation could provide more stability of the vertebral body and reduce stress concentration of internal fixation system compared to the traditional P4 fixation.

Keywords: Burst fracture · Thracolumbar spine · Cortical bone trajectory · Pedicle screw · Biomechanics · Injured level fixation

© Springer Nature Singapore Pte Ltd. 2017
M. Fei et al. (Eds.): LSMS/ICSEE 2017, Part I, CCIS 761, pp. 104–112, 2017.
DOI: 10.1007/978-981-10-6370-1_11

1 Introduction

Thoracic lumbar burst fracture is so common that much attention has been paid to the choice of surgical treatment. In spite of the great controversy [1, 2], internal fixation is still a preferred method to cure fractures which have surgical indications. After years of development, traditional pedicle screw-rod system was considered a mature method of treatment for such fractures after repeated testing [3, 4]. The treatment effects were more satisfactory [5, 6]. However, its growth curves were accompanied by the common complications and its limitations [7, 8]. Some accidents in the operation and difficulties for revision surgery also made doctors unprepared. Therefore, it is imperative to develop a safe and effective alternative way for traditional pedicle nailing method in case of need.

Santoni et al. [9] put forward a new kind of pedicle nailing method using cortical bone tunnel, in which the entry points were more inferomedial and the tunnel direction were more cephalad compared to other methods. They concluded that there was no obvious difference between CBT method and the traditional method in the intensity of nailing, and nailing anatomy is the basis of CBT method. Keitaro Matsukawa et al. [10] used the CT images of 100 persons to measure CBT canal of 470 vertebrae by 3D reconstruction software to get CBT channel diameter, length, roll angle of vertebra, and the reverse gantry of vertebral level. They summed up that the pedicle shape and axis angle of pedicle in different lumbar segment were different, but the data of CBT were similar in addition to the diameter. This means that CBT nailing direction and length varies a lot due to vertebral levels. This might reduce incidence of complication. But there are few reports on CBT application in burst fracture of thoracolumbar spine.

Finite element analysis method was applied in this paper to focus on influence of CBT techniques to postoperative vertebral stability of fixed lumbar spine and load distribution of internal fixation, so as to provide objective and theoretical basis for the application of CBT techniques.

2 Materials and Methods

We selected 192 slices raw data (Dicom format) of CT scan (64 spiral computed tomography scanner, Philips, Holland), scan slice 0.6 mm. The thoracolumbar spine was from a 32-year-old healthy male volunteer (height: 173 cm, weight: 75 kg), without history of spinal trauma, malformation or low back pain. The CT data was input into Mimics14.11 (Materialise, Belgium). STL format files of cortical bone, cancellous bone, annulus fibrosus and nucleus pulposus of thoracic lumbar segment (T12-L2) were obtained respectively through threshold segmentation, dynamic regional growth, cavity filling and Boolean operation steps with the help of the image segmentation tools. Then these files were input into Geomagic Studio12 (Geomagic, USA), after hole filling and deburring treatment, they were re-imported into Mimics. Then we got volume meshes, and cdb files of these parts directly from the corresponding mask (Hexahedral 8-point type). Then the cdb files were put into ANSYS Workbench 14.5 (Ansys, USA), assembled in the FE model module, and connected to the Static Structural module, where physical properties, elasticity modulus, contact condition of corresponding parts were defined. After ligament simulation, we determined thrust surface and fixed surface, set test stress and torque with directions, and planned the action time and the step length. Under mechanical condition and environment similar to classical biomechanical experiments in vitro, we finished the validation of the intact thoracic lumbar spine model.

The data of L1 vertebral body was input singly to Mimics, which was changed into fracture model with the use of the cutting function (Fig. 1). Then we got the corresponding vertebral fracture (cortical and cancellous bone) cdb files (Hexahedral 8-point type) just like the above steps, which were applied in the thoracic lumbar burst fracture model.

Fig. 1. Intact thoracic lumbar spine model and fractured model

Pedicle screw system and CBT screw system (Shangdong Weigao Biotech Co. Ltd) consisted of six pedicle screws (5.5 mm in diameter, 50 mm in length), two CBT screws (4.0 mm in diameter, 40 mm in length) and two connecting roots (6.0 mm in diameter, 100 mm in length). The simulation was completed by Solidworks 2012 software. After the formation of their stl files, we used Mimics to get their cdb documents directly from the mask body grid (Hexahedral 8-point type).

Surgery simulation system was contributed to the assembling of pedicle screw or CBT screw internal fixation system on fracture model. Traditional pedicle screw fixation and injured vertebral insertion models were made respectively: (1) 4 pedicle screws cross segment and 2 rods (P4); (2) 4 pedicle screws, 2 CBT screws at injured vertebrae and 2 rods (P4C2); (3) 6 pedicle screws and 2 rods (P6) (Fig. 2). We simulated normal weight status (NW), bending forward (BF), stretch (ST), rotation right (RR) and lateral bending right state (BR) respectively, and then calculated. Recorded results of the corresponding fixed system were compared in the total deformation of all nodes of the fracture spine model and the stress distribution of all nodes of fixation models of three groups above. The material properties are shown in table [11] (Table 1).

Fig. 2. Simulation of traditional pedicle screw fixation (P4, P6) and P4C2

Table 1. Material properties of human thoracolumbar spine

Materials	Young's modulus(MPa)	Poisson's ratio
Cortical bone	12000	0.3
Cancellous bone	100	0.2
Nucleus pulposus	1	0.49
Anulus fiber	400	0.3
Cortical bone (fractured)	1000	0.3
Cancellous bone (fractured)	10	0.3
Screw fixation	110000	0.3

Additional attachment included the interspinous ligament, supraspinous ligament, intertransverse ligament, anterior longitudinal ligament, posterior longitudinal ligament was simulated as incompressible spring element (Fig. 3), with axial stiffness (longitudinal stiffness) of 70000 N/m and axial damping (longitudinal damping) is 7000 N · s/m [12].

Fig. 3. The bone tunel of CBT screw and the internal fixation model with ligaments

3   Results

After the pre-treatment of finite element, contact mode of T12-L2 was obtained, the L1 burst fracture model and internal fixation model were also built based on it (Fig. 3). We got 133296 nodes (fractured spine model), 30365 nodes (P4), 39564 nodes (P4C2), 42282 nodes (P6) totally.

3.1   Model Validation

We compared ROM (range of angular movement) of virtual model in the study with classic biomechanical study in vitro of thoracolumbar segment specimen to validate the model (Fig. 4). After comparison, the finite element model of our study performanced similar to the specimen model in vitro biomechanical study of Markolf [13] in angular displacement - the applied torque gradient curve, results shown in figure. As a result, this finite element model had high reliability and it could be used in the biological mechanics simulation study effectively.

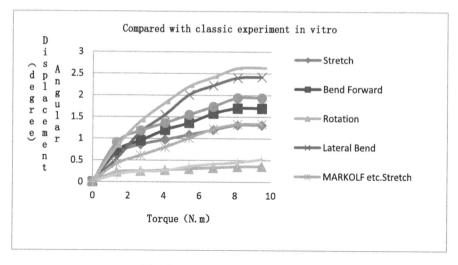

Fig. 4. The validation of the model

Mechanics Analysis

A. Total Deformation

The results of total deformation of five models were calculated out from Workbench static structural analysis system according to the previous steps: NW, BF, ST, RR, BR, and they were exported in excel format, which were the total deformation data of all nodes of the fracture model. The data of total deformation of all nodes in three kinds fixed fracture models was chosen. And then five kinds of model including intact model, burst fracture model, and three kinds of different internal fixation model were imported into statistics software SPSS 19. The mean and standard deviation were calculated, and went through paired sample T test. P values were obtained (Table 2).

Table 2. Total deformation of vertebrae nodes (mm, $\bar{X} \pm S$)

Group State	* Intact	**Fractured	P4	P4C2	P6
NW	0.037±0.036	0.239±0.273	0.058±0.044	0.046±0.037	0.040±0.035
BF	0.048±0.041	0.588±0.690	0.143±0.133	0.126±0.116	0.102±0.097
ST	0.157±0.096	0.293±0.209	0.152±0.090	0.124±0.076	0.090±0.059
RR	0.081±0.052	0.447±0.513	0.090±0.068	0.075±0.058	0.069±0.053
BR	0.103±0.070	0.287±0.275	0.121±0.086	0.092±0.076	0.073±0.065

By statistical comparison, we found average total deformation of fracture model were greater than the complete model in each movement and its standard deviation. And average total deformation and its standard deviation of fracture model were significantly decreased after using three kinds of internal fixation; The total deformation of all nodes of the fracture spine model of P4C2 during normal weight status, and rotation (right), bending forward, stretch, lateral bending (right) state, was less than the fracture spine model node group of P4, and larger than the fracture spine model node group of P6.

B. Equivalent Stress (von Mises)

Three kinds of internal fixation model were calculated in the workbench Static Structural analysis system to get the results according to the previous steps (Figs. 5, 6 and 7). Data of equivalent stress was exported in excel format. We selected the data of all nodes in fixation models for observation. Three sets of data of internal fixation models in the equivalent stress were imported into statistics software SPSS 19 in sequence according to five kinds of state: state of NW, BF, ST, RR, BR. The mean and standard deviation were calculated, and P values were obtained through the single factor analysis of variance (ANOVA) (Table 3).

Table 3. Equivalent stress of nodes of internal fixation (MPa, $\bar{X} \pm S$)

Group \ State	* P4	P4C2	P6
NW	5.850 ± 7.829	3.947 ± 5.736	3.096 ± 4.367
BF	13.045 ± 15.877	8.222 ± 10.609	11.260 ± 16.359
ST	6.709 ± 11.734	4.268 ± 7.498	3.744 ± 6.530
RR	8.344 ± 10.716	5.553 ± 7.889	4.539 ± 6.200
BR	6.854 ± 10.547	5.499 ± 7.101	4.149 ± 6.374

* Control group: P4   Experimental group: P4C2 and P6 $P < 0.001$

The equivalent stress of all nodes of internal fixation system of P4C2 during normal weight status, rotation (right), bending forward, stretch and lateral bending (right) state was smaller than that of P4, but bigger than that of P6 (except BF state).

We took BF state for example to explain the difference of equivalent stress in the three fixed models. In P4C2 model, the stress was more smoothly distributed in the screw-rod fixation than the P4 and P6 model (Figs. 5, 6 and 7). The data was compared by SNK test and we found it had obvious differences between three models (Table 4).

Table 4. The data of three models in bend forward state under Student-Newman-Keuls (SNK) test

Group	N	Alpha = 0.05		
		1	2	3
P4C2	39,564	8.22225798		
P6	42,282		11.26014106	
P4	30,365			13.04458581
Significance		1.000	1.000	1.000

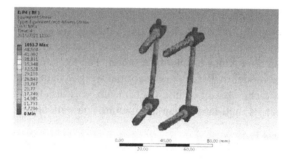

Fig. 5. Equivalent Stress of P4 model in BF state

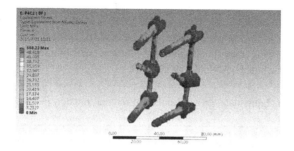

Fig. 6. Equivalent Stress of P4C2 model in BF state

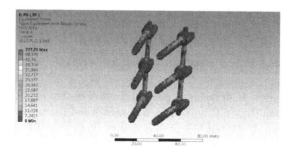

Fig. 7. Equivalent Stress of P6 model in BF state

4 Discussion

Traumatic fractures in thoracic lumbar segment (T10–L2) were the most common, and burst fractures in thoracic lumbar segment were representative of high energy damage which accounted for about 10% to 20% [14]. It is controversial for internal fixation of spinal fracture of osteoporosis, and the mainstream of research about it turns to the specialty of pullout resistance and anti-loosening of internal fixation. Solid mechanics experiment has been golden standard for comparing fixation or fusion treatment methods. Luis Perez-Orribo [15] considered that bilateral CBT nailing method provided similar stability to bilateral pedicle screw system in the specimen, regardless of intervertebral fusion. The CBT method could theoretically reduce the damage when compared with the traditional method.

It is a kind of simulation experiments for finite element analysis, and the reliability of the experimental results depends on the accuracy of model simulation. We had compared with the traditional classical experiments in vitro, and the finite element model seems to act the same as the classic experiments of mechanics, and the movement state had played similar angular displacement curve. So we believed the reliability of the model,and its well application to simulate the target segment of spinal and play similar biomechanics characteristics. After all, the finite element model is a kind of virtualization of the real state and a product of the development of computer technology. So we need further research to prove whether it can lead a right direction to represent real human spine and show believable results.

Single axial stress was equivalent to the multiple axis stress, and we make a uniaxial tensile force equivalent to six components of stress to the deformable body. The equivalent stress explores difference of stress on materials in all directions, and is mainly used to describe the phenomenon of stress concentration. In this paper, the result can reflect difference in equivalent stress of material of the three internal fixation model in the fracture model under stress. Objectively it could give expression to concentration degree of stress in internal fixation model, and indirectly reflect their difference in ability of load dispersion. Thus it could provide the reference to evaluate the stability and the service life of the fixation. This research cited the classic literature and used its material assignment only, but could not show all the properties of materials, such as yield, plastic deformation properties. So there was a certain gap from the real situation, and we will take into account in future.

The anterior column was not affected by the method of CBT. The diameter of the cortical bone tunnel was less and the length was shorter than that of the pedicle screws, so the compression of the cortical bone were smaller than that of the pedicle screws. Bone cutting can be also reduced compared with the pedicle screw method because of smaller screws and more content of cortical bone. CBT screws can also be inserted into vertebral bodies where the pedicle screws are already existed; thus, the CBT method is valuable in repeat surgery of same vertebrae.

Mechanics research based on fresh specimens is an ideal method of spinal biomechanics research. We used pre-processing and finite element analysis software to simulate the biomechanical state of spine, and calculate the corresponding results. Four kinds of simple motion state were tested in this study, and we should research more complex integrated motion, such as composite movement and its corresponding composite stress state. We will deepen the research and make the model perfect to obtain more accurate results.

5   Conclusion

CBT technology for injured vertebra fixation could provide more stability of the vertebral body and reduce stress concentration of internal fixation system when compared to the traditional P4 fixation. But it was still inferior to the P6 method in the two aspects.

Conflict of Interest
The authors declare that they have no conflict of interest.

Acknowledgment. The work was supported by National Natural Science Foundation of China (61273024 and 61673226).

References

1. Woodall Jr., J.W.: Evidence for the treatment of thoracolumbar burst fractures. Curr. Orthop. Pract. **23**(3), 188–192 (2012)
2. Canbek, U., Karapinar, L.: Posterior fixation of thoracolumbar burst fractures: Is it possible to protect one segment in the lumbar region? Eur. J. Orthop. Surg. Traumatol. **24**(4), 459–465 (2014)
3. Gaines Jr., R.W.: The use of pedicle-screw internal fixation for the operative treatment of spinal disorders. Bone Joint Surg. **82-A**(10), 1458–1476 (2000)
4. Ruf, M., Harms, J.: Pedicle screws in 1-and 2-year-old children: technique, complications, and effect on further growth. Spine **27**(21), 460–466 (2002)
5. McCormack, T., Karaikovic, E., Gaines, R.W.: The load-sharing classification of spine fractures. Spine **19**(15), 1741–1744 (1994)
6. Gelb, D., Ludwig, S.: Successful treatment of thoracolumbar fractures with short-segment pedicle instrumentation. J Spinal Disord. Tech. **23**, 293–301 (2010)
7. Esses, S.I., Sachs, B.L., Dreyzin, V.: Complications associated with the technique of pedicle screw fixation. A selected survey of ABS members. Spine **18**(15), 2231–2238 (1993)
8. Saita, K., Hoshino, Y., Kikkawa, I., et al.: Postlerior spinal shortening for paraplegia after vertebral collapse cause by osteopomsi. Spine **25**(21), 2832–2835 (2000)
9. Santoni, B.G., Hynes, R.A., McGilvray, K.C., et al.: Cortical bone trajectory for lumbar pedicle screws. Spine J. **9**, 366–373 (2009)
10. Matsukawa, K., Yato, Y.: Morphometric measurement of cortical bone trajectory for lumbar pedicle screw insertion using computed tomography. J. Spinal Disord. Tech. **26**(6), 248–253 (2013)
11. Shih, S.-L., Chen, C.-S., Lin, H.-M., et al.: Effect of spacer diameter of the dynesys dynamic stabilization system on the biomechanics of the lumbar spine: a finite element analysis. J. Spinal Disord. Tech. **25**(5), 140–149 (2012)
12. Ozgur, V., Mehmet, S.E., Levent, A., et al.: Biomechanical Evaluation of Syndesmotic Screw Position: a finite element analysis. J. Orthop. Trauma **28**(4), 210–215 (2014)
13. Markolf, K.L.: Deformation of the thoracolumbar intervertebral joints in response to external load: a biomechanical study using autopsy material. J. Bone Join Surg. Am. **54**(3), 511–533 (1972)
14. Wood, K.B., Li, W.: Management of thoracolumbar spine fractures. Spine J. **14**, 145–164 (2014)
15. Perez-Orribo, L., Kalb, S., et al.: Biomechanics of lumbar cortical screw-rod fixation versus pedicle screw-rod fixation with and without interbody support. Spine **38**(8), 635–641 (2013)

Current Solutions for the Heat-Sink Effect of Blood Vessels with Radiofrequency Ablation: A Review and Future Work

Zheng Fang[1], Bing Zhang[1,2(✉)], and Wenjun Zhang[1,2,3(✉)]

[1] CISR Lab, Tumor Ablation Group, East China University of Science and Technology, Shanghai, China
{bing.zhang84,chris.zhang}@usask.ca
[2] Division of Biomedical Engineering, University of Saskatchewan, Saskatoon, Canada
[3] Department of Mechanical Engineering, University of Saskatchewan, Saskatoon, Canada

Abstract. Radiofrequency ablation (RFA) as an alternative treatment to the conventional open surgery is the most popular minimally invasive thermal therapy, and it is widely used in clinic today. One of the most important limits for the RFA in clinic is the difficulty to deal with the heat-sink effect of blood vessels, as it causes the difficulty of control the RFA process and consequently the coagulation size of RFA is decreased considerably (empirically, the coagulation size is less than 3 cm with a single RFA electrode). This paper reviews the literature of the current solution for the heat-sink effect due to large blood vessels and suggests future work for finding more effective solutions.

Keywords: Radiofrequency ablation · Heat-Sink effect · Blood vessels · Incomplete ablation

1 Introduction

RFA, as a heat-mediated modality, has become an accepted treatment option for focal primary and secondary malignancies in some organs including the liver, lung, kidney, bone, and adrenal glands [1, 2]. So far, RFA treatment has achieved a competitive success rate compared to the conventional surgery treatment for small tumors (<3 cm in diameter) [3]. One of the major limits with RFA is the small size of coagulation [4]. It is noted that the coagulation zone also includes the 0.5 to 1-cm margin area of the healthy tissue adjacent to the tumor for the purpose of eliminating microscopic foci of disease and off-setting the possibility of incomplete tumor destruction [5]. The coagulation size is affected by heterogeneity of the tissue composition, which causes differences in the tumor tissue density and subsequently differences in the electrical and thermal conductivity; certainly, heterogeneity causes less coagulation [6]. Reduction of the coagulation size is further caused by the proximity of tumors to large blood vessels, as large blood vessels serve as a heat sink [7]. Goldberg et al. and Poch et al. [8, 9] demonstrated that the blood flow in blooded vessels is highly responsible for reduction of the coagulation size. It is noted that the electrical conductivity of blood is about five

© Springer Nature Singapore Pte Ltd. 2017
M. Fei et al. (Eds.): LSMS/ICSEE 2017, Part I, CCIS 761, pp. 113–122, 2017.
DOI: 10.1007/978-981-10-6370-1_12

times higher than that of the hepatic tissue, which means a low-resistance pathway to diffuse away heat [10].

This paper aims to provide a comprehensive review of the current solutions to the large blood vessel problem with RFA and to propose some new solution to the problem if the current solutions are found not satisfactory.

2   The Heat Transfer Principle of RFA with Large Blood Vessels

A general configuration of the RFA system is: an RF generator, an RF electrode (active tip and insulated shaft), several grounding pads, and several electrical wires [11]. With this configuration, a current field is established between the RF electrode and the grounding pads and an alternating current (375–500 kHz) is produced by the RF generator [12]. In the current field, ions (sodium, potassium, and chloride etc.) inside the biological tissue around the RF electrode are forced to interact with each other and to move back and forth rapidly along the direction of the alternating current. Then, the friction-induced heat will be generated due to the interaction between the ions and the electrons. Tissues around the RF electrode are heated to the temperature of approximately 50–100 °C, and the temperature in this range causes irreversible damages of the protein coagulation of cells [13].

The heat transfer inside the target tissue during the RFA procedure is governed by the so-called Pennes' bio-heat transfer equation with an outside energy source, which is expressed by [14–16]:

$$\rho c \frac{\partial T(x,t)}{\partial t} = \nabla \cdot k \nabla T(x,t) + \rho_b c_b \omega_b \left(T_b - T(x,t) \right) + Q_{hs}(x,t) + Q_m(x,t) \quad x \in \wedge \quad (1)$$

where $\rho \left(kg/m^3 \right)$ is the density, c (J/kg · K) is the specific heat, $T(\mathbf{x}, t)(K)$ is the temperature, k (W/m · K) is the thermal conductivity, $\rho \left(kg/m^3 \right)$ is the blood density, c_b (J/kg · K) is the specific heat of the blood, $\omega_b(1/s)$ is the blood perfusion, $T_b(K)$ is the temperature of the blood entering the tissue, $\mathbf{x} = \{x, y, z\}$ in the Cartesian coordinate system, $\wedge$ denotes the analyzed spatial domains, $Q_m(x, t) \left(W/m^3 \right)$ is the energy generated due to metabolic processes, and $Q_{hs}(x, t) \left(W/m^3 \right)$ is the spatial heat generated by the RF electrical current. It is to be noted that Eq. (1) does not capture the heat transfer of the tissue in the presence of large blood vessels in vicinity to the tissue [17].

For the situation in the presence of large blood vessels, the heat transfer of blood flow in the vessels may not be neglected. To take into account the large blood vessel in vicinity to the tissue, a convection boundary to the tissue should be considered based on the Newton law of cooling as Eq. (2) [18].

$$-\mathbf{n} \cdot (-k \nabla T(\mathbf{x}, t)) = h_b \left(T_b - T(\mathbf{x}, t) \right) \quad (2)$$

where h_b is the convection heat transfer coefficient of the blood to the tissue.

A constant heat transfer coefficient between vessel and tissue was assumed under the fully developed flow in the vessel, and the coefficient is expressed by [18, 19]:

$$h_b = Nu_D k_b/D \tag{3}$$

where Nu_D is the local Nusselt number, k_b is the thermal conductivity of blood, D is the vessel diameter.

For Nu_D, it can be approximated, within 3.5%, by [20]:

$$Nu_D = 4 + 0.48624 \ln^2[Re \cdot Pr \cdot D/(18 \cdot L)] \tag{4}$$

$$Re = \rho_b V_b D/\mu \tag{5}$$

where Re is the Reynolds number, Pr is the Prandtl number, L is the vessel length, μ is the viscosity of blood, and V_b is the average blood velocity.

3   The Influence of the Heat-Sink Effect

Based on studies, the thermal coagulation obtained ex vivo is larger than those in vivo [21]. The principle of causing smaller coagulation in vivo is that the blood flow in vessels is able to bring heat quantities transferred from active tip away which aggravates the heat lost. The act of heat convection between the tissue and the blood flow is termed the heat-sink effect [8]. Lu et al. and Shih et al. [22, 23] demonstrated that the heat-sink effect starts to occur with a blood vessel (<2 mm in diameter) and obviously occurs with a blood vessel (<3 mm in diameter), whereas, when vessel diameter was less than 2 mm, the short-duration and high-intensity heating scheme could overcome the heat-sink effect of the blood vessels. Therefore, the vessels exceeding 3 mm in diameter are defined as medium or large vessels. Pillai et al. [24] demonstrated the differences between the heat-sink and the heat-sink absent and the results are tabulated in Table 1. A huge influence of the heat-sink effect was illustrated, which causes the dramatic decrease of coagulation size. Furthermore, the recurrence rate in the vicinity of medium or large liver vessels is 36.5% compared to 6.3% without neighboring vessels [25]. Besides, the heat-sink effect can cause an irregular shape of coagulation zone closed to large vessels, which enhances the difficulty in ablating tumors completely and the early local recurrence in the tissue near the vessels [26].

Table 1. Comparison of the tumor-ablated parameters with the heat-sink and without the heat-sink in the perfused calf liver.

Parameter	Heat-sink absent	Heat-sink	Difference	Type of electrode
Lateral dimension (mm)	40 ± 3.2	28 ± 2.2	30%	Expandable electrode
Longitudinal dimension (mm)	50 ± 3.2	31 ± 3.2	38%	
Volume (cm³)	170 ± 12	100 ± 12	41%	
Mass (g)	120 ± 14	52 ± 11	56%	

4   The Current Physical Solution to the Heat-Sink Effect

For the heat-sink effect, an efficient method to solve this problem is still missing. In clinic, one of solution is the mechanical occlusion. And another way is to design an appropriate electrode, because a suitable structure of the electrodes is able to reduce the heat-sink effect effectively [24].

Table 2. In-vivo RFA results in liver.

Study	Electrode	L (cm)	D (mm)	N	T (min)	Ds (cm)	Di (cm)	V (cm³)
Goldberg et al. [31]	PE	3.0	1.3	3	6	1.5 ± 0.1	3.4 ± 0.2	N/A
Cha et al. [32]	ICE	2.0	1.5	30	12	2.0 ± 0.33	2.62 ± 0.39	5.76 ± 2.89
Pereira et al. [33]	PE	1.5	1.7	4	20	2.30 ± 0.94	5.85 ± 1.5	31.5 ± 16.8
	ME	4.0[a]	2.5	4	15	3.44 ± 0.21	3.10 ± 0.62	16.2 ± 7.3
Hirakawa et al. [34]	ME	3.0[a]	1.5	14	15.12	2.8 (3.1–6.0)[b]	4.0 (3.0–5.0)[b]	N/A
Lee et al. [35]	CE	2.5	N/A	14	18	3.49 ± 0.8	3.77 ± 1.01	33.08 ± 13.4 1
Yu et al. [21]	BE	3.0	1.7	2	10	1.61 ± 0.33	3.21 ± 0.51	No data

Note.-All data are mean ± SD (standard deviation) of central coagulated tissue.

PE is the plain electrode. ICE is the internally cooled electrode. PE is the perfusion electrode. ME is the multitoned expandable electrode. CE is the cluster electrode. BE is the bipolar electrode. L is the length of electrode active tip. d is the diameter of electrode. N is the number of ablation. T is the ablation time. Ds is the coagulation tissue short-axis diameter. Di is the length of coagulation along the electrode. V is the approximate volume of coagulation.

[a]maximum diameter of expanded shape.

[b]data are median (range).

4.1   New Designed Electrode

As the development of the technique of RFA, some new designed electrodes were invented for enlarging the size of coagulation, such as internally cooled electrode [27], expandable electrode [28], cluster (or multiple) electrode [29], perfusion electrode [30], and bipolar electrode [25]. All of those new designed electrodes, although, had enlarged the coagulation size (Table 3), their special characteristics also need to be further investigated to show the capability to overcome the heat-sink effect. For example, when an electrode abuts to a medium or large vessel, it is necessary to know the optimal characters about the relative position between the electrode and the vessel for increasing coagulation size [9] (Table 2).

Table 3. The pros and cons of current physical solutions to heat-sink effect

Solutions	Advantages	Disadvantages
New designed electrode	It increases the power deposition and decreases the heat-sink effect	It is difficult to design an optimal electrode
Mechanical occlusion	It decreases the heat-sink effect	Itincreases the difficulty andinvasiveness of RFA

Besides, Rossi et al. [36] discovered the reason avoiding the heat-sink effect efficiently by the hook tips is that expandable electrode can retract or redeploy the hooks after slightly rotating the electrode to change the hook position, or by completely repositioning the electrode until hook tips reach the killing temperature. Unfortunately, there is a challenge for an expandable electrode system to precisely deploy multiple electrodes simultaneously during percutaneous procedures and to modify the overall survival rate or the disease-free survival rate with less treatment sessions [37].

There is a tendency to design a novel electrode with a dynamic geometry, such as expandable electrode and spiral bipolar electrode, because it is able to adjust the geometry to contact a wider range of tissue and avoid or reduce the heat-sink effect, such as multitined expandable electrode with difference structures, bipolar-expandable electrode etc. [21, 38, 39]. Therefore, electrode structure is a main factor affecting the RFA treatment, because a pretty electrode structure can deliver more energy and far away. Taking into consideration the medium or large vessels during the process of electrode design seems to be able to reduce the heat-sink effect efficiently [38]. It is meaningful to design a new electrode for the purpose of reducing the heat-sink effect in future.

4.2 Temporary Vessel Occlusions by the Mechanical Occlusion

Temporary vessel occlusion as an alternative method for reducing the effect of heat sink is also used in the current procedure of RFA [40]. In modern clinical trials, surgeons usually prefer the occlusion with inflatable balloons to the Pringle maneuver to assist the RFA procedure [41]. Because, compared with the Pringle maneuver, the balloon occlusion is simple, safe, and highly feasible to avoid the effect of heat-sink [42]. However, for the method of inflatable balloon, some hepatic vessels cannot be occluded completely such as variant vascular anatomy, irregular shape, or stenosis of hepatic artery. Those problems can be solved by injecting a mixture sponge particles or contrast material [43].

Therefore, the hepatic vessel occlusion weakens the influence of the heat-sink effect for achieving a larger coagulation size and a more regular coagulation shape [44]. For the temporary vessel occlusion, the safety of mechanical occlusion had been approved in clinical practice and one-year survival rate was 44%, which was better than hepatectomy (28%) [45]. But, the difficulty and complexity of RF procedure are increased dramatically after the vessel occlusion [46].

In conclusion, all of those solutions obviously enlarged the size of coagulation (exceed the restriction of 3 cm in diameter) for liver tumors abut to medium or large vessels. The principle of mechanical occlusion is to decrease the heat loss. However, the mechanical occlusion increases the difficulty and invasiveness of RFA [46]. For the development of RF electrode, it aims to increase power deposition for enlarging coagulation, especially for that under the heat-sink effect of medium and large vessels. But, it is still difficult for an RF electrode having those characters simultaneously: (1) it can be inserted into body easily; (2) it is able to be expanded to maintain maximum contact with tissue; and (3) it is able to avoid the heat sink effect as much as possible [38].

5   RFA Computer Model with the Heat-Sink Effect

RFA computer models are often used to assist clinical treatments recently. RFA procedures are simulated with computer models for planning, evaluating, and optimizing RFA therapies. Meanwhile, RFA computer models also benefit to investigating the complex processes of RFA and to minimizing treatment risks preoperatively [47]. For example, a counter current blood vessels model analyzes the robustness of the power deposition scheme [48]. Especially, the heat-sink effect as a main factor of higher tumor recurrence is added into RFA computer model for investigating and optimizing clinical treatments [7]. Therefore, many RFA computer models with the heat-sink effect were proposed to predict the temperature, tissue properties change, ablation time, and ablation results within the organ intraoperatively for improving protocol of tumor ablation as well as reducing the damage of surrounding normal tissue [15]. Huang et al. [7] simulated the influence of position relationship contained direction and relative distance between vessel and RF electrode. Haemmerich et al. [20] simulated the results of RF ablation abutting large vessels with different RF electrodes. Nonetheless, those models, until now, did not include all variables among tumors, electrodes and medium or large vessels to simulate. For example, the variable of blood flow is excluded in Haemmerich's model. RFA procedures is also hard to plan and interventionally guide due to the heat-sink effect of medium and large vessels [49].

6   Future Work

During RFA treatment, the heat-sink effect of blood vessels remarkably narrows the coagulation size and raises mortality. To address this problem, great efforts should be made in all aspects of RFA technique, especially in the following areas:

First, further understanding of the surrounding characters of RFA is needed. Specially, the vessel structure, vessel types, and blood characteristics are needed to improve the accuracy and veridicality of RFA computer model. For example, O'Rourke et al. [50] demonstrated the performance of heat-sink effect in cirrhotic livers. Nevertheless, an RFA computer model requires complete surrounding information to achieve the predictability and controllability of RFA procedures with the heat-sink effect. Therefore, mass scientific experiments are necessary to obtain the surroundings characters. Correspondingly, the RFA model based on surrounding information is able to optimize the parameters of tumor ablation procedure for weakening the heat-sink effect.

Second, improving RFA procedure control protocol avoids the heat-sink effect of medium and large vessels. The current approach to heat dissipation is to block blood vessels or slow down blood flow, but this increases the complexity of the RFA procedure and introduces some other complications. A RFA procedure control protocol achieving the decrease of heat-sink effect is necessary. Nevertheless, it is also lack of a standard control protocol, which can achieve the largest coagulation in RFA procedure and meanwhile decrease the heat-sink effect. Adding the condition of medium and large vessels into RFA control protocol is able to avoid or overcome the effect of heat-sink.

Meanwhile, the control protocols taking into consideration the heat-sink effect of blood vessels need to be further studied to find an optimal method.

Third, designing a new electrode is another tendency to overcome the heat-sink effect of medium or large vessels. The development of RF electrode from structure, polar principle, and control method promotes efficiency and safety of tumor treatment. Until now, the design of RF electrodes, such as expandable, cooled cluster, or other combination designs, is still hard to obtain a large coagulation zone (>3 cm in diameter) abutting medium or large vessels [51]. Because all designs of RF electrodes just purely aimed at enlarging the coagulation zone in the tumor and ignored the influence of medium and large vessels. Only very few studies talking about the new design of RF electrodes considering the heat-sink effect can be found in literature [52]. The heat-sink effect initially is set into the design process as a restriction in conceptual design level. Then the new design of electrode possibly overcome the problem of heat-sink effect.

7 Conclusion

Radiofrequency ablation technique is very safe and effective for small tumor (<3 cm in diameter). One of the reasons causing the failure of RFA in the treatment of large tumors is the heat-sink effect with medium and large vessels. The small coagulation, lower survival rate and incomplete ablations have been demonstrated in vivo due to the heat-sink effect of blood vessels. Therefore, the further development of RFA is necessary to overcome the heat-sink effect for promoting RFA to be a more favorable tumor ablation modality in clinic.

References

1. Ahmed, M., Goldberg, S.N.: Principles of radiofrequency ablation. J. Interv. Oncol., 23–37 (2012). Springer
2. Ni, Y., et al.: A review of the general aspects of radiofrequency ablation. Abdom. Imaging **30**, 381–400 (2005)
3. Yang, W., et al.: Ten-year survival of hepatocellular carcinoma patients undergoing radiofrequency ablation as a first-line treatment. World J. Gastroenterol. **22**, 2993 (2016)
4. Zhang, B., et al.: A review of radiofrequency ablation: Large target tissue necrosis and mathematical modelling. Phys. Med. **32**, 961–971 (2016)
5. Lee, J.M., et al.: Switching monopolar radiofrequency ablation technique using multiple, internally cooled electrodes and a multichannel generator: ex vivo and in vivo pilot study. Invest. Radiol. **42**, 163–171 (2007)
6. Rhim, H., et al.: Essential techniques for successful radio-frequency thermal ablation of malignant hepatic tumors. Radiographics **21**, S17–S35 (2001)
7. Huang, H.-W.: Influence of blood vessel on the thermal lesion formation during radiofrequency ablation for liver tumors. Med. Phys. **40**, 073303 (2013)
8. Goldberg, S., et al.: Radio-frequency tissue ablation: effect of pharmacologic modulation of blood flow on coagulation diameter. Radiology **209**, 761–767 (1998)

9. Poch, F.G., et al.: The vascular cooling effect in hepatic multipolar radiofrequency ablation leads to incomplete ablation ex vivo. Int. J. Hyperth., 1–8 (2016)

10. Dodd III, G.D., et al.: Effect of variation of portal venous blood flow on radiofrequency and microwave ablations in a blood-perfused bovine liver model. Radiology **267**, 129–136 (2013)

11. Zhang, B., et al.: Evaluation of the current radiofrequency ablation systems using axiomatic design theory. Proc. Inst. Mech. Eng. H **228**, 397–408 (2014)

12. Ahmed, M., et al.: Image-guided tumor ablation: Standardization of terminology and reporting criteria—a 10-year update. J. Vasc. Interv. Radiol. **25**, 1691–1705, e1694 (2014)

13. Goldberg, S.N.: Radiofrequency tumor ablation: principles and techniques. Eur. J. Ultrasound **13**, 129–147 (2001)

14. Zhang, B., et al.: Study of the relationship between the target tissue necrosis volume and the target tissue size in liver tumours using two-compartment finite element RFA modelling. Int. J. Hyperth. **30**, 593–602 (2014)

15. Zhang, B., et al.: Numerical analysis of the relationship between the area of target tissue necrosis and the size of target tissue in liver tumours with pulsed radiofrequency ablation. Int. J. Hyperth. **31**, 715–725 (2015)

16. Zhang, B., et al.: A new approach to feedback control of radiofrequency ablation systems for large coagulation zones. Int. J. Hyperth. **33**, 367–377 (2017)

17. Hariharan, P., et al.: Radio-frequency ablation in a realistic reconstructed hepatic tissue. J. Biomech. Eng. **129**, 354–364 (2007)

18. Consiglieri, L., et al.: Theoretical analysis of the heat convection coefficient in large vessels and the significance for thermal ablative therapies. Phys. Med. Biol. **48**, 4125 (2003)

19. Nakayama, A., Kuwahara, F.: A general bioheat transfer model based on the theory of porous media. Int. J. Heat Mass Transf. **51**, 3190–3199 (2008)

20. Haemmerich, D., et al.: Hepatic bipolar radiofrequency ablation creates coagulation zones close to blood vessels: a finite element study. Med. Biol. Eng. Comput. **41**, 317–323 (2003)

21. Yu, J., et al.: A comparison of microwave ablation and bipolar radiofrequency ablation both with an internally cooled probe: results in ex vivo and in vivo porcine livers. Eur. J. Radiol. **79**, 124–130 (2011)

22. Lu, D.S., et al.: Effect of vessel size on creation of hepatic radiofrequency lesions in pigs: Assessment of the "heat sink" effect. AJR Am. J. Roentgenol. **178**, 47–51 (2002)

23. Shih, T.-C., et al.: Cooling effect of thermally significant blood vessels in perfused tumor tissue during thermal therapy. Int. Commun. Heat Mass **33**, 135–141 (2006)

24. Pillai, K., et al.: Heat sink effect on tumor ablation characteristics as observed in monopolar radiofrequency, bipolar radiofrequency, and microwave, using ex vivo calf liver model. Medicine **94**, e580 (2015)

25. Lehmann, K.S., et al.: Minimal vascular flows cause strong heat sink effects in hepatic radiofrequency ablation ex vivo. J. Hepatobiliary Pancreat. Sci. **23**, 508–516 (2016)

26. de Baere, T., et al.: Radiofrequency ablation of lung metastases close to large vessels during vascular occlusion: preliminary experience. J. Vasc. Interv. Radiol. **22**, 749–754 (2011)

27. Cha, J., et al.: Radiofrequency ablation using a new type of internally cooled electrode with an adjustable active tip: An experimental study in ex vivo bovine and in vivo porcine livers. Eur. J. Radiol. **77**, 516–521 (2011)

28. Ito, N., et al.: Bipolar radiofrequency ablation: development of a new expandable device. Cardiovasc. Intervent. Radiol. **37**, 770–776 (2014)

29. Yoon, J.H., et al.: Monopolar radiofrequency ablation using a dual-switching system and a separable clustered electrode: Evaluation of the in vivo efficiency. Korean J. Radiol. **15**, 235–244 (2014)

30. Lee, J., et al.: Radiofrequency ablation in pig lungs: in vivo comparison of internally cooled, perfusion and multitined expandable electrodes. Br. J. Radiol. (2014)
31. Goldberg, S.N., et al.: Tissue ablation with radiofrequency: effect of probe size, gauge, duration, and temperature on lesion volume. Acad. Radiol. **2**, 399–404 (1995)
32. Cha, J., et al.: Radiofrequency ablation zones in ex vivo bovine and in vivo porcine livers: Comparison of the use of internally cooled electrodes and internally cooled wet electrodes. Cardiovasc. Intervent. Radiol. **32**, 1235–1240 (2009)
33. Pereira, P.L., et al.: Radiofrequency ablation: In vivo comparison of four commercially available devices in pig livers. Radiology **232**, 482–490 (2004)
34. Hirakawa, M., et al.: Randomized controlled trial of a new procedure of radiofrequency ablation using an expandable needle for hepatocellular carcinoma. Hepatol. Res. **43**, 846–852 (2013)
35. Lee, E.S., et al.: Multiple-electrode radiofrequency ablations using Octopus® electrodes in an in vivo porcine liver model. Br. J. Radiol. **85**, e609–e615 (2014)
36. Rossi, S., et al.: Percutaneous treatment of small hepatic tumors by an expandable RF needle electrode. AJR Am. J. Roentgenol. **170**, 1015–1022 (1998)
37. Choi, D., et al.: Percutaneous radiofrequency ablation for recurrent hepatocellular carcinoma after hepatectomy: long-term results and prognostic factors. Ann. Surg. Oncol. **14**, 2319–2329 (2007)
38. Lau, L., Han, Y.-L.: Exploring a novel heating probe design for tumor ablation. J. Med. Device **10**, 030930 (2016)
39. Huo, Y.R., et al.: "Edgeboost": a novel technique to extend the ablation zone lateral to a two-probe bipolar radiofrequency device. Cardiovasc. Intervent. Radiol. **39**, 97–105 (2016)
40. Šubrt, Z., et al.: Temporary liver blood-outflow occlusion increases effectiveness of radiofrequency ablation: an experimental study on pigs. Eur. J. Surg. Oncol. **34**, 346–352 (2008)
41. Sobczyński, R., et al.: Transoesophageal echocardiography reduces invasiveness of cavoatrial tumour thrombectomy. Wideochir Inne Tech Maloinwazyjne **9**, 479 (2014)
42. de Baere, T., et al.: Hepatic malignancies: percutaneous radiofrequency ablation during percutaneous portal or hepatic vein occlusion 1. Radiology **248**, 1056–1066 (2008)
43. Rossi, S., et al.: Percutaneous radio-frequency thermal ablation of nonresectable hepatocellular carcinoma after occlusion of tumor blood supply. Radiology **217**, 119–126 (2000)
44. Ahmed, M., et al.: Principles of and advances in percutaneous ablation. Radiology **258**, 351–369 (2011)
45. Yamada, R., et al.: Hepatic artery embolization in 120 patients with unresectable hepatoma. Radiology **148**, 397–401 (1983)
46. Goldberg, S.N., et al.: Percutaneous radiofrequency tissue ablation: does perfusion-mediated tissue cooling limit coagulation necrosis? J. Vasc. Interv. Radiol. **9**, 101–111 (1998)
47. Rossmann, C., et al.: Platform for patient-specific finite-element modeling and application for radiofrequency ablation. Visual Image Proc. Comput. Biomed. **1** (2012)
48. Huang, H.-W., et al.: A robust power deposition scheme for tumors with large counter-current blood vessels during hyperthermia treatment. Appl. Therm. Eng. **89**, 897–907 (2015)
49. Audigier, C., et al.: Challenges to validate multi-physics model of liver tumor radiofrequency ablation from pre-clinical data. Comput. Biomech. Med., 29–40 (2015)
50. O'Rourke, A.P., et al.: Current status of liver tumor ablation devices. Expert Rev. Med. Devices **4**, 523–537 (2007)

51. Lim, D., et al.: Effect of input waveform pattern and large blood vessel existence on destruction of liver tumor using radiofrequency ablation: Finite element analysis. J. Biomech. Eng. **132**, 061003 (2010)
52. Wang, Z., et al.: Bi-component conformal electrode for radiofrequency sequential ablation and circumferential separation of large tumours in solid organs: development and in-vitro evaluation. IEEE Trans. Biomed. Eng. **64**, 699–705 (2017)

Extraction Technique of Spicules-Based Features for the Classification of Pulmonary Nodules on Computed Tomography

Xingyi He, Jing Gong, Lijia Wang, and Shengdong Nie[(✉)]

Institute of Medical Imaging Engineering, University of Shanghai for Science and Technology, Shanghai 200093, People's Republic of China
nsd4647@163.com

Abstract. To avoid the deformation of spicules surrounding pulmonary nodules caused by the classic rubber band straightening transform (RBST), we propose a novel RBST technique to extract spicules-based features. In this paper, the run-length statistics (RLS) features are extracted from the RBST image, in which a smooth circumference with a suitable radius inside the nodule is proposed as the border of transformed object. An experimental sample set of 814 images of pulmonary nodules was used to verify the proposed feature extraction technique. The best accuracy, sensitivity and specificity achieved based on the proposed features were 79.4%, 66.5%, 89.2%, respectively, and the area under the receiver operating characteristic curve was 87.0%. These results indicate that the proposed method of feature extraction is promising for classifying benign and malignant pulmonary nodules.

Keywords: Pulmonary nodules · Feature extraction · The rubber band straightening transform · Spicules-based features

1 Introduction

Since spiculation is one of the most important medical signs for the clinical diagnosis of benign and malignant pulmonary nodules (potential manifestation of lung cancer at early stage [1]), extraction technique of spicules-based features based on computed tomography (CT) scans has been widely studied [2–5]. The RBST technique introduced by Sahiner et al. [6] is an effective way to analyze the spiculation of nodules. According to the RBST technique, spicules that grow out radially from nodules can be transformed along approximately straight lines in the vertical direction and be tiled perfectly [6–8]. Therefore, it has to result in the deformation of the spicules, because of the rough borders of the transformed objects.

In order to suggest a solution for the deformable problem of spicules caused by the rough object border, we propose a novel RBST algorithm able to extract spicules-based features, in which the border of the transformed object is set to a smooth circumference with a suitable radius inside nodules.

© Springer Nature Singapore Pte Ltd. 2017
M. Fei et al. (Eds.): LSMS/ICSEE 2017, Part I, CCIS 761, pp. 123–131, 2017.
DOI: 10.1007/978-981-10-6370-1_13

2 Methods

In this section, we introduce the proposed spicules-based feature extraction methodology based on a novel RBST technique. The flow diagram is presented in Fig. 1. There are four details to be improved during this process: the determination of the object border, computation of normals, computation of pixel values, and the determination of well-suited sampling interval. Finally, the RLS analysis is done in order to extract features from the RBST image.

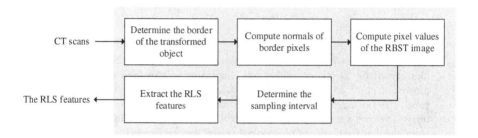

Fig. 1. The flow diagram of extracting features

2.1 Determine the Border of the Transformed Object

It is important to determine a suitable object border, because the inclusion of the RBST image will be determined by the object border. We consider a circumference with a suitable radius inside the nodule as the object border in this paper. There are two advantages: (1) avoiding the deformation of the spicules caused by the rough border; and (2) simplifying the calculation of normal vectors. The location and the radius of the circumference, which play an important role in deciding whether all of spicules can be included into the RBST image, are determined by the following strategies: firstly, the centroid of the circumference denoted by (X, Y) is considered as the center of the nodule to locate the position of the circumference; secondly, we set the circumference's radius denoted by r as the minimum value of the distance denoted by d from the centroid of the nodule to the surface of the nodule, in case the loss of spicules occurs, as illustrated in Fig. 2(a). However, when the length of spicules is too jagged, it is possible to include too much non-spicules region into the RBST image if the minimum d is equal to the r.

For this problem, we set experimentally the r as the half of the average of the d when the length of spicules is too jagged, as illustrated in Fig. 2(b). The jagged degree of spicules is measured by the standard deviation of the d and it is found experimentally that when the standard deviation of the d is equal to 3 mm, not only a number of non-spicules region can be removed from the RBST image, but also spicules can be included as many as possible. Therefore, the value of the r is determined finally according to the Eq. (1). Let (x_j, y_j) denote the position of the border pixel j on the original image, and the x_j and y_j can be calculated by Eq. (2), where the θ_j denotes the deviating degree on the Cartesian plane.

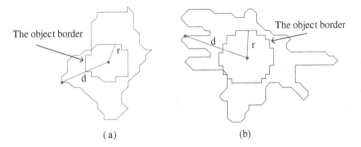

Fig. 2. Examples of the object borders inside the nodules

$$r = \begin{cases} \min(d), & \mathrm{std}(d) \leq 3; \\ \frac{1}{2}\mathrm{mean}(d), & \mathrm{std}(d) > 3. \end{cases} \tag{1}$$

$$\begin{bmatrix} x_j \\ y_j \end{bmatrix} = r \cdot \begin{bmatrix} \cos(\theta_j) \\ \sin(\theta_j) \end{bmatrix} + \begin{bmatrix} X \\ Y \end{bmatrix}. \tag{2}$$

2.2 Compute Normal of Border Pixels

Since the object border in this paper is a circumference as mentioned above, the normal direction to the object actually is equal to the direction from the border pixel to the centroid of the circumference. For a given border pixel which places the x_j and y_j coordinates on the original image, the normal direction $n(j)$ through the pixel j can be calculated by Eq. (3).

$$n(j) = \frac{1}{r}\left(\begin{bmatrix} x_j \\ y_j \end{bmatrix} - \begin{bmatrix} X \\ Y \end{bmatrix} \right). \tag{3}$$

2.3 Compute Pixel Values of the RBST Image

Afterwards, the value of each pixel on the RBST image is calculated by linear interpolation algorithm in this paper. For the pixel point (i, j) which places the ith row and the jth column on the RBST image, its normal line $n(j)$ on the original image is through the border pixel j, and let P(x, y) denote the pixel value which places the x and y coordinates on the original image. The value denoted by $p(i, j)$ at the pixel point (i, j) is calculated according to the closest pixel on the original image, as presented in Eq. (4), where the Δx_j and Δy_j denote the sampling intervals of the normal direction $n(j)$ in the horizontal and vertical direction, respectively.

$$p(i, j) = P\big(\lfloor x_j + \Delta x_j \cdot (i - 1)\rfloor, \lfloor y_j + \Delta y_j \cdot (i - 1)\rfloor\big). \tag{4}$$

Two examples of the original images and their RBST images about benign and malignant nodule were presented in Fig. 3.

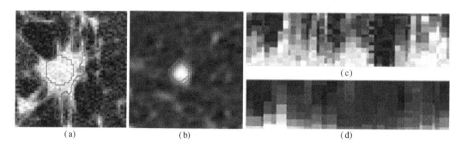

Fig. 3. The RBST images of benign and malignant nodules: in (a) and (c), the red line represents the circumference inside the nodule, and the blue one represents the nodule's boundary. The RBST image of (a) with the spicules was presented in (c), while the RBST image of (b) without the spicules was presented in (d). (Color figure online)

2.4 Determine the Sampling Intervals

It is necessary to decide appropriate sampling intervals Δx_j and Δy_j. In this paper, the Δx_j and Δy_j are determined by the pixel spacing of the RBST image, which is denoted by PX. For the circle object whose normal direction through the border pixel j is $n(j)$, the sampling intervals along the normal direction $n(j)$ (Δx_j and Δy_j) are calculated by Eq. (5). In addition, the number of sampling points on the circumference denoted by N and the θ_j mentioned above are also determined by PX, as presented in Eqs. (6), and (7). The number of the columns of the RBST image and PX are adjusted in this paper for the best values.

$$\begin{bmatrix} \Delta x_j \\ \Delta y_j \end{bmatrix} = PX \cdot n(j). \tag{5}$$

$$N = \frac{2\pi r}{PX}. \tag{6}$$

$$\theta_j = j \cdot \frac{2\pi}{N}. \tag{7}$$

2.5 Extract the RLS Features

Eleven RLS features [9, 10] are extracted in this paper according to the method presented by Way T.W. et al. [8], including gray-level uniformity (GLN), high gray-level run emphasis (HGRE), low gray-level run emphasis (LGRE), long run emphasis (LRE), long run high gray-level emphasis (LRHGE), long run low gray-level emphasis (LRLGE), run length non-uniformity (RLN), run percentage (RP), short run emphasis (SRE), short run high gray-level emphasis (SRHGE), long run low gray-level emphasis

(SRLGE). In the paper, the runs which are sets of consecutive pixels with the same gray-level intensity along the vertical direction represent either the spicules or the background surrounding the nodule. Feature extraction algorithm proposed in this paper is outlined in Algorithm 1.

Algorithm1. Feature Extraction Based on the RBST technique

---

1) Input the image of candidate nodule in the middle layer.
2) Set parameters *PX* (increases from 0.05mm to 0.5mm) and the number of the columns of the RBST image (increases from 20 to 60).
3) Calculate the centroid of the circumference according to the center of the nodule.
4) Calculate the radius of the circumference according to the Equation (1).
5) Get the number of sampling points in the circumference according to the Equation (6).
6) Calculate the position of each sampling point according to the Equation (2), (7).
7) Calculate the normals of sampling points according to the Equation (3).
8) Calculate the pixel values of interpolating points along assigned normals according to the Equation (4), (5).
9) Extract 11 RLS features from the RBST image.

---

3 Results and Discussion

In this section, the classification results which were obtained by using Linear Discriminant Analysis (LDA) classifier and Random Forest (RF) classifier are presented to evaluate the proposed method. In order to assess the performance of the proposed method, some compared classifications [7, 8, 11] which were using the same sample were carried out in this paper. Four common evaluation measures were employed: Accuracy (ACC), Sensitivity (SE), Specificity (SP) and the receiver operating characteristic curve (ROC) listed separately for the following results because of its description based on two-dimensional. We collected a total of 814 lung nodules (462 benign cases and 352 malignant cases) from the Lung Image Database Consortium and Image Database Resource Initiative (LIDC-IDRI) public database [12, 13]. It may be not enough to validate the effectiveness of the proposed approach when sample size is insufficient. Therefore, we applied 10-fold cross-validation method to validate the classification performance in order to make full use of these 814 samples. All classifications were based on the MATLAB 2014 platform.

3.1 The Results Under Different Parameters

The classification was performed based on the RLS features described in Sect. 2 and its area under the ROC curve (AUC) under the different parameters is presented in Fig. 4.

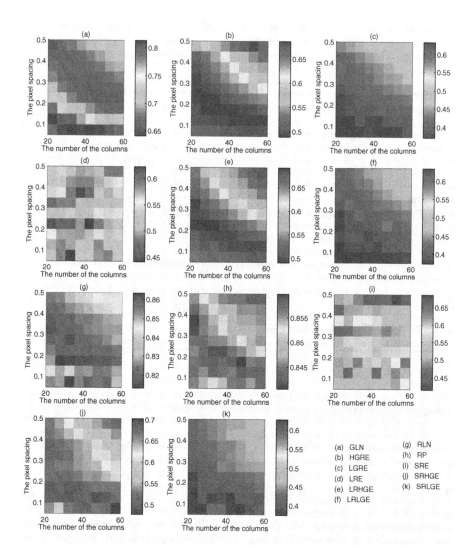

Fig. 4. The AUC values of 11 RLS features under the different parameters

The effect of two parameters on classification results was explored in this study. The one was the pixel spacing of the RBST image and the other one was the number of the columns of the RBST image denoted by *NC*. We adjusted the former ranging from 0.05 mm to 0.5 mm, and adjusted the latter ranging from 20 to 60. From these classification results presented in Fig. 4, we observed that the results are considerably different when setting different parameters. When it comes to features of GLN, HGRE, LGRE, LRHGE, LRLGE, RLN, RP, SRHGE, SRLGE, the value of AUC decreased either when the pixel spacing and the width of the band being too small or when the both being too large. It is because that it is bad for these features that too less spicules or too much background are included into the RBST image. If the value of the pixel spacing and the number of the columns of the RBST image are too large, the RBST image will not only lose some details of spicules, but also include much useless background. On the contrary, if these both parameters are too small, then the RBST image may be not enough to include too much details. Besides, we observed that the AUC values of LRE and SRE features are irregularly with the pixel spacing and the number of the columns changing. The reason causing this may be their calculations which are according to the length of the runs. Since the short runs and the long runs complement each other in the same column of the RBST image, as a result, the calculation of these two features are affected by the complementary relationship between the short runs and the long runs. Each of the RLS feature with the maximum AUC value whose parameters were recorded in Table 1 was used to classify the nodules in the following experiments.

Table 1. The maximum AUC values and the relative parameters of 11 RLS features

Features	AUC (%)	Parameters		Features	AUC (%)	Parameters	
		PX (mm)	NC			PX (mm)	NC
GLN	81.1	0.1	30	RLN	86.3	0.2	25
HGRE	68.6	0.3	20	RP	86.0	0.4	20
LGRE	63.3	0.1	20	SRE	68.6	0.55	60
LRE	62.3	0.1	30	SRHGE	70.3	0.1	55
LRHGE	69.2	0.25	25	SRLGE	62.7	0.15	20
LRLGE	63.4	0.1	30				

3.2   Comparisons of the Proposed Features with Others

The proposed features were compared with other relevant features [7, 8, 11], and the classification results obtained by using LDA classifier and RF classifier were presented in Table 2. Meanwhile, the performance on the two classifiers is displayed by the ROC curves, as shown in Fig. 5. For the proposed features, the LDA classifier achieved an accuracy of 79.4%, and the RF classifier achieved an accuracy of 80.7%. Besides, by analyzing area under ROC curves, we find that the proposed features are better, no matter what classifiers are used. These results indicated that the proposed features are more promising to distinguish malignant pulmonary nodules from benign ones. It is because that the improved RBST technique in this paper is capable of avoiding the deformation of the spicules occurred during the classic RBST technique. At the same time, the RLS statistics has a more powerful ability to analyze all kinds of streaky structures on the image. Although the proposed spicules-based feature extraction method is effective to classify pulmonary nodules, its limitations still exist. For example, there features are extracted from 2-dimensional slice images, so that they cannot characterize all of spicules growing in 3-dimensional space. Therefore, promoting the proposed method to 3-dimensional space should be well considered and studied in our further study.

Table 2. The classification results obtained by using LDA and RF classifiers

Classifiers	Features	ACC (%)	SE (%)	SP (%)	AUC (%)
LDA	The proposed	79.4	66.5	89.2	87.0
	Reference [7]	77.0	63.3	86.5	86.0
	Reference [8]	71.6	58.0	77.9	82.1
	Reference [11]	73.3	65.1	80.0	80.5
RF	The proposed	80.7	76.1	84.2	86.9
	Reference [7]	76.1	76.4	83.3	85.9
	Reference [8]	75.2	68.8	80.1	80.6
	Reference [11]	72.5	65.9	77.5	79.3

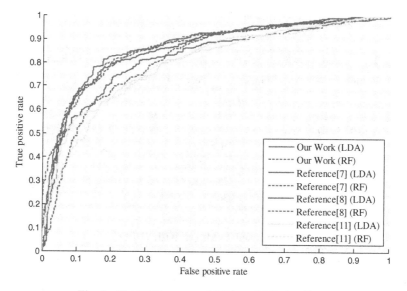

Fig. 5. The ROC curves of LDA and RF classifiers

4 Conclusions

In this study, we have developed spicules-based features based on the improved RBST algorithm. The results obtained were evaluated by the accuracy, sensitivity, specificity and the area under the receiver operating characteristic curve, and their values obtained were respectively 79.4%, 66.5%, 89.2%, and 87.0%. The comparisons with other reports indicate that the proposed features are beneficial to improve the classification performance of benign and malignant pulmonary nodules.

Acknowledgments. This work was partially funded by the National Natural Science Foundation of China under Grant (No. 60972122) and the Natural Science Foundation of Shanghai under Grant (No. 14ZR1427900).

References

1. Chowdhry, A.A., Mohammed, T.L.H.: Assessment of the solitary pulmonary nodule: an overview. In: Ravenel, J. (ed.) Lung Cancer Imaging, pp. 39–48. Springer, New York (2013). doi:10.1007/978-1-60761-620-7_4
2. Chen, H., Zhang, J., Xu, Y., et al.: Performance comparison of artificial neural network and logistic regression model for differentiating lung nodules on CT scans. Expert Syst. Appl. **39**, 11503–11509 (2012)
3. Cheng, J.Z., Ni, D., Chou, Y.H., et al.: Computer-aided diagnosis with deep learning architecture: applications to breast lesions in us images and pulmonary nodules in CT scans. Sci. Rep. **6**, 24454 (2016)
4. Ao, D.C.F., Silva, A.C., de Paiva, A.C., et al.: Computer-aided diagnosis system for lung nodules based on computed tomography using shape analysis, a genetic algorithm, and SVM. Med. Biol. Eng. Comput. **55**, 1129–1146 (2017)
5. Silva, E.C.D., Silva, A.C., de Paiva, A.C.D., et al.: Diagnosis of lung nodule using Moran's index and Geary's coefficient in computerized tomography images. Pattern Anal. Appl. **11**, 89–99 (2007)
6. Sahiner, B., Chan, H.P., Petrick, N., et al.: Computerized characterization of masses on mammograms: the rubber band straightening transform and texture analysis. Med. Phys. **25**, 516–526 (1998)
7. Way, T.W., Hadjiiski, L.M., Sahiner, B., et al.: Computer-aided diagnosis of pulmonary nodules on CT scans: segmentation and classification using 3D active contours. Med. Phys. **33**, 2323–2337 (2006)
8. Zhang, G., Xiao, N., Guo, W.: Spiculation quantification method based on edge gradient orientation histogram. In: International Conference on Virtual Reality and Visualization, pp. 86–91. IEEE Press, New York (2014)
9. Dasarathy, B.V., Holder, E.B.: Image characterizations based on joint gray level—run length distributions. Pattern Recogn. Lett. **12**, 497–502 (1991)
10. Brodić, D., Amelio, A., Milivojević, Zoran N.: Classification of the scripts in medieval documents from balkan region by run-length texture analysis. In: Arik, S., Huang, T., Lai, W.K., Liu, Q. (eds.) ICONIP 2015. LNCS, vol. 9489, pp. 442–450. Springer, Cham (2015). doi:10.1007/978-3-319-26532-2_48
11. Way, T.W., Sahiner, B., Chan, H.P., et al.: Computer-aided diagnosis of pulmonary nodules on CT scans: improvement of classification performance with nodule surface features. Med. Phys. **36**, 3086–3098 (2009)
12. Armato, S.G., McLennan, G., Bidaut, L., et al.: The Lung Image Database Consortium (LIDC) and Image Database Resource Initiative (IDRI): a completed reference database of lung nodules on CT scans. Med. Phys. **38**, 915–931 (2011)
13. Opulencia, P., Channin, D.S., Raicu, D.S., et al.: Mapping LIDC, RadLex, and lung nodule image features. J. Digit. Imaging **24**, 256–270 (2011)

Dynamical Characteristics of Anterior Cruciate Ligament Deficiency Combined Meniscus Injury Knees

Wei Yin[1], Shuang Ren[2], Hongshi Huang[2], Yuanyuan Yu[2], Zixuan Liang[2], Yingfang Ao[2(✉)], and Qiguo Rong[1(✉)]

[1] Department of Mechanics and Engineering Science, College of Engineering, Peking University, Beijing 100871, China
`qrong@pku.edu.cn`
[2] Beijing Key Laboratory of Sports Injuries, Institute of Sports Medicine, Peking University, Third Hospital, Beijing 100191, China
`yingfang.ao@vip.sina.com`

Abstract. It has been commonly believed that concomitant meniscus injuries may alter the dynamical condition of knee joint. The aim of this study was to analyze dynamical characteristics of ACLD knees with or without meniscus deficiency during level walking. The results indicated that meniscus plays an important role in bearing knee rotation moment. Additionally, the deficiency of meniscus could affect the dynamical condition of ACLD knees, especially during mid-stance phase and mid-swing. Future studies should focus on dynamical characteristics during those phases and related muscles.

Keywords: ACL deficiency · Meniscus injury · Dynamic analysis

1 Introduction

Anterior Cruciate Ligament (ACL) rupture accounts for 40% of sport injuries. ACL is an important restraint against anterior tibial translation, medial translation, and internal rotation, working in concert with other soft tissues and bony geometry to guide tibio-femoral kinematics during knee motion. ACL injuries are commonly combined with other soft tissue injuries, such as meniscus, collateral ligament and cartilage injuries. The medial and lateral menisci deform along the tibiofemoral joint line to guide tibio-femoral contact and improve load distribution and play a pivotal role in maintaining joint health after ACL injury, with or without reconstruction. Recent research, which was made by the Swedish National Knee Ligament Registry [1], showed 40% of ACL ruptures were combined with meniscus injures. Those who chose conservative treatment for ACL injury still suffered high incidences of meniscus injury, 40%, 60% and 80% in one year, five years and ten years, respectively.

Since ACL injury results in various kinematic alterations in daily activities, such as level walking, ascending/descending, jogging or lunging, meniscal load and movement may also be altered, possibly compromising knee joint load distribution and contact congruity. Former research had shown that, after ACL injury, posterior horn of lateral meniscus suffered much higher load, which would bring high risk of tears. Some

© Springer Nature Singapore Pte Ltd. 2017
M. Fei et al. (Eds.): LSMS/ICSEE 2017, Part I, CCIS 761, pp. 132–139, 2017.
DOI: 10.1007/978-981-10-6370-1_14

biomechanical studies found out that, during level walking gait, patients who suffered ACL injury combined with meniscus injury showed greater internal rotation torques of wounded knees than did those patients without meniscus injury, especially mid-stance gait phase. However, those studies only focused on the change of biomechanical condition. They could not provide doctors with succinct advice on conservative treatment or rehabilitation, though they might explain the pathology of combined meniscus injury.

Based on these shortcomings, the purpose of this study was to apply a musculoskeletal model which contained all the main muscles of human lower limbs to simulate dynamical condition of knee joint. We hypothesized that with concomitant meniscus injury, ACL deficient knees will present different knee torques and contract forces with unaffected side.

2   Materials and Methods

17 unilateral ACL deficient male patients' gait data was collected by a motion capture system. Customized musculoskeletal models were built in AnyBody Modeling System, based on subjects' morphological parameters. Knee contact force, such as proximal distal force and axial moment, and main muscle force were simulated in AnyBody Modeling System. Knee axial moments and contact forces were compared by statistical methods. The statistical analysis was performed with commercially available software.

2.1   Collection of Gait Data

17 patients were recruited in this approach, whose average body mass index was 25.5 ± 3.0 kg/m^2. Among those patients, 7 had isolated ACL injuries (Group I), 5 had combined ACL and medial-lateral meniscus injuries (Group II), 2 had combined ACL and medial meniscus injuries (Group III), 3 had combined ACL and lateral meniscus

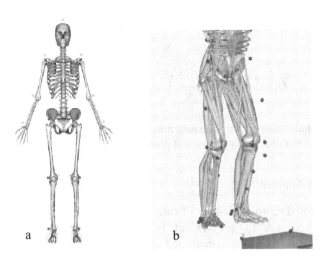

Fig. 1. a. Marker set model. b. musculoskeletal model in AnyBody Modeling System

injuries (Group IV). Though many kinds of tears were found in meniscus injuries, such as posterior horn meniscal root tear, horizontal cleavage tear or a large radial tear, we did not specify details of meniscal tear types because of the limited participants included. Morphological data was measured before gait test, including height, weight and foot size. After applying a marker model based on Helen Hayes Marker Set Model (Fig. 1a), the patients were asked to do level walking at normal speed of approximately 1.2 m/s, as well as collecting gait data and ground reaction force with the help of Vicon (Vicon, Oxford, Oxford Metrics Ltd.) and two force plates (AMTI, Advanced Mechanical Technology Inc., Watertown, Massachusetts, USA; 1000 Hz), respectively.

2.2 Dynamical Simulation of Musculoskeletal Model

AnyBody Modeling System provides a musculoskeletal model, containing all the main muscles of human, especially lower limb muscles [2]. The "LegTLEM" model, Twente Lower Extremity Model (TLEM), which was employed in this research, consisting of 159 muscles and 6 joint degrees of freedom, has been validated to deliver very convincing results.

Based on the min/max criterion, it can simulate human motion accurately and efficiently [3]. Muscle recruitment in inverse dynamics is the process of determining which muscle forces balance a given external load. The equilibrium equations for a musculoskeletal system can be organized on the following form:

$$\mathbf{Cf} = \mathbf{r} \tag{1}$$

$$Minimize\ G(\mathbf{f}^{(\mathbf{M})}) \tag{2}$$

$$0 \le f_i^{(M)}, i \in \{1, \dots, n^{(M)}\} \tag{3}$$

where $\mathbf{f}$ is a vector of muscle and joint force, $\mathbf{r}$ is a vector of external and inertia forces and $\mathbf{C}$ is a matrix of equation coefficients. G represents the tactic of recruiting muscles of central nervous system. The min/max criterion can be described as the following form:

$$G(\mathbf{f}^{(\mathbf{M})}) = \max\left(\frac{f_i^{(M)}}{N_i}\right), i = 1, \dots, n^{(M)} \tag{4}$$

where N_i is normalization factor. The min/max criterion could not only guarantee simulation efficiency, but also fit physiological phenomena.

2.3 Statistical Analysis

A two-way repeated measure analysis of variance was employed to analyze the biomechanical parameters of ACLD and unaffected contralateral knees. The two within factors were the knee status (ACLD and unaffected knee) and gait phase (equally divided into ten parts). The knee joint dynamic results during walking were normalized by body weight and height, into each gait cycle (from heel strike 0% to heel strike 100%). The level of statistical significance was set at P < 0.05. The statistical analysis was performed using commercially available software (Microsoft Office Excel 2010, Redmond, WA, USA).

3   Results

3.1   Rotation Moment

Similar trends of changes in knee rotation moment were observed in three groups (I, II and IV, Fig. 2). Firstly, compared with unaffected side, at heel strike (HS, 0%) and before next HS (95%), Group IV showed statistical increase, and Group II showed statistical greater moment at 60% and 70%. Additionally, Group III presented statistical higher moment than unaffected side at 10% and Group I had statistical increase at 70%.

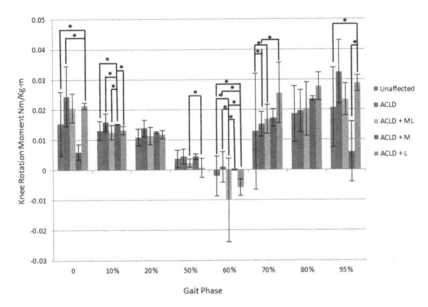

Fig. 2. Knee rotation moment among 5 groups. Unaffected = the unaffected side knee, ACLD = anterior cruciate ligament deficient patients with isolated ACL deficiency (Group I), + ML = combined ACL and medial-lateral meniscus injuries (Group II), + M = combined ACL and medial meniscus injuries (Group III), + L = combined ACL and lateral meniscus injuries (Group IV). * Statistically significant different (P < 0.05). Some phases that did not have any statistically significant difference between each group were not shown.

3.2   Flexion Moment

The changes in the flexion moment between affected and unaffected knees after ACL injury showed similar trends (Fig. 3). Group I had statistical difference with unaffected side at HS, 20% and 70% of gait cycle. At 70%, meanwhile, Group II and Group IV also presented statistical difference with unaffected side. Surprisingly, at 10%, 20%, 50% and 80%, Group I showed statistical difference with Group II.

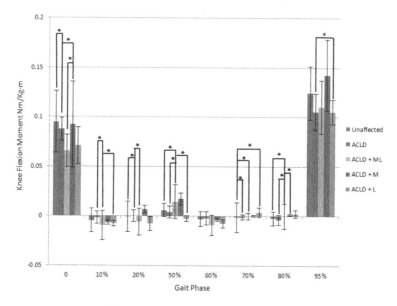

Fig. 3. Knee flexion moment among 5 groups. * Statistically significant different (P < 0.05). Some phases that did not have any statistically significant difference between each group were not shown.

3.3 Anteroposterior Force

Group I, II, IV and unaffected side showed similar trends of changes (Fig. 4). Group II presented statistical difference with unaffected at 30%, 50%, 60% and 70% of gait cycle. More statistical differences were found between Group I and II at 30%, 40%, 50%, 60% and 95%. Besides, Group I presented statistical difference with unaffected side.

3.4 Mediolateral Force

All groups showed similar trends of mediolateral force (Fig. 5). Statistical differences were found between Group I and unaffected side at 40% and 90% of gait cycle. Moreover, there were also statistical differences between Group II and unaffected side at 40% and 60%. Additionally, Group I presented statistical differences with Group II at 30% and 90%.

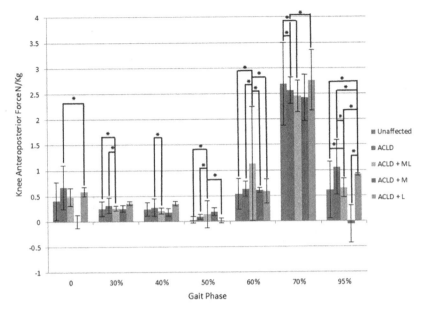

Fig. 4. Knee anteroposterior force among 5 groups. * Statistically significant different (P < 0.05). Some phases that did not have any statistically significant difference between each group were not shown.

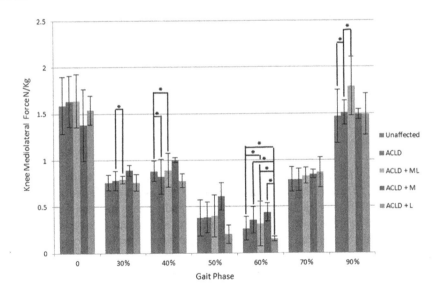

Fig. 5. Knee mediolateral force among 5 groups. * Statistically significant different (P < 0.05). Some phases that did not have any statistically significant difference between each group were not shown.

3.5 Proximodistal Force

It was also observed that all groups had similar trends of changes in knee proximodistal force (Fig. 6). No statistical differences were found between Group I and unaffected knees, or Group II and unaffected knees. However, Group III presented statistical differences with all the other groups at 10%, 30% and 40% of gait cycle, as well as with Group I and unaffected side at HS. Furthermore, Group I showed statistical difference with Group II at 90%.

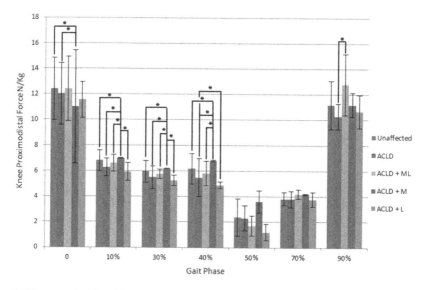

Fig. 6. Knee proximodistal force among 5 groups. * Statistically significant different (P < 0.05). Some phases that did not have any statistically significant difference between each group were not shown

4   Discussion

Dynamical characteristics of knee joint during level walking were investigated in patients with ACLD knees with or without meniscus injuries and unaffected contralateral knees. Our initial hypothesis was confirmed, as our finding demonstrated that statistical differences were found in knee rotation moment, flexion moment, anteroposterior force and mediolateral force among injured knees with different meniscus status in ACLD knees.

Both in vivo and in vitro simulated loading experiments have shown that the meniscus plays a greater role in contributing to stability in an ACLD knee than in an ACL intact knee. Kinematical research has reported that increased rotational motion during the gait is observed in the ACL deficient knees combined with meniscal injuries [4]. In our study, Group II showed higher rotation moment than unaffected side and Group I at 60% of gait when the toe off ground. Moreover, at 70%, the time between toe off and mid-swing, Group II also had statistical greater moment than Group I and unaffected side. Those phenomena could be explained by the absence of medial and lateral meniscus. Without the limitation of meniscus, knee needed to bear greater rotation moment to maintain knee stability. Though group III and IV presented similar changes, they could not be useful to approve the theory since they were lack of subjects, which was the largest shortcoming of our research. Group II also showed statistical difference with unaffected side at stance phase (10%–50% of gait cycle) in flexion moment and mediolateral force. Therefore, combined meniscus injuries could also change the dynamical condition of knee joint during stance phase.

When toe off ground (60% of gait cycle), Group II showed higher anteroposterior force than unaffected knees and group I. In contrast, at 70%, Group II presented less anteroposterior force than unaffected knees, as well as Group I showed less force either. This situation might attribute to some related muscles and more works should be done in the future. Moreover, at 90% of gait cycle, Group II showed greater mediolateral and proximodistal forces than Group I, which was related to the basic functions of meniscus, load distributing and keeping knee joint stability.

5   Conclusions

In conclusion, our findings indicated that meniscus plays an important role in bearing knee rotation moment. The deficiency of meniscus could affect the dynamical condition of ACLD knees, especially during stance phase and mid-swing. Future studies should focus on dynamical characteristics of stance and mid-swing phase. Additionally, related muscles should also be included in future study, as they contributed to different trends of some knee contact forces.

Acknowledgments. The study was financially supported by the Beijing Natural Science Foundation (No. 7172120).

References

1. Andernord, D., Bjornsson, H., Petzold, M., Eriksson, B.I., Forssblad, M., Karlsson, J., Samuelsson, K.: Surgical predictors of early revision surgery after anterior cruciate ligament reconstruction: results from the Swedish national knee ligament Register on 13,102 patients. Am. J. Sports Med. **42**(7), 1574–1582 (2014)
2. Horsman, M.K., Koopman, H.F.J.M., Van der Helm, F.C.T., Prosé, L.P., Veeger, H.E.J.: Morphological muscle and joint parameters for musculoskeletal modelling of the lower extremity. Clin. Biomech. **22**(2), 239–247 (2007)
3. Damsgaard, M., Rasmussen, J., Christensen, S.T., Surma, E., De Zee, M.: Analysis of musculoskeletal systems in the AnyBody Modeling System. Simul. Model. Pract. Theor. **14**(8), 1100–1111 (2006)
4. Harato, K., Niki, Y., Kudo, Y., Sakurai, A., Nagura, T., Hasegawa, T., Otani, T.: Effect of unstable meniscal injury on three-dimensional knee kinematics during gait in anterior cruciate ligament-deficient patients. Knee **22**(5), 395–399 (2015)

Medical Apparatus and Clinical Applications

A Survey of the State-of-the-Art Techniques for Cognitive Impairment Detection in the Elderly

Zixiang Fei[1], Erfu Yang[1(✉)], David Li[2], Stephen Butler[3], Winifred Ijomah[1], and Neil Mackin[4]

[1] Department of Design, Manufacture and Engineering Management,
University of Strathclyde, Glasgow G1 1XJ, UK
{zixiang.fei,erfu.yang,w.l.ijomah}@strath.ac.uk
[2] Strathclyde Institute of Pharmacy and Biomedical Sciences,
University of Strathclyde, Glasgow G4 0RE, UK
david.li@strath.ac.uk
[3] School of Psychological Sciences and Health,
University of Strathclyde, Glasgow G1 1QE, UK
stephen.butler@strath.ac.uk
[4] Capita plc, London SW1H 0XA, UK
Neil.Mackin@capita.co.uk

Abstract. With a growing number of elderly people in the UK, more and more of them suffer from various kinds of cognitive impairment. Cognitive impairment can be divided into different stages such as mild cognitive impairment (MCI) and severe cognitive impairment like dementia. Its early detection can be of great importance. However, it is challenging to detect cognitive impairment in the early stage with high accuracy and low cost, when most of the symptoms may not be fully expressed. This survey paper mainly reviews the state of the art techniques for the early detection of cognitive impairment and compares their advantages and weaknesses. In order to build an effective and low-cost automatic system for detecting and monitoring the cognitive impairment for a wide range of elderly people, the applications of computer vision techniques for the early detection of cognitive impairment by monitoring facial expressions, body movements and eye movements are highlighted in this paper. In additional to technique review, the main research challenges for the early detection of cognitive impairment with high accuracy and low cost are analysed in depth. Through carefully comparing and contrasting the currently popular techniques for their advantages and weaknesses, some important research directions are particularly pointed out and highlighted from the viewpoints of the authors alone.

Keywords: Cognitive impairment · Mild cognitive impairment (MCI) · Dementia · Early detection · Computer vision techniques · Literature review

© Springer Nature Singapore Pte Ltd. 2017
M. Fei et al. (Eds.): LSMS/ICSEE 2017, Part I, CCIS 761, pp. 143–161, 2017.
DOI: 10.1007/978-981-10-6370-1_15

1 Introduction

There are a growing number of elderly people in the UK. A significant number of people over 65 years old experience dementia. Dementia is a progressive cognitive impairment which may cause the impairment in many cognitive domains such as memory ability and language ability. Huge amounts of money is spent on the healthcare for the dementia patients every year. In addition, the number of people who have dementia is still increasing and it is estimated that there will be about 80 million dementia patients worldwide by 2040 [1]. As shown in Fig. 1, the possibility of the occurrence of dementia is often related to the age. With the age increasing, there is a higher possibility for elderly people to develop a dementia.

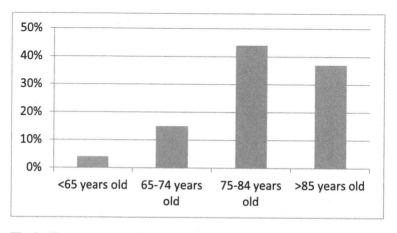

Fig. 1. The percentages of population aged 65 or over who have dementia [2]

Mild cognitive impairment (MCI) is an intermediate stage between the expected cognitive decline of normal aging and the more serious decline like dementia. For most patients with MCI, the normal activities of their daily life are not affected notably [3, 4]. Diagnosing the MCI is similar to diagnosing mild dementia, because they have similar symptoms [5]. Currently, there are many techniques available to detect cognitive impairment including MCI and more severe cognitive impairment like dementia. Detecting the occurrence of the cognitive impairment in an early stage is of great importance.

It is critical to early detect the symptoms of cognitive impairment, monitor the progress of the disease, and provide a collective care for the elderly people with cognitive impairment, especially for those living in low-income community. However, current methods for the detection of cognitive impairment have some weaknesses and cannot achieve this goal completely. For example, some methods are not suitable for the early detection of MCI and other methods may have weakness due to the high expenses. This survey mainly describes several currently popular techniques used in the detection of cognitive impairment, compares, and contrasts their advantages and weaknesses. As shown in Fig. 2, the techniques surveyed in this paper can be generally classified into cognitive tests, neuroimaging and computer vision. The cognitive test techniques include instruments-based and computer-based cognitive tests. There are also two kinds of neuroimaging techniques, i.e. magnetic resonance imaging (MRI) and metabolic positron emission tomography (FDG-PET). Computer vision techniques include body motion, activities, eye movements and facial expressions.

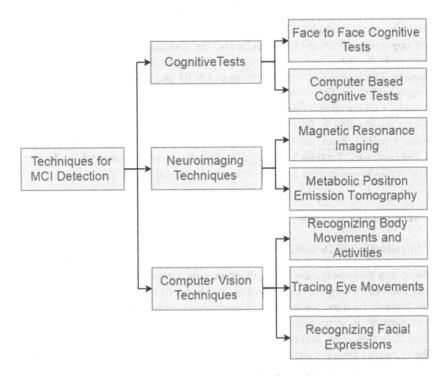

Fig. 2. Main techniques for the detection of cognitive impairment

There are also some relevant survey papers about detecting cognitive impairment. Latha et al. presented a review of 22 traditional face-to-face cognitive tests which are widely used in memory clinics in the aspect of level and quality of evidence [6]. On the other hand, Katherine et al. carried out a review for 18 computerized screening tools in the aspect of the test validity, reliability, comprehensiveness and usability [7]. They discussed the advantages and disadvantages of the computerized screening tools compared to the traditional face to face cognitive tests. In addition, some researchers reviewed the studies of using the magnetic resonance imaging (MRI) technique in diagnosing the cognitive impairment. The MRI technique can be used to detect the biochemical abnormalities in the regions of the brain that are related to cognitive impairment [8, 9]. Besides, Freitas et al. reviewed the studies about using the abnormalities in eye movement patterns to detect the cognitive impairment [10]. However, most of the related survey papers were focused on using one type of technique to detect cognitive impairment. In order to comprehensively review the state-of-the-art techniques currently used to detect cognitive impairment, compare and contrast their advantages and weaknesses, a literature survey is carried out in this paper.

This paper is organized as follows. Section 1 has a general overview and sets out the outline for the literature review. The background on dementia and MCI is further introduced in Sect. 2. Section 3 analyses some main research challenges for the early detection of cognitive impairment with high accuracy and low cost.

The techniques of using the cognitive test to detect the cognitive impairment have been widely used in clinics. Therefore, it is detailed in Sect. 4. These techniques include cognitive and functional instruments, and computerized cognitive testing.

Section 5 describes the neuroimaging techniques for detecting the cognitive impairment. The techniques include the use of MRI and metabolic positron emission tomography (FDG-PET).

Computer vision techniques with body motion, activities, eye movement and facial expression are detailed in Sect. 6, where Sect. 7 is for some discussions, where some important research directions are summarised. In this part, all the methods are compared and contrasted for their advantages and weaknesses in greater details. In Sect. 8, a conclusion is made to give a short summary of the paper.

2 Background Review

2.1 Introduction to Dementia

Dementia is a progressive cognitive disorder which may include the impairment in memory and other cognitive domains or skills. Dementia has some specific neuropathy logical changes, such as extracellular parenchymal lesions and intraneuronal. Because the investigation in neuropathy cannot be carried out when the patients are alive, dementia detection often involves a probabilistic diagnosis [11].

The dementia patients may have many symptoms which are different between individuals. The patients with the different types of dementia may also have different symptoms. Normally, the brain will have some changes related to the disease before the symptoms occur. For many patients, the initial symptom is that they are unable to remember new information. Then, other symptoms may appear. There are some typical symptoms for all the dementia patients such as language problem, movement problem, recognition problem, poor reasoning and judgement, and changes in personalities as well [2].

Dementia is caused by the damage of the brain cells. As a result, the brain cells cannot operate normally and communicate with each other. The affected brain parts will cause many different cognitive problems. In addition, the damage to different parts of the brain is related to the different types of dementia [12].

The common types of dementia are Alzheimer's Disease, Vascular Dementia, Dementia with Lewy Bodies and Fronto-Temporal Dementia [2]. Among them, the most common type of dementia is Alzheimer's disease. 60% to 80% of the dementia patients belong to this type of dementia [13].

2.2 Mild Cognitive Impairment

Mild cognitive impairment (MCI) is an intermediate stage between the expected cognitive decline of the normal aging and the more serious decline like dementia, which is greater than the decline expected for the individual's education and age. For the most patients with MCI, the normal activities of their daily life will not be affected greatly. Among the people with MCI, half of them can remain stable in their cognitive situation, but other half of the people may eventually become the patients with dementia. As a result, the MCI can be considered as the risky state for dementia [3, 14]. Especially, the MCI patients with the impairment in episodic memory, verbal abilities, associative recognition impairment and visual-spatial function may have a higher risk to develop into the state of dementia [15, 16].

3 Research Challenges

Although there are many methods and techniques developed to diagnose the cognitive impairments, few of them can provide the effective and low-cost solutions for the early detection of cognitive impairments. In the following, we highlight some major research challenges on the early detection of cognitive impairment.

To begin with, because different people may have different symptoms in the early stage, it is generally difficult to diagnose the cognitive impairment in the early stage. In the MCI stage, most of the symptoms may not be fully expressed. The MCI patients may be able to live independently and act like normal people. There are only minor abnormalities or differences between the MCI patients and healthy people. In order to find the abnormalities and differences, all the cognitive domains may need to be tested such as the reaction time, the memory and attention issues and the processing speed [17].

Over the past decades, many different methods with various techniques have been developed to diagnose the severe cognitive impairment. However, they may not be sensitive enough to detect the mild cognitive decline. For example, the Mini-Mental Status Examination (MMSE) cognitive test is one of the most widely used cognitive methods to detect the severe cognitive impairment. However, it has been reported that only 18% of the MCI subjects can be detected by the MMSE cognitive test [18].

In addition, it is also a challenge to provide a low-cost and effective cognitive impairment detection solution for elderly people, especially for those living in low-income community. For example, it is expensive to detect the cognitive impairment using the MRI-based method. Furthermore, for cognitive tests, professional neurophysiologists are needed to carry out the clinical tests for the elderly people, which suggests that a lot of time and cost are necessary.

4 Using Cognitive Tests to Detect the Cognitive Impairment

4.1 Traditional Cognitive Tests to Detect the Cognitive Impairment

It is difficult to diagnose the MCI because of the high variability of the patients with MCI. Nowadays, the MCI is often detected by the situations reported by self or family members or by the written and task-based neurological assessments. The cognitive tests are effective and widely used in clinics. There are many kinds of cognitive screening tools such as the Mini-Mental Status Examination (MMSE), the Montreal Cognitive Assessment (MoCA), and the Functional Activities Questionnaire (FAQ). Different cognitive tests have been designed to test the cognitive impairment of various types at different stages, so the cognitive test to be used needs to be selected carefully according to the different situations.

Alex conducted a survey of 34 cases in which the cognitive tests using the MMSE were performed in clinics [19]. The MMSE cognitive test has a good accuracy for the diagnosis of cognitive impairment. From [19], it suggests that the MMSE can have a sensitivity of about 80% in detecting the cognitive impairment. However, the MMSE cognitive test has several limitations. To begin with, the MMSE cognitive test has a ceiling effect for the very mild disease such as MCI. Therefore, the MMSE cognitive test has the limited value in the diagnosis of MCI and distinguishing Alzheimer's disease against MCI. In addition, the MMSE cognitive test has a floor effect for the patients with serious cognitive impairment. Moreover, special care and adjustment are needed for the people with poor education and non-English language speaking. At last, the MMSE cognitive test also has other limitations such as its length and non-linearity.

The Montreal Cognitive Assessment (MoCA) is another cognitive screening tool which can be used to diagnose the cognitive impairment. It can often be completed in about 10 min. Ziad et al. compared the MoCA with the MMSE in detecting the mild cognitive impairment [18]. In the experiment, there were 94 MCI patients, 93 mild Alzheimer's disease patients and 90 healthy controls. The experiment results showed that the MoCA was better than the MMSE in detecting the MCI and mild dementia. It has been reported that the MMSE detected 18% of the MCI subjects, whereas the MoCA detected 90% of the MCI subjects. Furthermore, as shown in Fig. 3, for the MoCA cognitive test, the difference between the mean score of the healthy controls and

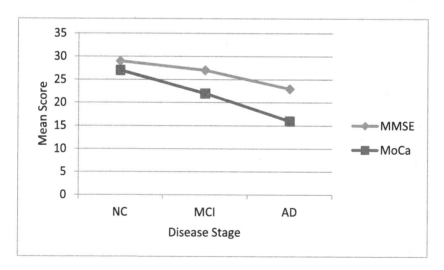

Fig. 3. Mean score for healthy control (NC), MCI patients (MCI) and dementia patients (AD) in MMSE and MoCA cognitive tests [18]

that of the dementia patients is bigger in comparison to the MMSE cognitive test. As a result, the MCI patients and the dementia patients may be diagnosed more easily using the MoCA cognitive test. The MoCA also had a better performance in detecting the mild dementia and excluding the elderly normal controls for about 87%. The MoCA cognitive test also has more words and fewer learning trials compared to the MMSE cognitive test. In summary, the different cognitive screening tools have been designed for the different levels of the cognitive impairment. The MoCA is better to test the patients with the mild cognitive impairment and the mild dementia. However, the MMSE is designed to diagnose the more serious cognitive impairment like dementia.

Latha et al. presented a review of 22 traditional face-to-face cognitive tests such as the Mini-Mental State Examination (MMSE), 6-Item Cognitive Impairment Test (6CIT) and the Hopkins Verbal Learning Test (HVLT) [6]. All these tests are designed to be completed in less than 20 min and are suitable for memory clinics. Latha reviewed these cognitive tests in level, quality of evidence and the types of cognitive impairment which they were suitable for. The conclusion of the review suggests that practitioners should choose the appropriate cognitive tests, because the aims of the different cognitive tests may be different from each other.

4.2 Computer Based Cognitive Tests to Detect the Cognitive Impairment

Nowadays, many cognitive tests are also available on computers. These computer based cognitive tests are convenient and provide professional help for a wide range of people. However, they have disadvantages such as the lack of the established psychometric standards and bad user-computer interface. There are mainly two types of the computerized cognitive tests: one type is converted from the existing cognitive instruments; another type is based on the development of novel computerized screening tools [7].

Many computer based cognitive tests are converted from the existing cognitive instruments. For instance, P. Garda et al. proposed a counter propagation network based network (CPN) using the combination of neuropsychological scales which are cognitive and functional instruments such as Geriatric Depression Scale (GDS), the Mini-Mental Status Examination (MMSE) and the Functional Activities Questionnaire (FAQ) [20]. The experiments were carried out by using the public Alzheimer's disease Neuroimage Initiative (ADNI) dataset. Kondou et al. also proposed a prototype of an automatic system for detecting the cognitive impairment [21]. The system uses the five-cog test, which is usually carried out manually. The system uses the image processing techniques to evaluate the cognitive functions of the users by recognizing the characters and patterns entered on a touch panel.

In order to compare and contrast the traditional cognitive tests with the computer based cognitive tests, many researchers have conducted the reviews of the existing cognitive tests and the existing computer based cognitive tests. Katherine et al. presented a review for the computerized screening tools which were used to detect the cognitive situation and to diagnose the mild cognitive impairment for elderly people [7]. In the research, 18 screening tools were reviewed in the aspect of the test validity, reliability, comprehensiveness and usability.

Katherine also discussed the advantages and disadvantages of the computerized screening tools compared to the traditional cognitive tests [7]. To begin with, the computer based cognitive tests have the advantages in record accuracy, speed and sensitivity of the responses and standardized format. These kinds of the characteristics are quite important in detecting the minor changes in the cognitive situation for the patients with the mild cognitive impairment and mild dementia. In addition, compared to the traditional cognitive tests, the computerized screening tools have the advantages in cost saving and alleviating the need of professional neurophysiologists. Moreover, the computerized cognitive tests can be available from the internet and provide the support for a large population of elderly people. However, there are also some disadvantages for the computerized screening tools. For instance, one disadvantage is the lack of the established psychometric standards. In addition, some computerized screening tools may have problems in bad computer-person interface and user interactions.

Thomas made a review for the use, development and performance of some computer based cognitive tests [22]. Both of the traditional cognitive tests and computer based cognitive tests have their own characteristics. The traditional cognitive tests are more sensitive and flexible. For instance, the situations such as the moods and motivation of the patients will be taken into consideration. On the other hand, the computerized screening tools often take less time and money. The computerized screening tools are very accurate and fast, e.g., timing can be done in milliseconds. The computer based screening tools also have some disadvantages. For example, some information generated during the tests may not be interpreted correctly. Furthermore, the results obtained from the computer based screening tools are not very sensitive. In other words, if someone does well in the computer based tests, it means that he does not have cognitive problems. However, someone does not do well in the computer based tests cannot make sure that he has dementia or other cognitive diseases. Therefore, clinical diagnosis may still be needed to accurately diagnose.

5   Using Neuroimaging Techniques to Diagnose Cognitive Impairment

There are two kinds of neuroimaging techniques which are widely used in clinics for the early detection of the cognitive impairment, i.e. magnetic resonance imaging (MRI) and metabolic positron emission tomography (FDG-PET) [23]. These two techniques are both used to detect the abnormalities in brain regions for the early detection of the patient with the cognitive impairment. Because the brain of the patient with the cognitive impairment may have the abnormalities before the symptoms appear, neuroimaging is quite useful for the early detection of the cognitive impairment when the clinical symptoms have not fully expressed [2]. However, diagnosing the cognitive impairment is expensive and unfriendly to those living in low-income community.

Herholz et al. presented a novel indicator of the FDG-PET with the automated voxel-based procedure to diagnose the cognitive impairment [24]. The data was collected from large amounts of the participants consisting of 110 normal healthy controls and 395 patients with the cognitive impairment. The results showed that all the regions of the brain related to the cognitive impairment were affected in the beginning of the disease.

On the other hand, many researchers reviewed the studies of using the MRI in diagnosing the cognitive impairment. Yuanyu et al. conducted a review of the studies about the magnetic resonance imaging [9]. The MRI technique can be used to detect the biochemical abnormalities in the regions of the brain that are related to the cognitive impairment. As a result, the MRI can be used for the early detection of the cognitive impairment and monitoring the disease progress. Amy et al. also reviewed the studies about the relationship between the cognitive problems with the neurometabolite values [8]. They have made a conclusion that the magnetic resonance spectroscopy (MRS) can be used to note the longitudinal changes and detect the cognitive problems such as dementia in an early stage.

To develop the MRI techniques in the diagnosis of the cognitive impairment, Stefan et al. proposed a novel automatic system to diagnose the cognitive impairment [25]. The system could be used to distinguish the normal aging from the cognitive impairment. In this research, the support vector machine (SVM) was used in MRI to detect the different disease states. The experiment showed that 96% of the patients were classified for the disease types correctly using the brain images.

6 Using Computer Vision to Detect the Cognitive Impairment

6.1 Detecting the Cognitive Impairment with Eye Movement

Abnormalities in eye movement patterns may be associated with the deficits in the patients with the cognitive impairment. As a result, the patients with the cognitive impairment can be diagnosed by analysing their eye movements. The patients with the cognitive impairment may have the deficits in static spatial contrast sensitivity, visual attention, shape-from-motion, colour, visual spatial construction, and visual memory. In addition, some researchers have found that several characteristics of the eye movement patterns may be particularly useful to diagnose the cognitive impairment. These characteristics are such as fixation duration, re-fixations, saccade orientation, pupil diameter smooth, pursuit movements and saccadic inhibition [10, 26]. They have also designed experiments to detect the cognitive impairment by eye movements [27–29]. However, most of the related investigations are carried out in dementia patients and the investigations in MCI patients are insufficient [10]. Furthermore, to improve the precision in measurements, eye trackers are often needed.

Yanxia et al. worked on the linkage between the cognitive situations obtained from the cognitive tests and the eye movement patterns collected when the participants were watching videos. They tried to find the specific eye movement patterns which were related to the cognitive impairment and then used these eye movement patterns for the automatic cognitive assessment [27]. From the experiments, they found that the participants with the cognitive impairment would have slower saccade motion and longer fixation time on average. However, the results might be limited because of the small number of samples obtained during the experiments.

In order to obtain the eye movements, eye trackers are often used to record the eye movement patterns while the patients with the cognitive impairment are doing some tasks such as watching images or videos on computers [10].

In particular, there are some eye movement experiments used to detect the cognitive impairment. One possible experiment example is the visual paired comparison (VPC) task. In the experiment, the participant is asked to select the picture which she saw previously. The eye tracker will be used to track the eye movement when the participant is doing this task. The eye movement pattern will be used to diagnose if the participant has a cognitive impairment or not.

Some researchers used the VPC task to find the difference in the eye movements between the patients with the cognitive impairment and the normal people. For instance, Dmitry et al. used the VPC task to detect the memory impairment in the MCI patients [28]. The VPC task would test the participants if they could figure out the novel visual stimuli with the repeated visual stimuli. As a result, the healthy controls would watch the novel images for more time.

Gerardo et al. proposed a novel method to detect the cognitive problems by the eye movements in the activities of reading proverbs [29]. In the experiments, there were 20 participants with Alzheimer's disease and 40 healthy participants. By the experiments, they have found that compared with the participants with Alzheimer's disease, the healthy participants would have the longer gaze periods for words with predictability. It showed that the participants with Alzheimer's disease had the impairment in word predictions.

6.2   Detecting the Cognitive Impairment by Recognizing Activities

When the cognitive decline occurs, it will have the impacts on doing their daily activities for elderly people [30]. As a result, the recognition of their activities has the potential for the detection of cognitive impairment. In particular, the systems with these techniques can be used to monitor the cognitive changes and disease progress for the elderly people automatically.

Sabrina et al. proposed a RGBD camera monitoring system to monitor the patients with the cognitive impairment using neural networks. The system can monitor the patients in home environment during daily activities which are in the Direct Assessment of Functional Status (DAFS) index. DAFS proposed by Zanetti et al. is a standardized observation-based checklist that is used to evaluate the cognitive situation for the elderly people with cognitive impairment [31]. The system was focused on monitoring the activities of brushing teeth and grooming hair. The system could detect the abnormalities during the activities to find out the cognitive situation of the elderly people. The result of the experiment showed that the system was able to recognize the daily activities and calculate the DAFS score to detect the occurrence of the cognitive decline. There were some problems that it could not recognize the objects during the activities and the kinds of the activities could be recognized was limited.

Aleksandar et al. also proposed a system to monitor the cognitive decline by recognizing the activities using the fusion of machine vision technique and RPID technique [32]. The system could provide an early warning of the disease by analysing the changes in their behavioural patterns during daily activities. The result for monitoring the patients would be provided to caregivers or doctors. This system could reduce the workload of the caregivers by monitoring the cognitive situations of the patients automatically. In [32], cooking and memory card game were selected to be as two daily activities monitored. However, this research was still in progress and there was no experiments carried out yet.

The kinds of activities could be recognized were limited in the both of the cases above. The various kinds of daily activities should be able to be monitored and recognized to know the cognitive situation of the elderly people in order that the MCI could be detected at the beginning. Vincent et al. proposed a system which could recognize 8 kinds of activities for the people with the cognitive impairment [33]. The activities were such as feeding birds, preparing and eating meal, making tea, playing a CD and so on. This system recognized the activities using the videos recorded by wearable devices. It was made up of the fusion of the object detectors and the location detectors. For example, to detect the activities of preparing and eating meal, the objects such as bowls, spoons and food were involved and the place such as kitchen was involved. The overall performance showed a good result. However, the recognition accuracy for some activities was low such as the recognition of making a phone call. The reason was that the phone was always near the ear and the wearable device might not be able to detect the phone. In addition, making phone calls might happen in many different places.

In order to improve the accuracy for recognizing the activities in complex environment and in long videos, locating the place and time that the activities take place is important. Konstantinos et al. did a research on monitoring and recognizing the activities for the elderly people [34]. The research focused on the improvement of the performance for recognizing the Activities of Daily Living (ADL). They proposed a novel algorithm that would locate the Activities of Daily Living both in time and space. It would find the start and end time of the activities and where they took place.

It may be difficult to monitor the patients during their daily activities all the day to find the abnormal body movements. In spite of monitoring the patients in daily activities, a method for detecting the cognitive problem when playing an interactive physical game was proposed in [35]. The cognitive situation can be evaluated by the video recorded with the body motion sensor when the elderly people are playing the game. The system can record the parameters of the body motion. In [35], 5 kinds of games were developed using the Kinect to detect if the elderly people had problems of the mild cognitive impairment.

6.3 Detecting the Cognitive Impairment by Facial Expression

The people with the cognitive impairment have the impairment in recognizing the facial expressions [36–38]. Meanwhile, the facial expressions of the people with the cognitive impairment also show the difference compared to normal people. As a result, the facial expressions also have potential to diagnose the cognitive impairment by observing and monitoring the facial expressions. However, there are few investigations in detecting the cognitive impairment by monitoring the facial expressions.

The difference of the facial expressions between the patients with the cognitive impairment and the normal people was discussed in many research papers [39, 40]. Keith et al. researched on the facial muscle activities for the patients with Alzheimer's Diseases (AD) [39]. In the experiment, the groups of patients with the AD and the groups of the normal controls were asked to watch some emotion-eliciting images while the facial muscles were researched. Smiles are often related to the zygomatic activities and frowns are often related to the corrugator activities. The experiment results showed that the AD patients showed more corrugator activities when viewing the negative images. Furthermore, there was also a big difference for zygomatic activities between the AD patients and the normal people. As shown in Fig. 4, the AD patients showed the maximum zygomatic activities when they were watching the negative images and minimum activities for the positive ones.

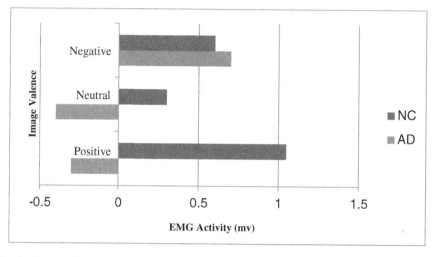

Fig. 4. Zygomatic change scores when viewing positive, neutral and negative images for the normal control (NC) and the patients with Alzheimer's Diseases (AD) [39]

The patients with the cognitive impairment have the impairment to control facial muscles and to express their feelings. Julie et al. researched on the facial expressions of the patients with the cognitive impairment [40]. In the experiment, the patients with the cognitive impairment and normal controls were asked to watch videos, when the facial expressions were recorded. In addition, they were asked to express their feelings in the conditions of spontaneous expressions, amplification and suppression of emotions. The results showed that for the patients with the cognitive impairment, the behavioural amplification of the expressed emotion was affected. However, the subjective feelings and suppression of emotions were not affected.

However, there are also some difficulties to use the facial expressions to diagnose the cognitive impairment. To begin with, the facial expressions may be different between individuals and would be affected by the premorbid personality [41]. The influence of the premorbid personality on the facial expressions was discussed in Carol's paper. They used the research data of the facial expressions collected during family visits to the patients with the cognitive impairment. They found that the avoidant attached individuals would show less the positive facial expressions than the securely attached individuals. In addition, the patients with the premorbid hostility would show more negative facial expressions. Moreover, the facial expression behaviours would be affected by culture and were different between the western people and the eastern people [42].

Furthermore, the facial expressions of the patients with the cognitive impairment will be affected by apathy [43]. Ulrich pointed out that with the progression of the disease, the total facial expressions would not decrease. Instead, the total facial expressions would increase, but there would be less specific facial expressions.

In order to use the facial expressions to diagnose the cognitive impairment automatically, computer vision technologies will play an important role in this field. There are many good survey papers about dealing with the facial expressions problems using computer vision techniques [44–46]. They mainly compared and contrasted the different techniques to carry out the major steps for the facial expression recognition such as face localization, face registration, face representation, face features extraction and face expression recognition.

For instance, Basel et al. reviewed the studies of the facial expression recognition and paid attention to the sections such as face normalization, dynamics and intensity for the facial expressions [44]. Georgia et al. mainly reviewed the computer vision techniques for the facial expression in the aspects of 3D or 4D facial expressions [45]. Evangelos et al. reviewed the state of the art computer vision techniques for the facial expressions and laid the emphasis on the part of facial representation [46]. In addition, they focused on discussing the difficulties in the process of the facial expression recognition such as the problems related to illuminations, the variations in head position, the errors in registration and so on.

Meanwhile, many researchers worked on the improvement of the automatic facial expression recognition techniques. For instance, Andre et al. proposed a novel solution for the recognition of expressions using a convolutional neural network and applying the pre-processing techniques to extract the specific features [47]. Murari et al. proposed a method to recognize the facial expressions by extracting the facial features using the higher-order Zernike moments and classification with an ANN based classifier [48]. Ashok et al. had a survey on the automatic recognition of the facial expressions and compared with the human visual system that deals with the same problem [49]. Some technologies such as deep learning will also improve the accuracy of the facial expression recognition [50].

7 Discussion and Research Directions

In this paper, many current methods and techniques used to diagnose and detect the cognitive impairment have been reviewed. There have been some big changes and progresses in the field of detecting and diagnosing the cognitive impairment in recent years. However, all these existing techniques have their own weaknesses. The current methods for diagnosing the cognitive impairment are still unsatisfying [51]. New techniques in particular need to be developed to meet the needs of the early detection of the cognitive impairment with the high accuracy and low cost.

Regarding the weaknesses of the existing techniques, both the traditional face to face cognitive tests and computer-based cognitive tests have the problems with diagnosing the mild cognitive impairment. More specifically, some personal information like age, education and personality will influence the test result and need to be taken into consideration carefully [52]. Moreover, the different cognitive tests are often suitable for the different types and stages of the cognitive impairment. As a result, the cognitive test to be used needs to be selected carefully according to the different situations. In addition, for the traditional face to face cognitive tests, the professional neurophysiologists are needed to carry out the cognitive tests for the patients. Meanwhile, for the computer based cognitive tests, if the elderly people would like to take the tests on their own, some basic computer skills are often required.

Although diagnosing the cognitive impairment with the neuroimaging techniques are widely used in clinics, the major weakness of them is the high expenses. Besides, the neuroimaging techniques are often used to diagnose the cognitive impairment only when the patients are found that they have serious cognitive problems.

Diagnosing the cognitive impairment by eye movements, facial expressions and daily activities are still in developing stage. Using the eye movements to diagnose the cognitive impairment needs eye trackers to improve the precision in measurements. In addition, the effective eye movement experiments need to be designed precisely.

Table 1. Compare and contrast several techniques to detect the cognitive impairment

Techniques	Characteristics	Weaknesses	Setting	Cost
Face to face cognitive tests	Test the cognitive domains such as memory ability finish in less than 20 min	Unfriendly to people with poor education background dependent on professional neurophysiologists	Memory clinics	Medium
Computer based cognitive tests	Test several cognitive domains available through internet	Requirement of computer skills problems from poor computer-human interface	Homes	Low or free
Neuroimaging techniques	Able to detect the abnormality in the brain when clinical symptoms are not fully expressed	Suitable for detailed diagnosis, not for early detection	Clinics	High
Recognising body movements and activities	Able to detect the cognitive decline	Limited types of activities that can be recognised difficulties in detecting minor abnormalities	Homes	Low or free
Tracing eye movements	Able to detect deficits in visual attention and so on	Need eye tracker need improvements in effective eye movement experiments	Labs	Medium
Recognizing facial expressions	Notice facial abnormalities	No developed system	Homes	Low or free

Meanwhile, the existing systems to detect the cognitive situations by the recognition of the activities are not effective and intelligent enough and the types of activities can be recognized are very limited. Besides, the systems which can detect the cognitive impairment by using facial expressions are still in progress. The major characteristics of these techniques are also compared and contrasted in Table 1.

Regarding the research directions, they can be roughly divided into three aspects: effective cognitive tests, advanced neuroimaging techniques for clinics situations, and automatic cognitive impairment detection system for family use.

The automatic cognitive impairment detection systems are suitable for the use in home situations for the early detection of the cognitive impairment and monitoring the disease progress. The systems need to be low cost, convenient and user friendly so that the systems can be used widely, but it may not be very precise as devices in clinics. The system should be able to detect the abnormalities related to the cognitive problems. The cameras from mobile phones could be used to monitor the eye movements, facial expressions and activities of elderly people. The system can work as an app on the mobile phones.

The future direction of the cognitive tests will probably be computer based cognitive tests, which can be easily available from internet. The input methods of the computer based cognitive tests will not be limited to keyboards and touch panels but will also include cameras which can record the eye movement, facial expression and body movement. The cognitive tests need to be able to diagnose the cognitive situations for the different types and stages of the cognitive impairment and can be completed in several minutes.

As the most direct and effective methods to diagnose the cognitive impairment in clinics, the future direction of the neuroimaging techniques should have a high accuracy and low cost. In addition, the effective automatic system to diagnose the cognitive impairment with the neuroimaging techniques needs to be developed.

8 Conclusion

This survey paper has reviewed several currently popular techniques for the early detection of cognitive impairment and discussed their advantages and weaknesses. These techniques included cognitive test, neuroimaging and computer vision. The cognitive test techniques had the traditional face to face cognitive tests and computerized cognitive tests. Two kinds of neuroimaging techniques were reviewed, i.e., magnetic resonance imaging (MRI) and metabolic positron emission tomography (FDG-PET). Computer vision techniques with body motion, eye movement and facial expression were highlighted as they have potentials to build cost-effective solutions for early detection of cognitive impairments.

This paper has also analysed and discussed about the future research directions in the area of diagnosing and early detecting the cognitive impairment. The first future direction is about the automatic cognitive impairment detection systems that use multiple information sources such as eye movements, facial expressions and body motions. These systems are expected to monitor the progress of the disease automatically and they are suitable for family use. The second future direction is about the advanced neuroimaging techniques to diagnose the different types of cognitive impairment with a higher accuracy and low cost for clinic situations. The third direction is about the improvement of the existing cognitive tests which can be carried out regularly with low cost and be easily available for the elderly living in the low-income communities.

Acknowledgements. This research is funded by Strathclyde's Strategic Technology Partnership (STP) Programme with CAPITA (2016–2019). The authors thank Dr Neil Mackin (CAPITA mentor) and Miss Angela Anderson (the STP's coordinator) for their support. The contents including any opinions and conclusions made in this paper are those of the authors alone. They do not necessarily represent the views of CAPITA plc.

References

1. Ferri, C.P., Prince, M., Brayne, C., Brodaty, H., Fratiglioni, L., Ganguli, M., Hall, K., Hasegawa, K., Hendrie, H., Huang, Y., Jorm, A., Mathers, C., Menezes, P.R., Rimmer, E., Scazufca, M.: Global prevalence of dementia: A delphi consensus study. Lancet **366**, 2112–2117 (2005). doi:10.1016/S0140-6736(05)67889-0
2. Gaugler, J., James, B., Johnson, T., Scholz, K., Weuve, J.: 2016 Alzheimer's disease facts and figures. Alzheimer's Dement. **12**, 459–509 (2016). doi:10.1016/j.jalz.2016.03.001
3. Wu, N.S.C., Ho, K.-S.: Mild cognitive impairment. Hong Kong Pract. **31**, 36–40 (2009)
4. Winblad, B., Palmer, K., Kivipelto, M., Jelic, V., Fratiglioni, L., Wahlund, L.-O., Nordberg, A., Bäckman, L., Albert, M., Almkvist, O., Arai, H., Basun, H., Blennow, K., De Leon, M., Decarli, C., Erkinjuntti, T., Giacobini, E., Graff, C., Hardy, J., Jack, C., Jorm, A., Ritchie, K., Van Duijn, C., Visser, P., Petersen, R.C.: Mild cognitive impairment - Beyond controversies, towards a consensus: Report of the international working group on mild cognitive impairment. J. Intern. Med. **256**, 240–246 (2004). doi:10.1111/j.1365-2796.2004.01380.x
5. Petersen, R.C.: Mild cognitive impairment as a diagnostic entity. J. Intern. Med. **256**, 183–194 (2004). doi:10.1111/j.1365-2796.2004.01388.x
6. Velayudhan, L., Ryu, S.-H., Raczek, M., Philpot, M., Lindesay, J., Critchfield, M., Livingston, G.: Review of brief cognitive tests for patients with suspected dementia. Int. Psychogeriatr. **26**, 1247–1262 (2014). doi:10.1017/S1041610214000416
7. Wild, K., Howieson, D., Webbe, F., Seelye, A., Kaye, J.: Status of computerized cognitive testing in aging: A systematic review. Alzheimer's Dement. **4**, 428–437 (2008). doi:10.1016/j.jalz.2008.07.003
8. Ross, A.J., Sachdev, P.S.: Magnetic resonance spectroscopy in cognitive research. Brain Res. Rev. **44**, 83–102 (2004). doi:10.1016/j.brainresrev.2003.11.001
9. Hsu, Y.-Y., Du, A.-T., Schuff, N., Weiner, M.W.: Magnetic resonance imaging and magnetic resonance spectroscopy in dementias. J. Geriatr. Psychiatry Neurol. **14**, 145–166 (2001)
10. Pereira, M.L.F., Von Zuben, A.C.M., Aprahamian, I., Forlenza, O.V.: Eye movement analysis and cognitive processing : detecting indicators of conversion to Alzheimer' s disease. Neuropsychiatr. Dis. Treat. **10**, 1273–1285 (2014). doi:10.2147/NDT.S55371
11. Dubois, B., Feldman, H.H., Jacova, C., Cummings, J.L., DeKosky, S.T., Barberger-Gateau, P., Delacourte, A., Frisoni, G., Fox, N.C., Galasko, D., Gauthier, S., Hampel, H., Jicha, G. A., Meguro, K., O'Brien, J., Pasquier, F., Robert, P., Rossor, M., Salloway, S., Sarazin, M., de Souza, L.C., Stern, Y., Visser, P.J., Scheltens, P.: Revising the definition of Alzheimer's disease: A new lexicon. Lancet Neurol. **9**, 1118–1127 (2010). doi:10.1016/S1474-4422(10)70223-4

12. Dementia – Signs, Symptoms, Causes, Tests, Treatment, Care | alz.org. http://www.alz.org/what-is-dementia.asp
13. Blennow, K., de Leon, M.J., Zetterberg, H.: Alzheimer's disease. Lancet **368**, 387–403 (2006). doi:10.1016/S0140-6736(06)69113-7
14. Nestor, P.J., Scheltens, P., Hodges, J.R.: Advances in the early detection of alzheimer's disease. Nat. Rev. Neurosci. **10**, S34 (2004). doi:10.1038/nrn1433
15. Arnáiz, E., Almkvist, O.: Neuropsychological features of mild cognitive impairment and preclinical Alzheimer's disease. Acta Neurol. Scand. Suppl. **107**, 34–41 (2003)
16. Troyer, A.K., Murphy, K.J., Anderson, N.D., Craik, F.I.M., Moscovitch, M., Maione, A., Gao, F.: Associative recognition in mild cognitive impairment: Relationship to hippocampal volume and apolipoprotein E. Neuropsychologia **50**, 3721–3728 (2012). doi:10.1016/j.neuropsychologia.2012.10.018
17. Gualtieri, C.T.: Dementia screening in light of the diversity of the condition. J. Insur. Med. **36**, 298–309 (2004)
18. Nasreddine, Z.S., Phillips, N.A., Bédirian, V., Charbonneau, S., Whitehead, V., Collin, I., Cummings, J.L., Chertkow, H.: The montreal cognitive assessment, MoCA: A brief screening tool for mild cognitive impairment. J. Am. Geriatr. Soc. **53**, 695–699 (2005). doi:10.1111/j.1532-5415.2005.53221.x
19. Mitchell, A.J.: A meta-analysis of the accuracy of the mini-mental state examination in the detection of dementia and mild cognitive impairment. J. Psychiatr. Res. **43**, 411–431 (2009). doi:10.1016/j.jpsychires.2008.04.014
20. Báez, P.G., Viadero, C.F., Espinosa, N.R., Pérez Del Pino, M.A., Suárez-Araujo, C.P.: Detection of mild cognitive impairment using a counter propagation network based system. An e-health solution. In: 2015 International Workshop on Computational Intelligence for Multimedia Understanding, IWCIM 2015 (2015)
21. Kondou, Y., Kawasumi, M., Yamamoto, O., Yamada, M., Yamamoto, S., Nakanno, T.: Study of early screening method of dementia and its systemization. In: Proceedings of the 11th IAPR Conference on Machine Vision Applications, MVA 2009 (2009)
22. Gualtieri, C.T.: Dementia screening using computerized tests. J. Insur. Med. **36**, 213–227 (2004)
23. Mosconi, L., Brys, M., Glodzik-Sobanska, L., De Santi, S., Rusinek, H., de Leon, M.J.: Early detection of Alzheimer's disease using neuroimaging. Exp. Gerontol. **42**, 129–138 (2007). doi:10.1016/j.exger.2006.05.016
24. Herholz, K., Salmon, E., Perani, D., Baron, J.-C., Holthoff, V., Frölich, L., Schönknecht, P., Ito, K., Mielke, R., Kalbe, E., Zündorf, G., Delbeuck, X., Pelati, O., Anchisi, D., Fazio, F., Kerrouche, N., Desgranges, B., Eustache, F., Beuthien-Baumann, B., Menzel, C., Schröder, J., Kato, T., Arahata, Y., Henze, M., Heiss, W.-D.: Discrimination between Alzheimer dementia and controls by automated analysis of multicenter FDG PET. Neuroimage **17**, 302–316 (2002). doi:10.1006/nimg.2002.1208
25. Klöppel, S., Stonnington, C.M., Chu, C., Draganski, B., Scahill, R.I., Rohrer, J.D., Fox, N. C., Jack Jr., C.R., Ashburner, J., Frackowiak, R.S.J.: Automatic classification of MR scans in Alzheimer's disease. Brain **131**, 681–689 (2008). doi:10.1093/brain/awm319
26. MacAskill, M.R., Anderson, T.J.: Eye movements in neurodegenerative diseases. Curr. Opin. Neurol. **29**, 61–68 (2016). doi:10.1097/WCO.0000000000000274
27. Zhang, Y., Wilcockson, T., Kim, K.I., Crawford, T., Gellersen, H., Sawyer, P.: Monitoring dementia with automatic eye movements analysis. In: Czarnowski, I., Caballero, A.M., Howlett, R.J., Jain, L.C. (eds.) Intelligent Decision Technologies 2016. SIST, vol. 57, pp. 299–309. Springer, Cham (2016). doi:10.1007/978-3-319-39627-9_26

28. Lagun, D., Manzanares, C., Zola, S.M., Buffalo, E.A., Agichtein, E.: Detecting cognitive impairment by eye movement analysis using automatic classification algorithms. J. Neurosci. Methods **201**, 196–203 (2011). doi:10.1016/j.jneumeth.2011.06.027

29. Fernández, G., Castro, L.R., Schumacher, M., Agamennoni, O.E.: Diagnosis of mild Alzheimer disease through the analysis of eye movements during reading. J. Integr. Neurosci. **14**, 121–133 (2015). doi:10.1142/S0219635215500090

30. Iarlori, S., Ferracuti, F., Giantomassi, A., Longhi, S.: RGBD camera monitoring system for Alzheimer's disease assessment using recurrent neural networks with parametric bias action recognition. In: IFAC Proceedings Volumes (IFAC-Papers Online) (2014)

31. Zanetti, O., Frisoni, G.B., Rozzini, L., Bianchetti, A., Trabucchi, M.: Validity of direct assessment of functional status as a tool for measuring Alzheimer's disease severity. Age Ageing. **27**, 615–622 (1998). doi:10.1093/ageing/27.5.615

32. Matic, A., Osmani, V.: Technologies to monitor cognitive decline: A preliminary case study. In: 2009 3rd International Conference on Pervasive Computing Technologies for Healthcare - Pervasive Health 2009, PCT Health 2009 (2009)

33. Buso, V., Hopper, L., Benois-Pineau, J., Plans, P.-M., Megret, R.: Recognition of activities of daily living in natural "at home" scenario for assessment of Alzheimer's disease patients. In: 2015 IEEE International Conference on Multimedia and Expo Workshops, ICMEW 2015 (2015)

34. Avgerinakis, K., Briassouli, A., Kompatsiaris, I.: Activity detection and recognition of daily living events. In: Proceedings of the 1st ACM International Workshop on Multimedia Indexing and Information Retrieval for Heathcare, MIIRH 2013, Co-located with ACM Multimedia 2013 (2013)

35. Chen, Y.T., Hou, C.J., Huang, M.W., Dong, J.H., Zhou, J.Y., Hung, I.C.: The design of interactive physical game for cognitive ability detecting for elderly with mild cognitive impairment. In: IFMBE Proceedings (2015)

36. Sapey-Triomphe, L.-A., Heckemann, R.A., Boublay, N., Dorey, J.-M., Hénaff, M.-A., Rouch, I., Padovan, C., Hammers, A., Krolak-Salmon, P.: Neuroanatomical correlates of recognizing face expressions in mild stages of Alzheimer's disease. PLoS One **10**, e0143586 (2015). doi:10.1371/journal.pone.0143586

37. Yang, L., Zhao, X., Wang, L., Yu, L., Song, M., Wang, X.: Emotional face recognition deficit in amnestic patients with mild cognitive impairment: Behavioral and electrophysiological evidence. Neuropsychiatr. Dis. Treat. **11**, 1973 (2015). doi:10.2147/NDT.S85169

38. Varjassyová, A., Hořínek, D., Andel, R., Amlerova, J., Laczó, J., Sheardová, K., Magerová, H., Holmerová, I., Vyhnálek, M., Bradáč, O., Geda, Y.E., Hort, J.: Recognition of facial emotional expression in amnestic mild cognitive impairment. J. Alzheimer's Dis. **33**, 273–280 (2013). doi:10.3233/JAD-2012-120148

39. Burton, K., Kaszniak, A.: Emotional experience and facial expression in Alzheimer's disease. Aging. Neuropsychol. Cogn. **13**, 636–651 (2006). doi:10.1080/13825580600735085

40. Henry, J.D., Rendell, P.G., Scicluna, A., Jackson, M., Phillips, L.H.: Emotion experience, expression, and regulation in alzheimer's disease. Psychol. Aging. **24**, 252 (2009). doi:10.1037/a0014001

41. Magai, C., Cohen, C.I., Culver, C., Gomberg, D., Malatesta, C.: Relation between premorbid personality and patterns of emotion expression in mid- to late-stage dementia. Int. J. Geriatr. Psychiatr. **12**, 1092–1099 (1997). doi:10.1002/(SICI)1099-1166(199711)12:11<1092::AID-GPS690>3.0.CO;2-X

42. Jack, R.E., Garrod, O.G.B., Yu, H., Caldara, R., Schyns, P.G.: Facial expressions of emotion are not culturally universal. Proc. Natl. Acad. Sci. **109**, 7241–7244 (2012). doi:10.1073/pnas.1200155109. USA

43. Seidl, U., Lueken, U., Thomann, P.A., Kruse, A., Schröder, J.: Facial expression in Alzheimer's disease: Impact of cognitive deficits and neuropsychiatric symptoms. Am. J. Alzheimers. Dis. Other Demen. **27**, 100–106 (2012). doi:10.1177/1533317512440495
44. Fasel, B., Luettin, J.: Automatic facial expression analysis: A survey. Pattern Recogn. **36**, 259–275 (2003). doi:10.1016/S0031-3203(02)00052-3
45. Sandbach, G., Zafeiriou, S., Pantic, M., Yin, L.: Static and dynamic 3D facial expression recognition: A comprehensive survey. Image Vis. Comput. **30**, 683–697 (2012). doi:10.1016/j.imavis.2012.06.005
46. Sariyanidi, E., Gunes, H., Cavallaro, A.: Automatic analysis of facial affect: A survey of registration, representation, and recognition. IEEE Trans. Pattern Anal. Mach. Intell. **37**, 1113 (2015). doi:10.1109/TPAMI.2014.2366127
47. Lopes, A.T., de Aguiar, E., De Souza, A.F., Oliveira-Santos, T.: Facial expression recognition with Convolutional Neural Networks: Coping with few data and the training sample order. Pattern Recogn. **61**, 610–628 (2017). doi:10.1016/j.patcog.2016.07.026
48. Mandal, M., Poddar, S., Das, A.: Comparison of human and machine based facial expression classification. In: International Conference on Computing, Communication and Automation, ICCCA 2015 (2015)
49. Samal, A., Iyengar, P.A.: Automatic recognition and analysis of human faces and facial expressions: a survey. Pattern Recogn. **25**, 65–77 (1992)
50. Taigman, Y., Yang, M., Ranzato, M., Wolf, L.: DeepFace: closing the gap to human-level performance in face verification. In: Proceedings of the IEEE Computer Society Conference on Computer Vision and Pattern Recognition (2014)
51. Mukadam, N., Cooper, C., Kherani, N., Livingston, G.: A systematic review of interventions to detect dementia or cognitive impairment. Int. J. Geriatr. Psychiatr. **30**, 32–45 (2015). doi:10.1002/gps.4184
52. Petersen, R.C., Stevens, J.C., Ganguli, M., Tangalos, E.G., Cummings, J.L., DeKosky, S.T.: Practice parameter: Early detection of dementia: Mild cognitive impairment (an evidence-based review). Neurology **56**, 1133–1142 (2001)

Automatic Measurement of Blood Vessel Angles in Immunohistochemical Images of Liver Cancer

Hongbin Zhang[1], Kun Zhang[1], Li Chen[2], Jianguo Wu[1(✉)],
Peijian Zhang[1], and Huiyu Zhou[3]

[1] School of Electrical Engineering, Nantong University, Nantong, China
1951299040@qq.com, zhangkun_nt@163.com,
{wu.jg, zhang.pj}@ntu.edu.cn
[2] Department of Pathology, Medical College,
Nantong University, Nantong, China
bll@ntu.edu.cn
[3] School of Electronics, Electrical Engineering and Computer Science,
Queen's University Belfast, Belfast, Northern Ireland, UK
h.zhou@ecit.qub.ac.uk

Abstract. This paper presents a method for automated measurement of vascular angle in immunohistochemical images of liver cancer. Firstly, Colour Deconvolution is used to conduct staining separation on a H&E-stained immunohistochemical image, and then blood vessels are segmented using an improved Otsu algorithm. Then the standard SURF algorithm is used to select feature points of the image, and then these feature points are divided into two equal groups according to the distance between individual feature points and the far left (or right) feature point. Finally, a standard least squares method is used to fit two lines using the two groups of points. When the linear deviation of the fitting result based on the two groups of feature points is significant, it is necessary to adjust the belonging of the points of the two groups, and then the two sets are fitted again respectively till the correlation coefficients of the two fitted lines are greater than the predefined threshold, meaning that the measurement of the blood vessel angle in the immunohistochemical map is completed. Compared with the experts' results, our proposed technique results in better accuracy. It is worthy to point out that, to our knowledge, our system is the first one that conducts automated measurement of blood vessel angle of immunohistochemistry.

Keywords: Immunohistochemical image · Color deconvolution · Image segmentation · Feature extraction · Least square

---

H. Zhang and K. Zhang—These authors contributed equally to this work and should be considered co-first authors.

M. Fei et al. (Eds.): LSMS/ICSEE 2017, Part I, CCIS 761, pp. 162–172, 2017.
DOI: 10.1007/978-981-10-6370-1_16

1   Introduction

Modern medicine shows the occurrence of liver cancer is a multi-factor, multi-step biological process. Clinically, when liver cancer progresses, vascular segments may become irregular and displaced, and liver capillaries can be deformed (e.g. hook-shaped). Therefore, we need to develop a method to identify the deformed blood vessels and then calculate the vascular angle. The stage immunohistochemical(IHC) technique [1] has been used to improve the accuracy of the clinical pathological diagnosis, and become an important evaluation standard for differentiating benign and malignant tumors. However, the current automated IHC analysis method is not enough to solve all the morphological problems of blood vessels, when only statistical information such as a blood vessel's length, area, aspect ratio and so on. Therefore, the automated angle measurement algorithm proposed in this paper is of great significance as it helps understand the growth and metastasis of malignant tumors [2]. (Fig. 1)

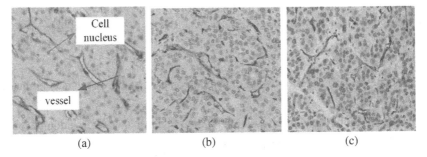

Fig. 1. IHC images of liver cancer (magnification of 400×) respectively for the well-segmented, moderately segmented and poorly segmented stage.

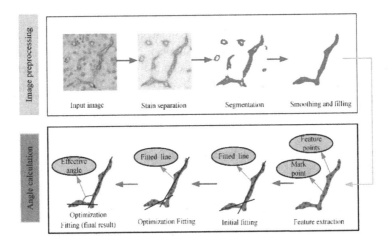

Fig. 2. Flowchart of image preprocessing and angle calculation.

The image registration method proposed in this paper is using image features. In general, this paper introduces a method for automated identification of blood vessels. This method, based on the selection of the characteristic points of images, performs line fitting and then derives the angle between lines to represent the deflection of a blood vessel [3].

2 Method

According to the characteristics of IHC images, our proposed algorithm includes the following steps: Firstly, a multi-colored IHC image passes through the Color Deconvolution step [4, 5], then the corresponding single colored image is obtained. An improved OTSU algorithm [6, 7] is used to obtain coarse image segmentation of potential vessels, then the smoothed boundaries and holes is solved through morphological processing [8, 9]. The standard SURF algorithm [10, 11] is used to extract feature points of the potential blood vessels, and a least square method [12] is used for fitting points of clouds to determine the position of two straight lines. Finally, using the common geometric relationship, we can calculate the vascular angle (Fig. 2).

2.1   Linear Fitting

We extract the coordinate information of feature points of blood vessel, and number all the N feature points:

(1) According to the coordinate positions of the feature points, we randomly choose the first endpoint, denoted as P_1;
(2) We calculate the distance between the endpoint P_1 and the other $N-1$ feature points. The 2nd endpoint P_2 is the one that has the shortest distance to the first endpoint. We then calculate the distance between P_2 and the other $N-2$ points respectively. When $|P_iP_2| + |P_iP_1|$ has the minimum value, i denoted as P_3. We perform such an iteration till all the feature points are marked.

When N is an odd number, the feature point with the number $(N+1)/2$ is the marked point. However, when N is an even number, N/2 or (N/2) + 1 is taken as the marker. The N feature points of the blood vessel are divided into two equal sets S_1 and S_2 (when N is an even number, $S_1 = \{P_1, P_2, \ldots, P_N/2\}$, $S_2 = \{P_N/2+1, \ldots, P_N\}$. When N is an odd number, $S_1 = \{P_2, \ldots, (P_N-1)/2\}$, $S_2 = \{(P_N+1)/2, \ldots, P_N\}$). The least square fitting is performed to obtain two straight lines L_1 and L_2. More details follow:

① When fitting the two straight lines L_1 and L_2, if the correlation coefficient (an indicator of the good fitting) is greater than 0.8, we determine the two straight lines as the outputs;
② Otherwise, denoted as L_x, the marked points moves away a position randomly, leading to $S_1 = \{P_1, P_2, \ldots, P_N/2, P_N/2+1\}$, $S_N = \{P_N/2+1, \ldots, P_N\}$ we perform line fitting again. If the correlation coefficient ρ becomes smaller, the marked point moves to the opposite direction, leading to $S_1 = \{P_1, P_2, \ldots, P_N/2 - 1.2 - 1\}$, $S_2 = \{P_N/2, P_N/2+1, \ldots, P_N\}$.

Once the two straight line positions have been determined, we calculate the blood vessel angle formed by the intersection of the two straight lines.

Linear fitting method simulation

Input: All the N characteristic points in a vessel

Output: The angle of the blood vessel

1. divide these feature points into two equal sets S_1 and S_2

2. S_1, S_2 least squares straight line fitting, denoted as L_1, L_2

3. calculate the correlation coefficients of lines L_1 and L_2, denoted as ρ_1, ρ_2

4. **if** $\rho_1 > 0.8$ and $\rho_2 > 0.8$, **then**

5. calculate the angle between lines L_1 and L_2 **end**

6. **else if**

7. mark point move a point to the left (or right), constitute two new set S_1', S_2'

8. S_1 S_2 least squares straight line fitting, denoted as L_1, L_2

9. calculate the correlation coefficients of lines L_1 and L_2, denoted as ρ_1, ρ_2

10. **if** $\rho_1 > 0.8$ and $\rho_2 > 0.8$, **then**

11. calculate the angle between lines L_1 and L_2 **end**

12. **else if**

13. $(\rho_1 + \rho_2) >$ initial $(\rho_1 + \rho_2)$ **then**

14. the marker moves a point in the same direction, constitute two new set S_1', S_2'

15. **repeat** steps 8-11

16. **else if** $(\rho_1 + \rho_2) <$ initial $(\rho_1 + \rho_2)$ **then**

17. the marker moves a point in the opposite direction, constitute two new set S_1', S_2'

18. **repeat** steps 8-11.

The processing steps of the proposed linear fitting scheme can be found in Fig. 3.

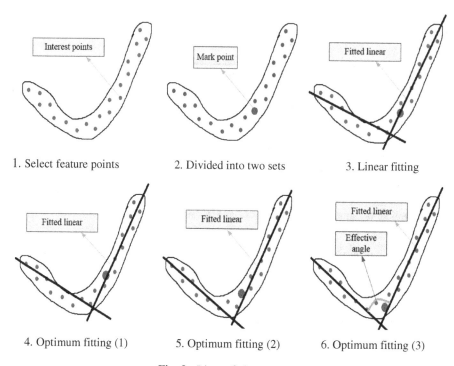

1. Select feature points   2. Divided into two sets   3. Linear fitting

4. Optimum fitting (1)   5. Optimum fitting (2)   6. Optimum fitting (3)

Fig. 3. Linear fitting process.

2.2 Proposed Framework

Step1: Staining Separation

The color deconvolution algorithm [4, 5] is to obtain independent information for each color. In order to get a correct color absorption factor, the orthogonal transformation of the RGB is needed, which can be achieved by the color deconvolution method.

Step 2: Segmentation

In this paper, the optimal threshold criterion function of the improved Otsu algorithm is:

$$T^* = Arg \max_{0 \leq T \leq L-1} (\sigma_B^2) \tag{1}$$

Step 3: Feature Extraction

The extraction of SURF [10, 13] feature points can determine the location of interest points through calculating the Hessian matrix [14] determinant of the local extreme points. Thus, the determinant of the Hessian matrix can be reduced to:

$$Det(H) = D_{xx}D_{yy} - (wD_{xy})^2 \tag{2}$$

Where W is the weight coefficient, which is appropriately given as 0.9.

Step 4: Linear Fitting

Linear fitting [3] is commonly achieved by a least square method. In statistics, the correlation coefficient is used to measure the degree of correlation between the degree of closeness.

$$\rho_{XY} = \frac{Cov(X, Y)}{\sqrt{D(X)} \sqrt{D(Y)}} \tag{3}$$

In the above formula, $Cov(X, Y)$ is the covariance of X and Y, $D(X)D(Y)$ are the variance of X and Y, respectively.

When $|\rho_{XY}| > 0.8$, the relationship between X and Y is known as highly relevant;

Step 5: The Angle Calculation

$$\alpha = \arctan \left| \frac{k_2 - k_1}{1 + k_1 k_2} \right| \tag{4}$$

where k_1 and k_2 were two straight lines.

3 Experiments and Results

In this paper we study the IHC images of 4 grades of liver cancer, each of which contains 20 cases. The samples come from 82 patients with liver cancer surgery in the Affiliated Hospital of Nantong University from January to September 2016. There were 68 males and 14 females, aged from 31 to 70, with an average age of (54.7 ± 10.5) years. No interventional therapy has been applied before the operation.

All of them had CPA examination within one week before the surgery, and postoperative pathology and immunohistochemistry after the surgery. All the pathological sections were treated with CD34 monoclonal antibody, H&E staining and IHC staining, whilst IHC shows positive, and we adopted a Olympus DP27 camera, which has the exposure time of 1/260 s, the horizontal and vertical resolution is 72dpi, and the image size is 1224 × 960 with 400 × magnification.

Clinically, according to the different levels of cell differentiation, malignant tumors are often divided into three groups: highly differentiated tumors (falling in 135–180 degree), moderately differentiated tumors (falling in 105–135 degree), poorly differentiated tumors (falling in 75–105 degree).

Each stage of the clinical study is conducted by the single factor analysis of variance. Bivariate correlation analysis is performed in different pathological stages. The t experiment is performed to compare the differences in a group, and the variance analysis is used to compare across groups, where $P < 0.05$ indicating statistical validity. All the data is analyzed using SPSS10 for windows.

3.1   IHC Image Preprocessing

For the stained IHC image, the extraction of blood vessels is performed by using the classical color deconvolution method in Fig. 4.

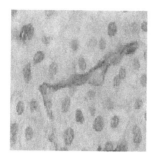

Fig. 4. Color deconvolution for staining separation (poorly differentiated)

3.2   Vascular Extraction

The determination of segmentation thresholds is the key factor to determine the accuracy of feature extraction, combining with the improved OTSU algorithm. Figure 5 illustrates the procedure of applying the improved OTSU algorithm.

Fig. 5. Comparison of several typical threshold segmentation methods, among them, they were obtained using percentile [15], shanbhag [16] and Otsu methods [6].

It can be seen from Fig. 5, immunohistochemical images only contain the characteristic area of blood vessels (the target feature) and the background area. After a binary operation, we obtain Fig. 6. which is an effective in this case.

3.3   Feature Extraction

We use the standard SURF algorithm [10, 11] to deal with the results of the three pathological stages as shown in Fig. 7.

Fig. 6. Illustration of smoothing and filling.

Fig. 7. Illustration of SURF feature extraction

3.4 Least Square Fitting

After having applied the least square [3, 17] method, we have the results shown in Fig. 8.

Fig. 8. Preliminary linear fitting results.

Fig. 9. The marker points are randomly moved by one unit and re-fitted

As can be seen from Fig. 8, one of the fitting lines has significant deviation. This requires the moving of the marked point and a proper re-fitting operation accordingly.

From Fig. 9, if the marked point is moved to right, the fitting line's deviation increases significantly, therefore we continue to move the marked point to the left. This

Fig. 10. We move the marked point to the opposite direction and then re-fit.

Fig. 11. We move the marked point in this direction and then re-fit

results in Fig. 10. By calculating the linear correlation coefficient, we can see that the deviation of the line fitting is dramatically reduced and the position of the marked point moves. We will obtain Fig. 11.

The angle of the blood vessel can be calculated accurately. However, there are some errors in the fitting of the feature points, and it is not easy to observe in the statistical analysis. Therefore, the method of classifying the blood vessel angle in each image is calculated so as to analyze the result by using the statistical principle [18].

The meaning of the P value is simply that the source of the difference is the possibility of sampling error (random error). Usually the definition of P is less than 0.05 that have statistical significance, meaning when the possibility of differences can be used to explain the sampling error of less than 5%, we would think that this has nothing to do with the sampling error, but caused by the factors of the test [13]. The results of Fig. 12 show that the experimental results are statistically significant, and the method proposed in this paper can produce reasonably correct vascular angles.

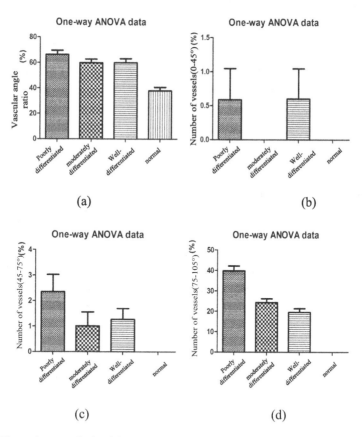

Fig. 12. The variance analysis of the number of vessels in different stages [19]. The value of P is a probability, which reflects the probability of occurrence of an event. Statistics based on the significance of the P value obtained by the significance test [20], the general P < 0.05 is significant, P < 0.01 is very significant, among them, (a) p < 0.0001; (b) p = 0.0026; (c) p < 0.0001; (d) p < 0.0001.

4   Conclusion

This paper introduces a new method of measuring vascular angles as the indicator of different stages of liver cancer. IHC images of liver cancer were processed through Color Deconvolution. The feature points of the region of interest were extracted using SURF, fitted by a least square method. Most of the blood vessels shown in the IHC images satisfied the requirements. However, for a small number of complex blood vessels, the use of the current method cannot produce satisfactory results. In the future work, we intend to closely look into the segmentation stage. We also compare our proposed method against the other state of the art technologies.

Acknowledgements. This work was financially supported by the Natural Science Foundation of Jiangsu Province, China under Grant No. BK20170443. Nantong Research Program of Application Foundation under Grant No. GY12016022, and Dr. H. Zhou is supported by UK EPSRC under Grant EP/N011074/1, and Newton Advanced Fellowship under Grant NA160342.

References

1. Bhat, P., Singh, N.D., Leishangthem, G.D., et al.: Histopathological and immunohisto-chemical approaches for the diagnosis of Pasteurellosis in swine population of Punjab. J. Vet. World. **9**, 989–995 (2016)
2. Cherni, M.A., Sayadi, M.: AI tools in medical image analysis: efficacy of ANN for oestrogen receptor status assessment in immunohistochemical staining of breast cancer. Int. J. Biomed. Eng. Technol. **12**, 60–83 (2013)
3. Berg, B.A.: Least square fitting with one explicit parameter less. J. Comput. Phys. Commun. **200**, 254–258 (2016)
4. Ruifrok, A.C.: Comparison of quantification of histochemical staining by Hue-Saturation-Intensity (HSI) transformation and color deconvolution. J. Appl. Immuno-histochem. Mol. Morphol. **11**, 85–91 (2004)
5. Onder, D., Zengin, S., Sarioglu, S.: A review on color normalization and color deconvolution methods in histopathology. J. Appl. Immunohistochem. Mol. Morphol. **22**, 713–719 (2014)
6. Chen, Z., Tu, Y.: Improved image segmentation algorithm based on OTSU algorithm. J. Int. J. Advancements Comput. Technol. **4**, 206–215 (2012)
7. Xiaodan, C., Li, S., Hu, J., Liang, Y.: A survey on Otsu image segmentation methods. J. Comput. Inf. Syst. **10**, 4287–4298 (2014)
8. Nudthakarn, K., Angkhana, J., Sirithan, J., Supatra, J., Ron, S.: Hydroxyapatite nanoparticles formed under a wet mechanochemical method. J. Biomed. Mater. Res. Part B Appl. Biomater. **105**, 679–688 (2017)
9. Shaaban, K.S., Abo-naf, S.M., Elnaeim, A.M.A., Hassouna, M.E.M.: Studying effect of MoO3 on elastic and crystallization behavior of lithium diborate glasses. J. Appl. Phys. A Mater. Sci. Process. **123**, 457 (2017)
10. Calamante, F., Jacques-Donald, T., Kurniawan, N.D., et al.: Super- resolution track-density imaging studies of mouse brain: Comparison to histology. J. Neuroimage **59**, 286–296 (2012)
11. Zhang, L., Dong, Y., Pu, J.: Object recognition based on SURF. J. ICIC Express Lett. Part B Appl. **6**, 259–264 (2015)

12. Xie, B., Bose, T.: Partial update least-square adaptive filtering. Synth. Lect. Commun. **7**, 1–115 (2014)
13. Yin, M.Y., Guan, F., Ding, P.: Implementation of image matching algorithm based on SIFT features. J. Appl. Mech. Mater. **602–605**, 3181–3184 (2014)
14. Lingyao, M., James, C.S.: Efficient computation of the Fisher information matrix in the em algorithm. In: 2017 51st Annual Conference on Information Sciences and Systems (2017)
15. Upadhyay, K., Asthana, A., Tiwari, N.: Determination of nimesulide in pharmaceutical and biological samples by a spectrophotometric method assisted with the partial least square method. Res. Chem. Intermed. **39**, 3553–3563 (2013)
16. Reeves, A.P., Liu, S., Xie, Y.: Image segmentation evaluation for very-large datasets. In: Conference on Medical Imaging–Computer-Aided Diagnosis, vol. 9785, San Diego, CA (2016)
17. Prasad, D.K., Leung, M.K.H., Quek, C.: ElliFit: an unconstrained, non-iterative, least squares based geometric ellipse fitting method. J. Pattern Recogn. **46**, 1449–1465 (2013)
18. Fan, W., Zhao, D., Wang, S.: A fast statistics and analysis solution of medical service big data. In: Proceedings - 2015 7th International Conference on Information Technology in Medicine and Education ITME, pp. 9–12 (2016)
19. Kapitány, K.: Barsi: deriving hierarchical statistics by processing high throughput medical images. IFMBE Proc. **50**, 32–35 (2015)
20. Shin, Y., Choi, Y., Lee, W.J.: Integration testing through reusing representative unit test cases for high-confidence medicalsoftware. J. Comput. Biol. Med. **43**, 434–443 (2013)

A Novel Segmentation Framework Using Sparse Random Feature in Histology Images of Colon Cancer

Kun Zhang[1], Huiyu Zhou[2], Li Chen[3], Minrui Fei[4], Jianguo Wu[1], and Peijian Zhang[1(✉)]

[1] School of Electrical Engineering, Nantong University, Nantong, China
zhangkun_nt@163.com, {wu.jg,zhang.pj}@ntu.edu.cn
[2] School of Electronics, Electrical Engineering and Computer Science,
Queen's University, Belfast, UK
h.zhou@ecit.qub.ac.uk
[3] Department of Pathology, Medical College,
Nantong University, Nantong, China
bll@ntu.edu.cn
[4] School of Mechatronic Engineering and Automaton,
Shanghai University, Shanghai, China
mrfei@staff.shu.edu.cn

Abstract. In this paper, we present a novel segmentation framework for glandular structures in Hematoxylin and Eosin stained histology images, choosing poorly differentiated colon tissue as an example. The proposed framework' target is to identify precise epithelial nuclei objects. We start with staining separate to detect all nuclei objects, and deploy multi-resolution morphology operation to map the initial epithelial nuclei positions. We proposed a new bag of words scheme using sparse random feature to classify epithelial nuclei and stroma nuclei objects to adjust the rest nuclei positions. Finally, we can use the boundary of optimized epithelial nuclei objects to segment the glandular structure.

Keywords: Poorly differentiated glandular · Colon tissue · Color-deconvolution · Multi-resolution morphology · Sparse random feature

1 Introduction

Motivation: In this paper, we address the problem of segmenting challenging glandular structures in colon histology images. Glandular structures are important for the diagnosis of several epithelial cancers. Glands in epithelial tissue, which are difficult to be differentiated from other tissues, normally consist of lumen structure surrounded by epithelial nuclei objects at the boundary, which can be used as a strong cue for the extraction of glandular structures [1, 2]. This problem becomes more significant while there are other tissue constituents, such as stroma nuclei and cytoplasm, around the glands.

© Springer Nature Singapore Pte Ltd. 2017
M. Fei et al. (Eds.): LSMS/ICSEE 2017, Part I, CCIS 761, pp. 173–180, 2017.
DOI: 10.1007/978-981-10-6370-1_17

Related Work: Existing methods for glandular structure segmentation can be categorised into texture and structure based method. In texture approaches, Farjam et al. [3] used Gaussian filters to extract texture features from glandular structures. For structural approaches, Naik et al. [4] used a level set method to segment lumen areas in a gland. Nguyen et al. [5] employed the prior knowledge about glandular constituents in order to extract glandular regions. Gunduz-Demir et al. [6] proposed object graphs for the segmentation of glandular structures.

Contributions: In this work, we propose a novel framework to segment glandular structures in Hematoxylin and Eosin (H&E) stained histology images, choosing colon tissue as an example. The proposed framework starts with nuclei identification. We deploy a color-deconvolution method to find the nuclei position. Then, a multi-resolution morphology operation is investigated to interpret the epithelial nuclei spatial distribution. In this study, we assume that epithelial nucleus cannot be fully separated from stromal nucleus. Therefore, we deploy a sparse random features based Bag of words model to classify these nucleus. Finally, we use the boundary of epithelial nuclei for the final segmentation.

2   Sparse Random Matrix Optimization

Recently, random projection feature has shown promise for complex classification [7, 8]. It uses random matrix to project high level features to low dimension with the promise of core features preserved. A simple example is illustrated in Fig. 2, which shows the reconstruction of an ideal texture map based on random projection. The reconstruction results from different dimensions of projection are shown in Fig. 2(b), (c) and (d), respectively. With random projection, the original ideal texture is well reconstructed.

Based on the principle of distance preservation [9], Gaussian random matrix [9] and sparse random matrices [10, 11] have been sequentially proposed for random projection. Although significant progress has been made, the Gaussian random matrix and sparse random matrices still have prominent limitations: Gaussian random matrix causes highly computing for its dense distribution and it is difficult to construct Gaussian distribution via hardware, on the other hand the sparser matrix tends to yield weaker distance preservation. It has been proven that the irrelevance of matrices column vector is highly related to the projecting performance. In this paper, we use the angle between matrix neighbor column vectors to indicate the projecting performance. The first row of Table 1 is the minimum and maximum angles of the matrix, and the second row is the difference value of the minimum and maximum angles. The difference value is smaller the projecting performance is better. From Table 1, the random matrix and sparse random matrix are better than Gaussian random matrix.

Table 1. Parameters comparison under three random matrices

Gaussian random matrix	Sparse random matrix	Very sparse random matrix
71.42/107.34	81.04/100.93	82.04/110.93
35.92	19.89	28.89

Furthermore, we propose a new method to optimize sparse random matrix. Firstly, we build an empty matrix U, the length of U is λk, is k signal length and λ is positive integer. Secondly, we generate uniform random value to U from the range of 0 and 1. Thirdly, we recursively scan every value, if the value is larger than $5/6$, the value is reset as "1.732", if the value is smaller than $1/6$, the value is reset as "−1.732", other values are reset as "0". Fourthly, calculate the angle column by column, if the angle is small than threshold, assemble the corresponding column to a new $d \times k$ matrix. Specifically, the optimize algorithm is to form a new matrix by selecting the column vectors which angle is small than threshold. The new matrix not only satisfy RIP theorem but also to achieve irrelevance of matrices column vector.

Figure 3 shows the signal recover ability based on six random matrices. All optimization method based curves are the right side of the original method based curves, which means the signal recover ability of optimization methods are more useful than basic ones. Meanwhile, the ability of Gaussian and sparse random matrices are almost the same. To achieve the convenience of hardware design, we use optimization sparse random matrix in this paper.

3   Poorly Differentiated Glandular Segmentation Framework

Figure 1 summarizes the poorly differentiated glandular segmentation framework. Given an H&E image (Fig. 1a), nuclei locations are represented by color-deconvolution. (Figure 1b), necessary to perform morphology operation and random feature based classification to generate glandular contour (Fig. 1c).

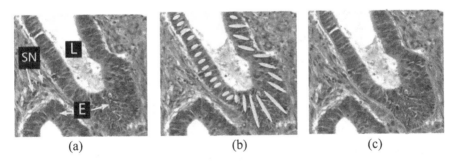

(a)   (b)   (c)

Fig. 1. A sample colon histology image showing various components (epithelial nucleus or E, stromal nucleus or SN, lumen or L)

3.1   Nuclei Identification

We employ a color-deconvolution method [12] to extract the Hematoxylin channel from the image. By thresholding the Hematoxylin channel using Otsu's threshold [13], we obtain a binary image corresponding to the approximate locations of nuclei in the image (Fig. 4).

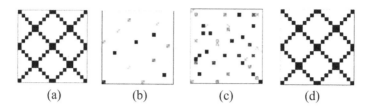

Fig. 2. Random projection based ideal texture reconstruction: (a) original 20×20 ideal texture image, (b) reconstruction using 50 RP impact factors, (c) reconstruction using 100 RP impact factors, (d) reconstruction using 150 RP impact factors.

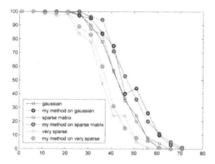

Fig. 3. The signal recover ability based on six random matrices.

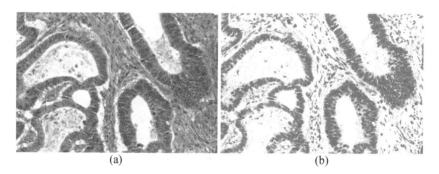

Fig. 4. (a) input image, (b) staining separated image using color-deconvolution.

3.2 Epithelial Layer Determination

Epithelial layers are most often formed by thick and solid objects such as epithelial nucleus. Therefore, a map describing the solidity and the connectedness of nuclei objects is calculated using this formula :

$$S_{map} = \frac{1}{R_s \times s_{max}} \sum_{r=1}^{R_s} \sum_{s=1}^{s_{max}} J_s((O_r(I_N))) \tag{1}$$

Where $O_r(X)$ is the morphological opening of image X with a disk of radius r, J_s is a removal operator of connected components with size smaller than s. R_s is the maximal radius and s_{max} the maximal connected component size. The use of these operations at different levels, i.e. with different values of r and s, agrees the variation in size of epithelial layers. The obtained map shows epithelial layers objects with higher solidity values than connective tissue nuclei objects (Fig. 5).

3.3 Epithelial Layer Optimization

Within the framework of RP based coding histogram coefficients in Fig. 6, for each RIRS feature vector [14] of the image, we can use RP to produce a compact and sparse representation, where only the entries corresponding to the same textons will have non-zero values, while the other entries in a compact vector are zeros. Then we can form the histogram coefficient as a new feature.

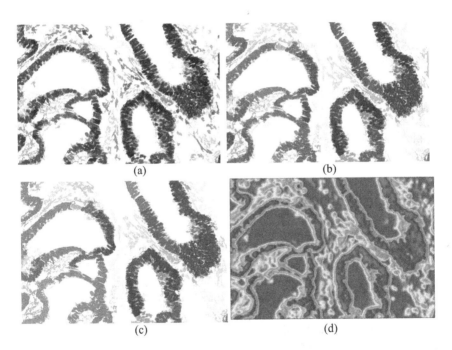

(a)

(b)

(c)

(d)

Fig. 5. (a)–(c) Epithelial layers identification after multi-resolution morphology operation, (d) Nuclei objects retained are overlayed in blue (Color figure online).

Figure 7 shows two nucleus of different classes (epithelial nuclei and stroma nuclei) and their 400 dimension histogram features. We can see that the histogram features of different classes are very different. For instance, in the upper two histograms, the features around "50", "100" and "350" show strong differences and it is convenient for the later classification. At the same time, when texton number changes, the two histograms for the same image patch also show a sufficient difference. The upper

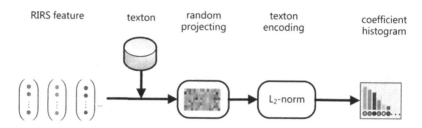

Fig. 6. Random projection based coding histogram coefficients. Best viewed in colour.

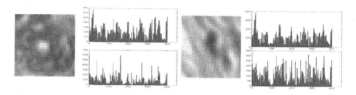

Fig. 7. epithelial nuclei and stroma nuclei images and their encoding histogram features (upper histogram – 9 textons and lower histogram – 25 textons). Best viewed in colour (Color figure online).

histograms are built from 9 textons and the lower histograms are constructed from 25 textons, and it is clear that the similarity of the two histograms based on 25 texton is much lower than 9 textons based histograms, and it is easy to find more discriminating features for classification.

4   Results and Discussion

To assess the effectiveness of our approach, the experiment was conducted on a public dataset of poorly differentiated H&E glandular tissue sections. The dataset, provided by Warwick university hospitals in the context of Gland Segmentation Challenge [15].

To quantitatively evaluate our segmentation results, the true positive (TP), false positive (FP), and false negative (FN) pixels are calculated. We use two quantitative criterions for the evaluation of our segmentation algorithm on the dataset. Table 2 shows the comparative quantitative performance of the proposed approach against 4 other methods in the literature; active contour [16], object-graphs [17], Morphology [18], and RIRS-bow [14], based on 3 quantitative measures: the precision also referred

Table 2. Quantitative evaluation of the segmentation methods.

Method	PPV	TRP	ACC
Active contour [16]	0.85	0.80	0.73
Object-graph [17]	0.91	0.87	0.77
Morphology [18]	0.89	0.86	0.78
RIRS-bow [14]	0.93	0.89	0.75
Proposed	0.85	0.87	0.81

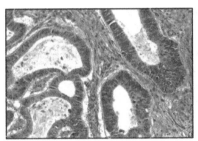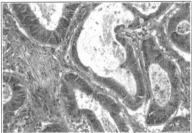

Fig. 8. Examples of segmentation results: image from Warwick university dataset. Best viewed in colour (Color figure online).

to as positive predictive value PPV $= \frac{TP}{TP + FP}$, the sensitivity referred to as the true positive rate TPR $= \frac{TP}{TP + FN}$, and the accuracy ACC $= \frac{TP}{TP + FP + FN}$. Figure 8 is examples of segmentation results.

5   Conclusion

This work presents a new gland segmentation approach based on gland structural feature. This method begin with staining separation to detect nuclei objects, then advance multi-resolution morphological operator is applied to map the initial epithelial nuclei objects. After that, a developed sparse random feature is used to classify the rest nuclei objects from epithelial nuclei class and stroma nuclei class. Finally, the boundary of epithelial nuclei objects is the core clue to segment the gland.

Acknowledgements. This work was financially supported by the Natural Science Foundation of Jiangsu Province, China under grant No. BK20170443. Nantong Research Program of Application Foundation under Grant No. GY12016022 and Dr. H Zhou is currently supported by UK EPSRC under Grant EP/N011074/1, and Newton Advanced Fellowship under Grant NA160342.

References

1. Gleason, D.F., Mellinger, G.T.: Prediction of prognosis for prostatic adenocarcinoma by combined histological grading and clinical staging. J. Urol. **167**(2), 953–958 (2002)
2. Hamilton, S.R., Aaltonen, L.A.: International Agency for Research on Cancer, World Health Organization, and others, Pathology and genetics of tumours of the digestive system (2000)
3. Farjam, R., Soltanian-Zadeh, H., Jafari-Khouzani, K., Zoroofi, R.A.: An image analysis approach for automatic malignancy determination of prostate pathological images. Cytometry Part B Clin. Cytometry **72**(4), 227–240 (2007)
4. Naik, S., Doyle, S., Madabhushi, A., Tomaszewski, J., Feldman, M.: Gland segmentation and gleason grading of prostate histology by integrating low-, high-level and domain specific information. In: Workshop on Microscopic Image Analysis with Applications in Biology (2007)

5. Nguyen, K., Jain, A.K., Allen, R.L.: Automated gland segmentation and classification for gleason grading of prostate tissue images. In: 2010 20th International Conference on Pattern Recognition (ICPR), pp. 1497–1500. IEEE (2010)
6. Gunduz-Demir, C., Kandemir, M., Tosun, A.B., Sokmensuer, C.: Automatic segmentation of colon glands using objectgraphs. Med. Image Anal. **14**(1), 1–12 (2010)
7. Varma, M., Zisserman, A.: A statistical approach to material classification using image patches. IEEE Trans. Pattern Anal. Mach. Intell. **31**(11), 2032–2047 (2009)
8. Zhang, J., Marszalek, M., Lazebnik, S., Schmid, C.: Local features and kernels for classification of texture and object categories: a comprehensive study. Int. J. Comput. Vis. **73** (2), 213–238 (2007)
9. Liu, L., Fieguth, P.: Texture classification from random features. IEEE Trans. Pattern Anal. Mach. Intell. **34**(3), 574–586 (2012)
10. Indyk, P.: Sparse recovery using sparse random matrices. In: López-Ortiz, A. (ed.) LATIN 2010. LNCS, vol. 6034, p. 157. Springer, Heidelberg (2010). doi:10.1007/978-3-642-12200-2_15
11. Lu, W., Li, W., Kpalma, K., et al.: Sparse matrix-based random projection for classification. Comput. Sci. **12**, 581–607 (2013)
12. Macenko, M., Niethammer, M., Marron, J., Borland, D., Woosley, J.T., Guan, X., Schmitt, C., Thomas, N.E.: A method for normalizing histology slides for quantitative analysis. In: International Symposium on Biomedical Imaging, vol. 9, pp. 1107–1110 (2009)
13. Otsu, N.: A threshold selection method from gray-level histograms. IEEE Trans. Syst. Man Cybern. **9**(1), 62–66 (1979)
14. Zhang, K., Crookes, D., Diamond, J., Fei, M., Wu, J., Zhang, P., Zhou, H.: Multi-scale colorectal tumour segmentation using a novel coarse to fine strategy. In: British Machine Vision Conference (BMVC) (2016)
15. Sirinukunwattana, K., Pluim, J.P., Chen, H., et al.: Gland segmentation in colon histology images: the glas challenge contest. Med. Image Anal. **35**, 489 (2016)
16. Cohen, A., Rivlin, E., Shimshoni, I., Sabo, E.: Memory based active contour algorithm using pixel-level classified images for colon crypt segmentation. Comput. Med. Imaging Graph. **43**, 150–164 (2015)
17. Gunduz-Demir, C., Kandemir, M., Tosun, A., Sokmensuer, C.: Automatic segmentation of colon glands using object-graphs. Med. Image Anal. **14**(1), 112 (2010)
18. Cheikh, B.B., Bertheau, P., Racoceanu, D.: A structure-based approach for colon gland segmentation in digital pathology. In: SPIE Medical Imaging, vol. 9791, p. 97910J (2016)

Surgical Timing Prediction of Patient-Specific Congenital Tracheal Stenosis with Bridging Bronchus by Using Computational Aerodynamics

Juanya Shen[1], Limin Zhu[1], Zhirong Tong[1], Jinfen Liu[1], Mitsuo Umezu[2], Zhuomin Xu[1(✉)], and Jinlong Liu[1,3(✉)]

[1] Department of Cardiothoracic Surgery, Shanghai Children's Medical Center, Shanghai Jiao Tong University School of Medicine, 1678 Dongfang Road, Shanghai 200127, China
zmxyfb@163.com, jinlong_liu_man@163.com
[2] Center for Advanced Biomedical Sciences, TWIns, Waseda University, 03C-301, ASMeW Lab 2-2 Wakamatsucho, Shinjuku, Tokyo 162-8480, Japan
[3] Institute of Pediatric Translational Medicine, Shanghai Children's Medical Center, Shanghai Jiao Tong University School of Medicine, 1678 Dongfang Road, Shanghai 200127, China

Abstract. Congenital tracheal stenosis (CTS) has a high clinical mortality in neonates and infants. Although the procedure of slide tracheoplasty (STP) applied over the years, it is still a challenge for clinicians to predict the surgical timing of the CTS correction. In the present study, we studied on three-dimensional (3D) aerodynamic analysis of an original tracheal model from a specific patient with CTS and bridging bronchus (BB) and four new reconstructed models. We constructed a 3D patient-specific tracheal model based on CT images and applied computer-aided design (CAD) to reconstruct four models to imitate the stenosis development of CTS. Average pressure drop (APD), wall shear stress (WSS) and velocity streamlines were calculated to analyze local aerodynamic characteristics for the evaluation of airflow at the inspiration phase and expiration phase, respectively. We found APD, WSS and AEL decreased during the respiration with the decrease of stenosis. Three abnormal gradients in APD were observed between the main stenosis of trachea arrived at 80% and 60%. This implied the surgical correction may be required when the main stenosis reached 60%. The combination of CAD and aerodynamic analysis is a potential noninvasive tool for surgical timing prediction in the management of patient-specific correction of CTS.

Keywords: Congenital tracheal stenosis · Computational fluid dynamics · Aerodynamics · Computer-aided design · Airflow

1 Introduction

Congenital tracheal stenosis (CTS) that would endanger respiratory function of children is a critical and life-threatening disease. In view of the multiformity of clinical

---

J. Shen and L. Zhu—Co-first author

© Springer Nature Singapore Pte Ltd. 2017
M. Fei et al. (Eds.): LSMS/ICSEE 2017, Part I, CCIS 761, pp. 181–190, 2017.
DOI: 10.1007/978-981-10-6370-1_18

manifestation, complexity of tracheal abnormalities and its rarity, a seasonable and efficient treatment of CTS is a great challenge to surgeons [1–3]. CTS with bridging bronchus (BB) is the most challenging type of CTS, which sometimes causes of high morbidity and mortality. At present, surgeons usually determine the surgical timing of CTS with BB depend on their own experience in clinical medicine and equipment inspection, which is hard to avoid the occurrence of misdiagnosis [3]. Recent years, the technology of image processing and numerical analysis [4] has played an increasingly important role in the areas of medical diagnoses and therapies and largely improves the possibility of pre-operative evaluation of aerodynamic outcomes of CTS [5–7]. Therefore, it is necessary to combine multiple disciplines to make a timely and effective approach for the patient with CTS.

In the present study, our research focused on the analysis of patient-specific tracheal aerodynamics. Three-dimensional 3D model with CTS before surgery was reconstructed. We defined it as the original model with 60% stenosis. According to the model, we imitated four stages of stenosis development (models with non-stenosis, 20% stenosis, 40% stenosis and 80% stenosis, respectively). Computational fluid dynamics (CFD) as a noteworthy method was used to simulate the tracheal aerodynamics. Average pressure drop (APD), wall shear stress (WSS) and velocity streamlines at the inspiration phase and expiration phase were calculated to evaluate the local aerodynamic characteristics for the analysis of the impairment of airflow in trachea. The major objection of the present study is to introduce our methods for the surgical timing prediction of CTS by computational aerodynamics.

2   Materials and Methods

2.1   Patient Information Acquisition

In the present study, a male little child was diagnosed with PAS which was associated with CTS, complete tracheal rings and BB, when he was 15-month-old. With the informed agreement from the parents and the local institutional review board and regional research ethics committee of Shanghai Children's Medical Center (SCMC) Affiliated Shanghai Jiao Tong University School of Medicine our research was agreed to be carried out.

As to image acquisition equipment, 16-slice multi-detector row enhanced CT scanner (Bright Speed Elite, GE Medical System, General Electric, America) was utilized to acquire the patient-specific CT images. Scanning parameters included the thickness of each slice 0.625 mm and CT image resolution 512×512 pixels. CT images were preserved in DICOM format.

2.2   Generation of Geometric Models

Medical imagining software, Materialise®-Mimics18.0, was used to compile and reconstruct the patient-specific tracheal model. By this step, a patient-specific model was visualized and transformed into 3D geometry from 2D images. After with an exact measurement in the original CT image, a 3D tracheal model of the 2-generation airway (bronchial tree) with high accuracy was reconstructed. The 3D tracheal model after surface smoothing is shown in Fig. 1, which can be observed clearly with three stenosis: Main Stenosis, Stenosis 1 and Stenosis 2. In the present study, we only considered the impact of changes in the geometric size of the main stenosis in trachea. We defined this patient-specific model as the original model, Model 2, and exported it in the most common geometric output format, stereo-lithography interface format (STL) for virtual design of different stages of stenosis development.

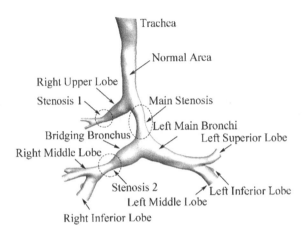

Fig. 1. Reconstruction of 3D patient-specific tracheal geometry

2.3   Model Rebuilding

The narrowest cross-sectional area of stenosis and the normal tracheal area can be determined by the imaging techniques. To calculate the stenosis ratio, we defined r as follows [8]:

$$r = \frac{S_N - S_S}{S_N} \times 100\% \tag{1}$$

S_N meant the normal tracheal area, and S_S expressed the narrowest cross-sectional area of main stenosis.

According to this definition, the stenosis ratio of patient's original model was about 60%. The original STL-formatted tracheal model was imported into computer-aided design (CAD) software Materialise®-3matic 10.0 and divided into three parts to geometrical design for the simulation of four stages of stenosis development as shown in Fig. 2. Then we modified the main stenosis to 0%, 20%, 40%, 80%, respectively to simulate four stages of stenosis development. Last, we integrated each part into a new model respectively. The Fig. 2 shows total five models in the present study: an original model (Model 2) with the stenosis ratio of 60%, Model 1 with the stenosis ratio of 80%, Model 3 with the stenosis ratio of 40%, Model 4 with the stenosis ratio of 20% and Model 5 with non-stenosis.

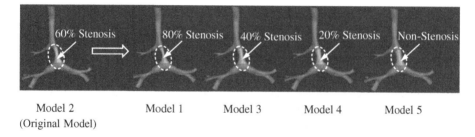

Model 2   Model 1   Model 3   Model 4   Model 5
(Original Model)

Fig. 2. Computer-aided design for model reconstruction

2.4   CFD Analysis

Governing equations of airflow. The airflow in trachea was assumed to be Newtonian and incompressible flow. Simulations were performed by solving the governing equations of the airflow including Navier-Stokes (N-S) and continuity equations [9], which were described as follows:

$$\frac{\partial}{\partial t}(\rho u_i) + \frac{\partial}{\partial x_j}(\rho u_i u_j) = -\frac{\partial p}{\partial x_i} + \frac{\partial}{\partial x_j}\left[u\left(\frac{\partial u_i}{\partial x_j} + \frac{\partial u_j}{\partial x_i}\right)\right] \tag{2}$$

$$\frac{\partial \rho}{\partial t} + \frac{\partial}{\partial x_j}(\rho u_j) = 0 \tag{3}$$

where ρ was the air density, μ was the viscosity, t was the time, u_i and u_j were the velocity vector of a point, p was the pressure, $i, j = 1, 2, 3$, x_i and x_j meant coordinate axes. We assumed the airflow with the constant density and viscosity in the present study ($\rho = 1.161$ kg/m^3 [4], $\mu = 1.864 \times 10^{-5}$ kg/m s [10]).

The Reynolds number, a dimensionless number, was used to characterize fluid flow, as defined in Eq. (4) [11]:

$$Re = \frac{\rho v d}{\mu} \tag{4}$$

where ρ and μ were air density and viscosity, v represented the velocity of airflow, and d was the characteristic length such as the diameter of the airway. We calculated the maximum Reynolds number was approaching to 3500. It suggested that the airflow in trachea should be the turbulence flow. In the present study, Wilcox k-ω model [12] was considered to obtain the turbulence viscosity for modeling the Reynolds stress, which was verified perfectly for airway flow simulations by CFD [13].

Mesh generation. In this step, we applied the grid-generation software, ANSYS®-ICEM 14.5, where five smoothed models with different stenosis ratio were imported. There were two types of grids in the fluid domain: three-layer body-fitted hexahedral grids and tetrahedral grids covering the remainder of the domain. Hereinto, an average nodal space of hexahedral grids increased by a ratio of 1.2. Grid independence verification was performed in order to find the optimal number of grids for the efficient calculation. Table 1 lists the number of total elements and total nodes for each model.

Table 1. Mesh information for each model

	Total elements	Total nodes
Model 1 (80% Stenosis)	1,207,424	453,344
Model 2 (60% Stenosis)	1,229,095	459,218
Model 3 (40% Stenosis)	1,093,932	394,447
Model 4 (20% Stenosis)	1,232,199	460,201
Model 5 (Non-stenosis)	1,242,588	463,383

Boundary conditions and calculations. To imitate more practical airflow in the physiological conditions of trachea, we considered that the inlet boundary was extended 20 times of its diameter to develop velocity profile. Due to the serious CTS, ventilator (Maquent, Servo-I, Sweden) was applied in this patient. Figure 3 showed the data of the mass-flow from ventilator after the amendment in one respiration cycle. In addition, the outlet boundaries for the tracheal domain were extended 40 times of each diameter to obtain sufficient pressure recoveries, and a constant pressure of 0 atm was set at each outlet [14]. The tracheal walls, including the extended boundary walls, were assumed to be rigid with no-slip conditions.

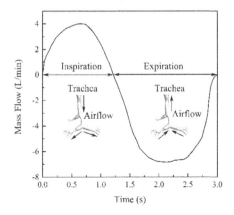

Fig. 3. Airflow mass flow rate in one respiration cycle

All of the above boundary conditions were set in the finite volume solver package, ANSYS®-FLUENT 14.5, which was utilized in simulate transient airflow of each tracheal model. The second-order upwind scheme and the semi-implicit (SIMPLE) method were chosen to solve the N-S equations. The convergence criteria were set to 10^{-5} for each time step.

3   Results

The stenosis ratios of Model 1, Model 2, Model 3, Model 4 and Model 5 were 80%,. 60%, 40%, 20% and 0%. Among these models, Model 2 with 60% stenosis was original patient-specific trachea model and Model 5 with non-stenosis was normal trachea model without CTS used as control group. In this study, average pressure drop (APD) of the main stenosis, wall shear stress (WSS) and streamlines of airflow in each tracheal model, were calculated to evaluate the outcomes of five models varying different degrees of stenosis.

3.1   Average Pressure Drop

Airway resistance and ventilation efficiency can be objectively measured by APD. Therefore, APD of each main stenosis at the phase of inspiration and expiration were respectively calculated in Fig. 4. Through comparisons of these values of each model's APD, it was obvious that APD decreased nonlinearly whether at the phase of inspiration or expiration with the decreasing degree of stenosis. And assuming under the same respiratory cycle, it is normal that the value of APD at the phase of inspiration was lower than that at the phase of expiration in each model. This indicated that its workload of expiration was heavier, and it's easier to lead to airway closure and airflow obstruction at this phase. Particularly there was a sharp decrease obtained between Model 1 and Model 2. And the change of the values of APD between two adjacent models was relatively steady in Model 3, Model 4 and Model 5.

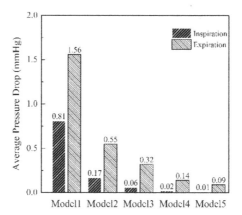

Fig. 4.   Average pressure drop at main stenosis at the phase of inspiration and expiration

3.2   Wall Shear Stress

WSS of tracheal wall was regarded as a measurement of the interaction between the air flow and the boundary of the airway. Many studies have demonstrated that much high WSS would damage to the tracheal wall and further cause many diseases. In Table 2, the closer the color of the tracheal wall was to the blue, the smaller the value of WSS was. If it is closer to the red, the value of WSS increased. As we can see, all the highest WSS of each model occurred in main stenosis. With the decrease of stenosis ratio, change of WSS between Model 1 and Model 2 is particularly evident, and there was no significant difference between other models. In comparison with expiration phase, WSS in inspiration phase was much lower.

Table 2. Wall shear stress (WSS) and aerodynamic analysis (streamlines)

	WSS		Streamline	
	Inspiration	Expiration	Inspiration	Expiration
Model 1	Airflow	Airflow	Airflow	Airflow
Model 2	Airflow	Airflow	Airflow	Airflow
Model 3	Airflo	Airflow	Airflow	Airflow
Model 4	Airflow	Airflow	Airflow	Airflow
Model 5	Airflow	Airflow	Airflow	Airflow
Legend	WSS (Pa) 0 1 2 3 4 5 6		Velocity (m/s) 0 4 8 12 16 20	

3.3 Streamlines

Table 2 displayed the streamlines at the maximum velocity of airflow at the inspiration phase and expiration phase as well. The higher value of velocity was observed in the main stenosis of the model 1. With the stenosis decreased, the velocity of airflow in the main stenosis slowed down. Accompanied by an increase of stenosis at the phase of inspiration, the size of turbulence became lager and the velocity of airflow was very small, which indicated resistance in the area of BB was getting stronger. And large rotating flow of each model was created area of BB to trachea at the phase of expiration. The velocity varied from Model 1 to Model 2 significantly, which was evidenced from another side to indicate it is critical timing for surgical correction.

4 Discussion

CTS is one of congenital tracheal diseases characterized by uncommon, complex, and potentially life-threatening, which could lead to pulmonary infection and respiration insufficiency. In the wake of the development in science and technology, the survival of CTS has greatly improved. One reason may be that the methods of diagnosis are becoming more diverse and accurate, such as bronchoscopy, bronchography and multi-detector computed tomography (MDCT) [15, 16], which could only provide the static imagine modalities of the main airway and is unavailable to the dynamic information. Surgeons usually depend on the results of above methods of diagnosis and their own previous experience to choose the surgical timing for the children with CTS. However, the optimal and exact timing for surgical intervention is still controversial. If it can be predicted, the operative survival rate of CTS would be definitely greatly improved.

In the present study, we utilized medical imagining software to generate 3D patient tracheal model with CTS from high resolution CT 2D images to 3D models, and CAD to rebuild four models with different stenosis based on the original 3D stenosis model. The aerodynamics of five models was calculated by using CFD. Several parameters including APD, WSS and streamlines were calculated to estimate aerodynamics of five models. Airway resistance and efficiency of ventilation can be measured objectively by APD. In the outcomes of APD, the general trend was that the values of APD declined as the degree of stenosis decreased, showing a nonlinear reducing. It was a direct proof of the degree of stenosis on the APD great impact and reasonable that airway resistance became lower as well as ventilation efficiency was getting higher with decrease of the stenosis ratio. The stenosis ratio was positively correlated with APD and the value of APD when in an inspiration phase was lower than that when in an expiration phase in each model, which were consistent with clinical practice. Moreover, there was a sharp decrease obtained between Model 1 and Model 2. Abnormal increase of APD disclosed that normal work of respiratory muscle can't meet the physiological needs. It would lead the patient to have symptoms like breathing difficulties, wheezing or even suffocation. It suggested surgical correction was necessary for the investigated patient with 60% tracheal stenosis.

WSS was regarded as a measurement of an interaction between the moving fluid and the solid wall, which was displayed in Table 2. High value of WSS appeared in main stenosis parts, which would cause additional destruction of the bronchial epithelium. Through the longitudinal observation from top to bottom, WSS in main stenosis gradually decreased both in the inspiratory phase and expiration phase. As stenosis ratio declined, airflow resistance became lower, which led to decreases WSS and strengthened interaction between the airflow and the wall of trachea in two phases, respectively. It was observed that WSS of main stenosis in the inspiratory phase were lower respectively than that at the expiration phase due to higher workload in the expiration phase. Similarly, change of WSS between Model 1 and Model 2 is particularly evident. The velocity varied from 80% stenosis to 60% stenosis significantly along with the decrease of stenosis ratio, which was evidenced that 60% stenosis is an inflection point for this patient from another side. Therefore, we should attach importance to this critical inflection point that can predict the critical timing of surgery.

5 Conclusion

Development of tracheal stenosis have great effects on the outcomes of aerodynamics. The combination with the technique of CFD and CAD is a potential tool for surgical timing prediction to minimize the surgical risk. APD, WSS, and streamlines are critical parameters for evaluation aerodynamics characteristics of airflow in trachea.

Acknowledgement. We have been genuinely appreciative of the support of the National Nature Science Foundation of China (No. 81602818, P.I.: Limin Zhu and No. 81501558, P.I.: Jinlong Liu), the Project-sponsored by the Scientific Research Foundation for the Returned Overseas Chinese Scholars, State Education Ministry (No. 20144902, P.I.: Jinlong Liu), the Fund of The Shanghai Committee of Science and Technology (No. 15411967100, P.I.: Limin Zhu and No. 14411968900, P.I.: Jinlong Liu) and the Biomedical and Engineering (Science) Inter-disciplinary Study Fund of Shanghai Jiaotong University (No. YG2014MS63, P.I.: Jinlong Liu).

References

1. Chung, S.R., Yang, J.H., Jun, T.G., et al.: Clinical outcomes of slide tracheoplasty in congenital tracheal stenosis. Eur. J. Cardio-Thorac. Surg. **47**(3), 537–542 (2015)
2. Grillo, H.C., Wright, C.D., Vlahakes, G.J., et al.: Management of congenital tracheal stenosis by means of slide tracheoplasty or resection and reconstruction, with long-term follow-up of growth after slide tracheoplasty. J. Thorac. Cardiovasc. Surg. **123**(1), 145–152 (2002)
3. Hofferberth, S.C., Watters, K., Rahbar, R., et al.: Management of congenital tracheal. Stenosis. Pediatr. **136**(3), e660–e669 (2015)
4. Mimouni-Benabu, O., Meister, L., Giordano, J., Fayoux, P., Loundon, N., et al.: A preliminary study of computer assisted evaluation of congenital tracheal stenoses: a new tool for surgical decision-making. Int. J. Pediatr. Otorhinolaryngol. **76**, 1552–1557 (2012)
5. Chen, F.L., Horng, T.L., Shih, T.C.: Simulation analysis of airflow alteration in the trachea following the vascular ring surgery based on CT images using the computational fluid dynamics method. J. Xray Sci. Technol. **22**, 213–225 (2014)
6. Cebral, J.R., Summers, R.M.: Tracheal and central bronchial aerodynamics using virtual bronchoscopy and computational fluid dynamics. IEEE Trans. Med. Imaging **23**(8), 1021–1033 (2004)

7. Gemci, T., Ponyavin, V., Chen, Y., et al.: Computational model of airflow in upper 17 generations of human respiratiory tract. J. Biomech. **41**(9), 2047–2054 (2008)
8. Liu, J.L., Itatani, K., Shiurba, R., et al.: Image-based computational hemodynamics of distal aortic arch recoarctation following the Norwood procedure. In: International Conference on Biomedical Engineering and Informatics. IEEE, 318—323. (2011)
9. Detta, R.A.K., Ducharme, N.G., Pease, A.P.: Simulation of turbulent airflow using a CT based upper airway model of a racehorse. J. Biomech. Eng. **130**(3), 13 (2008)
10. Chien-Yi, H., Liao, H.M., Tu, C.Y., et al.: Numerical analysis of airflow alteration in central
11. Chang, H.K., Mortola, J.P.: Fluid dynamic factors in tracheal pressure measurement. J. Appl. Physiol. Respir. Environ. Exerc. Physiol. **51**(1), 218–225 (1981)
12. Wilcox, D.C.: Reassessment of the scale determining equation for advanced turbulence models. J. Am. Inst. Aeronaut. Astronaut. **26**(11), 1299–1310 (1988)
13. Mylavarapu, G., Murugappan, S., Mihaescu, M., Kalra, M., Khosla, S., Gutmark, E.: Validation of computational fluid dynamics methodology used for human upper airway flow simulations. J. Biomech. **42**, 1553–1559 (2009)
14. Zhu, L., Liu, J., Zhang, W., et al.: Computational aerodynamics of long segment congenital tracheal stenosis with bridging bronchus. In: Asian Control Conference, pp. 1–5. (2015)
15. Baden, W., Schaefer, J., Kumpf, M., Tzaribachev, N., Pantalitschka, T., Koitschev, A., Ziemer, G., Fuchs, J., Hofbeck, M.: Comparison of imaging techniques in the diagnosis of bridging bronchus. Eur. Respir. J. **31**(5), 1125–1131 (2008)
16. Zhong, Y.M., Jaffe, R.B., Zhu, M., Gao, W., Sun, A.M., Wang, Q.: CT assessment of tracheobronchial anomaly in left pulmonary artery sling. Pediatr. Radiol. **40**(11), 1755–1762 (2010)

Finite Element Analysis and Application of a Flexure Hinge Based Fully Compliant Prosthetic Finger

Suqin Liu[1], Hongbo Zhang[1(✉)], Ruixue Yin[1], Ang Chen[2], and Wenjun Zhang[1,2(✉)]

[1] Complex and Intelligent Research Center, East China University of Science and Technology, Shanghai, People's Republic of China
morgenvera@163.com, wjz485@mail.usask.ca
[2] Department of Mechanical Engineering, University of Saskatchewan, Saskatoon, Canada

Abstract. Prosthetic hand is usually made by rigid body mechanism with ropes and pulleys. Such a hand is not "soft" to patients or to objects to be manipulated by the hand. In this paper, the concept of compliant mechanism is applied to prosthetic finger. The main challenge in designing and constructing such a finger lies in the design of flexure hinge. First, a fully compliant finger with a monolithic structure and flexure hinge was built. Then, finite element analysis for the compliant finger was implemented, and the results were compared with the experimental result to verify the design. Finally, the complaint finger was applied in a prosthetic hand design and worked excellent with the hand.

Keywords: Flexure hinge · Compliant finger · FEA · Prosthetic hand

1 Introduction

Human hand is a crucial tool for perceiving and manipulating objects in the external word [1]. When a hand is lost, there is a high desire to have a replacement, which is called prosthesis. The general function requirement (FR) of a prosthesis hand is of course the same function of a real hand, and the general condition requirement (CR) of a prosthesis hand is that a prosthesis hand should resemble a real hand (CR1) and should not interfere with other organs of the body (CR2).

The conventional prosthetic finger is built with "rigid" parts or links connected through revolute joints such as pins [2–9]. Rigid mechanisms have some disadvantages: (1) additional efforts to assemble parts together. (2) additional efforts to maintain the hand. (3) friction is presented in kinematic joints, resulting in wear in structure, backlash and damping in movement. (4) imprecision in motion and force transfer. Unlike rigid mechanisms, a compliant mechanism obtains at least one of its mobility from the deformation of its flexible members [10], transferring input motion/force to output motion/force.

Figure 1(a) shows a rigid four-bar mechanism, where all the links are rigid and connected through pin joints. In Fig. 1(b), short thin compliant segments (i.e., compliant joints or flexure hinges) are used to form a compliant mechanism. Figure 2(a) shows a fully compliant mechanism, while Fig. 2(b) is a partially compliant mechanism [2].

© Springer Nature Singapore Pte Ltd. 2017
M. Fei et al. (Eds.): LSMS/ICSEE 2017, Part I, CCIS 761, pp. 191–198, 2017.
DOI: 10.1007/978-981-10-6370-1_19

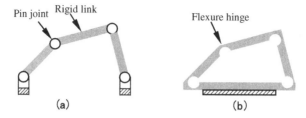

Fig. 1. Four-bar mechanism: (a) four-bar rigid mechanism, (b) four-bar compliant mechanism [2]

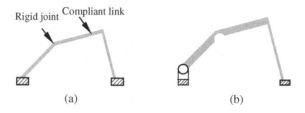

Fig. 2. Four-bar compliant mechanism: (a) four-bar fully compliant mechanism, (b) four-bar partially compliant mechanism [2]

A monolithic structure in the prosthetic finger, where a fully compliant mechanism is used, proves to be preferable due to its softness [11] as well as low cost in manufacturing. To the best of our knowledge, fully compliant mechanisms with a monolithic structure and lumped compliance were only found in mechanical grippers [12, 13] but not in prosthesis hands.

Based on our previous study [14], a compliant finger based on flexure hinges has been established. The main work of the previous paper 'Flexure hinge based fully compliant prosthetic finger' were as follows: (1) the design process for a fully compliant prosthetic finger based on cylinder body flexure hinges was claimed; (2) an evaluation system for the conceptual level design of the finger was developed. In this paper, based on the established flexure hinge model, further study was carried on to simulate on the complaint finger and apply the finger in a prosthetic hand design. The remainder of the paper is organized as follows. In Sect. 2, we outline the design process of a compliant finger with flexure hinges based on human anatomy. In Sect. 3, modeling and analysis of FEM are presented, and the simulation results are compared with the experiment results. The application of the complaint finger in a prosthetic hand prototype is also included. Section 4 gives a conclusion with discussion of future work.

2   Finger Model and Finite Element Analysis

2.1   Compliant Finger Model

For finger modeling, GB-T 10000-1988 and GB-T 16252-1996 standards were referred. The finger measurement data include length of finger, PIP and DIP as well as width of

PIP and DIP. It is necessary to simplify the parameters in the actual modeling process. The simplifying process was as follows: (1) ignore the changes of finger's thickness and take the width of PIP as the finger's width; (2) ignore the size differences among the 5 fingers and set the sizes of five fingers the same. Finally, the modeling data for the finger was: length 72 mm and diameter 17 mm. According to the human anatomy, the adjacent segments length ratio of each finger is about 0.618 [15] and they meet the Fibonacci sequence, see Fig. 3(a). A finger prototype was built based on this rule, and deep ellipse flexure hinge was used in the design, see Fig. 3(b). Figure 3(c) shows the printed finger.

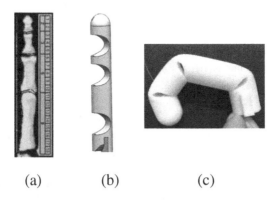

(a)   (b)   (c)

Fig. 3. Finger's model and prototype: (a) relation of finger segments' length, (b) finger model based on Ergonomics, (c) printed finger with deep elliptic flexure hinge

2.2 Finger Movement Patterns

There are two kinds of movement modes when human finger bends, as shown in Fig. 4. Figure 4(a) is free bending movement, where the joint angles ψ_1 and ψ_2 satisfy $\psi_1:\psi_2$ = constant. This mode can complete the fingertip grasp and side griping tasks. Figure 4(b) and (c) are finger's enveloped movements, and the enveloping motion is mainly used for adaptive grasping of irregularly shaped objects. For envelope motion, the joint angles ψ_1 and ψ_2 depends on the surface shape of the grasped object. The angle relation in Fig. 4(b) satisfies $\psi_1 < \psi_2$, and Fig. 4(c) satisfies $\psi_1 > \psi_2$ [16].

Like human fingers movement patterns, the compliant finger's bending process can be seen in Fig. 5. Figure 5(a) was the initial state, where the finger was relaxed with slightly curve. After force was applied, the PIP began to bend (Fig. 5(b)). Then the PIP sharply curved and DIP was less bended. Finally, DIP sharply bended. The bending pattern of the compliant finger was similar to that of human finger, which can realize the envelope movement and adaptive grasping.

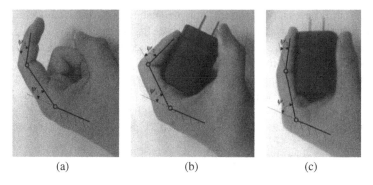

(a)   (b)   (c)

Fig. 4. Finger movement patterns: (a) free motion ($\psi_1:\psi_2$ = constant), (b) enveloping motion ($\psi_1 < \psi_2$), (c) enveloping motion ($\psi_1 > \psi_2$)

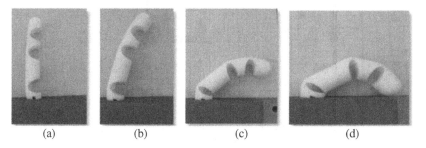

(a)   (b)   (c)   (d)

Fig. 5. Finger bending process: (a) the initial state, (b) PIP begins to bend first, (c) PIP is sharply bended and DIP is less bended, (d) DIP sharply bends

The proposed compliant finger had the following advantages: (1) manufacturing and assembly process simple; (2) good bending performance; (3) low weight; (4) bending pattern similar to human finger.

2.3   FEA and Experiment

As an efficient numerical calculation and analysis method, finite element analysis has become an important method of innovative research and design in engineering field. In this paper, ANSYS 14.0 will be used to simulate the finger model, analyze the stress, strain and displacement in each direction when the finger is bent. Comparing with the experimental results, the simulation results can help with future model optimization.

To decrease the modeling difficultly, finger constraints and load application method need to be simplified in FEM: (1) Assuming that the force point is concentrated at the rope hole of the DIP when the finger is pulley by the rope; (2) Assuming that PIP remains fixed when the finger is bent. Based on the above simplifications, the finite element simulation of finger bending process is as follows:

1. Export the .x_t finger model file from Solidworks and then import it into ANSYS to get the geometry model;

2. Add material: Add PolyFlex™ performance parameters to the library and assign it to the finger model;
3. Define contact: set the bottom of the finger to a fully fixed constraint;
4. Mesh: Select the program automatically divide the grid, the grid size is set to 1 mm;
5. Apply force: apply a concentration force at top of DIP. 5 groups of simulation were made with forces of 0.5 N, 1 N, 1.5 N, 2 N, 2.2 N (2.2 N was the maximum force needed to fully bend the finger);
6. Solve and result analysis.

3   Results and Discussions

3.1   FEA and Experiment Results

Figures 6 and 7 show the simulation results of finger displacement and stress under 1 N concentration force. Figure 6 showed the compliant finger's overall displacement and X, Y, Z direction of sub-displacement. According to the direction of the displacement vector and FEM analysis results, DIP produced the largest displacement and PIP produced the smallest displacement. Figure 7 presented the finger stress distribution status, according to the stress analysis results, the maximum stress located in the finger's hinges with stress concentration phenomenon, but still within the safety allowable range.

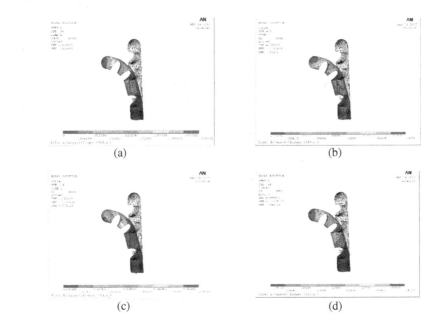

(a)   (b)

(c)   (d)

Fig. 6. Finite element result of finger's displacement: (a) displacement vector sum, (b) X-component of displacement, (c) Y-component of displacement, (d) Z-component of displacement

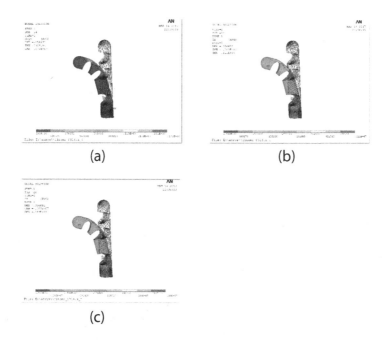

Fig. 7. Finite element result of finger's stress: (a) von Mises Stress, (b) X-component of stress, (c) Y-component of stress

In the finite element analysis, finger displacement can be obtained by changing the applied load. In order to verify the simulation results, the flexure hinge performance experiments were repeated for the compliant finger with 0.5 N, 1 N, 1.5 N, 2 N, and 2.2 N tension force, and five corresponding actual force-displacement relationship can be obtained. The force-displacement curves for experiment and simulation can be obtained by inputting the displacement vectors into Origin, as shown in Fig. 8. It can be seen that there were some errors between experiment and simulation results, and maximum error was 19.68%. But the overall trend was consistent, which verified the accuracy of the simulation model and method.

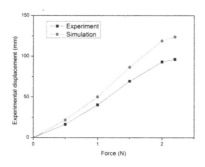

Fig. 8. Displacement of experiment and simulation result

3.2   Compliant Finger in Prosthetic Hand

Compliant finger with flexure hinge has been used in our prosthetic hand, see Fig. 9. The finger has proved to own excellent bending performance and easily manufacturing properties with low light. Adaptive grasping can be realized with this finger, and small, medium or large objects can be easily manipulated with this design.

Fig. 9. Grasping experiments of various sizes of objects with complaint finger

4   Conclusion and Future Work

This paper proposed a fully compliant finger with a monolithic structure based on CB flexure hinge. A soft material PolyFlex™ was used to fabricate the model with 3D printing technology. Various flexure hinges were designed, and the bending properties of them were compared. Then an evaluation system was brought up, and bending performance, appearance and mass were considered three evaluation criteria with their own weights. Deep elliptic flexure hinge was found to be the best with the score 4.4 of 5. Later, we applied the conclusion to a prosthetic finger design. FEA and experiment verification were implemented. Then the compliant finger was applied in a prosthetic hand with excellent performance.

Future work will include considering the flexure hinge's rebounding time as an evaluation criteria. Furthermore, FEA simulation can be improved by optimizing the model.

References

1. Koganezawa, K., Ito, A.: Artificial hand based on the planetary gear system-realization of daily utility motion of a hand with minimum actuators. In: 2013 IEEE International Conference on Mechatronics and Automation (ICMA), pp. 645–650 (2013)
2. Cao, L.: On advancing the topology optimization technique to compliant mechanisms and robots. Ph.D., University Saskatchewan, Saskatoon, Canada (2015)
3. Laliberté, T., Birglen, L., Gosselin, C.: Underactuation in robotic grasping hands. Mach. Intell. Robot. Control **4**, 1–11 (2002)

4. Dollar, A.M., Howe, R.D.: The SDM hand as a prosthetic terminal device: a feasibility study. In: IEEE 10th International Conference on Rehabilitation Robotics, ICORR 2007, pp. 978–983 (2007)

5. Carrozza, M.C., Suppo, C., Sebastiani, F., Massa, B., Vecchi, F., Lazzarini, R., et al.: The SPRING hand: development of a self-adaptive prosthesis for restoring natural grasping. Auton. Robots **16**, 125–141 (2004)

6. Hirose, S., Umetani, Y.: The development of soft gripper for the versatile robot hand. Mech. Mach. Theory **13**, 351–359 (1978)

7. Fukaya, N., Toyama, S., Asfour, T., Dillmann, R.: Design of the TUAT/Karlsruhe humanoid hand. In: Proceedings of 2000 IEEE/RSJ International Conference on Intelligent Robots and Systems, IROS 2000, pp. 1754–1759 (2013)

8. Montambault, S., Gosselin, C.M.: Analysis of underactuated mechanical grippers. J. Mech. Des. **123**, 367–374 (2001)

9. Kyberd, P., Pons, J.: A comparison of the oxford and manus intelligent hand prostheses. In: Proceedings of IEEE International Conference on Robotics and Automation, ICRA 2003, pp. 3231–3236 (2003)

10. Kozuka, H., Arata, J., Okuda, K., Onaga, A., Ohno, M., Sano, A., et al.: A compliant-parallel mechanism with bio-inspired compliant joints for high precision assembly robot. Procedia CIRP **5**, 175–178 (2013)

11. Howell, L.L.: Compliant Mechanisms. Wiley, Hoboken (2001)

12. Boudreault, E., Gosselin, C.M.: Design of sub-centimetre underactuated compliant grippers. In: ASME 2006 International Design Engineering Technical Conferences and Computers and Information in Engineering Conference, pp. 119–127 (2006)

13. Doria, M., Birglen, L.: Design of an underactuated compliant gripper for surgery using nitinol. J. Med. Devices **3**, 011007 (2009)

14. Liu, S.Q., Zhang, H.B., Yin, R.X., Chen, A., Zhang, W.J.: Flexure hinge based fully compliant prosthetic finger. In: 2016 SAI Conference (2016)

15. www.goldennumber.net/human-hand-foot/

16. Sheng, X., Hua, L., Zhang, D., et al.: Design and testing of a self-adaptive prosthetic finger with a compliant driving mechanism. Int. J. Humanoid Rob. **11**(3), 1450026 (2014)

Improvement of Acoustic Trapping Capability by Punching Specific Holes on Acoustic Tweezers

Haojie Yuan and Yanyan Liu[✉]

School of Mechatronic Engineering and Automation, Shanghai University,
Shanghai 200072, China
yyliu2014@shu.edu.cn

Abstract. It is found that small particles can be successfully manipulated by the acoustic tweezers. This paper presents a method to improve the acoustic trapping capability by punching specific round holes on two vibrating V-shaped metal strips of the acoustic tweezers. A particle is trapped under the sharp edges of metal strips with some specific round holes. Its trapping capability is improved under certain conditions compared with the original acoustic tweezers. A finite element model is developed to calculate the acoustic radiation force. The effects of the radius, the number and the arrangement of the round holes on the acoustic radiation force on the top surface of the particle are discussed. It is found that the acoustic radiation force increases obviously when the radius of the hole is more than a certain magnitude by changing the vibrational mode of the acoustic tweezers. With the increase of number and the row in vertical direction of the round holes, the acoustic radiation force acting on the particle increases correspondingly.

Keywords: Acoustic radiation force · Acoustic manipulation · Acoustic tweezers · Finite element method

1 Introduction

The acoustic forces excited by kilohertz or megahertz frequency can reach a magnitude that is in the range of small particle's weight, so that these can be used to manipulate the particles. Acoustic manipulation technology has practical applications in bioengineering [1], nanofabrication [2], energy efficient systems and clean utilization of coal [3], with the advantages of less damage and selectivity to samples [4, 5] and simple device structures. The acoustic radiation force used to collect, separate and transport particles is generated from the spatial non-uniformity of the kinetic and potential energy density around manipulated samples. The different motivation models and device structures utilized to generate acoustic radiation forces have been proposed and investigated by many authors until the present day. Wu [6] presented acoustical tweezers using two collimated focused ultrasonic beams propagating along opposite direction, which can trap latex particles of 270 μm diameter and clusters of frog eggs. S. B. Q. Tran [7] developed an acoustical tweezers using a slight frequency modulation of two ultrasound emitters to handle particles in a very controlled manner. Courtney [8] used an

© Springer Nature Singapore Pte Ltd. 2017
M. Fei et al. (Eds.): LSMS/ICSEE 2017, Part I, CCIS 761, pp. 199–207, 2017.
DOI: 10.1007/978-981-10-6370-1_20

electronically controlled acoustic tweezer consisting of a circular 64-element ultrasonic array to trap 45 and 90-μm-diameter polystyrene spheres in larger regions. Kang [9] proposed an acoustics-vortex-based trapping model of acoustic tweezers and the maximum trapping force acting on a 13-μm polystyrene sphere in the produced acoustic vortex was 50.0 pN. Lee [10] calculated acoustic radiation force on a lipid sphere in a 100-MHz focused Gaussian field to demonstrate the acoustic tweezer effect near the focus. Shi [11] presented an active patterning technique of acoustic tweezers using standing surface acoustic wave (SSAW) to manipulate and pattern cells and micro-particles. Hu [12] proposed an ultrasonic transducer to collect small particles by acoustic radiation surface of shaped ultrasonic tweezers. Other significant work in this area was reported [13–16].

In the above listed research, the acoustic trapping capability in various acoustic tweezers has been proposed and discussed. In order to manipulate heavier particle, the greater acoustic radiation force is required to be generated in practical applications. In this work, we proposed a method to increase the acoustic radiation force by optimizing a simple structure with two vibrating V-shaped metal strips [17]. The acoustic radiation force is generated by the leakage of a standing wave ultrasonic field in a triangular air gap and is opposite to the acoustic leakage direction. By punching symmetrical round holes at the bottom of the metal strips, it is found that the acoustic trapping capability increases with the change of ultrasonic field compared to the original structure. To calculate the acoustic radiation force with different strips situations in a complicated sound field, the analysis of finite element method (FEM) combined with the theory of acoustic radiation force is applied. The effects of the radius of holes, the position of holes at the bottom of the V-shaped strips and the arrangement of holes are clarified in detail.

2   Computational Model and Analysis Method

The original structure of the acoustic tweezers proposed by the authors' research group [18] is shown in Fig. 1. The metal strips made of aluminum have the shape and size shown in Figs. 1(a) and (b).

In the experiment, two identical aluminum strips are clamped to a Langevin trans-ducer through a 10-mm-diameter hole in the upper part of the strips. The upper part is a rectangular aluminum plate with a size of $40 \times 45 \times 1.5$ mm. The lower part of the structure is a V-shaped aluminum strip with a length of 99 mm, width of 22.5 mm and thickness of 1.5 mm at the top. The lower part aluminum strip is tapered off from the upper end to the lower end and the thickness of strip at the tip is around 200 μm. As a consequence, a triangular air gap is formed between the two V-shaped strips with a thickness of 1.3 mm at the tip. A flexural vibration is excited in the aluminum strips by the ultrasonic transducer with a resonance frequency of 25.3 kHz. An acoustic radiation force used to suck the particles to the lower end of strips is generated by the sound filed near the lower end of the gap, which is formed by the flexural vibration [18].

To maximize the trapping capability, a round hole which is located at the center of the bottom of each metal strip is punched on each strip by changing the vibration mode, compared to the original structure. As shown in Fig. 1(c), it is 1 mm from the bottom of

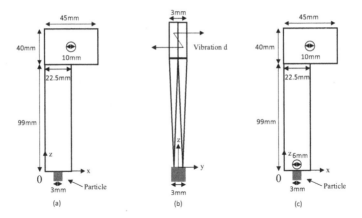

Fig. 1. Structure and size of acoustic tweezers. (a) Shape and size of the metal strip of original acoustic tweezers. (b) Air gap formed by the two V-shaped strips. (c) Shape and size of improved acoustic tweezers.

round hole with a radius of 3 mm to the bottom edge of the aluminum strip. The center of the hole is on the central line of the lower part of the metal strip in xz-plane.

The following integration over the surface of the object is used to calculate the acoustic radiation force F on a rigid immovable object in an ultrasonic field in ideal fluid [19].

$$F = \left\langle \iint_S (K - U)nsdS \right\rangle. \tag{1}$$

where the notation $\langle \ \rangle$ denotes time average over one period, K is the kinetic energy destiny, U is the potential energy density and $\mathbf{n}$ is the outward normal unit vector of the surface. The following formulas are used to calculate the kinetic and potential energy densities K and U [20].

$$K = \frac{\rho_0 v^2}{2}. \tag{2}$$

$$U = \frac{p^2}{2\rho_0 c_0^2}. \tag{3}$$

where ρ_0 and c_0 are the density of and sound speed in the fluid, v is the velocity and p is the sound pressure.

Our computation of acoustic radiation force is implemented by the FEM software COMSOL Multiphysics for the sound field surrounding the $3 \times 3 \times 3$ mm cube particle under the two vibrating sharp edges in air. The excitation frequency f applied to the structure model is 25.3 kHz, which is the resonance frequency of the transducer in the experiment. A prescribed displacement condition is added to the upper part of metal plates and then the amplitude of the y-direction vibration displacement (0-peak) d is set to 10 μm. In the computational model, an air domain with a size of $25 \times 20 \times 15$ mm is added as the sound field, where all the sound filed boundaries are radiation boundary. A cube particle with six sound hard boundaries is located below the bottom vibrating edges of the metal strips and the distance between the upper surface of the cube particle and the strip edges is 0.05 mm. Figure 2(a) shows the mesh of the structure mode in 3-D FEM calculation, where the minimum mesh size is 0.05 mm and the maximum mesh size at the surface boundaries is 0.8 mm, which is around 0.58% of the wavelength.

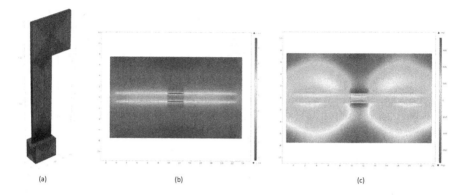

(a)   (b)   (c)

Fig. 2. 3-D FEM analyses of the structure model with a 3-mm-radius hole at the bottom area of the vibrating strips in air. (a) Mesh of the structure model. (b) The y-direction velocity on top surface of the particle. (c) The sound pressure on top surface of the particle.

The kinetic energy and potential energy of the original structure model with no round holes are calculated as the reference data. The computation process consists of two steps. In the first step, the velocity and sound pressure on top surface of the particle are solved by the acoustic module of the software shown in Figs. 2(a) and (b). In the second step, the kinetic and potential energy K and U is calculated by the Eqs. (2) and (3) through the post processing function of the software, which generate the acoustic radiation force. From the 3-D FEM results, it is known that $\iint_S \langle K \rangle dS$ and $\iint_S \langle U \rangle dS$ on the top surface of the cube particle are calculated to be 6.47×10^{-4} N and 1.87×10^{-5} N, respectively. The acoustic radiation force acting on a cube particle is determined by the force on the top surface of the particle and pointing upward, as the kinetic energy and potential energy on the side and bottom surfaces are less than 1% of that on the top surface.

3   Results and Discussion

As the cube particle is trapped at the center of the metal strip's edge in xz plane, the effect of radius of a single round hole on the acoustic radiation force is investigated. The hole's radius is ranging from 1 mm to 6 mm, which is on the center line of metal plates with 1 mm away from the bottom edges as shown in the Figs. 3(a) and (b). From the FEM computational results, it is found that the kinetic energy is almost 10 times the potential energy therefore the acoustic radiation force is mainly determined by the velocity on the top surface of the particle. If the vibrational velocity at the bottom edges of the metal strips increases, the acoustic radiation force is also strengthened. The y-direction velocities at the center line on inner side of one metal plate is shown in Fig. 3(c) when the radius of hole is less than 2.5 mm, and the voids in the figure indicate the position of hole on the strip. It is seen with the increase of the radius of the hole, the velocity at bottom edges decreases correspondingly. The computational acoustic radiation force F of acoustic tweezers with a single hole of radius less than 4.5 mm is shown in Fig. 3(d). It is found that by punching a small hole (radius less than 4.5 mm) in the bottom area of strips can't increase the acoustic trapping capability and even decrease the radiation force. Figure 3(e) shows the y-directional velocities at the center line when the radius of the hole is larger than 5 mm. The vibrational mode changes significantly compared with Fig. 3(c). In the radius range of 5 ~ 6 mm, the acoustic radiation force is strengthened compared with the original acoustic tweezers, and it decreases as the radius increasing shown in Fig. 3(f). The acoustic radiation force reaches its maximum when the hole's radius is 5 mm, which is 4 times of that with no hole on the strip. The

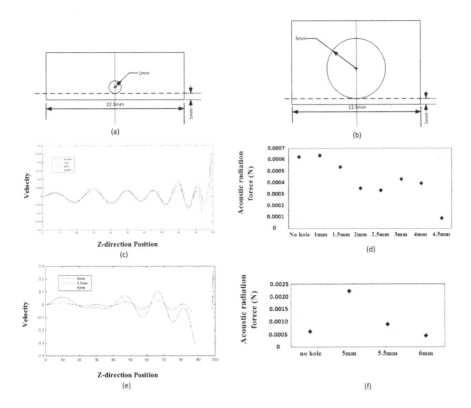

Fig. 3. The effect of hole's size on the vibration velocity and acoustic radiation force. (a - b) Position and size of a single hole. (c) The y-direction velocities at the center line of metal plates when the hole's radius is less than 2.5 mm. (d) The acoustic radiation force with hole's radius less than 4.5 mm. (e) The y-directional velocities at the center line with hole's radius larger than 5 mm. (f) The acoustic radiation force when the hole's radius is larger than 5 mm.

original acoustic tweezers is in flexural vibration. This vibration changes to be coupling with the radial vibration when punching a hole on the metal strip. The change in vibration mode causes the change of resonance of the acoustic tweezers, which leads to change in acoustic radiation force.

The effect of the number of the round hole on the acoustic radiation force is discussed in the Fig. 4. Two holes are symmetrically located about the center line of the lower metal plates and 1 mm away from each other as shown in Fig. 4(a). The vibrational modes and velocities of acoustic tweezers with two holes with radius more than 3.5 mm are shown in Fig. 4(b). The acoustic radiation force acting on the upper surface of the cube particle under two vibrating shape edges of the strips with two round holes is calculated by FEM software shown in Fig. 4(c). The radii of the round hole are all larger than 3.5 mm. The acoustic radiation force doesn't increase when the radius of hole is less than 3.5 mm compared with no holes case. In the condition that the radii of holes are 3.5 mm ~ 4.25 mm, the acoustic radiation forces generated by the two sharp edges of the strips with two round holes are obvious larger than that by the structures with only one hole and no hole. The acoustic radiation force reaches its maximum when the radii of two holes are 3.5 mm, which is 5.7 times of that with one hole with the radius of 5 mm on the strip, and 20 times of that with original acoustic tweezers.

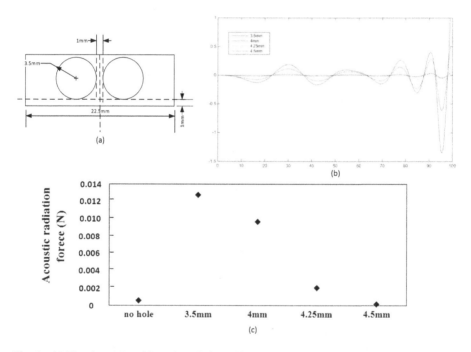

Fig. 4. (a) The size and position of two holes with radii of 3.5 mm. (b) The y-direction velocities at the center line of metal plates when the radii of two holes are more than 3.5 mm. (c) The calculated acoustic radiation force when the radii of two holes are more than 3.5 mm.

The effect of hole array on the acoustic trapping capability is also discussed in our works shown in Fig. 5. There are seven 1 mm-radius holes in a row on the bottom area of the metal strips and every single row is 1 mm away from the adjacent row shown in Fig. 5(a). The acoustic radiation forces is calculated for the different conditions with one, two and three rows of hole array on the strip, which is shown in Fig. 5(b). Adding more rows of round holes in the bottom area of metal strips can increase the velocity of the metal edges shown in Fig. 5(c), which leads to the significant improvement of the acoustic trapping capability. It is seen that from row number one to three, the acoustic radiation force on top surface of a particle increases correspondingly.

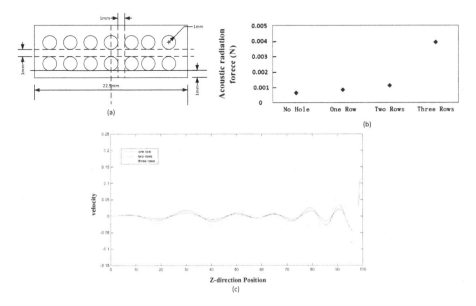

Fig. 5. (a) The size and position of two rows of hole array with radii of 1 mm. (b) The calculated acoustic radiation forces under the conditions of one, two and three rows of hole array on the metal strips. (c) The velocities of the center line with different rows of holes.

4   Conclusion

In summary, we proposed and explored a method to improve the acoustic trapping capability by punching specific round holes on acoustic tweezers, in which the acoustic radiation force is generated by the two vibrating V-shaped metal strips. Combining the FEM analysis and theory of acoustic radiation force, the effects of the radius of the round holes, the number and the arrangement of holes on the acoustic radiation force acting on particles are investigated by the COMSOL software. The acoustic radiation force is principal decided by the kinetic energy generated by the acoustic field, and the kinetic energy is influenced by the velocity of the metal strips. By punching the hole on the acoustic tweezers, the original flexural vibration couples with radial vibration in the direction of the radius of the hole, which causes the change in resonance of the structure.

When the hole's radius is less than 5 mm in only one hole condition, the trapping capability doesn't increase by punching a small round hole. When the hole's radius is larger than 5 mm, the velocity of metal strips and acoustic radiation force obviously increase compared with the original acoustic tweezers. The acoustic radiation force is maximum when the radius of hole is 5 mm. It is also found that the acoustic radiation force of acoustic tweezers with two holes on the bottom area of metal strips is significantly strengthened compared to that with original acoustic tweezers, if the radii of two holes are larger than 3.5 mm. By increasing the number of row of holes in vertical direction, there is obvious improvement on the acoustic radiation force even for the small radius holes.

Acknowledgement. This work is supported by Shanghai Young Eastern Scholar Talent Program (QD2015030).

References

1. Doblhoff-Dier, O., Gaida, T., Katinger, H., et al.: A novel ultrasonic resonance field device for the retention of animal cells. J. Biotechnol. Prog. **10**(4), 428–432 (1994)
2. Kosmala, A., Zhang, Q., Wright, R., et al.: Development of high concentrated aqueous silver nanofluid and inkjet printing on ceramic substrates. J. Mater. Chem. Phys. **132**(2–3), 788–795 (2012)
3. Ambedkar, B., Nagarajan, R., Jayanti, S.: Ultrasonic coal-wash for de-sulfurization. J. Ultrason. Sonochem. **18**(3), 718–726 (2011)
4. Castillo, J., Dimaki, M., Svendsen, W.E.: Manipulation of biological samples using micro and nano techniques. J. Integrative Biol. **1**(1), 30–42 (2009)
5. Lee, C., Lee, J., Lau, S.T, et al.: Single microparticle manipulation by an ultrasound microbeam. In: IEEE International Ultrasonics Symposium, pp. 849–852. IEEE Press (2010)
6. Wu, J.: Acoustical tweezers. J. Acoust. Soc. Am. **89**(5), 2140–2143 (1991)
7. Tran, S.B.Q., Marmottant, P., Thibault, P.: Fast acoustic tweezers for the two-dimensional manipulation of individual particles in microfluidic channels. J. Appl. Phys. Lett. **101**(11), 1109–1111 (2012)
8. Courtney, C.R.P., Demore, C.E.M., Wu, H., et al.: Independent trapping and manipulation of microparticles using dexterous acoustic tweezers. J. Appl. Phys. Lett. **104**(15), 154103 (2014)
9. Kang, S.T., Yeh, C.K.: Potential-well model in acoustic tweezers. J. IEEE Transactions on Ultrasonics Ferroelectrics & Frequency. Control **57**(6), 1451–1459 (2010)
10. Lee, J., Shung, K.K.: Radiation forces exerted on arbitrarily located sphere by acoustic tweezer. J. Acoust. Soc. Am. **120**(2), 1084–1094 (2006)
11. Shi, J., Ahmed, D., Mao, X., et al.: Acoustic tweezers: patterning cells and microparticles using standing surface acoustic waves (SSAW). J. Lab Chip **9**(20), 2890–2895 (2009)
12. Hu, J., Santoso, A.K.: A /spl pi/-shaped ultrasonic tweezers concept for manipulation of small particles. J. IEEE Transactions on Ultrasonics Ferroelectrics & Frequency. Control **51**(11), 1499–1507 (2004)
13. Gesellchen, F., Bernassau, A.L., Déjardin, T., et al.: Cell patterning with a heptagon acoustic tweezer–application in neurite guidance. J. Lab Chip **14**(13), 2266–2275 (2014)
14. Mitri, F.G.: Radiation force of acoustical tweezers on a sphere: The case of a high-order Bessel beam of quasi-standing waves of variable half-cone angles. J. Appl. Acoust. **71**(5), 470–472 (2013)

15. Hu, J., Yang, J., Xu, J.: Ultrasonic trapping of small particles by sharp edges vibrating in a flexural mode. J. Appl. Physics Lett. **85**(24), 6042–6044 (2004)
16. Hu, J., Xu, J., Yang, J., et al.: Ultrasonic collection of small particles by a tapered metal strip. J IEEE Trans Ultrason Ferroelectr Freq Control **53**(3), 571–578 (2006)
17. Liu, Y., Hu, J.: Trapping of particles by the leakage of a standing wave ultrasonic field. J. Appl. Phys. **106**(3), 034903 (2006)
18. Liu, Y., Hu, J., Zhao, C.: Dependence of acoustic trapping capability on the orientation and shape of particles. IEEE Trans. Ultrason. Ferroelectr. Freq. Control **57**(6), 1443–1450 (2010)
19. Morfey, C.L.: Dictionary of Acoustics. Academic Press, San Diego, pp. 14–15 (2000)
20. Hasegawa, T., Kido, T., Iizuka, T., et al.: A general theory of Rayleigh and Langevin radiation pressures. J. Acoust. Soc. Jpn. **21**(3), 145–152 (2000)

Bionics Control Methods, Algorithms and Apparatus

Motion Planning and Object Grasping of Baxter Robot with Bionic Hand

Xinyi Fei[1], Ling Chen[1(✉)], Yulin Xu[1], and Yanbo Liu[2]

[1] School of Mechatronics Engineering and Automation,
Shanghai University, Shanghai, China
lcheno@shu.edu.cn
[2] Shanghai Industrial Technology Institute, Shanghai, China

Abstract. Grasping and moving objects is a natural behavior in human daily life, whereas it turns into an enormous challenge with robots. To analyze the difficulty of grasping and moving target objects, a arm-hand system is performed with 7-DOF dual arms robot and bionic hand in this paper. A numerical method is proposed to solve the problem of arm motion planning. And a novel grasping strategy is proposed for enabling bionic hand to grasp efficiently. Finally, the effectiveness of the proposed methodology is demonstrated using both computer simulation and physical experiment.

Keywords: Motion planning · Grasping strategy · Robot arm-hand system

1 Introduction

Nowadays, robots are widely used in numerous industrial and scientific circumstances to replace humans as executors of some dangerous and complicated works. But in daily life, task like grasping objects are still an enormous challenge that robots face. Thus, some researches have been developed to find the solution of robot uses in daily life [1–4]. A dual-arm robotic system is presented to combine the motion planning with task assignment to have objects manipulated in cluttered environment [5]. Besides, objects grasping of different sizes and shapes with robot arms and hand coordination is shown in [6]. It is important to appropriately control the motion of robot arms while robots accomplish the target of grasping objects.

Object manipulation in cluttered environment is a task that humans perform daily. Different types of problems are found when humans are replaced with robots that have to perform this type of tasks. Problems like the motion planning, the computation of the object grasping, and the computation of a path to move the object from its initial configuration to a new goal configuration. This kind of problems are considered classics in the literature in robotics [7–9].

This work is supported by Science and Technology Commission of Shanghai (15411953500) and Science and Technology Commission of Shanghai Municipality under "Shanghai Sailing Program" (16YF1403700), Shanghai University Youth Teacher Training Assistance Scheme (ZZSD15088).

© Springer Nature Singapore Pte Ltd. 2017
M. Fei et al. (Eds.): LSMS/ICSEE 2017, Part I, CCIS 761, pp. 211–221, 2017.
DOI: 10.1007/978-981-10-6370-1_21

To enable the movement of robot arm from initial to the goal configuration, a methodology of motion control is proposed in this paper. A numerical algorithm based on Newton-Raphson method is illustrated to solve the inverse kinematics of robot manipulator. Unlike analytical solution, the numerical algorithm could easily figure out the joint angles. In addition, the purpose to grasp objects stably requires a strategy to adapt robot hand to the object pose. In order to accomplish the objective of having the objects grasped stably, the strategy of grasping different sized and shaped objects is proposed in details.

This paper presents an implementation of motion planning and object grasping of Baxter robot with bionic hand using methodology proposed. The simulation and experiment are established in ROS (Robot Operation System) [10]. MoveIt plugin [11] is used to study the performance of the robot arm-hand system. At the end of this paper, the result of manipulator experiment of grasping object is shown.

2 System Specification

In order to have the objects grasped as well as motion planning, a system is set up include three parts: 1. Baxter robot, 2. SHU-II bionic hand and 3. Kinect sensor. The SHU-II hand and Kinect sensor are attached to the robot with program run on PC. The details of components is illustrated in the following content.

2.1 Baxter Robot

Baxter robot is a humanoid robot as shown in Fig. 1 that includes a head-pan with a screen, located on the top of torso, can rotate in the horizontal plane followed by a torso based on a movable pedestal and two 7-DOF (degree of freedom) arms installed on left/right arm mounts respectively. Each arm has 7 rotational joints and 8 links, as well as an interchangeable gripper which can be installed at the end of the arm [12]. But the gripper is not operating well and unsteady to grasp objects. Thus, an bionic hand named SHU-II hand [13] is adopted. And because of its anti-collision design, Baxter robot is s able to sense a collision at a very early time instant, before it hits badly onto a subject. Thus it is a good choice to fulfill safe human-robot interaction.

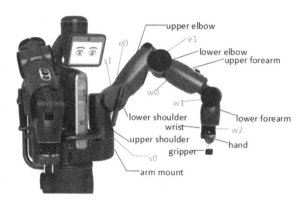

Fig. 1. Baxter robot with 7-DOF and the structure

2.2 SHU-II Hand

The SHU-II hand is equipped with force sensor that is used to grasp objects. Figure 2 shows the prototype of the SHU-II robot hand consisting of 5 under-actuated fingers and each finger is controlled by a DC motor with a gear box. A servo operates the thumb to be positioned parallel to the other four forefingers. A palm is designed to carry five fingers including their motors, servo and encoders. The thumb has four joints with four-degrees-of-freedom (DOF) and each finger has three joints with 3-DOF, which means it has sixteen degrees of freedom like real human hand. The size of the artificial hand is 1.1 times the human hand size.

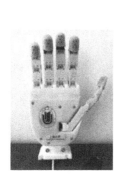 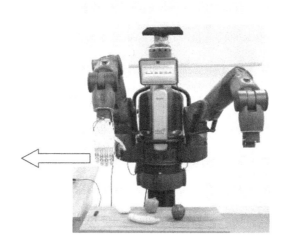

Fig. 2. SHU-II hand   **Fig. 3.** Baxter with arm-hand

The under-actuated hand with anthropomorphic motion can adaptively grasp objects with different sizes and shapes like the human hand. Because of its higher integration, it can be very easy to install and test on the other platform.

2.3 Vision

Total coordinated system is consists of robot, SHU-II hand, and the vision system. This vision system provides the object information that is obtained by Kinect sensor showed in Fig. 3. The object image processing goes as follows: 1. Information of depth and RGB colors of image is collected by Kinect. 2. Information is segmented by the image depth. 3. Target identification based on objects' shapes and colors. 4. Kinect based frame is converted to the robot based frame. Hereafter the robot gets the 3D coordinates of objects of the robot frame.

2.4   Operation Flow

The main controller is operated on ROS (Robot Operation System), and consists of three threads. One is ROS operation thread called Main thread. It gets the command from user then starts the main process. The forward kinematics of robot arm is computed, then the initial value of trajectory is set which called the original position. The main thread gets the object position from second thread Vision thread, and convert the Kinect based frame to Robot based one. Hereafter, serial data is sent to the SHU-II hand then it will grasp the object we chosen before. A feedback is returned once the grasp is complimented. Thus the robot arm begins to move. Place the object at the placed position (Fig. 4).

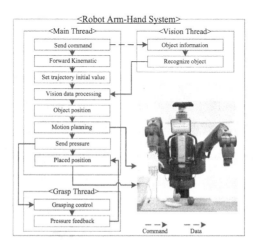

Fig. 4. Main flow of Robot arm-hand system

3   Motion Planning and Grasp Strategy

3.1   Motion Control of Robot Arm

The whole system is completed under the ROS system, and we obtain the DH parameters of Baxter from its URDF (Unified Robot Description Format) file [14]. The D-H parameters of robot left arm measured by the mechanical structure (Table 1) can be utilized to establish the D-H equation in formula (1):

$$
^{i-1}T_i = \begin{bmatrix} c\theta_i & -s\theta_i c\alpha_i & s\theta_i s\alpha_i & a_i c\theta_i \\ s\theta_{ii} & c\theta_i c\alpha_i & -c\theta_i s\alpha_i & a_i s\theta_i \\ 0 & s\alpha_i & c\alpha_i & d_i \\ 0 & 0 & 0 & 1 \end{bmatrix} \tag{1}
$$

Table 1. DH notation table of the left arm

i	$\theta_i^{(rad)}$	$d_i^{(m)}$	$a_i^{(m)}$	$\alpha_i^{(rad)}$
1	q_1	0.27	0.069	$-\pi/2$
2	$q_2 + \pi/2$	0	0	$\pi/2$
3	q_3	0.364	0.069	$-\pi/2$
4	q_4	0	0	$\pi/2$
5	q_5	0.375	0.01	$-\pi/2$
6	q_6	0	0	$\pi/2$
7	q_7	0.28	0	0

Where c and s represent the trigonometric of cos and sin respectively. Thus the position and orientation of the end effector described by coordinate in a Cartesian coordinate system can be calculated by each transformation matrix in formula (2):

$$
{}^0T_n = {}^0T_1\,{}^1T_2\cdots{}^{n-1}T_n =
\begin{bmatrix}
n_x & o_x & a_x & p_x \\
n_y & o_y & a_y & p_y \\
n_z & o_z & a_z & p_z \\
0 & 0 & 0 & 1
\end{bmatrix}
\tag{2}
$$

And $^{i-1}T_i$ could represents as A_i, so the formula (2) could be rewritten as formula (3):

$$
{}^0T_n = A_1 A_2 \cdots A_n
\tag{3}
$$

Forward kinematics analysis of Baxter robot determines workspace of the robot arms, from which we could obtain the jacobian matrix of robot. Jacobian matrix of robot will establish the relationship between joint motion and hand motion. And the relationship can be represented shortly as formula (4):

$$
D = JD_\theta
\tag{4}
$$

Where $D = [v \quad \varpi]^T$, $D_\theta = [\dot{q}_1 \cdots \dot{q}_n]^T$, and the formula showed above can be rewritten as formula (5):

$$
\begin{bmatrix} v \\ \varpi \end{bmatrix} =
\begin{bmatrix}
J_{11} & \cdots & J_{1n} \\
\vdots & \ddots & \vdots \\
J_{61} & \cdots & J_{6n}
\end{bmatrix}
\begin{bmatrix} \dot{q}_1 \\ \vdots \\ \dot{q}_n \end{bmatrix}
\tag{5}
$$

Where $[v \quad \varpi]^T$ is the linear velocity and angular velocity of robot hand which is equally to $[v_x \quad v_y \quad v_z \quad \varpi_x \quad \varpi_y \quad \varpi_z]^T$ while $[\dot{q}_1 \cdots \dot{q}_n]^T$ shows the angular velocity from different joints, n is the DOF of robot. For us n is equally to 7 as Baxter robot has 7-DOF. Hereafter $J = [J_1 \quad \cdots \quad J_n]$ is the differential kinematic relation of them. The jacobian matrix could be computed from its kinematic equation. With regard to a matrix:

$$J_i = \begin{bmatrix} J_{1i} \\ J_{2i} \\ J_{3i} \\ J_{4i} \\ J_{5i} \\ J_{6i} \end{bmatrix} = \begin{bmatrix} (p_6^i \times n_6^i)_z \\ (p_6^i \times o_6^i)_z \\ (p_6^i \times a_6^i)_z \\ (n_6^i)_z \\ (o_6^i)_z \\ (a_6^i)_z \end{bmatrix} \qquad (6)$$

The elements of it can be calculated one by one with specific equation and $(p \times n)_z = (p_x \times n_y - p_y \times n_x)$ where i varies from 1 to n.

J_i is the i_{th} column of jacobian J and it is based on the kinematic:

$$^i T_n = A_i A_{i+1} \cdots A_n = \begin{bmatrix} n_x & o_x & a_x & p_x \\ n_y & o_y & a_y & p_y \\ n_z & o_z & a_z & p_z \\ 0 & 0 & 0 & 1 \end{bmatrix} \qquad (7)$$

And p_6^i is present by formula (8):

$$p_6^i = [p_x \quad p_y \quad p_z]^T \qquad (8)$$

An analytical methodology of inverse kinematic computation for 7-DOF redundant manipulator may be difficult. But with the numerical methodology it can be solved easily. The problem of finding the inverse solution of the robot is considered as a problem of solving a set of nonlinear equations. Newton-Raphson method is an optional method for this problem. The problem rewritten as equations with joint angles unknown like:

$$f = \begin{bmatrix} f_1(x_1 \cdots x_n) \\ \vdots \\ f_n(x_1 \cdots x_n) \end{bmatrix} = \begin{bmatrix} 0 \\ \vdots \\ 0 \end{bmatrix} \qquad (9)$$

Where x_1 to x_n is the angles of different joints. From the equation mentioned above, we have:

$$J = f' = \begin{bmatrix} \frac{\partial f_1}{\partial x_1} & \cdots & \frac{\partial f_1}{\partial x_n} \\ \vdots & \ddots & \vdots \\ \frac{\partial f_6}{\partial x_1} & \cdots & \frac{\partial f_6}{\partial x_n} \end{bmatrix} \qquad (10)$$

It means jacobian matrix consists of elements with respect to the partial differential of function f_1 to f_6.

And finally the joint angles will be calculated:

$$\begin{bmatrix} q_{1(k+1)} \\ \vdots \\ q_{n(k+1)} \end{bmatrix} = \begin{bmatrix} q_{1(k)} \\ \vdots \\ q_{n(k)} \end{bmatrix} + J^{-1} Err \tag{11}$$

$q_{i(k+1)}$ is the i_{th} joint angle obtained after $k+1$ times. J^{-1} is the inverse matrix of jacobian J.

$$Err = f(x_k) \tag{12}$$

Represents the difference between the current value and the target value after the k times iteration. And the final solution will be obtained when the difference is close to 0.

3.2 Grasping Strategy

The robot grasping of different sized and shaped objects is a complex problem to be solved [15]. A strategy of objects manipulation is proposed in this part. Because the shape of any object can be viewed as a combination of several special ones like sphere and cuboid, the methodology to grasp these specific objects could be generally extended.

Just like human beings seize things, the grasping pose of robot hand is changed on the basis of postures of different objects. And the approach to grasp is sometimes distinct with same thing of changed pose. The whole task could be divided into two aspects: 1. the position robot hand moved and 2. the pose of robot hand to grasp the objects stably. Strategy to solve the problems is performed after the analysis of objects structure.

Sphere and cuboid is taken for the analysis of objects. By the experience human hand seizes, the grasping point is always on the longer side of object. As shown in Fig. 6, the Z axis represents the hand approach direction to the object, and X axis is from the paper outward which matters the pose to grasp.Y axis is vertical to the plane shaped by X and Z axis. The height and width of object are represented as H and W in Fig. 5. $d_{\max}$ and $d_{\min}$ is defined as the biggest and least of H and W.

$$\begin{cases} d_{\max} = \max\{H, W\} \\ d_{\min} = \min\{H, W\} \end{cases} \tag{13}$$

As mentioned before, the Z axis known as approach direction is parallel with the normal vector of plane shaped by the direction vector of high and wide side which is obtained from the vision process. Similar to Z axis determined, the direction vector of Z axis and $d_{\max}$ confirm the one of X axis.

Some objects have the similar H and W like sphere, it's not necessary to choose which pose to grasp. Thus, a criterion named d_{cr} is defined to divide the objects into different groups. If $\Delta d = d_{\max} - d_{\min}$ is smaller than d_{cr}, the pose is not defined. The

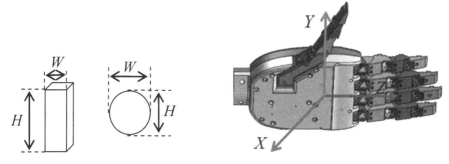

Fig. 5. The height and width of object   Fig. 6. Coordinate of hand

parameter of d_{cr} is decided by d_{max} and the grasp range d_{rg} of bionic hand as shown in formula (14).

$$d_{cr} = \begin{cases} d_{max}, & d_{max} \leq d_{rg} \\ d_{rg}, & d_{max} > d_{rg} \end{cases} \quad (14)$$

Because of the shape and size of SHU-II hand, there is an offset attached to the grasp point. The strategy to grasp cuboid and sphere is simulated in Fig. 7.

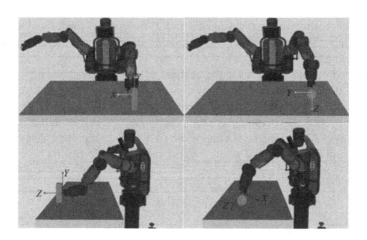

Fig. 7. Grasp simulation

4   Experiment

4.1   Motion Planning

When the Cartesian position and orientation of robot are known, we can calculate the 7 joint angles based on the algorithm illustrated in Sect. 3. The simulation result in Fig. 8 shows the accuracy of the solution and well plan of the trajectory.

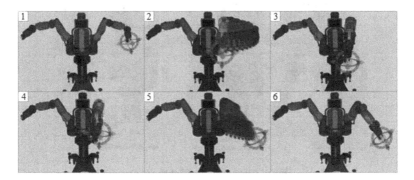

Fig. 8. Motion plan of robot arm

4.2   Object Grasping

The arm-hand system comprises Baxter robot, Kinect sensor of vision and SHU-II hand. The whole system is operated on ROS and the movement of manipulator is decided by algorithm mentioned in Sect. 3. Strategy to grasp is used to grasp the object stably. The experiment selects the green apple and cola jar as targets. The Figs. 9 and 10 shows

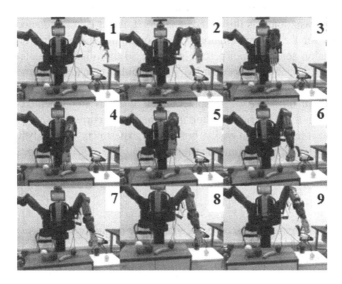

Fig. 9. Baxter robot recognizes and grasps green apple

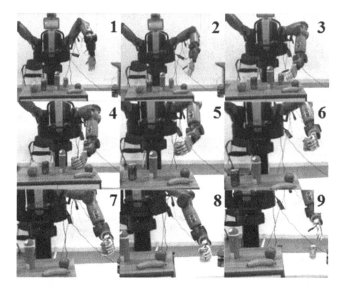

Fig. 10. Baxter robot recognizes and grasps cola jar.

the validity of the Motion Planing and Object Grasping of Baxter Robot with SHU-II hand.

5   Conclusion

In this paper, a robot arm-hand system which is comprised of a 7-DOF humanoid robot, a bionic hand and vision sensor is shown. We have proposed an effective numerical method to accomplish the task of robot arm motion planning. And the strategy for bionic hand to grasp object is illustrated. The target object is grasped successfully. The main difficulty for the implementation is that sometimes the position of grasp object obtained by vision sensor is not precise enough. So the offset of bionic hand is necessary to grasp the object successfully. The simulation and experiment show the validity of the robot arm-hand system.

Now the main objective of grasping target object is implemented. The motion planning based on the numerical method is verified to be correct. Using them we could build a robot arm-hand system easily, but further experiment and analysis of grasping strategy need to be carried out in our future work.

References

1. Taylor, R.H.: Planning and execution of straight line manipulator trajectories. IBM J. Res. Develop. **23**(4), 424–436 (1979)
2. Hirano, Y., Kitahama, K., Yoshizawa, S.: Image-based object recognition and dexterous hand/arm motion planning using RRTs for grasping in cluttered scene. In: IEEE/RSJ International Conference on Intelligent Robots and Systems. pp. 2041–2046. IEEE (2005)

3. Calinon, S., Guenter, F., Billard, A.: On learning, representing, and generalizing a task in a humanoid robot. IEEE Trans. Syst. Man Cybern. Part B **37**(2), 286–298 (2007)
4. Ciocarlie, M.T., Allen, P.K.: Hand posture subspaces for dexterous robotic grasping. Int. J. Robot. Res. **28**(7), 851–867 (2009)
5. Rodriguez, C., Suarez, R.: Combining motion planning and task assignment for a dual-arm system. In: IEEE/RSJ International Conference on Intelligent Robots and Systems, pp. 4238–4243. IEEE (2016)
6. Cheon, S., Ryu, K., Oh, Y.: Object manipulation using robot arm-hand system. In: 10th International Conference on Ubiquitous Robots and Ambient Intelligence (2013)
7. Shin, S., Kim, C.: Human-like motion generation and control for humanoid's dual arm object manipulation. IEEE Trans. Ind. Electron. **62**(4), 2265–2276 (2015)
8. Ko, C.H., Lin, S.H., Chen, J.K.: Motion planning of multifingered hand-arm system with optimal grasping force. In: 10th International Conference on Ubiquitous Robots and Ambient Intelligence (2013)
9. Bae, J.H., Sekimoto, M., Arimoto, S.: Effect of Virtual Spring-Damper in Grasping and Object Manipulation of a Robotic Hand-Arm System. In: IEEE Xplore of International Joint Conference on SICE-ICASE 2006, pp. 2222–2226 (2006)
10. Information of ROS. http://wiki.ros.org
11. Information of MoveIt. http://moveit.ros.org
12. Baxter product Data sheet. http://sdk.rethinkrobotics.com
13. Xu, Y., Jiang, C., Yuan, J.: Compliance control for grasping with a bionic robot hand. In: Chinese Control and Decision Conference, pp. 5280–5285. IEEE (2016)
14. URDF file. http://wiki.ros.org/urdf
15. Mouri, T., Kawasaki, H., Ito, S.: Unknown object grasping strategy imitating human grasping reflex for anthropomorphic robot hand. J. Adv. Mech. Des. Syst. Manufac. **1**(1), 1–11 (2005)

Grasping Force Control of Prosthetic Hand Based on PCA and SVM

Jian Ren[1], Chuanjiang Li[1(✉)], Huaiqi Huang[2,3], Peng Wang[1], Yanfei Zhu[1], Bin Wang[1], and Kang An[1]

[1] The College of Information, Mechanical and Electrical Engineering,
Shanghai Normal University, Shanghai 201418, China
licj@shnu.edu.cn
[2] EPFL, 2002 Neuchâtel, Switzerland
[3] BFH, 2502 Biel, Switzerland

Abstract. This paper presents a control method of grasping force of prosthetic hand. Firstly, the correlated features of surface electromyogram (sEMG) signal that collected by MYO are calculated, and then principal component analysis (PCA) dimension reduction is processed. According to pattern classification model and sEMG-force regression model which based on support vector machine (SVM) to gain the force prediction value. In this approach, force is divided into different grades. The predicted force value is used as the given signal, and grasping force of prosthetic hand is controlled by a fuzzy controller, and combined with vibration feedback device to feedback grasping force value to patient's arm. The test results show that the method of prosthetic hand grasping force control is effective.

Keywords: MYO · PCA · sEMG-force · SVM · Fuzzy controller · Vibration feedback device

1 Introduction

The second national sample survey of disabled people shows that the number of amputation patients in China is as high as 2 million 260 thousand. A conservative estimate of the number of patients who need to install prosthetic hand up to more than 250 thousand people, therefore, it is necessary to study grasping force control in order to improve the functionality of prosthetic hand.

Currently, most research focus on identification of finger grasping mode, and few studies on force of prosthetic hand during grasping [1–3]. And some of the flexibility and operating sense of prosthetic hand are poor, many patients are reluctant to use the prosthetic hands. Commercial prosthetic hand adopts proportional control mode and determine the force directly according to the sEMG amplitude, but its accuracy is limited; Ahmet Erdemir established Del explicit model [4] of force-length-velocity-action potential, due to the complexity of human physiological structure, it is hard to establish an accurate model; Claudio Castellini used 10 electrodes to record sEMG data, but the more the number of electrodes, the more uncomfortable the patient.

© Springer Nature Singapore Pte Ltd. 2017
M. Fei et al. (Eds.): LSMS/ICSEE 2017, Part I, CCIS 761, pp. 222–230, 2017.
DOI: 10.1007/978-981-10-6370-1_22

This paper adopts MYO [5] to collect sEMG. Grasping force is controlled by sEMG, which has the advantages of low cost, no invasion, convenience and high reliability. In addition, sEMG containing the information of user's action intention can be used as driving or feedback signal to improve movement effect and most prosthetic hands are controlled by sEMG [6–9]. From the literature, we know that force is contributed by many muscles [10, 11], and MYO has 8 electrodes that can meet the demand to collect multiple muscles at the same time. In addition, MYO has the advantages that it is simple and convenient to wear, thus high acceptance rate. We set sEMG and force as input and output, using SVM to establish nonlinear regression relationship between them. This method does not need complex modeling process, and shows good prediction effect [12–14], which is more practical than other modeling methods [9, 15–18]. For grasping force control part, we divide force into different grades, set the predicted force value as the given signal, use fuzzy control to control grip strength of prosthetic hand, and have vibration feedback device to feedback force grade value to patient's arm, which improve patient's sense of using prosthetic hand and realize accurate grasping. The whole structure is shown in Fig. 1.

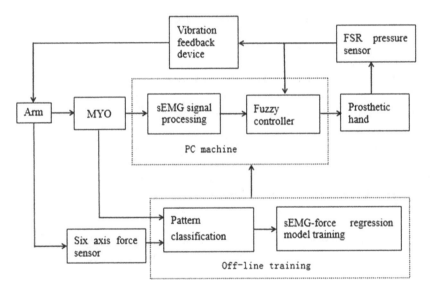

Fig. 1. The prosthetic hand grip control system diagram.

2 Methods

2.1 Signal Acquisition and Feature Extraction

In this method, the force is divided into eight different grades from 0 N to 16 N, and acquisition of sEMG of arm by MYO corresponding to each grade, and at the same time, acquisition of hand grip signal by a six axis force sensor [5], as shown in Fig. 2.

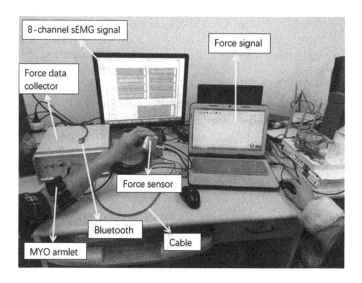

Fig. 2. sEMG and force signal acquisition.

According to time domain feature of sEMG to determine starting and ending time of movement, specific method is used to calculate the sum of Mean Absolute Value (MAV) of 8 channels sEMG, and then compare with the preset threshold value to judge the starting point and the ending point of action.

The accuracy of force prediction is related to features of sEMG, so it is essential to select suitable features [19, 20]. Most of sEMG analysis use time domain features, frequency domain features and other more complex features [21]. Features used in current study include four time domain features: MAV, Root Mean Square (RMS), Standard Deviation (SD), and Waveform Length (WL). MAV reflects the average intensity and the concentration of sEMG. RMS represents the contribution of each muscle organization in the process of the movement. SD can demonstrate the dispersion degree of a data set. WL can reflect the complexity of sEMG waveform, and the joint effect of sEMG amplitude, frequency and duration. Time domain calculation is simple, and that can ensure real-time grasping [22]. The calculation methods of each feature are as follows:

$$MAV_k = \frac{1}{N} \sum_{i=1}^{N} x(i) \tag{1}$$

$$RMS_k = \sqrt{\frac{1}{N} \sum_{i=1}^{N} x(i)^2} \tag{2}$$

$$SD_k = \sqrt{\frac{1}{N-1} \sum_{i=1}^{N} (x(i) - \mu)^2} \tag{3}$$

$$WL_k = \sum_{i=1}^{N-1} |x(i+1) - x(i)| \tag{4}$$

Where x (i) is the sEMG data of each sample, N is the number of data of each channel, μ is the average of N data, $k = 1, \ldots, M$, M is the number of channels.

2.2 PCA Dimension Reduction

Considering real-time, we adopt the method of PCA dimension reduction to reduce the computational complexity, shorten the computing time and improve grasping speed [23]. Between the beginning and ending time of one hand movement, a section of sEMG is intercepted as a one-dimensional signal sequence. Assuming that the number of the feature extracted is n, the dimension of the feature vector is $1 \times n$. Assuming $U \in R^{n \times k}$ is the dimension reduction matrix by PCA, the feature matrix multiplied by the dimension reduction matrix becomes $1 \times k$, namely from the original n-dimensional down to k-dimension. PCA can be used to analyze the main influencing factors from multiple sources and simplify complex problems [24].

2.3 Pattern Classification Model

To distinguish between 8 grades of grip strength, the pattern classification model is built, which is a 3-layer back propagation (BP) neural network. The number of input neurons is the dimension of feature matrix, denoted as K, the number of output neurons is the number of grip grades, denoted as C, the number of hidden layer neurons of H can be calculated by an empirical formula (5), and then adjust it according to the training precision. We can get weights and thresholds of pattern classification through training BP neural network model, and then save the weights and thresholds.

$$H = \sqrt{CK} \tag{5}$$

2.4 sEMG-Force Regression Model

In order to gain the grasping force value, we have to get force information from sEMG that is why to build sEMG-force model. The sEMG-force model is based on SVM, which uses sEMG data $x[x_1, x_2, \cdots \cdots x_n]$ (n element one dimensional vector) as input, and use force signal z as output to form several sample vectors $[x_1, x_2, \cdots \cdots x_n, z]$, then build the nonlinear regression relationship f from x to the target z: $z = f(x)$. The following is a detailed introduction: Nonlinear regression function is $f(x) = w \bullet \phi(x) + b$, in the constraint condition of $\sum_{i=1}^{l} (\alpha_i - \alpha_i^*) = 0$, to optimize formula (6).

$$L(w, b, \xi) = -\frac{1}{2} \sum_{i,j=1}^{l} (\alpha_i - \alpha_i^*)(\alpha_j - \alpha_j^*) K(x_i, x_j) + \sum_{i=1}^{l} (\alpha_i - \alpha_i^*) y_i$$
$$- \sum_{i=1}^{l} (\alpha_i + \alpha_i^*) \varepsilon \tag{6}$$

In which, $w = \sum_{i=1}^{l} (\alpha_i - \alpha_i^*) \phi(x_i)$, combined with Karush-Kuhn-Tucker optimality condition, we can get the nonlinear regression function:

$$f(x) = \sum_{i=1}^{l} \left(\alpha_i - \alpha_i^* \right) K\left(x_i, x \right) + b \tag{7}$$

Where $K(x_i, x_j)$ is a kernel function,which is the sample inner product of ϕ space by using Gaussian kernel function: $K(x_i, x_j) = \left\langle \varphi(x_i), \varphi(x_j) \right\rangle = \exp\left(-\gamma \left\| x_i - x_j \right\|^2 \right)$, where γ is kernel function parameter. α and α^* can be calculated from training sample space $\left\{ \left[x_1, z_1 \right], \cdots, \left[x_l, z_l \right] \right\}$.

2.5 Grasping Force Control

The signal is collected from one healthy subject, which include eight grades of grasping force read through a six-axis force sensor, and eight channels sEMG of sEMG signal gathered from MYO corresponding to eight grades. Extracting time domain features, then feature matrix is used as a sample for training pattern classification model, and feature matrix with force data are used to construct sample for sEMG-force regression model training, then the parameters of each model is saved and written to the PC. After the preparation below, patient can wear MYO to collect sEMG, read sEMG by PC machine and process sEMG as mentioned earlier. And then pattern classification model will recognize which grade the signal belongs to, and then the signal will be send to the corresponding grade of sEMG-force model to predict force value. What follows is that setting the predicted force value as the given signal to control grasping force of prosthetic hand by fuzzy control, and the size of grip strength of prosthetic hand induced by the FSR pressure sensor on prosthetic hand. Then vibration feedback device changed the size to force grade value, and feedback it to patient's arm, in order to make patient get the real grasping value grade and adjust grip strength according to it, so that can achieve accurate grasping.

3   Results

In order to reduce calculation time, ensure real-time grasping and improve force prediction effectiveness, comparison of pattern recognition rate of eight grades force among the number of sEMG channel, feature and PCA dimension reduction has been analyzed. By mapping eight channels of sEMG, we observed that there are three channels signal fluctuated obviously, which means these three channels signal relatively sensitive and ideal for grasping movement. The recognition of reading eight and three channels of sEMG, extracting four time domain features, and their eight grades force pattern recognition are shown in Fig. 3.

For the eight channels sEMG, result of four features is shown as the blue bars in Fig. 3, result of three features is shown in Fig. 4. The results show that for the eight channels signal, the selection and number of features have little effect on recognition rate. Probably because

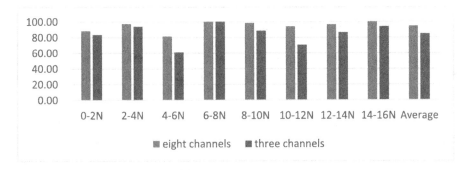

Fig. 3. Pattern recognition rate of different channel number using four features. The bars with different colors represent different channels of sEMG signal. The X-axis represents the magnitude of the forces at eight different levels, and Average represents average recognition rate of the eight grades. The Y-axis represents recognition rate (%). The X-axis and Y-axis in Figs. 4 and 5 have the same meaning. (Color figure online)

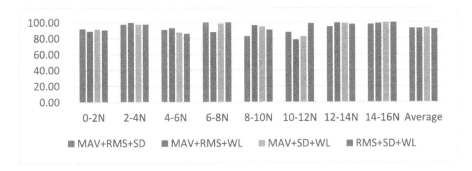

Fig. 4. Pattern recognition rate of eight channels with three features. The bars with different colors represent the recognition of combinations of different three features.

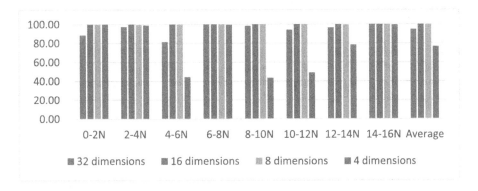

Fig. 5. Pattern recognition rate of eight channels with four features reduced by PCA to different dimensions. The bars with different colors represent the recognition of different dimensions.

For the eight channels sEMG, each channel four features are extracted, then we can get a 32-dimension feature matrix, and reduce dimension by PCA to 16, 8 and 4. Figure 5 shows the result of the four cases. Experimental results show that the recognition rate of 16 and 8 dimensions are better than that of 32 and 4 dimensions; In addition, the time of reducing dimension from 32 to 16 or 8 is the same, but the calculation of 16 dimensions is much larger than that of 8, so we choose to sacrifice a little bit of recognition rate in exchange for higher real-time grasping.

Through the above analysis: when grasping force control of prosthetic hand, we adopt the method of reducing dimension to 8 dimensions by PCA of the feature vector of eight channels and four features. And as shown in Fig. 6 is the prediction result of the grip strength of the eight grades. Which shows that the predicted value of each grade force is almost in the range of the corresponding force. On the whole, it displays a good force prediction effect. And therefore,

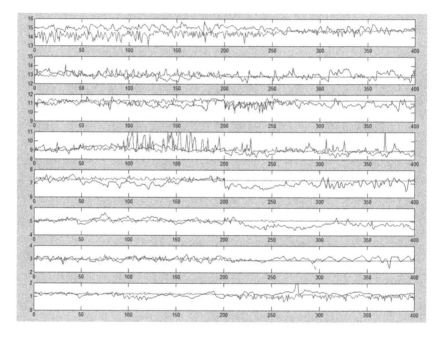

Fig. 6. The prediction results of the grip strength of the eight grades. The X-axis represents sampling point of force signal, and Y-axis represents eight grades of force size. The blue curve represents actual grasping force value, and the red curve represents predicted grasping force value. (Color figure online)

4   Conclusion

In this paper, we proposed a control method of hand grasping force of prosthetic hand based on PCA and SVM, using PCA to reduce dimension of time domain feature matrix, forecasting grasping strength by establishing regression model between sEMG and force based on SVM; The calculation of time domain is simple, in addition, PCA reduces the amount of computation and shortens calculation time, which provides real-time for the control of prosthesis grasping, and the method of establishing regression model by SVM is simple and also get a good prediction effect. Using fuzzy controller to control grasping force of prosthesis hand and vibration feedback device to feedback force value to patient's arm can improve the patient's sense of using prosthetic hand. The experiment shows that this approach is an effective control method of prosthesis hand grip.

References

1. Bingke, Z., Xiaogang, D., Hua, D.: Force estimation in different grasping mode from electromyography. J. Comput. Appl. **35**(7), 2109–2112 (2015)
2. Huatao, Z., Aiguo, S., Yan, G., Jiatong, B., Kui, Q.: Control method for myoelectric prosthetic hand's grip force based on neural network and fuzzy logic controller. Adv. Mater. Res. **403–408**, 2843–2847 (2011)
3. Fukuda, O., Tsuji, T., Kaneko, M., Otsuka, A.: A human-assisting manipulator teleoperated by sEMG signals and armmotions. IEEE Trans. Robot. Autom. **19**(2), 210–222 (2003)
4. Erdemir, A., Mclean, S.G., Herzog, W., van den Bogert, A.J.: Model-based estimation of muscle forces during movements. Clin. Biomech. **22**(2), 131–154 (2007)
5. Peidong, L., Chenguang, Y., Ning, W., Ruifeng, L.: A discrete-time algorithm for stiffness extraction from sEMG and its application in antidisturbance teleoperation. Discrete Dyn. Nat. Soc. **2016**(2), 1–11 (2016)
6. Zhijun, L., Baocheng, W., Fuchun, S., Chenguang, Y., Qing, X., Weidong, Z.: sEMG-based joint force control for an upper-limb power-assist exoskeleton robot. IEEE J. Biomed. Health Inf. **18**(3), 1043–1050 (2014)
7. Ng, G.Y.F., Zhang, A.Q., Li, C.K.: Biofeedback exercise improved the EMG activity ratio of the medial and lateral vasti muscles in subjects with patellofemoral pain syndrome. J. Electromyogr. Kinesiol. **18**(1), 128–133 (2008)
8. Li, G., Schultz, A.E., Kuiken, T.A.: Quantifying pattern recognition based myoelectric control of multifunctional transradial prostheses. IEEE Trans. Neural Syst. Rehabil. Eng. **18**(2), 185–192 (2010)
9. Kamavuako, E.N., Farina, D., Yoshida, K., Jensen, W.: Relationship between grasping force and features of single-channel intramuscular EMG signals. J. Neurosci. Methods **185**(1), 143–150 (2009)
10. Hanliang, Y., Chase, R.A., Mosby, B.S.: Atlas of hand anatomy and clinical implications. J. Hand Surg. Br. Eur. **29**(4), 409 (2004)
11. Choi, C., Kwon, S., Park, W., Lee, H., Kim, J.: Real-time pinch force estimation by surface electromyography using an artificial neural network. Med. Eng. Phys. **32**(5), 429–436 (2010)
12. Xinqing, W., Yiwei, L., Dapeng, Y., Shaowei, F., Hong, L.: Tracking control of real-time force for prosthetic hand based on EMG. J. Shenyang Univ. Technol. **34**(1), 63–67 (2012)
13. Dapeng, Y., Jingdong, Z., Li, J., Hong, L.: Force regression from EMG signals under different grasping patterns. J. Harbin Inst. Technol. **44**(1), 83–87 (2012)
14. Xinhua, D., Zengqiang, C., Zhuzhi, Y.: Function approximation research based on support vector machine. Comput. Eng. **32**(8), 52–54 (2006)
15. David, G., Lloyd, T.F.B.: An EMG-driven musculoskeletal model to estimate muscle forces and knee joint moments in vivo. J. Biomech. **36**(6), 765–776 (2003)

16. Gaofeng, W., Feng, T., Gang, T., Chengtao, W.: A wavelet-based method to predict muscle forces from surface electromyography signals in weightlifting. J. Bionic Eng. **09**(9), 48–58 (2012)
17. Asefi, M., Moghimi, S., Kalani, H., Moghimi, A.: Dynamic modeling of SEMG–force relation in the presence of muscle fatigue during isometric contractions. Biomed. Signal Process. Control **28**, 41–49 (2016)
18. Soo, Y., Sugi, M., Yokoi, H., Arai, T., Nishino, M., Kato, R., Nakamura, T., Ota, J.: Estimation of handgrip force using frequency-band technique during fatiguing muscle contraction. J. Electromyogr. Kinesiol. **20**(5), 888–895 (2010)
19. Englehart, K., Hudgins, B., Parker, P.A.: A wavelet-based continuous classification scheme for multifunction myoelectric control. IEEE Trans. Biomed. Eng. **48**(3), 302–311 (2001)
20. Kamavuako, E.N., Rosenvang, J.C., Bøg, M.F., Smidstrup, A., Erkocevic, E., Niemeier, M.J., Jensen, W., Farina, D.: Influence of the feature space on the estimation of hand grasping force from intramuscular EMG. Biomed. Signal Process. Control **8**(1), 1–5 (2013)
21. Dongmei, W., Xin, S., Zhicheng, Z., Zhijiang, D.: Feature collection and analysis of surface electromyography signals. J. Clin. Rehabil. Tissue Eng. Res. **14**(43), 8073–8076 (2010)
22. Yutao, J., Zhizeng, L.: Summary of EMG feature extraction. Chin. J. Electron Devices **30**(1), 326–330 (2007)
23. Yapeng, Z.: Research on face recognition system based on parallel pca algorithm. Comput. Knowl. Technol. **12**(19), 147–148 (2016)
24. Yuan, Z., Yanping, Z.: An algorithm of PCA and its application. Comput. Technol. Dev. **15**(2), 67–68 (2005)

Adaptive SNN Torque Control
for Tendon-Driven Fingers

Minrui Meng[1,2], Xingbo Wang[1,2(✉)], and Xiaotao Wang[2,3]

[1] College of Automation, Nanjing University of Posts and Telecommunications,
Wenyuan Road 9, 210046 Nanjing, China
`minruimeng@163.com`, `sinbowang@163.com`
[2] Shanghai Key Laboratory of spacecraft mechanism,
Shanghai 201108, China
`wangxtao1977@nuaa.edu.cn`
[3] College of Astronautics, Nanjing University of Aeronautics and Astronautics,
210016 Nanjing, China

Abstract. Tendon-driven robot manipulators are often used to actuate distal joints. The tendons allow the actuators to be located outside the fingers. Conventionally, the use of the tendons of the fingers allows for the significant reduction to the size and weight, in this case, which approximately similar to that of the human. To achieve the interaction with unstructured environments, a torque control system is presented based on the single neuron networks (SNN) in this paper. The torque control allows the system maintain proper torques on the joints. Meanwhile, this controller calculates actuator positions based on the error measured by the actual joint torques and desired joint torques. Simulations have been conducted on a tendon-driven finger model to demonstrate that the proposed controller can achieve the faster response, and then decrease overshoot comparing to a PI controller.

Keywords: Single neuron network · Torque control · Uncertain parameters · Tendon-driven fingers

1 Introduction

In the recent decades, many technologies have been made for the dexterous hand to perform complex tasks. The joints of the dexterous hand which like the humans' are often operated by actuators. Their implementation, however, in the system is limited by the size of them. Tendons can actuate fingers with the advantages of low inertia and friction.

In [1], each finger has three joints and the finger is actuated by only one actuator. Therefore, every joint must move connectedly together with the others. The connected motion among the joints can be implemented by a mechanism using the cross-coupling tendons. Traditionally, the n joints can be fully controlled by a minimum quantity of $n + 1$ tendons. But the coupling in their kinematics makes it difficult to design a controller for the tendon-driven fingers. In addition, the friction between the tendon and its routing path increases the uncertainty for dynamics modeling. Many control strategies have been proposed for the tendon-driven fingers.

© Springer Nature Singapore Pte Ltd. 2017
M. Fei et al. (Eds.): LSMS/ICSEE 2017, Part I, CCIS 761, pp. 231–241, 2017.
DOI: 10.1007/978-981-10-6370-1_23

Most tendon-driven fingers have applied force control through independent tension controllers on each tendon in the tendon space. The coupled kinematics of the tendons, however, causes a transient coupling of the controllers in their response. This problem can be resolved by alternatively framing the controllers in the joint space of the finger. A torque and stiffness control method of a tendon-driven finger in the joint space was introduced in [2]. The work demonstrates both the decoupled and faster response than an equivalent tendon-space formulation. The law was also demonstrated to operate greater speed and robustness than a PI controller.

For the force and the impedance controller, an approach has been proposed using tension sensors in [3]. The method is very useful with the sheaths in compensating joint torque errors, which were caused by large friction in sheaths. In many tasks, dexterity may be provided for the fingers of a hand for not allowing grasped object to slip. An impedance control scheme with multi-priority tasking was designed with a situation where enforces the impedance behavior of the grasped objects in [4]. The task of grasping objects was in keeping with governing the impedance response at the joints. Decoupled kinematics of fully-actuated tendon-driven fingers has been analyzed in [5] for dealing with the impedance control of the joints. And a new stiffness control law was proposed to control the fingertip force. In a case where the second priority joint space impedance operates with the null-space of the first priority Cartesian impedance of the end effectors, the multi-priority control law has been considered in [6]. The control law in the paper was shown to be a generalization of motions. And the impedance control laws were presented in the other literatures.

Although the PID controller has the ability to achieve the flexibility for the tendon-driven fingers, the challenge is to develop a controller which can achieve the desired performance while adapting to the uncertain environment. So improved control laws have being presented in these years.

Many adaptive control systems exist in the literatures for developing the impedance control of the tendon-driven fingers. A neuron adaptive controller learning the robot dynamics online and responding like the prescribed impedance model was presented in [7]. The system integrates the assist-as-needed (AAN) gait training algorithm based on the adaptive impedance control, which was proven practicable that the adaptive impedance control system is able to provide gait motion training in [8]. In [9], an impedance control method based on the position is proposed for a specific task. In that paper, the neural network estimated and compensated the uncertainty of the dynamic model. An adaptive robust controller, which generated by an impedance controller, is designed to track the desired position.

In this paper, a new torque controller based on the single neuron network (SNN) is proposed for the tendon-driven finger. Based on an error measured in terms of desired joint angles and actual joint angles, the torque controller calculates the desired joint torques. The proposed controller calculates actuator positions based on the error measured by actual joint torques and desired joint torques. The actuator position and rate-of-change of the length of the tendons are controlled by a PI controller and the adaptive SNN controller, respectively. New actuator position which is combined with the rate-of-change of the length of the tendons can be transformed into desired forces directly. The simple adaptive controller was proposed to increase the ability to adapt to the changing environment position or stiffness by adjusting the parameters of the SNN controllers.

The proposed algorithm has an implementation on top of the pure PI controller for controlling the tendon-driven fingers. The structure of the paper is as follows: In Sect. 2 the kinematics is reviewed for the tendon-driven finger. In Sect. 3 an adaptive torque controller in joint-space integrated with a torque controller is proposed for the tendon-driven finger. In Sect. 4 simulations have been conducted to verify the efficiency of the controller presented in this paper. Finally, concluding remarks are given in Sect. 5.

2   Kinematics of the Tendon Driven Finger

2.1   The Design of the Finger

A model of the tendon-driven finger in ADAMS consists of four DOFs (Degrees of freedom), as shown in Fig. 1. For the model shown here, the fingertip's motion depends on the coupled link, as shown in Fig. 1. The actuation system of the robot hand is remotely packaged in the forearm, which makes the size of the robot hand as large as a man's hand. Each unit of the actuation system consists of a brushless motor and a lead screw. The lead screw can convert rotary motion to linear motion. Each of the tendons connects the finger joint and the lead screw. The motors can drive the lead screws to control the finger motion through the tendons. Since the tendons can only transmit forces in tension, the number of tendons should be more than the DOFs. It turns out that only one tendon more than the number of the DOF is needed, so a three-DOF finger needs four tendons. Then $\tau = (\tau_1\ \tau_2\ \tau_3)^T$ denotes a vector of joint torques acting on the finger, and $x = (x_1\ x_2\ x_3\ x_4)^T$ is a vector of tendon positions.

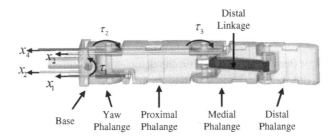

Fig. 1. The model of the tendon-driven finger

2.2   Kinematics

Let q and f represent joint angles vector and tendon tensions vector, respectively. For the above three-DOF finger, the relationship between the joint torques and the tendon tensions is given as follows,

$$\tau = Rf \tag{1}$$

where R is the mapping matrix translating the tendon tensions to the joint torques. The matrix consists of the joint radii data and must be full row rank.

Given the joint torques, the tendon tensions can be obtained from Eq. (1) as,

$$f = f_a + f_{in} \tag{2}$$

where f_a is the special solution of Eq. (1) which generates the joint torques, and f_{in} is the homogeneous solution and represents the internal tensions of the tendons which will not affect the torques of the finger. Both f_a and f_{in} are given as follows,

$$f_a = R^+\tau, \quad f_{in} = W^T t \tag{3}$$

where R is the pseudo-inverse of the matrix R, and W spans the right null space with the matrix R. The constant t is chosen for ensuring the tendon tensions are all positive.

As long as the tendons remain taut, tendons will be elastic. The relation between the joint motion and the actuator motion is governed by the same matrix, as follows,

$$\dot{x} = R^T\dot{q} + \dot{l} \tag{4}$$

where $\dot{x}$ indicates the tendon velocities in the actuator-side. $\dot{q}$ and $\dot{l}$ are the velocities of the joint angle and the rate-of-change of the tendons, respectively.

Integrating these velocities and assuming the constant R:

$$x = R^T q + \Delta l \tag{5}$$

where the Δl denotes a change in length relative to the zero-positions length. Then, modeling the tendons as linear springs with the stiffness k_t and assuming that the springs remain taut. Considering the un-stretched length of the tendons, the force f is proportional to the Δl:

$$f = k_t \Delta l \tag{6}$$

Of course, an external load may cause the actual forces to drop to zero or to reach excessive highs. But the forces will return to their original state once the load removes.

3 Adaptive SNN Torque Controller

3.1 The Structure of the Control System

Neural network is a powerful tool which may strengthen the ability to control the joint angles of the finger when it interacts with the uncertain environments. The classical PI controllers are the standard approaches and, in other words, the simple ones, which can also result in a fast-rise time. Unfortunately, overshoot may be caused of the response. For the three-DOF finger, four tendons are controlled to fully generate the torques of the joints together with the stiffness control. Each joint is supposed to have an independent torque. Every joint was expected to obey a desired stiffness relationship.

In this paper, a torque controller integrated with an adaptive SNN controller was presented for the tendon-driven finger. The block diagram of the controller is illustrated in Fig. 2.

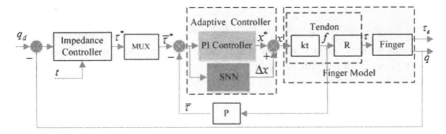

Fig. 2. The components of the control system

For the tendon-driven finger, joint motion and tendon motion are usually coupled. The coupled joint motions are derived by the coupled nature of tendon-driven manipulators. For a controller in tendon space, the coupled kinematics of the tendons can exhibit a transient coupling in their response. Accordingly, an ideal controller was presented here in joint space. The control system was built for the three-DOF finger as shown in Fig. 2.

First, the impedance controller takes the difference between the desired joint angle and the actual angle and passed down the desired joint torque. Second, the input of the PI controller is the difference between the desired joint torque and the actual joint torque and output is the position of the joint-side tendon. x represents the actuator positions which are the sum of the joint contribution plus Δx. Δx is the rate-of-change of the length of the tendon which defined as the output of the adaptive SNN controller. And the input of the adaptive SNN controller is similar to that of the PI controller. Finally, the output of the finger is the actual joint angle and the tendon tensions.

3.2 Impedance Controller

The impedance controller was proposed according to the joint position error. The joint torque would be proportional to the difference between the desired joint angle and the actual joint angle for the particular joint being controlled, as follows,

$$\tau = k(q_d - q) \tag{7}$$

where k translates the difference of the joint angle into the joint torque, acting as a stiffness matrix. q_d is the vector which represents the desired joint angles and q is the actual joint angle.

The combined signal will be sent to the converter to limiting the tendon tensions. First, it is necessary to converter the signal from joint space to the tendon space using the inverse matrix P^{-1} for the actuators to exert forces. Second, the matrix L provides the limiting term for the tensions in the tendon space to protect tendons. Third, it is necessary to use the matrix P to converter the signal from the tendon space to the joint space, which can be represented as follows,

$$\bar{\tau}^* = \begin{pmatrix} \tau^* \\ t^* \end{pmatrix} = PLP^{-1} \begin{pmatrix} \tau \\ t \end{pmatrix} \tag{8}$$

where $P = [R^T \quad W^T]^T$, $\bar{\tau} = [\tau^T \quad t^T]^T$, and $[\tau^{*T} \quad t^{*T}]$ is the column denoting the desired joint torques, t and t^* are the actual internal tendon tension signal and desired internal tendon tension signal respectively. The constant t was provided to ensure the positive tendon tensions.

3.3   PI Controller

The tendon position of the actuator-side is calculated based on the error between the desired joint torques and the actual joint torques as follows,

$$x^* = P^T(k_p + \frac{k_i}{s})(\bar{\tau}^* - \bar{\tau}) \tag{9}$$

where x^* is the output of the PI controller. Both k_p and k_i denote the proportional gain matrix and integration gain matrix respectively. According to Eqs. (3) and (8), the above equation can be rewritten as follows,

$$x^* = (k_p + \frac{k_i}{s})(R^T k(q_d - q) + W^T t - P^T Pf) \tag{10}$$

Unfortunately, the PI controller may cause overshoot in the transient response and this situation is quite undesirable for the motion of the tendon-driven finger. Moreover, the pure PI controller might not adjust itself when acting with the uncertain environment.

3.4   SNN Controller

To avoid the former problems, the new controller based on the adaptive SNN controller is presented here. The new controller does a good job of tracking the desired joint angles and little overshoot with the faster settle time. The adaptive SNN control law implements a compensator. According to the description above, x is expressed as,

$$x = x^* + \Delta x \tag{11}$$

For the sake of convenience, the control law is written as,

$$\Delta x = (K(w_1 x_1 + w_2 x_2 + w_3 x_3)) * (\bar{\tau}^* - \bar{\tau}) \tag{12}$$

where $[\bar{\tau}^* - \bar{\tau}]$ is the input of the controller, and w_1, w_2 and w_3 are the weights of the adaptive SNN controller. The matrix K is the proportion coefficient ($K > 0$). The output of the controller is denoted as Δx, and n is a number on behalf of the iterations.

Given $\Delta x = y(n) * (\bar{\tau}^* - \bar{\tau})$, the $y(n)$ will be calculated as follows,

$$y(n) = y(n-1) + K \sum_{i=1}^{3} w_i(n)x_i(n) \tag{13}$$

where n >= 2, K is the ratio coefficient of neuron network, which affects the performance of the system greatly. To ensure the convergence of the learning algorithm and the robustness of the control, the output of the controller may be improved as,

$$y(n) = y(n-1) + K \sum_{i=1}^{3} w_i'(n)x_i(n) \tag{14}$$

$$w_i'(n) = w_i(n) / \sum_{i=1}^{3} |w_i(n)| w \tag{15}$$

According to the dynamic of the finger and ignoring the some small terms, Eq. (11) can be rewritten as,

$$\Delta x = K(w_1x_1 + w_2x_2 + w_3x_3)(k(q_d - q) - Rf) \tag{16}$$

3.5 Finger Model

To start the analysis, we consider the equation of motion for the tendon-driven finger after simplifying. For our purposes now, assume that the external forces are zero.

$$M\ddot{q} + \eta = \tau \tag{17}$$

where M is the joint-space inertia matrix and η represents the sum of coriolus centripetal gravitational and frictional forces which are assumed to be zero.

According to Eqs. (5) and (6), the Eq. (1) can be rewritten as:

$$\tau = Rk_t(x - R^T q) \tag{18}$$

The analysis provides theoretical validation for the claims of the previous section. And the experimental validation will follow in the next section.

4 Simulation

4.1 Simulation Setups

In this section, simulation has been conducted to compare the performance of the proposed controller with that of the PI controller proposed in [10] for a low-inertia finger. The finger has four tendons and three independent DOF which actuated by the four tendons. For the sake of convenience, the coriolus, centripetal and gravitational terms were ignored, and the inertial matrix M is formulated for controlling the finger.

The parameters of the adaptive SNN controller are shown as follows,

$$\eta_P = 120,\ \eta_I = 0.5,\ \eta_D = 50,\ K = 0.000030$$

The initial values of the weights of the adaptive controller and that of the output of the SNN controller are given as,

$$w_1 = w_2 = w_3 = 0.5,\ y(0) = 0$$

4.2 Simulation Results

In this section, we conducted the simulation using MATLAB. The control law operated on the tendon-driven finger was proved available. The search algorithms are evaluated and the controller performances are presented along with simulations and experimental results.

The step response of each controller was tested using a step input of (q_1 q_2 q_3). In the experiment, the desired joint angle q_3 was set from 10 to 40°, and the other joint references were kept constant at 10°. The setting time of the medial joint dropped from 3 s to 1 s as shown in Fig. 5. The new controller produced little overshoot on the response of the three joints, but a slow rise time before 10 s. The comparison between the SNN controller and PI controller testing on the finger, including the yaw joint, the proximal joint and the medial joint was applied here. In Fig. 3, the settle time of the yaw joint increased from 12 s to 20 s, but the SNN controller generated little overshoot

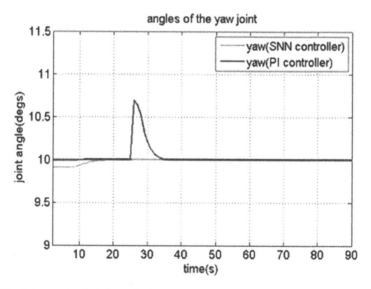

Fig. 3. The joint angle of the yaw joint for the two controllers under $q_1 = 10$ (Color figure online)

than the PI controller. The controller parameters are shown in the formal section. The performance maximized with the gains tuned suitably. The simulation results of the proximal joint are similar to that of the yaw joint. In Fig. 4, the PI controller produced the faster rise time, the overshoot, however, cannot be avoided. On the other word, the SNN controller produced little overshoots.

The clear phenomenon was produced on the medial joint, as the step response of the controllers was tested using a step input q_3 increased from 10 to 40°. Through our experiences, the new controller responded without obvious overshoot. The proposed controller, however, resulted in the slow rise time.

The simulation was performed to evaluate that the adaptive SNN controller is priority to the pure PI controller in controlling the finger when in uncertain environment. A classical PI controller is not the best choice for the tendon-driven fingers. Then a new impedance controller is the most common way to deal with the situation where the finger handles environmental interactions. When a step input of $(q_1\ q_2\ q_3)$ was commanded, the simulation results were shown in Figs. 3, 4 and 5. The red line shows the joint angle controlled by the adaptive SNN controller, and the blue line shows that controlled by the PI controller. The PI controller increased the overshoot, delayed the response time and made the motion in the finger unsightly. The adaptive SNN controller, however, decreased the overshoot in a great extent, significantly increased the speed of the response. The blue line shows the response controlled by the SNN controller and the red line shows that controlled by the PI controller.

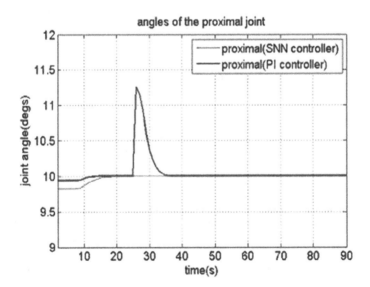

Fig. 4. The proximal joint response for the $q_2 = 10$ (Color figure online)

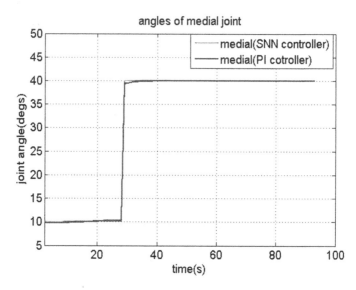

Fig. 5. The medial joint is controlled with the step input changed from 10° to 40° (Color figure online).

5   Conclusion

In this paper, a new torque control in the joint space for the tendon-driven finger was proposed based on the adaptive SNN control. The new torque controller can command the position of the finger by regulating the parameters of the adaptive SNN controller. Simulation results have shown that the proposed controller is more robust than the pure PI controller to the external disturbance on the finger.

References

1. Ozawa, R., Hashirii, K., Kobayashi, H.: Design and control of under-actuated tendon driven mechanisms. In: 9th IEEE International Conference on Robotics and Automation, pp. 1522–1527. IEEE Press, New York (2009)
2. Abdallah, M.E., Robert, P.J., Charles, W.W., Hargrave, B.: Applied joint-space torque and stiffness control of tendon-driven fingers. In: 10th IEEE-RAS International Conference on Humanoid Robots (Humanoids), pp. 74–79. IEEE Press, New York (2010)
3. Abdallah, M.E., Platt, R.J., Wampler, C.W.: Decoupled torque control of tendon-driven fingers with tendon management. Int. J. Robot. Res. **32**, 247–258 (2013)
4. Abdallah, M.E., Charles, W., Platt, J.R.: Object impedance control using a closed chain task definition. In: 10th IEEE-RAS International Conference on Humanoid Robots (Humanoids), pp. 26–274. IEEE Press, New York (2010)
5. Lee, Y.T., Choi, H.R., Chung, W.K., Youm, Y.: Stiffness control of a coupled tendon-driven robot hand. IEEE Control Syst. Mag. **14**, 10–19 (1994). IEEE Press, New York
6. Platt, R.J., Abdallah, M.E., Wampler, C.W.: Multiple priority impedance control. In: Proceeding of the IEEE International Conference on Robotics and Automation, pp. 6033–6038. IEEE Press, New York (2011)

7. Alqaudi, B., Modares, H., Ranatunga, I.: Model reference adaptive impedance control for physical human-robot interaction. Control Theor. Technol. **14**, 68–82 (2016). Springer, Heidelberg

8. Hussain, S., Xie, S.Q., Jamwal, P.K.: Adaptive impedance control of a robotic orthosis for gait rehabilitation. IEEE Trans. Cybern. **43**, 1025–1043 (2013). IEEE Press, New York

9. Huang, P., Meng, Z., Wang, D.: Impact dynamic modeling and adaptive target capturing control for tethered space robots with uncertainties. IEEE/ASME Trans. Mechatron. **21**, 2260–2271 (2016). IEEE Press, New York

10. Reiland, M.J., Platt, R., Charles, W.W.I., Abdallah, M.E., Hargrave, B.: U.S. Patent No. 8,060,250. U.S. Patent and Trademark Office, Washington, DC (2011)

Application of Human Learning Optimization Algorithm for Production Scheduling Optimization

Xiaoyu Li[1], Jun Yao[1(✉)], Ling Wang[1], and Muhammad Ilyas Menhas[2]

[1] School of Mechatronic Engineering and Automation, Shanghai University, Shanghai 200072, China
grandone0529@shu.edu.cn
[2] Department of Electrical (Power) Engineering, Mirpur University of Science and Technology, MUST, Mirpur AJ&K, Pakistan

Abstract. In this paper, Human Learning Optimal (HLO) algorithm is presented to solve the scheduling problem. HLO is a meta-heuristic search algorithm which is inspired by the process of human learning. Three learning operators are developed to generate new solutions and search for the optima by mimicking the learning behaviors of human. This new algorithm has been proved to be very effective in solving optimization problems. HLO is applied to solve an actual production scheduling problems in a dairy factory and the performance of HLO is compared with that of two other meta-heuristics algorithms, BSO-PSO and HGA. Comparison results demonstrate that HLO is a promising optimization algorithm.

Keywords: Human learning optimization · Meta-heuristic · Scheduling problem

1 Introduction

Scheduling optimization problems are wide ranging and plentiful, which people always find in various fields of engineering, scientific, economic management and so on [1]. However, optimization problems are becoming more and more complicated with the development of science and technology, and traditional gradient-based methods are inefficient and inconvenient for such problems as they require substantial gradient information, depend on a well-define starting point, and need a large amount of enumeration memory. Meta-heuristics mimic nature biological systems to solve various kinds of optimization problems. For instance, Genetic Algorithms (GAs) [2], Particle

J. Yao—This work was financially supported by the Science and Technology Commission of Shanghai Municipality of China under Grant (No.17511107002).

© Springer Nature Singapore Pte Ltd. 2017
M. Fei et al. (Eds.): LSMS/ICSEE 2017, Part I, CCIS 761, pp. 242–252, 2017.
DOI: 10.1007/978-981-10-6370-1_24

Swarm Optimization (PSO) [3], Ant Colony Optimization (ACO) [4], Tabu Search [5] are popular with optimizers. In recent years, more and more novel algorithms have been proposed to tackle optimization problems, such as the Binary Differential Evolution algorithm (BDE) [6], the Binary Artificial Bee Colony Algorithm (BABCA) [7], the Binary Bat Algorithm (BBA) [8], the Binary Flower Pollination Algorithm (BFPA) [9], the Binary Gravitation Search Algorithm (BGSA) [10], the Binary Simulated Annealing Algorithm (BSAA) [11], and the Bi-Velocity Discrete Particle Swarm Optimization (BVDPSO) [12].

To solve hard optimization problems more effectively and efficiently, new powerful meta-heuristics inspired by nature, especially by biological systems, must be explored, which is a hot topic in evolutionary computation community [13]. As is known to all, human beings are the smartest creatures in the earth and invent numbers of machines and tools to make our lives convenient, which indicates human beings are gifted to tackle many complicated problems. People master and improve skills through repeatedly learning which is similar to iteratively searching for best solutions by optimal algorithms. The process can be considered as an optimization of iterative process. For the example of learning Sudoku, a person may learn randomly due to the lack of prior knowledge or exploring new strategies (random learning), learn from his or her previous experience (individual learning), and learn from his or her friends and books (social learning) [14, 15]. Inspired by human learning mechanisms, a new meta-heuristic algorithm called human learning optimization (HLO) algorithm is presented. The performance of HLO to solve the continuous optimization problems such as a suit of numerical benchmark functions, deceptive problems, 0–1 knapsack problems [16, 17] has been proved. HLO will be applied to an actual production scheduling problems in this work.

The rest of the paper is organized as follows. Section 2 introduces the presented HLO in detail. In Sect. 3, HLO is applied to an actual production scheduling problems in a dairy factory and the results are compared with those of other meta-heuristics collected from recent works to validate its performance. Finally, Sect. 4 gives a conclusion of the work in this paper.

2 Human Learning Optimization Algorithm

2.1 Initialization

HLO adopts the binary-coding framework in which each bit corresponds to a basic component of knowledge to solve problems. Therefore, an individual, i.e. a candidate solution, is represented by a binary string as Eq. (1) which is initialized as "0" or "1" randomly assuming that there is no prior-knowledge of problems,

$$x_i = \begin{bmatrix} x_{i11} & x_{i12} & \cdots & x_{i1j} & \cdots & x_{i1M} \\ x_{i21} & x_{i22} & \cdots & x_{i2j} & \cdots & x_{i2M} \\ \vdots & \vdots & & \vdots & & \vdots \\ x_{ip1} & x_{ip2} & \cdots & x_{ipj} & \cdots & x_{ipM} \\ \vdots & \vdots & & \vdots & & \vdots \\ x_{iP1} & x_{iP2} & \cdots & x_{iPj} & \cdots & x_{iPM} \end{bmatrix} \tag{1}$$

$$x_{ipj} \in \{0,1\}, 1 \le p \le P, 1 \le j \le M$$

where xi is the ith individual, P is the number of product, and M is the number of machine i.e.

After all the individuals are initialized, the initial population of HLO is generated as Eq. (2). N is the number of individuals of the population.

$$X = \begin{bmatrix} x_1 & x_2 & \cdots & x_i & \cdots & x_N \end{bmatrix}^T, 1 \le i \le N \tag{2}$$

2.2   Learning Operators

Random Exploration Learning Operator. While learning to solve unfamiliar problems, people usually learn randomly because of lacking or forgetting knowledge. Simulating these phenomena, HLO performs the random exploration learning with some probability as Eq. (3)

$$x_{ij} = RE(0,1) = \begin{cases} 0, & 0 \le rand() \le 0.5 \\ 1, & else \end{cases} \tag{3}$$

Where $RE(0,1)$ is a random number in [0, 1).

Individual Learning Operator. Individual learning is defined as the ability to build knowledge through individual reflection about external stimuli and sources [18]. Every person learns in conscious or unconscious states, which is a fundamental requirement of existence. In HLO, an individual learns to solve problems by the individual learning operator based on its own experience which is stored in the Individual Knowledge Database (IKD) as Eqs. (4) and (5),

$$x_{ipj} = ik_{ipj} \tag{4}$$

$$IKD_i = \begin{bmatrix} ikd_{i1} \\ ikd_{i2} \\ \vdots \\ ikd_{ip} \\ \vdots \\ ikd_{iP} \end{bmatrix} = \begin{bmatrix} ik_{i11} & ik_{i12} & \cdots & ik_{i1j} & \cdots & ik_{i1M} \\ ik_{i21} & ik_{i22} & \cdots & ik_{i2j} & \cdots & ik_{i2M} \\ \vdots & \vdots & & \vdots & & \vdots \\ ik_{ip1} & ik_{ip2} & \cdots & ik_{ipj} & \cdots & ik_{ipM} \\ \vdots & \vdots & & \vdots & & \vdots \\ ik_{iP1} & ik_{iP2} & \cdots & ik_{iPj} & \cdots & ik_{iPM} \end{bmatrix} \tag{5}$$

$$IKD = \begin{bmatrix} ikd_1 & ikd_2 & \cdots & ikd_i & \cdots & ikd_N \end{bmatrix}^T$$

$$1 \leq i \leq N, 1 \leq i \leq N, 1 \leq p \leq P, 1 \leq j \leq M$$

where IKD_i is the individual knowledge database of person i which stands for the ith best solution of person I and P denotes the size of the IKDs.

Social Learning Operator. Social learning is a transmission of knowledge and skills through direct or indirect interactions among individuals. In the social context, people can learn from not only their own direct experience but also the experience of the other members, and therefore they can develop further their abilities and achieve the higher efficiency with an effective knowledge sharing. To possess the efficient search ability, the social learning mechanism is mimicked in HLO. Like human learning, each individual of HLO studies the social knowledge which is stored in the Social Knowledge Database (SKD) with some probability as Eqs. (6) and (7) when it yields a new solution,

$$x_{ipj} = sk_{pj} \tag{6}$$

$$SKD = \begin{bmatrix} sk_{11} & sk_{12} & \cdots & sk_{1j} & \cdots & sk_{1M} \\ sk_{21} & sk_{22} & \cdots & sk_{2j} & \cdots & sk_{2M} \\ \vdots & \vdots & & \vdots & & \vdots \\ sk_{p1} & sk_{p2} & \cdots & sk_{pj} & \cdots & sk_{pN} \\ \vdots & \vdots & & \vdots & & \vdots \\ sk_{P1} & sk_{P2} & \cdots & sk_{Pj} & \cdots & sk_{PN} \end{bmatrix} \tag{7}$$

$$1 \leq p \leq P, 1 \leq j \leq M$$

where the SKD is the best knowledge of the social collectivities. The S best individuals are selected as initial social knowledge database stored in the SKD. In the following iterated searching, the IKD and SKD will be updated if new knowledge is better than that in the IKD and SKD.

In summary, HLO yields a new solution by means of random exploration learning, individual learning and social learning with certain rates, which can be simplified and formulated as

$$x_{ipj} = \begin{cases} RE(0,1) & 0 \le rand \le pr \\ ikd_{ipj} & pr \le rand \le pi \\ skd_{pj} & else \end{cases} \tag{8}$$

where x_{ipj} is the productivity of the pth product on jth machine, pr is the probability of random exploration learning, (pi-pr) and (1-pi) represent the rates of individual learning and social learning, respectively.

2.3 Updating of the IKD and SKD

After individuals accomplish learning in each generation, the fitness values of new solutions that evaluated through the fitness function f(x) are obtained. If fitness values of new candidate solutions are better than the worst one in the IKD, or the dimension of individual knowledge which is stored in the IKD is less than N*P, the new candidate solutions will be saved in the IKD. In the same way, the SKD is updated. The SKD is updated by at least one solution every iteration in case of falling into local optimum.

3 Establishment of the Model and Experimental Results

3.1 Establishment of the Model

Scheduling problems have a major impact on the productivity of a manufacturing system, which can be described as follows. Given a number of tasks which must be carried out by some processors, it is required to find the best resource assignments and tasks sequencing, For example, the total completion time needs to be as small as possible. The study of production scheduling problem in this paper is based on actual production procedure in a dairy factory. The efficiency and capacity of each product on each machine are fixed and the demands for the output of dairy products in daily order are variant. The purpose of production scheduling is to distribute different kinds of dairy products to different kinds of equipment and minimize production time. The production scheduling model can be described as follows.

$$minT = max(t_1, t_2, \cdots, t_i, \cdots, t_m)$$

$$t_i = \sum_{j=1}^{n} \left(\frac{c_{ij}}{\eta_{ij} v_{ij}} + b \times tc_{ij} \right)$$

$$s.t. \quad c_i = \sum_{j=1}^{n} c_{ij} \tag{9}$$

where t_i is one day runtime of the ith machine, c_{ij} is the amount of the ith product on the jth machine, ηij is the Productivity of the ith product on jth machine, vij is the capacity of the ith product on jth machine, b is 0 or 1, If it needs to switch to other milk after the ith product, b is equal to 1 and if not, b is equal to 0. The number of ith production on the jth machine c_{ij} is between $minc_{ij}$ and $maxc_{ij}$ as Eq. (10)

$$\min c_{ij} \le c_{ij} \le \max c_{ij} \qquad (10)$$

In summary, the procedure of HLO for scheduling problems in a dairy factory can be concluded as follows:

Step 1: Set the parameters and initialize the population of HLO;

Step 2: Repair the initial solution as Eq. (10), calculate the fitness of initial individuals and initialize the IKDs and SKD ;

Step 3: If reaching the maximum generation, the results will be output; otherwise go to step 5.

Step 4: Generate new individuals by performing the three learning operators as Eq. (8);

Step 5: If the new individuals is feasible solution, go to step 7, otherwise repair the new solution as Eq. (10);

Step 6: Evaluate the fitness of each individual and update the IKD, SKD ;

Step 7: If reaching the maximum generation, the results will be output; otherwise repeat step 4-step 6.

3.2   Experimental Results and Discussions

Production scheduling based on HLO is carried out according to production plan of a certain day in a dairy factory to verify the effectiveness of HLO. 25 kinds of milk products need to be produced on the 10 machines. The efficiency and capacity of each product on each machine are show in Tables 1 and 2. Table 3 is orders of one day in a dairy factory.

Table 1. Capacity on each machine.

(piece/h)10^3	M1	M2	M3	M4	M5	M6	M7	M8	M9	M10
P1	18	20	18	29	29	29	39	39	39	39
P2	18	20	18	/	/	/	39	39	39	39
P3	18	20	18	29	29	29	/	/	/	/
P4	15	/	18	/	/	/	38	38	38	38
P5	/	/	18	29	29	29	35	35	35	35
P6	/	/	/	29	29	29	35	35	35	35
P7	/	19	/	29	29	29	/	/	/	/
P8	18	19	/	29	29	29	/	/	/	/
P9	18	16	16	/	/	/	35	35	35	35
P10	18	16	16	25	25	25	/	/	/	/
P11	16	18	17	/	/	/	30	30	30	30
P12	16	19	/	/	/	/	30	30	30	30

(*continued*)

Table 1. (*continued*)

(piece/h)10^3	M1	M2	M3	M4	M5	M6	M7	M8	M9	M10
P13	15	16	16	/	/	/	30	30	30	30
P14	18	20	/	25	25	25	/	/	/	/
P15	/	20	18	/	/	/	30	30	30	30
P16	/	18	20	25	25	25	/	/	/	/
P17	/	/	/	25	25	25	30	30	30	30
P18	16	/	18	25	25	25	/	/	/	/
P19	/	20	/	/	/	/	30	30	30	30
P20	/	/	/	20	20	20	30	30	30	30
P21	/	/	/	20	20	20	/	/	/	/
P22	/	/	/	20	20	20	30	30	30	30
P23	15	20	19	/	/	/	30	30	30	30
P24	/	/	/	/	/	/	30	30	30	30
P25	/	19	20	/	/	/	30	30	30	30

Table 2. Efficiency on each machine.

(0 ~ 1)	M1	M2	M3	M4	M5	M6	M7	M8	M9	M10
P1	0.6	0.7	0.65	0.8	0.78	0.62	0.54	0.48	0.72	0.45
P2	0.58	0.72	0.65	/	/	/	0.66	0.71	0.58	0.67
P3	0.42	0.68	0.68	0.62	0.71	0.68	/	/	/	/
P4	0.62	/	0.55	/	/	/	0.77	0.64	0.63	0.71
P5	/	/	0.77	0.55	0.53	0.56	0.62	0.65	0.48	0.51
P6	/	/	/	0.51	0.43	0.53	0.47	0.61	0.55	0.48
P7	/	0.63	/	0.55	0.61	0.43	/	/	/	/
P8	0.6	0.45	/	0.57	0.44	0.48	/	/	/	/
P9	0.62	0.57	0.64	/	/	/	0.61	0.65	0.45	0.61
P10	0.57	0.7	0.54	0.62	0.61	0.74	/	/	/	/
P11	0.44	0.63	0.76	/	/	/	0.55	0.65	0.59	0.61
P12	0.7	0.54	/	/	/	/	0.44	0.38	0.62	0.58
P13	0.55	0.33	0.46	/	/	/	0.51	0.56	0.77	0.49
P14	0.62	0.5	/	0.71	0.58	0.62	/	/	/	/
P15	/	0.73	0.73	/	/	/	0.68	0.55	0.57	0.61
P16	/	0.71	0.68	0.48	0.55	0.57	/	/	/	/
P17	/	/	/	0.58	0.59	0.44	0.47	0.6	0.49	0.52
P18	0.54	/	0.64	0.52	0.65	0.58	/	/	/	/
P19	/	0.82	/	/	/	/	0.71	0.62	0.65	0.55
P20	/	/	/	0.6	0.71	0.58	0.45	0.49	0.51	0.61
P21	/	/	/	0.7	0.68	0.69	/	/	/	/
P22	/	/	/	0.65	0.56	0.71	0.68	0.62	0.58	0.71
P23	0.32	0.45	0.55	/	/	/	0.54	0.51	0.43	0.47
P24	/	/	/	/	/	/	0.61	0.58	0.54	0.59
P25	/	0.66	0.43	/	/	/	0.54	0.47	0.42	0.51

Table 3. Orders of each production.

Product	P1	P2	P3	P4	P5	P6	P7	P8	P9	P10
Output (10^3)	100	150	200	80	60	56	50	8	100	40
Product	P11	P12	P13	P14	P15	P16	P17	P18	P19	P20
Output (10^3)	50	80	/	200	280	180	/	28	320	30
Product	P21	P22	P23	P24	P25					
Output (10^3)	10	50	/	40	/					

The performance of HLO is compared with the BSO-PSO (Brain Storm Optimization with Discrete Particle Swarm Optimization) [19] and HGA (Hybrid Genetic Algorithms) [20]. All experimental tests were implemented on a PC of Intel Core CPU i7-2700 K @ 3.50 GHz with 8 GB RAMs. A set of fair parameters are chosen for all three algorithms in this paper. For example, population size is set to 20 and iteration times are 300. Other diverse parameters that are used in HLO, BSO-PSO, and HGA are listed in Table 4.

Table 4. Parameters settings of HLO, BSO-PSO, and HGA.

Algorithms	Parameters		
HLO	$Pr = 0.01$		$Pi = 0.07$
BSO-PSO	$Pone = 0.6$		$Pinverse = 0.8$
HGA	$PC = 0.9$		$Pm = 0.02$

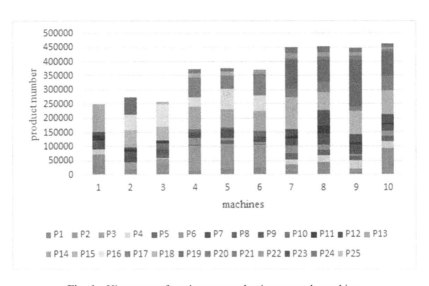

Fig. 1. Histogram of equipment production on each machine.

Histograms of task arrangement are given in Fig. 1 and histograms of runtime are given in Fig. 2 to display the task arrangement.

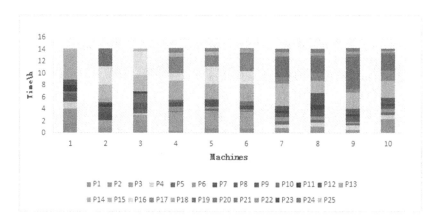

Fig. 2. Histogram of Runtime on each machine.

Figure 1 displays the task arrangement and illustrates that the task is evenly distributed to each machine. Figure 2 indicates that the run time of each machine is almost the same and no machine is idle. Therefore the utilization of machines are improved.

Table 5 shows that HLO, BSO-PSO and HGA find solutions respectively after running 30 times. Row "Best" denotes the best solution from each algorithm. Row "Mean" denotes the mean value of the total run solutions from each algorithm. Row "Worst" denotes the worst solution found from each algorithm. As we can see, all results from HLO in Table 5 are better than other two algorithms.

Table 5. Result of HLO, BSO-PSO, HGA.

Algorithms	HLO	BSO-PSO	HGA
Best	14.57	14.96	15.32
Mean	14.93	15.19	16.04
Worst	16.5	18.3	18.01

Figure 3 illustrates the iteration curves of the optimal algorithms in this paper. It is no surprise that the convergence speed of HLO is faster than both BSO-PSO and HGA.

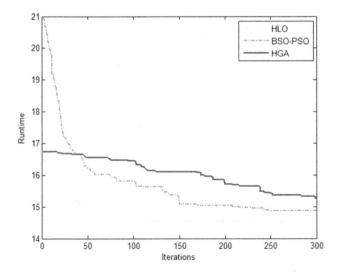

Fig. 3. The iteration curves of HLO, BSO-PSO and HGA.

4   Conclusion

In this paper, a novel human learning optimization algorithm (HLO) is presented which is inspired by the human learning process. In this method, three learning operations, i.e. the random learning operator, the individual learning operator, and the social learning operator, are developed by mimicking human learning behaviors to generate new solutions and search for the optimal solution of problems. The performance of this proposed method is validated by applying to an actual production scheduling problems in a dairy factory. The experimental results show that the performance of the proposed method is better than the performance of compared methods.

References

1. Mullen, R.J., Monekosso, D., Barman, S., et al.: A review of ant algorithms. J. Expert Syst. Appl. **36**(6), 9608–9617 (2009)
2. Elsayed, S.M., Sarker, R.A., Essam, D.L.: A new genetic algorithm for solving optimization problems. J. Eng. Appl. Artif. Intell. **27**(1), 57–69 (2014)
3. Fang, W., Sun, J., Chen, H., et al.: A decentralized quantum-inspired particle swarm optimization algorithm with cellular structured population. J. Inf. Sci. **330**, 19–48 (2016)
4. Prakasam, A., Savarimuthu, N.: Metaheuristic algorithms and probabilistic behaviour: a comprehensive analysis of Ant Colony Optimization and its variants. J. Artif. Intell. Rev. **45** (1), 1–34 (2015)
5. Qin, J., Xu, X., Wu, Q., et al.: Hybridization of tabu search with feasible and infeasible local searches for the quadratic multiple knapsack problem. J. Comput. Oper. Res. **66**, 199–214 (2016)
6. Chen, Y., Xie, W., Zou, X.: A binary differential evolution algorithm learning from explored solutions. Neurocomputing **149**, 1038–1047 (2015)

7. Zhang, X., Zhang, X.: A binary artificial bee colony algorithm for constructing spanning trees in vehicular ad hoc networks. Ad Hoc Netw. **58**, 198–204 (2017)
8. Mirjalili, S., Mirjalili, S.M., Yang, X.S.: Binary bat algorithm. Neural Comput. Appl. **25**(3–4), 1–19 (2014)
9. Rodrigues, D., Silva, G.F.A., Papa, J.P., et al.: EEG-based person identification through binary flower pollination algorithm. Expert Syst. Appl. **62**, 81–90 (2016)
10. Automatic channel selection in EEG signals for classification of left or right hand movement in Brain Computer Interfaces using improved binary gravitation search algorithm
11. Li, X., Ma, L.: Minimizing binary functions with simulated annealing algorithm with applications to binary tomography. Comput. Phys. Commun. **183**(2), 309–315 (2012)
12. Shen, M., Zhan, Z.H., Chen, W.N., Gong, Y.J., Zhang, J., Li, Y.: Bi-velocity discrete particle swarm optimization and its application to multicast routing problem in communication networks. IEEE Trans. Ind. Electron. **61**(12), 7141–7151 (2014)
13. Fister, J.I., Yang, X.S., Fister, I.: A brief review of nature-inspired algorithms for optimization. Elektroteh. Vestn. **80**(3), 1–7 (2013)
14. Wang, L., Ni, H., Yang, R., et al.: A simple human learning optimization algorithm. J. Commun. Comput. Inf. Sci. **462**, 56–65 (2014)
15. Wang, L., An, L., Pi, J., et al.: A diverse human learning optimization algorithm. J. Global Optim. **67**(1–2), 1–41 (2016)
16. Wang, L., Yang, R., Ni, H., et al.: A human learning optimization algorithm and its application to multi-dimensional knapsack problems. J. Appl. Soft Comput. **34**, 736–743 (2015)
17. Wang, L., Ni, H., Yang, R., et al.: An adaptive simplified human learning optimization algorithm. J. Inf. Sci. **320**, 126–139 (2015)
18. Forcheri, P., Molfino, M.T., Quarati, A.: ICT driven individual learning: new opportunities and perspectives. J. Educ. Technol. Soc. **3**(1), 51–61 (2000)
19. Hua, Z., Chen, J., Xie, Y.: Brain storm optimization with discrete particle swarm optimization for TSP. In: 2016 12th International Conference on Computational Intelligence and Security (CIS), pp. 190–193. IEEE (2016)
20. Kundakcı, N., Kulak, O.: Hybrid genetic algorithms for minimizing makespan in dynamic job shop scheduling problem. J. Comput. Ind. Eng. **96**, 31–51 (2016)

An Improved WKNN Indoor Fingerprinting Positioning Algorithm Based on Adaptive Hierarchical Clustering

Jian Li[1], Jingqi Fu[1(✉)], Ang Li[1], Weihua Bao[2], and Zhengming Gao[2]

[1] School of Mechatronic Engineering and Automation, Shanghai University,
NO. 149 Yanchang Road, JingAn District, Shanghai 200072, China
jqfu@staff.shu.edu.cn
[2] Shanghai Automation Instrumentation Co., Ltd., Shanghai, China

Abstract. Aiming at the dependence of the traditional indoor clustering positioning accuracy on the initial center and clustering number selection, an improved WKNN indoor fingerprint localization algorithm based on adaptive H clustering algorithm is proposed in this thesis. Specifically, an adaptive hierarchical clustering combined with positioning environment and fingerprint information without initial clustering center is introduced. At the same time, a RSSI information compensation method based on cosine similarity is proposed aiming at the problem of RSSI information packet loss for test nodes in complicated indoor location environment, with the result of positioning error decrease at test node by using cosine similarity between test nodes and fingerprint points to approximately compensate the missing RSSI information. The experimental results indicate that the proposed adaptive hierarchical clustering algorithm can divide the experimental area adaptively according to fingerprint information, meanwhile the proposed fingerprint information compensation method can decrease the positioning error of the test node with incomplete information, by which the average positioning error in the experimental environment is decreased to 0.78 m compared with other indoor positioning algorithms.

Keywords: Location fingerprint localization · Adaptive hierarchical clustering · Cosine similarity · Received signal strength indication

1 Introduction

With the comprehensive popularization of wireless communication technology, location-based service problems are attracting more and more attention [1]. However, traditional positioning systems such as GPS could not provide a good positioning service when used in such environment. Hence, exploring effective methods of indoor positioning has become a research focus [2].

Fingerprint location algorithm based on received signal strength indicator (RSSI) decreases the effects of the multi-path, non-line of sight and other environmental factors through the idea of mathematical statistics brought by the complexity of the internal structure [3], which is becoming the main method for indoor positioning

© Springer Nature Singapore Pte Ltd. 2017
M. Fei et al. (Eds.): LSMS/ICSEE 2017, Part I, CCIS 761, pp. 253–262, 2017.
DOI: 10.1007/978-981-10-6370-1_25

technology nowadays. However, the problem of higher computation and lower positioning efficiency arise due to the large amount of fingerprint data when applied in large-scale complicated indoor scenes. Therefore, it is necessary to design corresponding algorithms to achieve the rapid matching for target location.

In [4], the KNN fingerprint localization algorithm is proposed. The node is located by matching its position information with the fingerprints among the whole fingerprint space in the algorithm, but it is difficult to eliminate the interference of singular points. In [5], an enhanced weighted K nearest neighbor algorithm is proposed by changing the number of neighbors to be considered. Literature [6] uses the point sporadic intensity of the nearest neighbor to determine the reference point control network and in order to select the key parameter K of KNN algorithm dynamically. However, the KNN algorithm has high computational complexity and low matching efficiency. In [7], the method of K-means clustering is used to cluster the fingerprint points with matching area determined by matching the node with the sub-category. The computation is reduced, however, it is much easy to trap into the local optimum. In [8], an error beacon filtering algorithm is proposed, but the algorithm does not overcome the problem that the clustering results are sensitive to initial center values. In [9], a KNN algorithm based on fuzzy C-means clustering (FCM) is proposed. It is used to soften the candidate points, meanwhile the membership degree of each candidate point is determined and then clustered. In [10], a WLAN hybrid indoor location method based on FCM and artificial neural network (ANN) is proposed. FCM method is used to select the reference point affected by multipath effect, and ANN is used to approximate the coordinate position. However, the parameter values introduced by the hybrid algorithm are complicated and are difficult to get the optimal combination of parameters.

In order to solve problems mentioned above and the RSSI packet loss of the node to be positioned due to environmental noise in complicated indoor environment, an improved weighted K-nearest neighbor location algorithm based on adaptive hierarchical clustering (AHC-iWKNN) is proposed in Sect. 2. Adaptive hierarchical clustering is used to classify fingerprint space for the sake of avoiding parameter setting and initial clustering center selection in the algorithm. Meanwhile, a RSSI compensation method based on cosine similarity is proposed to reduce the positioning errors of nodes to be positioned with RSSI packet loss. In Sect. 3, the proposed algorithm is verified by the built-in indoor positioning system and the compare results are listed. Conclusions are introduced in Sect. 4.

2 Research on Improved Fingerprint Location Algorithm

The positioning nodes can be located by different methods base on fingerprint location. A novel fingerprints clustering method and an improved Weighted K-Nearest Neighborhood Matching Algorithm are introduced in this section.

2.1 Fingerprint Space Division Based on Adaptive Hierarchical Clustering

Hierarchical clustering algorithm is also called tree clustering algorithm. Unlike the K-means clustering, there is no need to determine the global optimal criterion function and the initial clustering center for hierarchical clustering algorithm. Instead, the threshold criteria is used which clustering according to the given distance threshold or cluster number. A threshold T is set based on human experience to stop the clustering iteration for cohesive hierarchical clustering. In this thesis, an adaptive hierarchical clustering method is proposed, which can generate the distance threshold adaptively with the assurance of difference among clusters by using the maximum distance in the distance similarity matrix as the distance threshold.

The main idea of clustering fingerprints based on adaptive hierarchical clustering is:

(1) At the initial layer, a fingerprint data x_i consisting of the position coordinates and the RSSI information is regarded as a cluster C_i.

$$C_i = \{x_i\}, \forall i. \tag{1}$$

(2) Calculate the Euclidean distance between any two fingerprint data:

$$d_{i,j} = D(C_i, C_j), \forall i,j. \tag{2}$$

Where C_i, C_j are fingerprint data in (2).

Then, generate the distance similarity matrix and find out the maximum distance (d_{max}) as the threshold T.

$$T = \max(d_{i,j}), \forall i,j. \tag{3}$$

In (3), $d_{i,j}$ is the Euclidean distance from fingerprint i to fingerprint j, calculated by formula (2).

As can be seen in Table 1, A, B, C, D, E, F, G are six fingerprint points with known coordinates, d(A → B) represents the Euclidean distance from fingerprint A to fingerprint B, which equals to d(B → A). In addition, d_{max} is assumed as d(B → D) in Table 1.

Table 1. Distance similarity matrix of 6 clusters.

	A	B	C	D	E	F
A	0	d(A→B)	d(A→C)	d(A→D)	d(A→E)	d(A→F)
B	d(B→A)	0	d(B→C)	d(B→D)	d(B→E)	d(B→F)
C	d(C→A)	d(C→B)	0	d(C→D)	d(C→E)	d(C→F)
D	d(D→A)	d(D→B)	d(D→C)	0	d(D→E)	d(D→F)
E	d(E→A)	d(E→B)	d(E→C)	d(E→D)	0	d(E→F)
F	d(F→A)	d(F→B)	d(F→C)	d(F→D)	d(F→E)	0

(3) Find two clusters (C_a and C_b) with the smallest distance in the range of the whole fingerprint database:

$$a, b = \arg \min_{i,j} d(i,j). \tag{4}$$

Then, merge C_a and C_b into a new cluster C_n and calculate the Euclidean distance between C_n and the other clusters except C_a and C_b. Finally, delete C_a and C_b.

$$C_n = \{C_a, C_b\}. \tag{5}$$

$$Restruture\left(\sum C_i\right) = \sum C_i - C_a - C_b + C_n. \tag{6}$$

For Table 1, find out the minimum distance (d_{min}) in the distance similarity matrix, assume d(D → F). Then merge two clusters with the smallest distance (D, F) into a new cluster G, generate a new distance similarity matrix between cluster G and remaining clusters, as shown in Table 2.

(4) Loop (3), compare T with the minimum distance (d_{min}) in each merging process and stop clustering until d_{min} is bigger than T.

Table 2. Distance similarity matrix of 5 clusters

	A	B	C	E	G
A	0	d(A→B)	d(A→C)	d(A→E)	d(A→G)
B	d(B→A)	0	d(B→C)	d(B→E)	d(B→G)
C	d(C→A)	d(C→B)	0	d(C→E)	d(C→G)
E	d(E→A)	d(E→B)	d(E→C)	0	d(E→G)
G	d(G→A)	d(G→B)	d(G→C)	d(G→E)	0

2.2 Improved Weighted K-Nearest Neighborhood Matching Algorithm

In the process of locating, beacon nodes are easy to be blocked due to indoor obstacles and frequent flow of personnel, which lead to serious interference in the signal strength or even no RSS from beacon nodes for test nodes. In order not to affect the positioning matching, for conventional methods, remaining RSS is applied to calculate the distance or attach the same value of the fingerprint data from the matching fingerprint point to signal strength values that the node to be located could not receive from the beacon nodes. For former method, locating the node with RSS packet loss easily results in matching misalignment. For the latter, test nodes with forced assignment are easy to lose the signal characteristics at the location of the node to be located, which will reduce the accuracy of matching.

In this thesis, a method of compensating test nodes RSS by cosine similarity is proposed, where a complete RSS vector which has similar characteristics of test nodes is obtained and fingerprint matching based on Euclidean distance is performed. Cosine similarity refers to measuring the cosine of the angle of two vectors to determine the degree of similarity between two vectors. According to the Euclidean inner product formula, the cosine of the angle between two phases which is also called the similarity of two vectors is calculated by:

$$similarity = \cos(\theta) = \frac{A \bullet B}{\|A\|\|B\|} = \frac{\sum\limits_{i=1}^{n} A_i \times B_i}{\sqrt{\sum\limits_{i=1}^{n} A_i}\sqrt{\sum\limits_{i=1}^{n} B_i}}. \tag{7}$$

Where A, B in (7) are n-dimensional vector.

The test nodes can receive the RSS signals from all beacon nodes without interference after beacon nodes deployed in the environment and the RSS vector of a test node should be contained by:

$$s^* = \{s_1, s_2, \cdots, s_N\}. \tag{8}$$

Where s^* is N-dimensional vector. N is the number of beacon nodes.

The number of RSS values in s^* reduces to $N - p$ when test nodes couldn't receive signals from p beacon nodes and the algorithm is introduced as follows:

(1) Adaptive hierarchical clustering of fingerprint data.

Define ψ as fingerprint space:

$$\psi = \begin{bmatrix} x_1 & y_1 & RSS_{1,1} & RSS_{1,2} & \cdots & RSS_{1,N} \\ x_2 & y_2 & RSS_{2,1} & RSS_{2,2} & \cdots & RSS_{2,N} \\ \vdots & \vdots & \vdots & \vdots & \ddots & \vdots \\ x_M & y_M & RSS_{M,1} & RSS_{M,2} & \cdots & RSS_{M,N} \end{bmatrix}. \tag{9}$$

In (9), M is the number of fingerprint points, N is the number of RSS values from all beacon nodes, and (x_i, y_i) represents the coordinate of a fingerprint point.

Clustering the fingerprint data through adaptive hierarchical clustering. First, calculate the Euclidean distance between any two fingerprint points:

$$D_{i,j} = \sqrt{(x_i - x_j)^2 + (y_i - y_j)^2}. \tag{10}$$

Where i, j in (10) are two different fingerprint point in fingerprint space ψ.

At the beginning of the algorithm, each fingerprint data is regarded as a cluster, the threshold T is obtained from the distance similarity matrix of all fingerprint data. Start clustering by merging two clusters with smallest distance calculated by (10) into a new cluster until the smallest distance is rather bigger than T. Then, calculate the RSS vector of the centroid of each cluster:

$$Centroid_f = \frac{1}{H} \sum_{i=1}^{H} (RSS_{i,1}, RSS_{i,2}, \cdots, RSS_{i,M}), f = 1, \ldots, k. \tag{11}$$

In (11), H is the number of fingerprints in a cluster, k is the number of clusters.

(2) Pretreatment for the vector of the test node (s^*).

All the fingerprint data in ψ is processed as follows: First, remove p corresponding RSS values of all fingerprint points and construct the new fingerprint space v. Calculate similarities between s^* ($N - p$ dimensional vector) and all fingerprint RSS vector in v by formula (7):

$$\cos(\theta) = [\cos(\theta_1), \cdots, \cos(\theta_M)]^T. \tag{12}$$

In (12), M is the number of fingerprint points.

Then, select the maximum p similarities from $\cos(\theta)$, calculate the mean value of p similarities $(\cos(\theta)_{mean})$ and record the corresponding p fingerprint points.

$$\cos(\theta)_{mean} = \frac{1}{p} \sum_{i=1}^{p} \max(\cos(\theta)). \tag{13}$$

In (13), $\cos(\theta)$ is the similarities vector calculated by formula (12).

Finally, regard missing p RSS values in s^* as unknown, meanwhile substitute corresponding p RSS vector that recorded above and $\cos(\theta)_{mean}$ into formula (7) and solve the p-dimensional equations, as can be seen in formula (14). Solutions of the equations approximately equal to p RSS values that s^* lacks so that the new complete RSS vector $s^{*\prime}$ could be constructed.

(3) Matching and location.

Calculate the Euclidean distance between $s^{*\prime}$ and $Centroid_f$:

$$L_j = \sqrt{\sum_{i=1}^{N} (s_i - RSS_i)^2}, j = 1, \ldots, k. \tag{14}$$

Where N is the RSS values of each fingerprint point received from all beacon nodes.

Then, compare the distance between the test node and k centroids, and the cluster that has smallest centroid distance becomes the cluster for test node. Finally, use weight K-nearest algorithm to obtain the estimated coordinate of the test node.

3   Experimental Results and Analysis

In order to verify the effectiveness of the improved algorithm proposed in this thesis, an indoor positioning system based on wireless sensor network is developed independently. The experiment is carried out in Room 305, Western Automation Building, Shanghai University. The area is a 5.7 m × 4 m rectangular laboratory, as is shown in Fig. 1, where there is the office furniture such as tables, chairs and bookcases. The positioning system contains a wireless sensor measurement node A1, 5 beacon nodes, a wireless sensor gateway and detection software in the PC. Measurement node A1 and 5 beacon nodes all contain the TI's CC2530 chip with wireless communication capabilities. In addition, the wireless sensor gateway (including TI's RF module CC2591 and AT91SAM9260 processor) is also integrated in measurement node A1. The experiment area is divided into 35 fingerprint locations and each of which has a 0.5 m × 0.5 m area. A fingerprint point of each location is randomly selected at each area. The positions of 20 test nodes which are randomly selected can receive RSSI values from 3 to 5 beacon nodes. 200 groups of RSS values of the fingerprint point are collected continuously in the offline phase and 20 groups of RSS values of the test nodes are acquired continuously during the online matching phase.

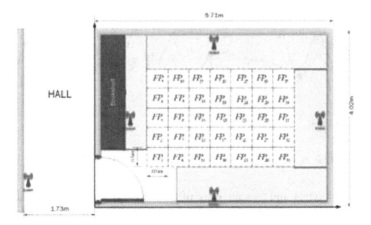

Fig. 1. Experimental area layout

3.1   Results and Analysis of Hierarchical Clustering Experiment

The proposed system is used to build the fingerprint space and MATLAB software is applied in splitting fingerprint points by K-means clustering and adaptive hierarchical clustering. According to reference [7], with the positioning accuracy assured, K-means clustering number is set for 4 on the basis of experimental area size in order to minimize the amount of calculation. The distance threshold T is 3.65 m according to the distance similarity matrix of fingerprints in experimental area.

   In order to further verify the clustering effectiveness of adaptive hierarchical clustering on fingerprint points, a positioning experiment for 20 test nodes (p1, p2, ..., p20) is conducted by using the above positioning system. Among these test nodes, p9, p10 and p18 lack the RSSI signals from node4 and node5. Three algorithms are KNN, Kmeans+WKNN, AHC+WKNN that can be applied to find the position of test nodes. Table 3 is the maximum errors and average errors of three algorithms for the positioning to test nodes. Figure 2 is the positioning error contrast curve of test nodes among three algorithms.

Table 3. Maximum errors and average errors of three algorithms

Algorithm	Average error (m)	Max error (m)
KNN	1.02	1.81
AHC+WKNN	0.84	1.44
Kmeans+WKNN	1.33	2.83

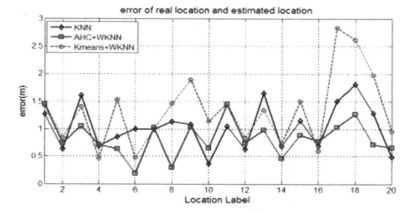

Fig. 2. Positioning error contrast curve of three algorithms

The maximum error and average error of the AHC+WKNN algorithm are 1.44 m and 0.84 m, as is shown in Fig. 2, with 1.39 m and 0.49 m errors decreased respectively when compared with that of the Kmeans+WKNN algorithm from Table 3. The experimental results show that the adaptive hierarchical clustering has a better performance in positioning than K-means clustering.

3.2   Results and Analysis of Improved Weight KNN Location Algorithm

In Fig. 2, for test nodes with incomplete RSSI signals, positioning errors which increase the average positioning error are eliminated more than 1 m except P10 by clustering algorithms. In this thesis, an improved weighted KNN location algorithm for the test nodes with RSS packet loss is proposed, in addition, in order to verify the performance of the proposed algorithm, the standard WKNN location algorithm and the improved WKNN location algorithm are used to estimate the position of test nodes after the hierarchical clustering for the fingerprint space. The positioning errors of the test nodes with RSS packet loss can be seen in Table 4 and Fig. 3 is the contrast curve of positioning errors between the WKNN algorithm and the improved WKNN algorithm after adaptive hierarchical clustering.

Table 4. Positioning errors of test nodes with incomplete RSSI signals

Test node	AHC+WKNN(m)	AHC+iWKNN(m)
P_9	1.04	0.35
P_{10}	0.64	0.45
P_{18}	1.26	0.99
All nodes	0.83	0.78

In Fig. 3, for test nodes with complete RSSI signals, positioning errors are same when using two algorithms respectively. As can be seen in Table 4, positioning errors for all the test nodes with incomplete RSSI signals are reduced, where the decreases are 66.3% at test node P9 and 21.4% at test node P18 respectively. In addition, the average positioning error is decreased from 0.84 m to 0.78 m. Experiments indicate the improved weighted KNN localization algorithm proposed in this thesis can improve the positioning accuracy of the nodes to be positioned with RSS packet loss.

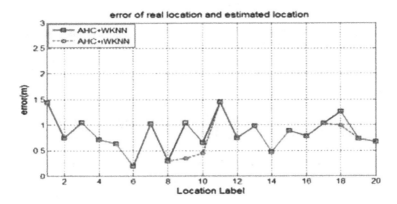

Fig. 3. Positioning error contrast curve of two algorithms

4 Conclusion

Aiming at the dependence of the traditional indoor clustering positioning accuracy on the initial center and clustering number selection, an adaptive hierarchical clustering algorithm is proposed in this thesis. Specifically, an adaptive clustering positioning is realized by the method of combining unnecessary of setting initial clustering center with hierarchical clustering and a method of obtaining the optimal clustering number adaptively according to the fingerprints information proposed in this thesis. Simulation results show that the average positioning error dropped to 0.84 m in the 5.7 m × 4 m experimental environment, achieving a good performance in clustering with the calculation amount increased. At the same time, in order to solve the problem of RSS packet loss for test nodes located in the indoor positioning process in complicated environment, a method to compensate RSSI signals of test nodes by cosine similarity is proposed, which improves anti-noise performance of algorithm in such environment with the result of improving the positioning accuracy further. Experimental results indicate the proposed algorithm can compensate the RSSI of test nodes reasonably in the case of test node packet loss, which optimizes the positioning effects and the average positioning error of all test nodes is reduced from 0.84 m to 0.78 m in the 5.7 m × 4 m experimental environment.

Acknowledgment. This work was financially supported by the Science and Technology Commission of Shanghai Municipality of China under Grant (No. 17511107002).

References

1. Gu, Y.Y., Lo, A., Neimegeers, I.: A survey of indoor positioning system for wireless personal network. IEEE Commun. Surv. Tutor. **11**(1), 13–32 (2009)
2. Xu, J., Luo, H., Zhao, F., Tao, R., Lin, Y.: Dynamic indoor localization techniques based on RSSI in WLAN environment. In: 2011 6th International Conference on Pervasive Computing and Applications (ICPCA), pp. 417–421 (2011)
3. Liao, X.-Y., Ke, H., Min, Y.: Research on improvement to WiFi fingerprint location algorithm. In: 10th International Conference on Wireless Communications, Networking and Mobile Computing (WiCOM 2014), Beijing, pp. 648–652 (2014)
4. Xu, Y., Zhou, M., Meng, W., Ma, L.: Optimal KNN positioning algorithm via theoretical accuracy criterion in WLAN indoor environment. In: Global Telecommunications Conference (GLOBECOM 2010). IEEE (2010)
5. Shin, B., Lee, J.H., Lee, T., Kim, H.S.: Enhanced weighted K-nearest neighbor algorithm for indoor Wi-Fi positioning systems. In: 2012 8th International Conference on Computing Technology and Information Management (NCM and ICNIT), Seoul, Korea (South), pp. 574–577 (2012)
6. Liu, Y., Wang, J.: Normalized KNN model based on geometric clustering fingerprint library. J. Wuhan Univ. (Inf. Sci. Ed.) (11), 1287–1292 (2014)
7. Altintas, B., Serif, T.: Improving RSS-based indoor positioning algorithm via k-means clustering. In: 17th European Wireless 2011 - Sustainable Wireless Technologies, Vienna, Austria, pp. 1–5 (2011)
8. Liu, L., Du, J., Guo, D.: Error beacon filtering algorithm based on K-means clustering for underwater Wireless Sensor Networks. In: 2016 8th IEEE International Conference on Communication Software and Networks (ICCSN), Beijing, pp. 435–438 (2016)
9. Sun, Y., Xu, Y., Ma, L., et al.: KNN-FCM hybrid algorithm for indoor location in WLAN. In: 2009 2nd International Conference on Power Electronics and Intelligent Transportation System (PEITS), vol. 2, pp. 251–254. IEEE, Zhangjiajie (2009)
10. Yubin, X., Zhou, M., Lin, M.: Hybrid FCM/ANN indoor location method in WLAN environment. In: 2009 IEEE Youth Conference on Information, Computing and Telecommunication, Beijing, pp. 475–478 (2009)

Short-Term Load Forecasting Model Based on Multi-label and BPNN

Xiaokui Sun$^{(\boxtimes)}$, Zhiyou Ouyang, and Dong Yue

Institute of Advanced Technology,
Nanjing University of Posts and Telecommunications, Nanjing 210046, China
260602812@qq.com

Abstract. With the rapid development of smart grid, the importance of power load forecast is more and more important. Short-term load forecasting (STLF) is important for ensuring efficient and reliable operations of smart grid. In order to improve the accuracy and reduce training time of STLF, this paper proposes a combined model, which is back-propagation neural network (BPNN) with multi-label algorithm based on K-nearest neighbor (K-NN) and K-means. Specific steps are as follows. Firstly, historical data set is clustered into K clusters with the K-means clustering algorithm; Secondly, we get N historical data points which are nearest to the forecasting data than others by the K-NN algorithm, and obtain the probability of the forecasting data points belonging to each cluster by the lazy multi-label algorithm; Thirdly, the BPNN model is built with clusters including one of N historical data points and the respective forecasting load are given by the built models; Finally, the forecasted load of each cluster multiply the probability of each, and then sum them up as the final forecasting load value. In this paper, the test data which include daily temperature and power load of every half hour from a community compared with the results only using BPNN to forecast power load, it is concluded that the combined model can achieve high accuracy and reduce the running time.

Keywords: BP neural network · K-means algorithm · K-nearest neighbor · Multi-label · Short-term load forecasting

1 Introduction

With the trend of smart grid, it is very important to forecast the user-side power load. High quality and accuracy load forecasting is essential in the smart grid [1, 2]. In recent years, the residents ladder price is implemented by the State promoting to save electricity and use electricity reasonably. The smart grid of China starts too late and the relatively technology is backward and it results in a number of power supply and demand imbalance [3]. In this case, the state encourages the power plants to generate more power. On the one hand, it meets the electricity consumption in some power-needed areas. On the other hand, it also leads to a lot of electricity vastly waste in many areas.

STLF usually refers to the load demand for the next 24 h. Its methods include time series analysis [5, 6], grey theory method [7] and neural network method [8]. Among

© Springer Nature Singapore Pte Ltd. 2017
M. Fei et al. (Eds.): LSMS/ICSEE 2017, Part I, CCIS 761, pp. 263–272, 2017.
DOI: 10.1007/978-981-10-6370-1_26

them, an effective mathematical model of the time series analysis method is difficult to be built and the prediction accuracy is generally not high. Grey prediction model is theoretically applicable to any nonlinear load forecasting and the differential equation is suitable for load forecasting with exponential growth trend, but it is difficult to improve the accuracy of fitting gray with other indicators of trend. The neural network usually trained by back-propagation (BP) of errors is one of the popular network architecture in use nowadays [9]. BPNN, due to its excellent ability of non-linear mapping, generalization and self-learning, has been proved to be widely used in engineering optimization field [10]. Due to too much training data, BPNN training time is too long [11]. Because the BP algorithm is essentially a gradient descent method, and the objective function to be optimized is very complex, therefore, there will be a "saw-tooth" phenomenon, which makes the BP algorithm inefficient [11]. Nowadays, in the trend of big data, we usually deal with mass of data. Therefore, we should minimize the size of the training data without losing effective data before using the BPNN. [13] uses the BPNN with K-means clustering algorithm to forecast power load. It regards the center of clustering as the label and just train the BPNN model with the one cluster. So, it is possible to lose valid information in other clusters.

In this paper, our contribution is to propose the BPNN with multi-label algorithm based on K-NN and K-means clustering algorithm. We give different weight of each cluster for the forecasting points by Multi-label algorithm based on K-NN and K-means proposed by us and build models by BPNN. This method can avoid losing valid information of other clusters. Compared with the traditional BPNN algorithm, the combined algorithm can obtain better performance in accuracy and running time.

This paper is organized as follows. The introduction is described in Sect. 1. In Sect. 2, the BPNN is discussed. BPNN with multi-label algorithm based on K-NN and K-means is given in Sect. 3. The experiment on real data are given in Sect. 4. In Sect. 5, we draw conclusions and give insightful discussions.

2 BP Neural Network

The BP algorithm [14, 15] uses the gradient of the performance function to determine how to adjust the weights to minimize errors that affect performance.

Figure 1 shows a multi-layer feedforward network structure with d input layer neurons, l output layer neurons, q hidden layer neurons, where the threshold value of the j th neurons in the output layer is represented by θ_j, The threshold of the h th neurons is denoted by γ_h. The connection weight between the i th neurons of the input layer and the h neurons of the hidden layer is v_{ih}, and the connection weight between the h th neurons of the hidden layer and the j th neurons of the output layer is w_{hj}. $\alpha_h = \sum_{i=1}^{d} v_{ih} x_i$ represents the input received by the h th neurons in the hidden layer, and $\beta_j = \sum_{h=1}^{q} w_{hj} b_h$ represents the input received by the j th neurons in the output layer, where b_h is the output of the h th neurons in the hidden layer.

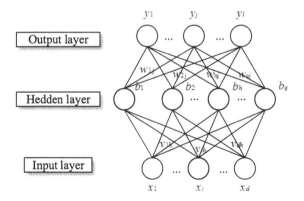

Fig. 1. Architecture of BP neural network

There are many forms for activation function, but the most popular is sigmoid function. In this paper, we use the sigmoid function as output function of the hidden neuron and the output neuron. The sigmoid function is as follows,

$$f(x) = \frac{1}{1 - e^{-x}} \tag{1}$$

For training sample (x_k, y_k), assuming that $\hat{y} = (\hat{y}_1^k, \hat{y}_2^k, \ldots, \hat{y}_l^k)$ is the output of the neural network, where

$$\hat{y}_j^k = f(\beta_j - \theta_j) \tag{2}$$

So we can obtain the mean square error E_k of the neural network in training sample,

$$E_k = \frac{1}{2} \sum_{j=1}^{l} (\hat{y}_j^k - y_j^k)^2 \tag{3}$$

The neuron gradient term is

$$g_j = \hat{y}_j^k (1 - \hat{y}_j^k)(y_j^k - \hat{y}_j^k) \tag{4}$$

The updating formula of w_{hj} is

$$\Delta w_{hj} = \eta g_j b_h \tag{5}$$

where η is the leaning rate.
The updating formula of θ_j is

$$\Delta \theta_j = -\eta g_j \tag{6}$$

The updating formula of v_{ih} is

$$\Delta v_{ih} = \eta e_h x_i \tag{7}$$

where $e_h = b_h(1-b)\sum_{j=1}^{l} w_{hj} g_j$, x_i is the input data.

The updating formula of γ_h is

$$\Delta \gamma_h = -\eta e_h \tag{8}$$

BPNN is an iterative learning algorithm that arbitrary parameter v update that are

$$v \leftarrow v + \Delta v \tag{9}$$

And the iterative process of w, θ and γ are the same to v. Updating w and v will consume a lot of time, and we reduce the running time by the mean of the combined model.

The goal of the BPNN is to minimize the cumulative error of the historical set. During the training process, historical set that differ greatly from the predicting load are involved in the training model, which not only increase the training time but also reduce the accuracy of the prediction. Therefore, in this paper, we introduce a method to reduce the impact of these irrelevant data sets.

3 BPNN with Lazy Multi-label Algorithm Based on K-NN and K-means

3.1 K-means Clustering Algorithm

K-means [16] is an unsupervised learning algorithm, and it is an algorithm for clustering data points by the mean values. The K-means algorithm divides similar data points into K clusters, that the K is a pre-setting value, and each cluster has an initial center which is selected from data set randomly. In this paper, Euclidean distance is used to represent the similarity between data. For instance, the Euclidean distance of $\vec{X} = (x_1, x_2, \ldots, x_n)$ and $\vec{Y} = (y_1, y_2, \ldots, y_n)$ is

$$d_{XY} = \sqrt{\sum_{i=1}^{n} (x_i - y_i)^2} \tag{10}$$

K-means algorithm general steps:

Step 1: select K data from the data set as the initial cluster center;

Step 2: calculating the Euclidean distance between each data and K cluster centers and dividing data into the nearest cluster center;

Step 3: recalculating the cluster center by mean values;

Step 4: calculating the standard measure function if it reaches the maximum number of iterations, and then stop, otherwise go to step 2.

After the iteration is completed, we can get K clusters which high similarity within cluster and low similarity between clusters. In this paper, K-means algorithm is used to divide the data set into K clusters. Then, these clusters are regarded as original labels which will be used in the next step.

3.2   Lazy Multi-label Algorithm Based on K-NN and K-means

In Sect. 3.1, the historical data set is clustered into K clusters. In [17], the cluster center is regarded as feature of each cluster, and the forecasting load are divided into the cluster whose cluster center is the shortest from forecasting load. Because the K-means clustering algorithm is an unsupervised learning algorithm, we cannot get the specific characteristics of each cluster. [17] will lose some information of other clusters. We solve this problem by the lazy multi-label algorithm based on K-NN and K-means.

ML-KNN, i.e. multi-label K-NN, was proposed in [17]. It is mainly used for pattern recognition. On this basis, we put forward a lazy multi-label algorithm based on K-NN and K-means. Specific description is following.

Define the result of K-means as.$C = \{C_1, C_2, \ldots, C_i, \ldots, C_k\}$, where C_i represents the i th cluster.

Step 1: compute and finding N train points which are the nearest distance from the forecasting load through the K-NN algorithm. The distance is computed by formula (10).

Step 2: count that each cluster contains the number of N training points and the result is $\vec{n}$, where $\vec{n} = (n_1, n_2, \ldots, n_i, \ldots, n_k)$, and n_i represents the number of N training points in the i th cluster.

Step 3: use lazy multi-label theory, we can obtain a set of weights $\vec{p}$, where $\vec{p} = (p_1, p_2, \ldots, p_i, \ldots, p_k) = (\frac{n_1}{N}, \frac{n_2}{N}, \ldots, \frac{n_i}{N}, \ldots, \frac{n_k}{N})$, and p_i is the probability that the forecasting load belongs to the i th cluster.

We can observe that the proposed algorithm will reduce some training data. We think that the reduced data are low related to the forecasting load and if they are used to build the BPNN model, it abates accuracy of the BPNN.

3.3   BPNN with Multi-label Algorithm Based on K-NN and K-means

As shown in Fig. 2, The date set is clustered into K clusters, such as $C_1, C_2, \ldots, C_i, \ldots, C_k$, by the K-means algorithm. In this paper, our historical data include temperature and power load and the temperature of the forecasting points can be known from weather forecasting. Therefore, the temperature of the forecasting points is regarded as the known element of forecasting data. Using the temperature, we can get the weights of each cluster by Multi-label based on K-NN and K-means, such as $p_1, p_2, \ldots, p_i, \ldots, p_k$. If p_i isn't zero, we build BPNN of the C_i and obtain forecasting power load of the C_i which multiply p_i is $Load_i$. Then we sum the forecasting load of each cluster, they are $Load_1, Load_2, \ldots, Load_i, \ldots, Load_k$, and get the forecasting load.

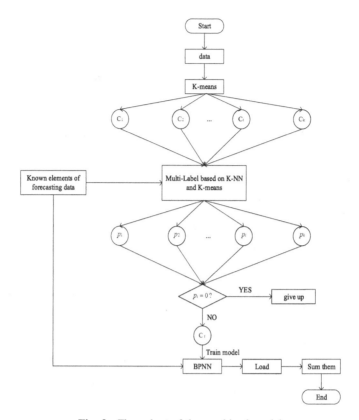

Fig. 2. Flow chart of the combined model

4 Experiment

4.1 Data Preprocessing

In this paper, the historical data set is a real data of a community from January to December. The data set has one temperature per day and have 48 load data points, that is, a load value point is the past half an hour. Temperature is the only information related to the power load in this data sets. We can get the future temperature information from weather forecasting web.

In order to facilitate observation, the community total daily power load is painted as Fig. 3. It can be seen from the Fig. 3 that the seasonal variation of the load is very obvious. The load of May to September compared to other months, there is a significant decline, but the volatility is periodic.

In order to speed up the convergence of clustering, avoid data annihilation and reduce the sensitivity of the singular data to the algorithm, the historical data needs to be normalized. There are many normalized methods, for instance, Min-Max scaling and Z-score standardization. In this paper, we use Min-Max scaling.

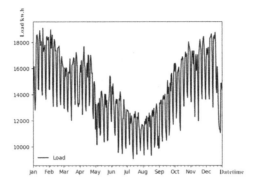

Fig. 3. Total power load

The method realizes the equal scaling of the historical data. The Min-Max scaling is following,

$$X_{norm} = \frac{X - X_{min}}{X_{max} - X_{min}} \tag{11}$$

where X_{norm} is the normalized data, X is the original data, and X_{max} and X_{min} are the maximum and minimum of the original data set.

There are two evaluation standards. One is RMSE which is root-mean-square error, the other is MAPE where is mean absolute percentage error.

$$RMSE = \sqrt{\frac{1}{n} \sum_{i=1}^{n} (x_{pi} - x_{ri})^2} \tag{12}$$

$$MAPE = \frac{1}{n} \sum_{i=1}^{n} |x_{pi} - x_{ri}| \times 100\% \tag{13}$$

where x_{pi} is the i th forecasting data, and x_{ri} is the i th real data. In this paper, we use the data of 1–201 days to forecast the load of the 202th day.

4.2 BPNN Experiment

The data is 49 dimensions containing temperature and 48 load points. BPNN is built with one input layer neuron, five hidden layer neurons and 48 output layer neurons. The temperature is the input layer neuron and the 48 load points is the output layer neurons. Then, normalize the data set by formula (11) and set the learning rate to 0.05, the training times to 1000 and the correct to 0.01.

We use the normalized data of 1–201 days to build BPNN model and forecast the load of the 202th day. This article uses python to implement it. The results of the 202th day's load are in Fig. 4 and the RMSE, MAPE and running time are in Table 1.

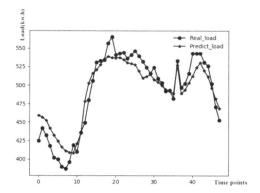

Fig. 4. BPNN forecasting load compared with real load

Table 1. BPNN results

Model	RMSE	MAPE(%)	Running time(s)
BPNN	18.11234	3.32004	1002.28238

4.3 BPNN with Lazy Multi-label Based on K-NN and K-means (BPNN-L-ML-K-K) Experiment

The theory and implementation steps of the algorithm have been introduced in Sect. 3.3 and BPNN is built with one input layer neuron, five hidden layer neurons and 48 output layer neurons. We set the learning rate to 0.05, the training times to 1000 and the correct to 0.01. The temperature is the input layer neuron and the 48 load points is the output layer neurons. We use the normalized data of 1–201 days to build BPNN-L-ML-K-K model and forecast the load of the 202th day.

Firstly, we normalize the data. Secondly, in this paper, the historical data are clustered into five clusters by K-means clustering algorithm, recorded as $C = \{C_1, C_2, C_3, C_4, C_5\}$. Thirdly, make $N = 60$, and count that each cluster contains the number of 60 training samples and the result is $\vec{n} = (0, 0, 22, 14, 24)$. Fourth, using lazy multi-label theory, we can obtain a set of weights $\vec{p} = \left(\frac{0}{60}, \frac{0}{60}, \frac{22}{60}, \frac{14}{60}, \frac{24}{60}\right) = (0, 0, 0.37, 0.23, 0.40)$. Finally, the BPNN is built with the non-zero weight clusters and the forecasting load is given each other, where the forecasting load is represented by L. The last forecasting load is $L_{total} = L \times \vec{p}^T$. The results of the 202th day's load are in Fig. 5 and the RMSE, MAPE and running time are in Table 2.

4.4 Comparison of Experimental Results

Table 3 shows that the BPNN-L-ML-K-K model performs better than BPNN in RMSE, MAPE and running time. Through the processing of clustering and multi-label algorithm, the historical data with high correlation with the predicting load are used to build the model and get the weights of different clusters. Some lower correlation data are deleted. Thus, the training time is reduced and the efficiency of the model is

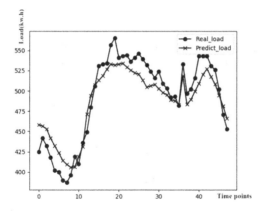

Fig. 5. BPNN-L-ML-K-K forecasting load compared with real load

Table 2. BPNN-L-ML-K-K effect

Model	RMSE	MAPE(%)	Running time(s)
BPNN-L-ML-K-K	16.42211	2.91566	469.55906

Table 3. Models comparison

Model	RMSE	MAPE(%)	Running time(s)
BPNN	18.11234	3.32004	1002.28238
BPNN-L-ML-K-K	16.42211	2.91566	469.55906
Improvement	9.3%	12.2%	53.2%

improved. The experimental results show that BPNN-L-ML-K-K can improve the accuracy of prediction.

5   Conclusion

In view of the characteristics of the STLF, we put forward a BPNN with multi-label based on K-NN and K-means model. Based on the analysis of a community power load, the forecast results show that the new model not only can improve the prediction accuracy, but also can reduce the running time. In the trend of big data, a lot of data has been collected and stored. We should use highly related data for us and delete irrelevant data. In this paper, the combined model can achieve it with better accuracy and less running time.

References

1. Javed, F., Arshad, N., Wallin, F.: Forecasting for demand response in smart grids: An analysis on use of anthropologic and structural data and short term multiple loads forecasting. Appl. Energy **96**, 150–160 (2012)
2. Borges, C.E., Penya, Y.K.: Fernandez I. Evaluating combined load forecasting in large power systems and smart grids. IEEE Trans. Industr. Inf. **9**(3), 1570–1577 (2013)
3. Beccali, M., Cellura, M., Brano, V.L.: Forecasting daily urban electric load profiles using artificial neural networks. Energy Convers. Manag. **45**(18), 2879–2900 (2004)
4. Hagan, M.T., Behr, S.M.: The time series approach to short term load forecasting. IEEE Trans. Power Syst. **2**(3), 785–791 (1987)
5. Papalexopoulos, A.D., Hesterberg, T.C.: A regression-based approach to short-term system load forecasting. IEEE Trans. Power Syst. **5**(4), 1535–1547 (1990)
6. Niu, D.: Adjustment gray model for load forecasting of power systems. IEEE Trans. on PWRS. **5**(4), 1535W–1547 (1990)
7. Chow, T.W.S., Leung, C.T.: Neural network based short-term load forecasting using weather compensation. IEEE Trans. Power Syst. **11**(4), 1736–1742 (1996)
8. Yu, S., Zhu, K., Diao, F.: A dynamic all parameters adaptive BP neural networks model and its application on oil reservoir prediction. Appl. Math. Comput. **195**(1), 66–75 (2008)
9. He, Y., Xu, Q.: Short-term power load forecasting based on self-adapting PSO-BP neural network model. In: IEEE 2012 Fourth International Conference on Computational and Information Sciences (ICCIS), pp. 1096–1099 (2012)
10. Wang, H., Shan, G., Duan, X.: Optimization of LM-BP neural network algorithm for analog circuit fault diagnosis. In: IEEE 2013 International Conference on Information Science and Technology (ICIST), pp. 271—274. Karnataka (2013)
11. Huang, H.C., Hwang, R.C., Hsieh, J.G.: A new artificial intelligent peak power load forecaster based on non-fixed neural networks. Int. J. Electr. Power Energy Syst. **24**(3), 245–250 (2002)
12. Comité, Francesco, Gilleron, Rémi, Tommasi, Marc: Learning multi-label alternating decision trees from texts and data. In: Perner, Petra, Rosenfeld, Azriel (eds.) MLDM 2003. LNCS, vol. 2734, pp. 35–49. Springer, Heidelberg (2003). doi:10.1007/3-540-45065-3_4
13. Frean, M.: The upstart algorithm: A method for constructing and training feedforward neural networks. Neural Comput. **2**(2), 198–209 (1990)
14. Lu, Y.W., Sundararajan, N., Saratchandran, P.: Performance evaluation of a sequential minimal radial basis function (RBF) neural network learning algorithm. IEEE Trans. Neural Netw. **9**(2), 308–318 (1998)
15. MacQueen, J.: Some methods for classification and analysis of multivariate observa-tions. In: Proceedings of the fifth Berkeley symposium on mathematical statistics and probability, pp. 281–297 (1967)
16. Huang, L., Cheng, H., Yi, Q.M.: OLoad forecast based on hybrid model with k-means clustering and BP neural network. Power Energy **1**, 56–60 (2016)
17. Zhang, M.L., Zhou, Z.H.: ML-KNN: A lazy learning approach to multi-label learning. Pattern Recogn. **40**(7), 2038–2048 (2007)

Hybrid Fx-NLMS Algorithm for Active Vibration Control of Flexible Beam with Piezoelectric Stack Actuator

Yubin Fang, Xiaojin Zhu$^{(\boxtimes)}$, Haotian Liu, and Zhiyuan Gao

School of Mechatronic Engineering and Automation, Shanghai University,
Shanghai 200072, People's Republic of China
mgzhuxj@shu.edu.cn

Abstract. Filtered-x Least Mean Square (FxLMS) algorithm is a meaningful adaptation algorithm used in the field of Active Vibration Control (AVC). Hybrid FxLMS algorithm, which is the combination of the feedforward structure and the feedback structure of FxLMS, has a better stability and could get the same performance with a lower filter order. In order to get a faster convergence speed, this paper adopts Normalized LMS (NLMS) algorithm to replace of LMS algorithm in the hybrid AVC system. To verify the Hybrid Fx-NLMS algorithm, this paper developed a simulation platform for active vibration control of a flexible beam with piezoelectric stack actuator using ADAMS and MATLAB SIMULINK. Simulation results show that the convergence speed and vibration suppression performance of the Hybrid Fx-NLMS algorithm are better than other traditional algorithms.

Keywords: Active Vibration Control · Fx-LMS algorithm · Hybrid Fx-NLMS algorithm · Flexible beam · Convergence analysis

1 Introduction

Active vibration control has received a great deal of attention due to its effectiveness at low frequencies and alternative passive absorption materials are relatively more expensive to deploy [1]. A AVC system is mainly classified into two categories: feed-forward AVC and feedback AVC. If combined the feed-forward with feedback control structure, we get a hybrid AVC method [2]. The hybrid AVC method can use a lower order filter to achieve the same performance [3]. The hybrid method is also clearly demonstrated an advantage over either simple feedforward AVC or feedback AVC method alone when there is significant plant noise.

In an AVC system, the secondary path is vital to the convergence of whole system and the control effect [4]. The FxLMS algorithm eliminates the effect of secondary path

X. Zhu—This work is supported by National Natural Science Foundation (NNSF) of China under Grant 51575328, 61503232. Mechatronics Engineering Innovation Group project from Shanghai Education Commission and Shanghai Key Laboratory of Power Station Automation Technology.

© Springer Nature Singapore Pte Ltd. 2017
M. Fei et al. (Eds.): LSMS/ICSEE 2017, Part I, CCIS 761, pp. 273–281, 2017.
DOI: 10.1007/978-981-10-6370-1_27

and then becomes one of the most commonly used algorithms in AVC also for its robust, low computational complexity, and easy to implement [5]. Although FxLMS algorithm is widely used but it has a slower convergence [6]. Numerous research works have been carried out on this problem. Various improvements were made to offset the defects of the FxLMS algorithm, such as the Filtered-x Normalized LMS (Fx-NLMS) [7], the Modified Filtered-x LMS (MFXLMS) [8], the Leaky FxLMS [9], etc.

In this paper, the hybrid filtered-x normalized LMS (Fx-NLMS) is applied to control the flexible beam with piezoelectric stack actuator. The hybrid FxLMS is presented in Sect. 2. The Sect. 3 briefs about the hybrid Fx-NLMS algorithm. The Sect. 4 shows the flexible beam module in ADAMS. The Sects. 5 and 6 give the simulation and conclusion.

2 Hybrid FxLMS Algorithm

In applications where adaptation is needed, the LMS algorithm is probably the most frequently used algorithm. The LMS algorithm, FxLMS algorithm, feedback FxLMS algorithm and hybrid FxLMS algorithm are briefly introduced as follow.

2.1 LMS Algorithm

As shown in Fig. 1, the adaptive filter $\mathbf{W}$ is fed with the input sequence $x(n)$. $y'(n)$ is the output of the filter $y(n)$ through the secondary path. The noise signal $d(n)$ compared with $y'(n)$, $e(n)$ is get. The goal is to adjusts the filter to minimize $e(n)$.

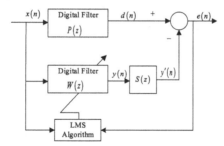

Fig. 1. The block diagram of LMS in feed-forward AVC system

From Fig. 1, $e(n)$ can be described as

$$e(n) = d(n) - S(n) * \left\{ \mathbf{W}^T(n)\mathbf{x}(n) \right\} \tag{1}$$

Where, n is time index, and $S(n)$ is impulse response of secondary path $S(z)$. And the filter coefficients expressed as

$$\mathbf{W}(n+1) = \mathbf{W}(n) + \mu x(n)e(n) \tag{2}$$

2.2   Feed-Forward FxLMS Algorithm

The block diagram for a single-channel feed-forward AVC system using the FxLMS algorithm is shown in Fig. 2 [1].

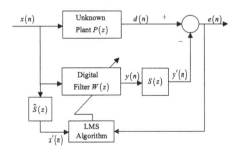

Fig. 2. The block diagram of feed-forward FxLMS

The unknown plant is represented by $P(z)$ that the vibration propagates. The vibration signal is represented by $d(n)$. The controller cancel this vibration signal through a signal $y'(n)$ that has a same amplitude but opposite phase to $d(n)$. After the $y'(n)$ mixed with $d(n)$, we get the error signal $e(n)$. Through updating the coefficients of filter, $e(n)$ is subsequently minimized.

The error signal can be expressed with Eq. (1), the same like LMS algorithm. Specifically, the weight updating equation is given by Eq. (3).

$$\mathbf{W}(n+1) = \mathbf{W}(n) + \mu x'(n)e(n) \tag{3}$$

2.3   Feedback FxLMS Algorithm

In real system, the reference signal $x(n)$ cannot be obtained directly sometimes. Then, the feedback FxLMS algorithm was proposed [10]. A block diagram for a single-channel feedback AVC system is presented in Fig. 3.

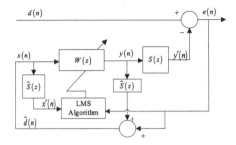

Fig. 3. The block diagram of feedback FxLMS

Compared to FxLMS algorithm, the obvious distinguish is to estimate the $x(n)$. From the diagram, it can be figured out that the estimated primary vibration signal $\widehat{d}(n)$ is generated by adding the output of filter $y(n)$ and the error signal $e(n)$. It is shown as:

$$\widehat{d}(n) = e(n) + y'(n) = e(n) + S(n) * \left\{ \mathbf{W}^T(n)\mathbf{x}(n) \right\} \tag{4}$$

$$x(n) \equiv \widehat{d}(n) \tag{5}$$

Except $x(n)$, the other parameters of feedback FxLMS algorithm are the same as that in FxLMS algorithm. The coefficients of the filter updating also as Eq. (3).

2.4   Hybrid FxLMS Algorithm

In feed-forward AVC system, to function efficiently, the primary signal must be highly correlated with the reference signal. However, it is difficult in complex practical application. And in feedback AVC systems, the cancellation effect only efficient in narrow band, and may be unstable on some frequencies. A coordination of the FxLMS and feedback FxLMS algorithm is called the hybrid FxLMS, as in Fig. 4 [11].

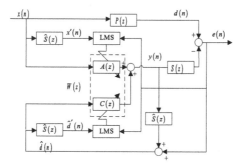

Fig. 4. The block diagram of hybrid FxLMS

In the hybrid FxLMS algorithm, there are two sensors to sample the corresponding signals, one is the reference sensor for FxLMS; the other one is error sensor to provide the error for synthesizing the reference signal. At the same time, the error signal is used for both the FxLMS algorithm and feedback FxLMS algorithm.

From Fig. 4, it can be figured out that the adaptive filter of FxLMS and feedback FxLMS were assumed as $A(n)$ and $C(n)$. And then, it is obvious that the control output is the sum of the output of $A(n)$ and $C(n)$. The length of $A(n)$ and $C(n)$ can be different. It is shown as

$$y(n) = \mathbf{A}^T(n)\mathbf{x}(n) + \mathbf{C}^T(n)\widehat{\mathbf{d}}(n) \tag{6}$$

And the error signal is:

$$e(n) = d(n) - S(n) * y(n) = d(n) - S(n) * \left\{ \mathbf{A}^T(n)\mathbf{x}(n) + \mathbf{C}^T(n)\widehat{\mathbf{d}}(n) \right\} \quad (7)$$

The hybrid FxLMS algorithm can get a lower order filter in the same performance with FxLMS algorithm or feedback FxLMS algorithm.

3   Hybrid Fx-NLMS Algorithm

In the process of convergence, a great step size may lead a fast convergence. But it may be divergent if the step-size is too large. Many variable step-size (VSS) LMS algorithms were proposed.

3.1   Normalized LMS Algorithm

The purpose of VSS-LMS is to adjust the step-size. In 1967, Nagumo proposed the Normalized LMS (NLMS) algorithm, which is considered as the first VSS modify [7].
The step-size of NLMS algorithm is shown as:

$$\mu(n) = \frac{\overline{\mu}}{\beta + R} = \frac{\overline{\mu}}{\beta + \mathbf{x}(n)^T \mathbf{x}(n)} \quad (8)$$

Where $\overline{\mu} > 0$ is a scalar, the maximum of $\overline{\mu}$ in some case is 2 [12]. The value of β is made quite small as possible.

3.2   Hybrid Fx-NLMS Algorithm

In the hybrid AVC system, replacing the LMS algorithm with the NLMS algorithm, it is obtained a hybrid Fx-NLMS algorithm. This algorithm update the step-size by

$$A(n+1) = A(n) + \left(\frac{\overline{\mu}}{\beta + \mathbf{x}^T \mathbf{x}} \right) x'(n)e(n) \quad (9)$$

$$C(n+1) = C(n) + \left(\frac{\overline{\mu}}{\beta + \widehat{\mathbf{d}}^T \widehat{\mathbf{d}}} \right) \widehat{d}'(n)e(n) \quad (10)$$

Where $A(n)$ is the filter coefficients of the feedforward Fx-NLMS algorithm, and $C(n)$ is the filter coefficients of the feedback Fx-NLMS algorithm.
The hybrid Fx-NLMS algorithm can get a greater step-size μ when the error $e(n)$ is great. Then, the convergence of this algorithm is faster than the traditional one.

4 Module of the Flexible Beam

To verify the effectiveness of the above algorithms, a flexible beam with piezoelectric stack actuator is built in ADAMS as a test bed. As shown in Fig. 5, the length, width and thickness of the flexible beam is L, W and T. Specifically, L is 1500 mm, W is 60 mm, and T is 15 mm. The piezoelectric stack actuator installed at the bottom of flexible beam. X_1 is the distance between the fixed ends of the flexible and the piezoelectric stack actuator. The disturbance signal is a force which exert at the X_2 position from the top of flexible beam. The X_1 is 450 mm, and the X_2 is 1050 mm. Also, the reference sensor fixed at the X_2 position in the bottom of flexible beam, and the error sensor fixed at the barycenter of flexible beam.

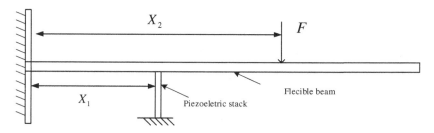

Fig. 5. Module of flexible beam with piezoelectric stack actuator

5 Simulation

To verify the effectiveness of the above algorithms, the flexible beam with piezoelectric stack actuator is setup as a test bed. In this section, both initial states of the flexible beam and the designed parameters of the algorithms are detailed for the realization. In what fellow, performance comparing between the six kind algorithms are designed to demonstrate the performances of the control schemes.

In the simulation of this paper, the disturbance force is a sinusoidal signal whose frequency is 99 Hz, and amplitude is 0.1 N. Figure 6 is the vibration observed at the beam without active control.

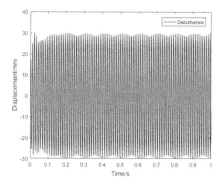

Fig. 6. Characteristics of disturbance signal

Figure 7(a) show the comparison of error signal with the control of FxLMS algorithm and Fx-NLMS algorithm; Fig. 7(b) show the comparison of error signal with the control of Feedback FxLMS algorithm and Feedback Fx-NLMS algorithm; Fig. 7 (c) show that the comparison of error signal with the control of Hybrid FxLMS and Hybrid Fx-NLMS. From the characteristics, it could be observed that the error signals of vibration by using NLMS algorithm have a better convergence speed than the traditional one and also have a satisfied steady-state error.

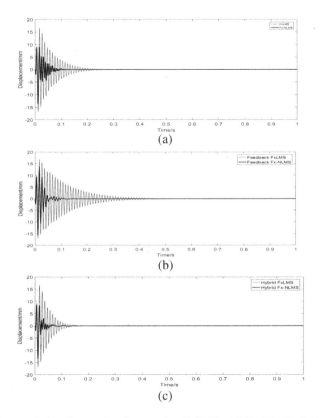

Fig. 7. (a) Characteristics of error signal comparing FxLMS with Fx-NLMS. (b) Characteristics of error signal comparing Feedback FxLMS with Feedback Fx-NLMS. (c) Characteristics of error signal comparing Hybrid FxLMS with Hybrid Fx-NLMS

To better analysis the effect of these above algorithms applying to the flexible beam with piezoelectric stack actuator, the error signals of vibration by using FxLMS algorithm, feedback FxLMS algorithm, hybrid FxLMS algorithm, Fx-NLMS algorithm, feedback Fx-NLMS algorithm and hybrid Fx-NLMS algorithm are shown in Fig. 8(a). For the comparison, original disturbance signal and error signals of vibration by using FxLMS is shown in Fig. 8(b).

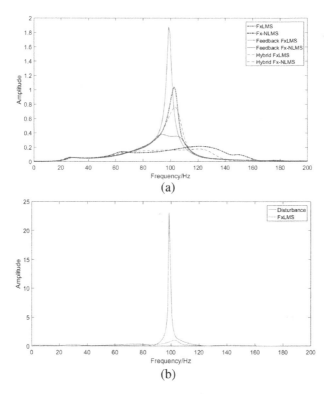

Fig. 8. (a) Characteristics of error signal with active control in frequency domain. (b) Characteristics of original disturbance signal and error signal with FxLMS algorithm in frequency domain

From Fig. 8(b), it could be observed that the effectiveness of the FxLMS is obvious. And in Fig. 8(a), it can be analysis based on the FxLMS. First, all the six adaptive algorithms are effective in the active vibration control of the simulation module of flexible beam in this paper. Second, the control performance of the above six algorithms are compared. Except feedback FxLMS algorithm, the amplitude of error signal with other four algorithms are all smaller than FxLMS'. And the amplitude of Fx-NLMS algorithm, feedback Fx-NLMS algorithm and hybrid Fx-NLMS algorithm are obvious smaller than FxLMS algorithm, feedback FxLMS algorithm and hybrid FxLMS algorithm as the normalized step size.

6 Conclusion

In this paper, the hybrid Fx-NLMS algorithm is adopted for vibration control of the flexible beam with piezoelectric stack actuator. The control effect between FxLMS algorithm, feedback FxLMS algorithm, hybrid FxLMS algorithm, Fx-NLMS algorithm, feedback Fx-NLMS algorithm and hybrid Fx-NLMS algorithm are compared and analyzed in terms of time and frequency domain.

References

1. Morgan, D.R.: History, applications, and subsequent development of the FXLMS algorithm [DSP History]. IEEE Sig. Process. Mag. **30**, 172–176 (2013)
2. Swanson, D.C.: Active noise attenuation using a self-tuning regulator as the adaptive control algorithm. In: INTER-NOISE and NOISE-CON Congress and Conference Proceedings, pp. 467–470. Institute of Noise Control Engineering, Newport Beach (1989)
3. Kuo, S.M., Morgan, D.R.: Active noise control: a tutorial review. Proc. IEEE **87**, 943–973 (1999)
4. Zhu, X., Gao, Z., Huang, Q.: Active vibration control for piezoelectric flexible structure using multi-channel FxLMS algorithm. J. Vibr. Measur. Diagn. **31**, 150–155 (2011)
5. Barkefors, A., Sternad, M., Brannmark, L.J.: design and analysis of linear quadratic gaussian feedforward controllers for active noise control. IEEE/ACM Trans. Audio Speech Lang. Process. **22**, 1777–1791 (2014)
6. Kar, A., Chanda, A.P., Mohapatra, S.: An improved filtered-x least mean square algorithm for acoustic noise suppression. Smart Innov. Syst. Technol. **27**, 25–32 (2014)
7. Nagumo, J., Noda, A.: A learning Method for system identification. IEEE Trans. Autom. Control **12**, 28–287 (1967)
8. Rupp, M., Sayed, A.H.: Two variants of the FxLMS algorithm. In: IEEE ASSP Workshop on Applications of Signal Processing to Audio and Acoustics, pp. 123–126. IEEE Press, New York (1995)
9. Elliott, S.J., Stothers, I.M., Nelson, P.A.: A multiple error LMS algorithm and its applications to active control of sound and vibration. IEEE Trans. Acoust. Speech Sig. Process. **35**, 1423–1434 (1987)
10. Popovich, S.S., Melton, D.E., Allie, M.C.: New adaptive multi-channel control systems for sound and vibration. In: INTER-NOISE and NOISE-CON Congress and Conference Proceedings, pp. 19–20. Institute of Noise Control Engineering, Toronto (1992)
11. Swanson, D.C.: Active noise attenuation using a self-tuning regulator as the adaptive control algorithm. In: INTER-NOISE and NOISE-CON Congress and Conference Proceedings, pp. 467–470. Institute of Noise Control Engineering, Newport Beach (1989)
12. Bismor, D., Czyz, K., Ogonowski, Z.: Review and comparison of variable step-size LMS algorithms. Int. J. Acoust. Vibr. **21**, 2–39 (2016)

Research of Model Identification for Control System Based on Improved Differential Evolution Algorithm

Li Zheng[1], Daogang Peng[1(✉)], Yuzhen Sun[1], and Sheng Gao[2]

[1] College of Automation Engineering, Shanghai University of Electric Power,
Shanghai 200090, China
pengdaogang@126.com
[2] Shanghai Power Equipment Research Institute, Shanghai 200240, China

Abstract. Differential evolution algorithm is a heuristic global search technology based on population, which has received extensive attention from the academic community. Evolution algorithm is applied to the identification and optimization of double-tank system in this article. Firstly, the paper introduces the basic principle of the system identification and differential evolution algorithm. Secondly, design the identification scheme of double-tank system based on differential evolution algorithm. Identify the system according to the data measured in the experiment. Based on the commonly used three models and combined with DE/rand/1/bin, the model structure which best complies with the original experimental data is selected, and the improved form of the difference algorithm is further studied on the basis of the model structure. A large number of experiments have been carried out, the algorithm in other references may only improve one of CR or F, and the two will be all compared in this paper. The results of comparative analysis show that the improved differential evolution algorithm is, to some extent, superior to the basic differential evolution algorithm on identification accuracy of double-tank.

Keywords: Differential evolution algorithm · System identification · Two-tank

1 Introduction

People's research on process behavior has been paid more and more attention. The mastery of process behavior has largely determined the performance of process control system, which has promoted the development of process modeling and identification. There are two ways to build a mathematical model: analytic method and system identification. This paper mainly studies the method of establishing the mathematical model of the process to be studied by system identification. System identification, a kind of science and technology, establishes the dynamic system model by examining the input and output data of the system. When faced with some systems which are more complex, the classic system identification method will be ineffective.

Evolutionary computation has many advantages, such as self-organization, self-adaptation, self-learning, and it is not limited by the type and nature of the problem to be studied. The differential evolution (DE) algorithm uses the population-based global

© Springer Nature Singapore Pte Ltd. 2017
M. Fei et al. (Eds.): LSMS/ICSEE 2017, Part I, CCIS 761, pp. 282–293, 2017.
DOI: 10.1007/978-981-10-6370-1_28

search strategy and uses real number coding and simple variation based on the difference, which effectively reduces the complexity of genetic operation. In addition, the DE algorithm has the ability of memory, it can dynamically track the current search situation of the algorithm and adjust its search strategy. The algorithm does not need to know the characteristic information of the problem studied in advance, so it can solve the optimization problem in some complicated environment. DE algorithm has been widely concerned and applied because of its obvious advantage in the continuous domain optimization problem, which has led to the upsurge in the field of evolutionary algorithm research.

2   Principle and Process of Differential Evolution Algorithm

The DE algorithm is usually composed of four operations: initialization, mutation, crossover and selection. Combined with the evaluation of the fitness value, the optimal solution can be approximated by repeated iterations. In general, the global optimization problem can be transformed into solving the minimum or maximum problem. The step of solving the minimum problem of the function is shown in Eq. (1).

$$\begin{cases} \min f(X), X = [x_1, x_2, \dots, x_D] \\ s.t.\ a_j \le x_j \le b_j,\ j = 1, 2, \dots, \dots D \end{cases} \quad (1)$$

In the formula (1), b_j is used to express the upper limit of x_j, and a_j is used to express the lower limit of x_j.

The process of differential evolution algorithm is shown as follows.

2.1   Initialize the Population

$$\{X_i(0)|X_i(0) = [x_{i,1}, x_{i,2}, \dots, x_{i,D}], i = 1, 2, \dots NP\} \quad (2)$$

$$\begin{cases} x_{i,j} = a_j + rand \cdot (b_j - a_j) \\ i = 1, 2, \dots, NP; j = 1, 2, \dots, D \end{cases} \quad (3)$$

$X_i(0)$ is used to express the i-th individual of the initial population, and $x_{i,j}$ is used to express the i-th individual of the j-th dimension.

2.2   Mutation Operation

$$V_i(g + 1) = X_{r_1}(g) + F \cdot (X_{r_2}(g) - X_{r_3}(g)) \quad (4)$$

$i \ne r_1 \ne r_2 \ne r_3,\ i = 1,2,\dots,NP,\ r_1,r_2,r_3$ are the random integers in the closed interval [1, NP], and g represents the current number of iterations. $X_i(g)$ Represents the i-th individual in the g-th iteration population. $V_i(g + 1)$ represents a new population generated after the mutation.

2.3 Cross Operation

$$u_{i,j}(g+1) = \begin{cases} V_{i,j}(g+1) \, if \, rand \, \leq \, CR \, or \, j = j_{rand} \\ x_{i,j}(g), \quad otherwise \end{cases} \tag{5}$$

$i = 1,2,...,NP, j = 1,2,...,D, U_i(g+1) = [u_{i,1},u_{i,2},...,u_{i,D}]$ represents the new population after the cross operation, j_{rand} is a random integer within the interval [1, D] (Fig. 1).

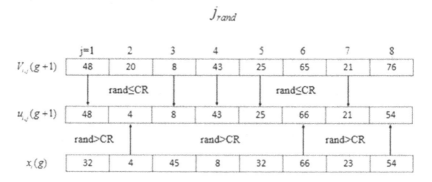

Fig. 1. Binomial cross process

2.4 Selection Operation

$$X_i(g+1) = \begin{cases} U_i(g+1), \, if \, f(U_i(g+1)) \leq f(X_i(g)) \\ X_i(g), \quad otherwise \end{cases} \tag{6}$$

2.5 Termination Condition

The judgment condition here is that the number of evolutionary iterations reaches its maximum value, that is to say, the operation of the algorithm can be stopped only if g (the number of iterations) is greater than G_m (the maximum number of iterations); otherwise, the algorithm will continue until the termination condition is satisfied.

3 Improved Differential Evolution Algorithm

3.1 New Mutation Operation

The differential vector of the parent is the most basic component of the mutation operation of the DE algorithm, and each differential vector consists of two different individuals of the parent. According to a variety of generating methods of mutation individual, a variety of different differential evolution algorithm has be formed. The equations for the DE/rand/1 and DE/best/1 are:

$$v_i^t = x_{r1}^t + F(x_{r2}^t - x_{r3}^t) \tag{7}$$

$$v_i^t = x_{best}^t + F(x_{r2}^t - x_{r3}^t) \tag{8}$$

In the Eqs. (7) and (8), $X_{r1}^t, X_{r2}^t, X_{r3}^t$ are random individuals different from each other, X_{best}^t is the best individual in the population, $F \in [0,1]$ is Scale factor. From Eq. (7) we can see that the mutation individual consists of three different random individuals. Since the base vector is a random individual, which do not need any fitness information, the mutation operation shown in Eq. (7) contributes to the diversity of the population and global search capability of differential evolution algorithm, at the same time, the convergence rate of the algorithm will be reduced. From Eq. (8) we can see that mutation individual X_{best}^t is the base vector of V_i^t, therefore, its local search ability is high, the precision is high, however, the fast convergence speed may cause the algorithm fall into the local optimal point. Combine the characteristics of these two different mutation patterns, take the effects of random individuals X_{r3}^t and optimal individuals X_{best}^t into account when make the mutation strategy of the mutation equation. In this paper, a new mutation strategy is adopted, the mutation equations are:

$$v_i^t = \lambda x_{r1}^t + (1 - \lambda)x_{best}^t + F(x_{r2}^t - x_{r3}^t) \tag{9}$$

$$\lambda = (T_{max} - t)/T_{max} \tag{10}$$

In Eq. (10), T_{max} is used to express the maximum number of iterations, t indicates the current number of iterations, $\lambda \in [0,1]$. If $\lambda = 1$, then Eq. (9) is equivalent to Eq. (7), which refers to DE/rand/1; If $\lambda = 0$, then Eq. (9) is equivalent to Eq. (8), which refers to DE/best/1. A well-performing algorithm generally requires strong global search capability at the initial stage of the search to find possible global optimizations as much as possible, and at the end of the search it should have strong local search capability. Combine the global search capabilities and local search abilities, then the accuracy and convergence rate of the algorithm can be improved. Therefore, the simulated annealing strategy is introduced, λ is set as the annealing factor, as it is shown in Eq. (10). In the process of running the DE algorithm, λ will be gradually reduced from 1 to 0, so that the weight of X_{r3}^t gradually reduces and the weight of X_{best}^t gradually increases, so as to ensure the strong global search and faster convergence rate and search accuracy of the differential evolution algorithm.

3.2 The Strategy of Adaptive Scaling Factor

In the standard differential evolution algorithm, the scaling factor F is generally a fixed value. The scaling factor F affects the degree of disturbance to the base vector, so the size of F will affect the convergence and convergence speed of the algorithm. Das and so on proposed F dynamic adjustment strategy, F gradually reduces with the increase of the number of iterations, so that the algorithm can make multi-directional exploration in the early stages of evolution, and in the late stages of evolution, the gradual reduction of the scaling factor is conducive to adjust the search direction of the vector, which contributes to the development in the region where the global optimal solution are. The following gives a strategy for adaptive scaling factor F, in which F decreases with the increase of the number of iterations:

$$F = F_{max} - \frac{t(F_{max} - F_{min})}{T_{max}} \tag{11}$$

3.3 The Strategy of Crossover Probability Factor

CR determines whether the test vector of the individual in the population is taken from the mutation vector or the target vector. When the CR is large, the information of the test vector will be taken from the mutation vector more, and when CR is small, the information of the test vector will be taken from the target vector more, so as to achieve the desired cross effect. When the fitness value of the mutation vector is better, the test vector should be taken from the mutation vector with a larger probability, and the value of CR should be increased. When the fitness value of the mutation vector is poor, the test vector should be taken from the mutation vector with a smaller probability, and the value of CR should be reduced. The crossover probability factor chosen in this paper is shown in Eq. (12).

$$CR = CR_{min} + \frac{t(CR_{max} - CR_{min})}{T_{max}} \tag{12}$$

4   Application and Verification of Improved DE Alogorithm

In order to verify the effect of model identification based on differential evolution algorithm, the process control experiment device is adopted, and the tank level is the controlled amount. Give the output data collected from the experiment back to the computer to study the dynamic characteristics of dual capacity tank. A total of two sets of experimental data were collected, the first group receives a 5% step signal, and the second group receives a 15% step signal.

From the literature [2] and the literature [3] we can see that the transfer function of the dual-capacity tank can be expressed as three models as shown in Table 1.

Table 1. Three models of the dual-capacity tank

Model 1	$G_1(s) = \dfrac{K}{(T_1 s + 1)(T_2 s + 1)}$
Model 2	$G_2(s) = \dfrac{K}{(T_1 s + 1)(T_2 s + 1)} e^{-\tau s}$
Model 3	$G_3(s) = \dfrac{K}{T s + 1} e^{-\tau s}$

In this paper, the precision index shown in Eq. (13) is chosen as the fitness function. y_p is the output value obtained from the simulation of the model, y is the real output data measured from the experiment.

$$J = \sum_{i=1}^{n} \frac{1}{2}(y_p - y)^2 \tag{13}$$

The differential evolution algorithm uses DE/rand/1/bin. After a large number of experiments, it is found that the influence of the range of the parameters to be identified on the performance of the algorithm is poor than the influence of the algorithm parameter on the performance of the algorithm. Therefore, after several tests, identify the upper limit of the range of parameters. The three models identified by the basic differential evolution algorithm are shown in Table 2.

Table 2. The three models of the double tank identified

Model 1	$G_1(s) = \dfrac{93.7977}{(64.0501s + 1)(291.6566s + 1)}$
Model 2	$G_2(s) = \dfrac{93.8236}{(59.9938s + 1)(296.2190s + 1)}e^{-0.9874s}$
Model 3	$G_3(s) = \dfrac{94.2152}{340.5260s + 1}e^{-42.7492s}$

From Table 3, the model 1 is also applicable to the second set of experimental data, and the accuracy is high and the time is short (Fig. 2).

Table 3. The model parameters identified by the second set of experimental data

	J	K	T/T$_1$ T$_2$	τ	Time
Model 1	0.0816	94.64	66.6 285	/	79.4
Model 2	0.0817	94.65	66.4286	0.1	128
Model 3	0.2939	97.19	330	47	123

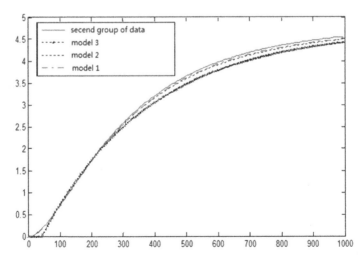

Fig. 2. Validation of model accuracy

The step response curve of model 3 identified by the first set of experimental data is far from the second set of experimental data, and the step response curve of model 1 is closer to the real experimental data. Therefore, model 1 is used as the object of further study (Fig. 3).

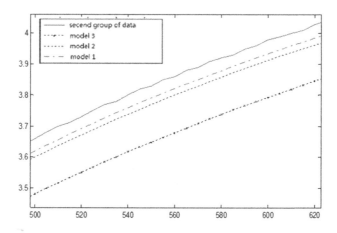

Fig. 3. Proximity between three models and experimental data

Dynamically changing the value of the parameter allows the parameters of the differential evolution algorithm to be adapted to the algorithm process automatically, while improving the mutation operation to apply it to the identification of the system model. The MATLAB experiments were carried out with the mutation operations shown in Eqs. (8) and (9). The PC system is win10 Professional Edition, and the MATLAB version is 2011b.

$$v_i^t = \lambda x_{r1}^t + (1 - \lambda)x_{best}^t + F(x_{r2}^t - x_{r3}^t) \tag{14}$$

$$\lambda = (T_{\max} - t)/T_{\max} \tag{15}$$

Now select the following two adaptive parameter strategy to identify.

$$CR = CR_{\min} + \frac{t(CR_{\max} - CR_{\min})}{T_{\max}} \tag{16}$$

$$F = F_{\max} - \frac{t(F_{\max} - F_{\min})}{T_{\max}} \tag{17}$$

4.1   Application of Improved Mutation Operator Algorithm

Under the premise that the population size NP, the maximum evolutionary algebra. T_{max} and the scaling factor F are not changed, the system is identified by the mutation strategy shown in Eqs. (14) and (15). The change of the optimal solution with the evolutionary algebra is shown in Fig. 4. It can be seen that the recognition accuracy based on the basic differential evolution algorithm and the recognition accuracy based on the improved mutation operator are not very different and the convergence rate of the improved differential evolution algorithm is superior to the basic differential evolution algorithm to a certain extent. It can be seen that the convergence rate of the differential evolution algorithm with improved mutation operator is faster than that of the basic mutation operator. So we can know that the performance of the improved mutation operator algorithm is better than that of the basic differential evolution algorithm.

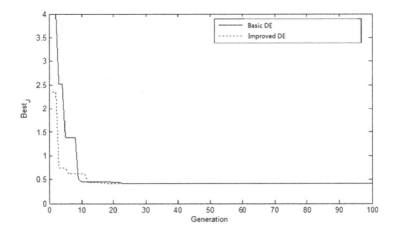

Fig. 4. Effect of mutation operator

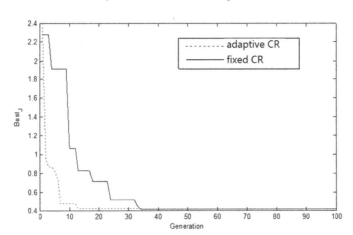

Fig. 5. Effect of cross factor strategy

4.2   Application of Improved Crossover Probability Factor Algorithm

Under the premise that the population size NP, the maximum evolutionary algebra T_{max} and the scaling factor F are invariant, we choose the crossover probability factor strategy as shown in Eq. (16) to identify the parameters of the mathematical model of the double-tank test system. The change of fitness value with evolutionary algebra is shown in Fig. 5.

From Fig. 5, the convergence rate of the differential evolution algorithm with the adaptive cross factor parameter strategy is faster than that of the fixed CR value. It can be seen that the performance of adaptive crossover factor algorithm is better than that of fixed cross factor algorithm.

4.3   Application of Improved Scaling Factor Algorithm

Similarly, under the premise that the population size NP, the maximum evolutionary algebra T_{max} and the cross factor CR are invariant, we choose the following adaptive scaling factor F strategy, in which F decreases with the increase of the number of iterations:

$$F = F_{max} - \frac{t(F_{max} - F_{min})}{T_{max}} \tag{18}$$

Now study the effect of adaptive parameter scaling factor, where t is the current number of iterations, parameters $F_{max} = 0.8$, $F_{min} = 0.2$. The change of the optimal solution with the evolutionary algebra t is shown in Fig. 6.

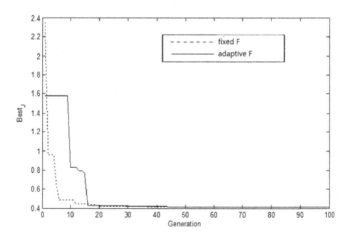

Fig. 6.   The effect of scaling factor strategy

It can be seen from Fig. 6 that the convergence rate of the differential evolution algorithm with the scaling factor parameter adaptive selection strategy is faster than that of the fixed F-value differential evolution algorithm. Which proves that the superiority of the differential evolution algorithm which adopts the adaptive scaling factor strategy.

4.4 Comparison of Identification Results of Double-Capacity Water Tank System Based on Improved DE Algorithm

The three improved DE algorithms are used to identify the parameters of the dual-tank system. After selecting the best parameters, three new transfer functions can be formed, and then the simulation is carried out by MATLAB. The step response curve is shown in Fig. 7.

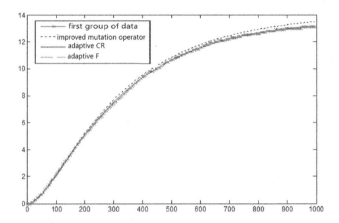

Fig. 7. Improved DE algorithm to identify the results of the response curve and experimental data comparison

It can be seen from Fig. 7 that among the three improved differential evolution algorithms, the result of the improved mutation operator strategy is relatively poor, while the accuracy of the other two improved differential evolution algorithms is similar to each other.

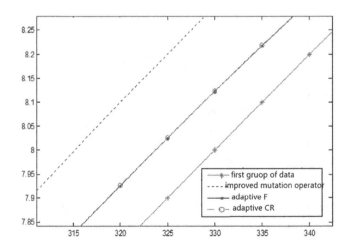

Fig. 8. Comparison of the results of the improved algorithm and the experimental data

After further enlarging the curve graph shown in Fig. 8, it can be seen that the accuracy of the improved F and adaptive CR differential evolution algorithm is similar to that of the improved CR, and then compare the recognition result of improved CR differential evolution algorithm with the recognition result of the basic differential evolution algorithm, the result is shown in Fig. 9.

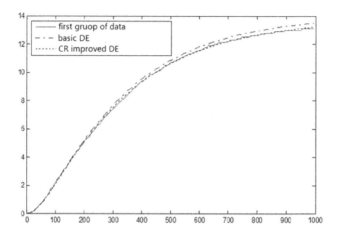

Fig. 9. Comparison of improved DE and basic DE identification results

From Fig. 9, we can see that the step response curve of the parametric model identified by the basic differential evolution algorithm is far from the first group of experimental data, and the step response curve of the parametric model identified by the improved differential evolution algorithm is almost overlaps with the curve of the experimental data, thus it can be seen that the improved differential evolution algorithm is superior to the basic differential evolution algorithm in the accuracy of the dual-tank model.

5   Conclusion

This paper introduces the differential evolution algorithm and the principle of improved differential evolution algorithm and its several different forms. The performance of improved differential evolution algorithm is verified by the case of double - capacity water tank. In this paper, the operation flow of the basic differential evolution algorithm is given and used in the model identification of the system. The influence of the parameters in the differential evolution algorithm on the performance of the algorithm is studied by repeated comparative experiments.

Because of the strong and rapid optimization ability of the differential evolution algorithm, In this paper, the differential evolution algorithm is used to identify the parameters of the model, and the individuals of differential evolutionary algorithms is composed of the parameters of the model to be identified. The results show that the improved differential evolution algorithm is superior to the basic evolutionary algorithm in terms of system identification.

DE algorithm has become a new research hotspot in the field of evolutionary computation, however DE algorithm still has a wide range of research space worthy of mining in its theoretical analysis, algorithm improvement and applied research. For example, we can consider the diversity of the population and the change of the individual fitness value in the process of the algorithm, and introduce the adaptive adjustment mechanism. When the population converges to a certain range near the optimal solution, take the reverse direction, and this will cause greater perturbation to mutation vector, which can increase the diversity of the population and prevent local optimum; we can take advantage of the high degree of parallelism in the operation of the DE algorithm to divided the initial population into multiple subpopulations, and the subpopulations are interconnected according to the von Neumann topology, and the migration mechanism allows the information to be shared periodically throughout the population; DE can also be combined with other algorithms with respective advantages within a certain scope of the solution to produce a new algorithm which surpass the parent algorithm.

Acknowledgments. This work was supported by Shanghai Science and Technology Commission Program (No. 16111106300, No. 17511109400 and No. 15510722100) and Engineering Research Center of Shanghai Science and Technology Commission Program (No. 14DZ2251100).

References

1. Guo, L.-H., Zhu, L.-H., Gao, W.: Characteristics and simulation of least squares system based on MATLAB. J. Xuchang Univ. **29**(2), 24–27 (2010)
2. Liu, Z.-Q., Wang, S.-X.: Study on modeling and PID control algorithm of liquid level control system for double-tank. J. Appl. Sci. **9**, 95–96 (2015)
3. Wang, Z.: Process Control Systems and Instruments. Mechanical Industry Press, Beijing (2006)
4. Zhang, C.: Differential Evolution Algorithm Theory and Application. Beijing University of Science and Technology Press, Beijing (2014)
5. Liu, Y., Gu, F.: Study on differential evolution algorithm. Technol. Square **03**, 20–23 (2013)
6. Li, Q., Guo, Y., Ying, P.: Comparison of water level model DE and PSO and LS identification of double-capacity water tank. Comput. Simul. **32**(11), 407–409 (2015)
7. Li, X.-L., Li, J., Shi, L.-H.: Construction and simulation of tank level system. J. Syst. Simul. **22**(4), 829–832 (2010)
8. Tasgetiren, M.F., Suganthan, P.N.: A multi-populated differential evolution algorithm for solving const rained optimization problem. In: IEEE Congress on Evolutionary Computation (2006)
9. Das, S., Suganthan, P.N.: Differential evolution: a survey of the state-of-the-art. IEEE Trans. Evol. Comput. **15**(1), 4–31 (2011)
10. Liu, J., Lampinen, J.: A fuzzy adaptive differential evolution algorithm. Soft. Comput. **9**(6), 448–462 (2005)
11. Vu, C.C., Bui, L.T., Abbass, H.A.: DEAL: a direction-guided evolutionary algorithm. In: Bui, L.T., Ong, Y.S., Hoai, N.X., Ishibuchi, H., Suganthan, P.N. (eds.) SEAL 2012. LNCS, vol. 7673, pp. 148–157. Springer, Heidelberg (2012). doi:10.1007/978-3-642-34859-4_15
12. Gong, W., Cai, Z., Ling, C.X.: Enhanced differential evolution with adaptive strategies for numerical optimization. IEEE Trans. Syst. Man Cybern. B Cybern. **41**(2), 397–413 (2011)

Multi-variety Fresh Agricultural Products Distribution Optimization Based on an Improved Cuckoo Search Algorithm

Wenqiang Yang[(✉)], Junpeng Xu, and Yongfeng Li

Henan Institute of Science and Technology, Xinxiang, China
yangwqjsj@163.com

Abstract. To minimize the losses of multi-variety perishable agricultural products, a mathematical model considering time sensitive feature of each perishable agricultural product is proposed. Meanwhile, a cuckoo search algorithm (CSA) is introduced to minimize the total losses of agricultural products. In view of poor exploration and exploitation ability of CSA, adaptive adjusting discovery probability and dynamic step-length is imposed to form an improved cuckoo search algorithm (ICSA). Finally, to verify the performance of the proposed algorithm, it is compared with cuckoo search and genetic algorithm (GA). Simulation results prove that the feasibility and superiority of the proposed algorithm.

Keywords: Multi-variety fresh agricultural products · Improved cuckoo search algorithm (ICSA) · Adaptive adjusting discovery probability · Dynamic step-length · Distribution optimization

1 Introduction

Agricultural cooperatives are booming in china, in order to realize the benefit maximization, diversification and anti-seasonal of agricultural products gradually taken over the sole of agricultural products from self-employed [1]. Meanwhile, as necessities of life, the safety of agricultural products has been the focus of attention. Therefore, how to optimize the distribution route to minimize the decay degree of fresh agricultural products which has become important in the development of agricultural cooperatives. Duo to this and the traditional distribution modes must bring about some changes. So far, domestic and foreign scholars have acquired some findings [2–7]. For instance, Min et al. [8] present a network-based food supply chain model, which focus on fresh produce and food deterioration. Rong et al. [9] provide a methodology to model food quality degradation and integrate food quality in decision-making on production and distribution in a food supply chain. Soysal et al. [10] model the inventory routing problem to account for perishability, explicit fuel consumption and demand uncertainty and present the applicability of the model on the fresh tomato distribution operations of a supermarket chain. Ahumada et al. [11] present an operational model that generates short term planning decisions for the fresh produce industry and helps the grower to maximize his revenues. Yang et al. [12] propose a non-linear programming model to

© Springer Nature Singapore Pte Ltd. 2017
M. Fei et al. (Eds.): LSMS/ICSEE 2017, Part I, CCIS 761, pp. 294–302, 2017.
DOI: 10.1007/978-981-10-6370-1_29

optimize logistics network layout for fresh agricultural products. Yang et al. [13] propose bi-level model based on aquatic product distribution. From above study, it would be concluded that much more researches are concerned about a single product than multi-variety products in the field of fresh agricultural products distribution. Actually, as more agricultural cooperatives spring up, multi-variety fresh agricultural products distribution would be the main trend in the future. Thus, it is necessary to study multi-variety fresh agricultural products distribution.

Fresh agricultural products distribution problem is one of complex combinatorial optimization problems and NP-Hard, which is difficult to solve properly by using traditional optimization methods. Cuckoo search algorithm is a meta-heuristic algorithm simulating the parasitism breed behavior of cuckoos and used for continuous nonlinear optimization problems [14, 15]. But up to now, CSA has not yet been applied to discrete combinatory optimization problems. Hereinafter, To solve such problems does require a series of changes on CSA to make it suitable for discrete problems.

2 Problem Description and Modeling

Suppose there is an agricultural cooperatives that produces tomatoes, cucumbers, watermelon and each production may have different demands for every distribution destination. Sketches of the fresh agricultural products distribution is shown in Fig. 1.

From Fig. 1, the vehicle that transports fresh agricultural products to every distribution destination in sequence and then returns back to agricultural cooperative.

Problem statement
Suppose a fresh agricultural products distribution task that has the following features.

(I) The number of the distribution destinations is m, the assigned locations are represented respectively as $(p_1, p_2, \ldots, p_m)$. Similarly, p_0 corresponding to input buffer or output buffer, then the distance between them can be expressed as $d_{ij}(i, j \in \{0, 1, 2, \cdots, m\})$.

(II) For each distribution destination, o_{it}, o_{ic} and $o_{iw}(i \in \{1, 2, \cdots, m\})$ are represented by orders of tomatoes, cucumbers and watermelon.

Definition 1: If the vehicle with agricultural products consecutively accesses p_i and p_j during distribution, $e_{ij} = 1$; Otherwise, $e_{ij} = 0$.

Definition 2: If the sub-route r belongs to one of the routes which the vehicle complete distribution tasks and p_i belongs to sub-route r, $g_{ir} = 1$; Otherwise, $g_{ir} = 0$.

Definition 3: If the task of one distribution destination is completed, $l_i = 0$; Otherwise, $l_i = 1$.

Definition 4: To quantify the decay degree of fresh agricultural products, the perishable character of tomatoes, cucumbers and watermelon are denoted respectively with w_t, w_c and w_w, which indicate how much of products is decayed in unit time. Based on prior experience, w_t, w_c and w_w are assigned to 0.3, 0.05 and 0.4, respectively.

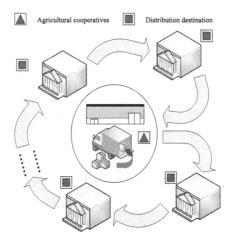

△ Agricultural cooperatives   ■ Distribution destination

Fig. 1. The sketches of the fresh agricultural products distribution

2.1 Mathematical Modeling

To deal with distribution optimization problems of perishable agricultural products, need to establish a mathematical model. Suppose that maximum load and speed of the vehicle is Q and v, respectively. When minimizing the decay degree of fresh agricultural products as objective function, the mathematical model can be described as

$$\min f(e) = \sum_{j=1}^{m} (d_{0j}/v) \cdot e_{0j} \cdot \phi(i) + \sum_{i=1}^{m} \sum_{j=1}^{m} (d_{ij}/v) \cdot e_{ij} \cdot \phi(i) + \sum_{i=1}^{m} (d_{i0}/v) \cdot e_{i0} \cdot \phi(i) \tag{1}$$

where, $\phi(i) = \sum_{i=1}^{m} (o_{it} \cdot w_t + o_{ic} \cdot w_c + o_{iw} \cdot w_w) \cdot l_i$.

s.t.

$$\sum_{j\in\{t,c,w\}} o_{ij} \leq Q \quad \forall i \in \{1,2,\cdots,m\} \tag{2}$$

$$\sum_{i=1}^{m} (\sum_{j\in\{t,c,w\}} o_{ij}) \cdot g_{ir} \leq Q \tag{3}$$

$$\sum_{i=1}^{m} (\sum_{j\in\{t,c,w\}} o_{ij}) \cdot g_{ir} \leq Q \quad \&\& \quad \sum_{i=1}^{m} (\sum_{j\in\{t,c,w\}} o_{ij}) \cdot g_{ir} + \sum_{j\in\{t,c,w\}\&\&p_i\notin r} o_{ij} > Q \tag{4}$$

$$\sum_{r} g_{ir} = 1 \quad \forall i \in \{1,2,\cdots,m\} \tag{5}$$

Equation (1) is the objective function. Constraints that are from Eqs. (2) to (5). Equation (2) requires the order of each distribution destination must not exceed maximum load of the vehicle. Equation (3) prevents the overload of the vehicle. Equation (4) makes sure that the vehicle is almost fully loading. Equation (5) ensures that the order of each distribution destination is completed by the vehicle at a time.

3   Improved Cuckoo Search Algorithm (ICSA)

Standard cuckoo search algorithm is a community intelligent optimization technique, which is originally used to solve optimization problems in continuous domain. First, for the sake of solving discrete combination optimization problems, a new kind of encoding schema needs to be adopted. Second, the random searching step and the fixed discovery probability can't coordinate the exploration and exploitation ability of the algorithm. Based on the conditions mentioned above, this paper makes some improvements to standard cuckoo search algorithm, thereinafter.

Solution Encoding and Decoding
To use cuckoo search algorithm conveniently, random numbers-encoded is adopted. However, fresh agricultural products distribution is discrete combination optimization problem, need to create a one-to-one mapping between solutions and cuckoo nests located in continuous domain. In this case, all dimensions of the cuckoo nest position is to be sorted in ascending order. Thus to each dimension corresponds to the distribution destination number. Meanwhile, one dimension at which the order totals approach or reach fully loading of the vehicle, insert here agricultural cooperatives number 0. This avoids the generation of infeasible solutions. As Fig. 2 indicates the detailed encoding process.

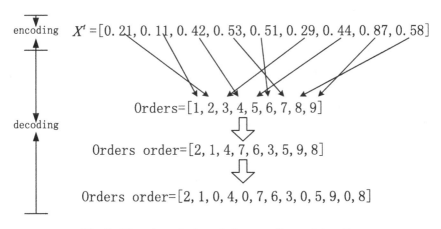

Fig. 2. The schematic for solution encoding and decoding

Adaptive Adjusting Recognization Probability

Discovery probability p_a can be understood as what extent parent individuals are reserved. Hence, the larger p_a, the more chance parent individuals are reserved. And yet, too large fixed discovery probability makes cuckoo search trap into local optima easily and have a low convergence accuracy. Meanwhile, too small fixed discovery probability leads to the blindness of search. To handle this problem, a method which adaptively adjusts discovery probability of each is defined as

$$p_a = p_a^{min} + (p_a^{max} - p_a^{min}) \cdot \frac{f_i}{f_{max}} \tag{6}$$

where $p_a^{min}, p_a^{max}, f_i, f_{max}$, refers to the minimum of p_a, the maximum of p_a, the fitness of the i th cuckoo nest and the maximum fitness of all cuckoo nests. This not only keeps the diversity of the population but good individuals.

Dynamic Step-Length

Standard cuckoo search algorithm adopts levy flight to get random steps, which lacks effective guide mechanism and makes it difficult to achieve balance between searching efficiency and solution accuracy during the searching process. For this reason, a strategy of adaptive step is proposed as

$$L_i^{(t+1)} = (X_i^{(t)} - X_{best}) \cdot \frac{N_a - N_c}{N_a} \tag{7}$$

where L_i^{t+1}, $X_i^{(t)}$, X_{best}, N_a and N_c refers to the step of the (t + 1)th iteration, the position of the i th cuckoo nest in the t th iteration, the best position of all the cuckoo nests until the t th iteration, the maximum number of iteration and the number of current iteration.

Representation of ICSA

Step 1. Initialize parameters: the generation counter t the number of cuckoo nests N_n, p_a^{min}, p_a^{max}, N_a, and generate N_n cuckoo nests;
Step 2. Calculate the fitness of N_n cuckoo nests based on Eq. (1) and find the best cuckoo nest X_{best};
Step 3. Global search: $t = t + 1$;
Step 4. Update the cuckoo nests according to Eq. (8);

$$X_i^{(t+1)} = X_i^{(t)} + L_i^{(t+1)} \tag{8}$$

Step 5. Compare cuckoo nests of the current generation and the previous generation. And preserve the better cuckoo nests.
Step 6. Generate random numbers r. If $r > p_a$, then establish new nests;
Step 7. Evaluate all the cuckoo nests, and find the best cuckoo nest X_{best};
Step 8. If $t \leq N_a$, and then go to step3; Otherwise, output the optimal solution.

4 Numerical Example and Analysis

To evaluate the performance of ICSA, the examples of multi-variety fresh agricultural products distribution problems are given in Fig. 3, where some comparisons are made with CSA and GA. In addition, simulations are made under the same condition, such as windows XP operating system, 2.19 GHz processor, 1.99 GB of RAM and development environment Matlab 7.0. Population size and the number of generations of all algorithms are 50 and 600, respectively. Moreover, for ICSA, $p_a^{min} = 0.15$, $p_a^{max} = 0.6$. For CSA, $p_a = 0.5$. For GA, crossover probability and mutation probability are set 0.8, 0.15, respectively. For the vehicle, $Q = 10t$, $v = 70$ kph.

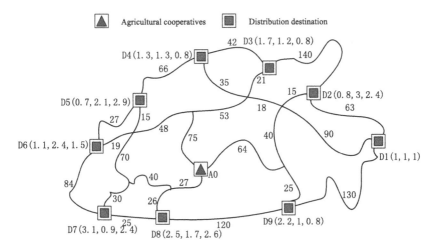

Fig. 3. Distribution nodes layout and customer demands

In Fig. 3, $Di(x, y, z)$ represents customer demands, of which i, x, y, z stand for distribution destination number, tomatoes, cucumbers and watermelon, respectively. The distance between nodes are marked along the arcs among nodes. Figure 4 gives the optimization results with ICSA, CSA and GA, respectively. In the meantime, to analyze the relation between $f(e)$ and driving time of vehicle, which are compared in Fig. 5. Moreover the number of generations on the performance of ICSA is shown in Fig. 6. To better understand the effect of order change on the decay degree, the sensitivity of each order parameters is surveyed in Fig. 7.

To further test stability and accuracy of ICSA which is relative to other two algorithms, benchmark instances from TSPLIB are adopted. Each algorithm runs thirty times, results are presented in Table 1 in which opt, n and dev indicate optimal value, the times of finding optimal, and deviation rate relative to optimal respectively.

As shown in Fig. 4 and Table 1, compared with CSA and GA, ICSA shows a better performance. Furthermore, CSA and GA trap into local optima easily. GA, by using mutation, and CSA, by using discovery probability, can reduce the chance of falling into local minimum to some extent. Nevertheless, the fixed mutation and discovery probability make them be blindness or randomness. For ICSA, strategy of adaptive

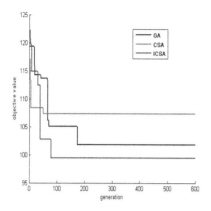

Fig. 4. Evolution comparison

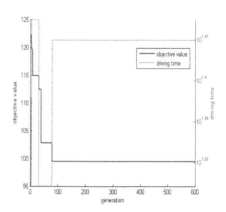

Fig. 5. The corresponding relation

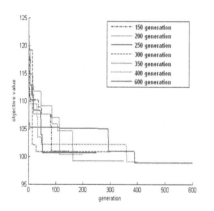

Fig. 6. Evolution comparison for ICSA

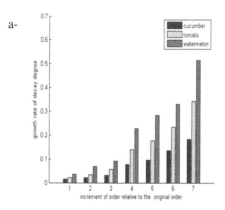

Fig. 7. Sensitivity of order parameters test

Table 1. Performance comparison among algorithms

TSP	Optimal (TSPLIB)	GA			CSA			ICSA		
		opt	n	dev	opt	n	dev	opt	n	dev
Dantzigz42	699	756	0	0.082	778	0	0.113	699	30	0
Eil51	426	435	0	0.021	441	0	0.035	426	30	0
Berlin52	7542	7629	0	0.012	7650	0	0.014	7542	30	0
St70	675	693	0	0.027	687	0	0.018	675	26	0
Eil76	538	642	0	0.193	676	0	0.257	538	23	0
Pr107	44303	44542	0	0.005	44617	0	0.007	44303	23	0
Gr120	6942	8807	0	0.269	9131	0	0.315	6942	16	0

djusting discovery probability presented in this paper improves the quality of the population, which can upgrade the accuracy of optimal solution. Furthermore, dynamic step-length are adopted during search, which greatly speed up the convergence. In addition, as seen in Table 1, the performance difference among algorithms is a little when the size of tasks is smaller; However, the performance of ICSA is much better than that of the other two when the size of tasks is larger. It further demonstrates that ICSA has strong global and local search ability, especially in a larger size. Besides, Fig. 6 provides that evolutionary generations is related to the scale of the problem. So it is possible that the optimal solution may not be found with too few generations, and computation time will increase with too many generations. Therefore, how to choose appropriate evolutional generations is a problem worth researching.

In order to understand the relationship between decay degree and driving time contrasted in Fig. 5, from which we can see that decay degree is not linearly proportional to driving time. It means that the smallest decay degree necessarily correspond to the shortest driving time. It is exactly what is different from other studies in this paper and ensure the safety of agricultural products more effectively. Finally, sensitivity of order parameters is tested in Fig. 7. It is clear that the decay degree growth rate increases with increment of order among cucumber, tomato and watermelon. And, the easier perishable the agricultural products, the more decay degree growth rate increases. It gives us some idea that the easier perishable agricultural products may be distributed firstly taking little account of transportation costs.

5 Conclusion

To optimize the distribution of a multi-variety fresh agricultural products, an ICSA is proposed in this paper. Firstly, strategy of adaptive adjusting discovery probability is introduced, which improves the quality of the population and upgrades the accuracy of optimal solution. Secondly, and dynamic step-length is adopted, which greatly speed up the convergence. Finally, the superiority of ICSA is demonstrated by the comparison with CSA and GA.

Acknowledgments. This work is supported by the scientific and technological project of Henan province under Grant No. 172102110031, and by the high-tech people project of Henan institute of science and technology under Grant No. 203010616001.

References

1. Ito, J., Bao, Z.S., Su, Q.: Distributional effects of agricultural cooperatives in China: exclusion of smallholders and potential gains on participation. Food Policy **37**(6), 700–709 (2012)
2. Reis, S.A., Leal, J.E.: A deterministic mathematical model to support temporal and spatial decisions of the soybean supply chain. J. Transp. Geogr. **43**, 48–58 (2015)
3. van Donselaar, K., van Woensel, T., Broekmeulen, R., et al.: Inventory control of perishables in supermarkets. Int. J. Prod. Econ. **104**(2), 462–472 (2006)

4. Ahumada, O., Villalobos, J.R., Mason, A.N.: Tactical planning of the production and distribution of fresh agricultural products under uncertainty. Agric. Syst. **112**, 17–26 (2012)
5. Santa, J., Zamora-Izquierdo, M.A., Jara, A.J., et al.: Telematic platform for integral management of agricultural/perishable goods in terrestrial logistics. Comput. Electron. Agric. **80**, 31–40 (2012)
6. Wang, L.J.: Grey dynamic programming model for grey game in perishable product's supply chains. J. Huazhong Univ. Sci. Technol. **37**(4), 89–92 (2009)
7. Ahumada, O., Villalobos, J.R.: Application of planning models in the agri-food supply chain: a review. Eur. J. Oper. Res. **196**(1), 1–20 (2009)
8. Min, Y., Anna, N.: Competitive food supply chain networks with application to fresh produce. Eur. J. Oper. Res. **224**(2), 273–282 (2013)
9. Rong, A., Akkerman, R., Grunow, G.: An optimization approach for managing fresh food quality throughout the supply chain. Int. J. Prod. Econ. **131**(1), 421–429 (2011)
10. Soysal, M., Bloemhof-Ruwaard, J.M., Haijema, R., et al.: Modeling an inventory routing problem for perishable products with environmental considerations and demand uncertainty. Int. J. Prod. Econ. **164**, 118–133 (2015)
11. Ahumada, O., Villalobos, J.R.: Operational model for planning the harvest and distribution of perishable agricultural products. Int. J. Prod. Econ. **133**(2), 677–687 (2011)
12. Yang, H.L., Ji, Y.F., Liu, F.F.: Layout optimization of logistics network nodes for fresh agricultural products. J. Dalian Marit. Univ. **36**(3), 47–49 (2010)
13. Yang, Z.Z., Mu, X., Zhu, X.C.: Optimization model of distribution network with multiple distribution centers and multiple demand points considering traffic flow change. J. Traffic Transp. Eng. **15**(1), 100–107 (2015)
14. Yang, X.S., Deb, S.: Engineering optimization by cuckoo search. Int. J. Math. Model. Numer. Optim. **1**(4), 330–343 (2010)
15. Rajabioun, R.: Cuckoo optimization algorithm. Appl. Soft Comput. **11**(8), 5508–5518 (2011)

Research on Indoor Fingerprint Localization System Based on Voronoi Segmentation

Ang Li, Jingqi Fu[✉], and Huaming Shen

Department of Automation, College of Mechatronics Engineering and Automation,
Shanghai University, No. 149, Yanchang Rd, Shanghai 200072, China
jqfu@staff.shu.edu.cn

Abstract. The location of entities in a smart indoor environments is an important context information. To this end, several indoor localization algorithm have been proposed with the received signal strength fingerprint (RSS-F) based algorithm being the most attractive due to the higher localization accuracy. However, RSS-F based localization accuracy is highly degraded on account of non-line-of-sight (NLOS) propagation in indoor or harsh environment. This thesis proposes an approach for NLOS self-monitoring and autonomous compensation. Firstly, the localization area is regionalized according to Voronoi Diagram. Then, the self-monitoring and autonomous compensation is realized by propagation environment similarity represented by the dynamic path attenuation index between the domains. The verification experiment results show that the proposed algorithms can adaptively identify the NLOS interference and accomplish compensation. Compared with other localization algorithm, the maximum error is reduced from 3.04 m to 1.71 m, the average error is reduced to 0.90 m, and the localization time is reduced to 2.113 s (contain 10 test point) compared with other tracking algorithm.

Keywords: Indoor localization · Voronoi diagram · Fingerprint algorithm

1 Introduction

With the rapid development of wireless technology and popularity of indoor wireless systems, the Indoor Location Based Service have penetrated into many aspects of modern life [1]. The Received Signal Strength Fingerprints (RSS-F) based indoor localization technology has become research focus of indoor localization because of the highly localization accuracy and hardware-independent [2, 3]. RSS-F based indoor localization accuracy is mainly depend on the RSS-F database, while the construction of a complete RSS-F database is extremely labor-extensive and time-consuming. On account of the RSS is vulnerable to the environment, the RSS-F based indoor localization is mainly used in the small ambient noise contained indoor environment [4]. In order to improve the efficiency of RSS-F database establishment and improve the ability of fingerprint localization algorithm to resist noise. The localization accuracy fluctuation caused by the RSS-F sparse degree and RSS-F establishment mechanism is discussed in [5]. In [6], a Gaussian regression RSS-F database generation mechanism is proposed.

© Springer Nature Singapore Pte Ltd. 2017
M. Fei et al. (Eds.): LSMS/ICSEE 2017, Part I, CCIS 761, pp. 303–312, 2017.
DOI: 10.1007/978-981-10-6370-1_30

A regional centroid algorithm with non-fixed path attenuation index is proposed in [7], but the online global optimization of the proposed algorithm requires a larger amount of calculation is not conducive to real-time improvement. The clustering algorithm which is widely used in image processing has a good classification ability [8–10], but the clustering performance highly depends on the initial point selection. Voronoi diagram has a good regional division and scalability, a Voronoi diagram based RSS-F database establishment algorithm is proposed in [11], the proposed algorithm divided the default RSS-F into two categories, the primary calibration points and the secondary calibration points, the secondary calibration points obtained based on the radio propagation model. But the proposed algorithm is sensitive to the real-time ambient noise, the localization error will be greatly increased if the RSS contains N interference. At present, the most widely used method is RSS filtering, for the nonlinear system which can be described by the state equation, the particle filter algorithm is the most suitable filter method [12]. In [13], a Pedestrian Dead-Reckoning [14] based filter algorithm is proposed. An unsupervised learning based particle filter algorithm is proposed in [15]. However, these algorithms increase the computational complexity and lower the real-time localization performance.

In view of the above problem, a Voronoi segmentation based RSS-F localization system is proposed to detect and compensate the NLOS effects. Voronoi Segmentation Based Fingerprint Localization (VBFL) algorithm is proposed in Sect. 2. The effect of NLOS propagation is compensated by the communication environment similarity between Voronoi Domains (VDs). The wireless indoor localization system based on RSS-F is set up in the third section to verify the proposed algorithm and draw some constructive conclusions.

2   Voronoi Based Indoor Fingerprint Localization Algorithm

The RSS-F based localization algorithm consists of two phases: the offline acquisition phase and online matching phase as shown in the Fig. 1. In the offline acquisition phase, the RSS corresponding to the Beacon Point (BP) label is obtained from the physical coordinates fixed BP at the preset access points (APs) by the mobile intelligent terminal to establish RSS-F database $\{(x_i, y_i), RSS_{i,1}, RSS_{i,2}, \cdots RSS_{i,m}\}$. (x_i, y_i) is the physical coordinates of BP. The quantity of BP in the localization system is m.

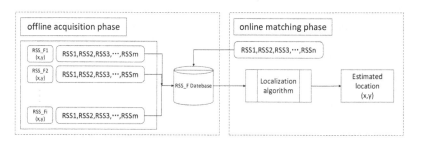

Fig. 1. Traditional fingerprint localization algorithm

As shown in the Fig. 1, the mapping between RSS and geographical location can be established since the RSS of APs have the features of stable, special and distinguishable [16]. In the online matching phase, a set of RSSs obtained by the mobile intelligent terminal are compared with the RSS-F database to obtain the estimation of the geographical position, then the static indoor localization or dynamic indoor target tracking can be achieved. The method of establishing the offline RSS-F includes the propagation model method and RSS feature method.

In this thesis, the mean value and the most value are chosen to construct the RSS-F database.

2.1 Voronoi Based Regionalization Processing

The view of "the solar system is composed of vortices" is proposed by Descartes in his book name "philosophy". In this book, the space can be decomposed into some convex domain, each convex domain is formed around a fixed star, this point of view provides a idea of multi-dimensional space domain segmentation. A two-dimensional space Voronoi diagram based regionalization preprocessing of RSS-F map is proposed to realize the regionalization of the localization area without the initial cluster point. Assuming the preset AP in the localization area is $(AP_1, AP_2, \cdots, AP_i)$, the amount of AP is i, and the coordinate of the AP is $\{(x_{AP1}, y_{AP1}), (x_{AP2}, y_{AP2}), \cdots, (x_{APi}, x_{APi})\}$. Simultaneously, $\{(x_{BP1}, y_{BP1}), (x_{BP2}, y_{BP2}), \cdots, (x_{BPm}, y_{BPm})\}$ is the coordinate of preset BPs.

As shown in the Fig. 2, the VD is constructed through adopting Delaunay triangulation algorithm with the nearest three AP points serving as the vertex. Assuming the vertex is $AP_{i1}, AP_{i2}, AP_{i3}$, the external round center of ΔAP_i is AP_{qi}. So that, AP_{qi} is a voronoi nucleus, the connection of nucleus is a Voronoi segmentation line, the closed polygon constituted by lines is a VD. The characteristics of the VDs that are adaptively divided according to the ADs satisfy the following items:

- Any VD contains only one domain nucleus.
- The Euclidean distance from the mobile intelligent terminal in a VD to the VD nucleus is smaller to any other VD nucleus.
- The AP addition, remove or movement only affect the adjacent VD.

The RSS-F of domain nuclear AP_i is $\{(x_i, y_i), RSS_{i,1}, RSS_{i,2}, RSS_{i,3}, \cdots RSS_{i,m}\}$, and the RSS received by the mobile intelligent terminal is $(RSS_1, RSS_2, RSS_3, \cdots RSS_m)$. The estimated location of the mobile intelligent terminal can be calculated by the matching function 1.

$$\inf(\text{gap}_i) = \min\left(\frac{1}{m}\sum_{n=1}^{m}|RSS_{i,m} - RSS_m|\right). \tag{1}$$

The RSS-F is generally collected in the static localization environment for intuitive reflection of the static propagation environment. According to the mapping characteristics, it can be deduced that in the ideal state (assuming that there is no measurement

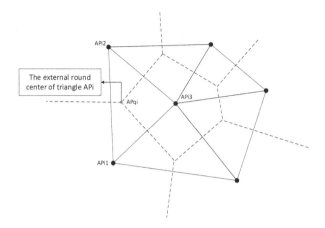

Fig. 2. The construction of the Voronoi Domain

noise and observed noise), if the received real-time RSS is closest to one RSS-F, the estimated location is closest to corresponding VD nucleus coordinate.

2.2 Regional Path Attenuation Index Based RSS-F Indoor Localization Algorithm

In the online matching phase, the propagation of RSS is in a dynamic environment. The NLOS propagation caused by the moving objects in the dynamic environment can result in a large degree of weakening of the signal strength. As the RSS-F localization algorithm mainly relies on the offline collection of RSS, the existence of NLOS will result in the degradation of the localization accuracy. Aiming at this problem, voronoi segmentation based indoor fingerprint localization algorithm is proposed in this thesis.

After the deployment of the BP, the Euclidean distance between AP and BP is a fixed value. Assuming $\left\{ \left(x_j, y_j \right), RSS_{j,1}, RSS_{j,2}, \cdots RSS_{j,m}, n_{j,1}, n_{j,2}, \cdots, n_{j,m} \right\}$ is the best-matching RSS-F of the jth gait in the RSS-F database, and the received RSS of the jth gait is $\left(RSS_1, RSS_2, \cdots RSS_m \right)$. Since the distance between the two gaits is small, the VD propagation environment of the previous gait can be the criterion of the next gait. If there exist NLOS propagation between AP_j and BP_2, then the $RSS_2 \ll RSS_{j,2}$. The path attenuation index between the two gaits toward the same BP is compared to judge if the following gait propagation contains the NLOS error. If there exit NLOS error, the later RSS propagation obeys the propagation environment of the previous VD. This algorithm uses the propagation environment similarity of the adjacent VD to realize the recognition and compensation of the NLOS interference. Combining the optimization algorithms proposed in the previous section, the specific steps of the improved fingerprint algorithm is shown as follows:

- Get the RSS of the jth gait $\left(RSS_1, RSS_2, \cdots RSS_m \right)$, assuming that the RSS-F map $\left\{ \left(x_i, y_i \right), RSS_{i,1}, RSS_{i,2}, \cdots RSS_{i,m}, n_{i,1}, n_{i,2}, \cdots, n_{i,m} \right\}$ have been established.

- The recognition of the NLOS:

$$n_{j,m} = \frac{\Pr(d_0) - RSS_m}{10 * \log_{10}\left(\frac{d_{j-1,m}}{d_0}\right)}. \tag{2}$$

Where RSS_m denote the real-time RSS of mobile intelligent terminal received from $BP_m . d_{j-1,m}$ is the Euclidean distance between AP_{j-1} and BP_m in the $(j-1)^{th}$ gait. If $\left|n_{j,m} - n_{j-1,m}\right| > k$, there exists NLOS in the propagation path between AP_i and BP_m. Where k is the threshold which can be determined according to the environment.

The compensation of the NLOS:

$$RSS_m = \Pr(d_0) - 10n_{j-1,m} \log_{10}\left(\frac{d_{j,m}}{d_0}\right). \tag{3}$$

Where $n_{j-1,m}$ denote the $BP_m - direction$ path attenuation index of the corresponding VD of the $(j-1)^{th}$ gait in the RSS-F database.

- Initial localization point generation method:

Regional path attenuation index based RSS-F generation based on Shadow Model [7]:

$$RSS_{q,ms} = \Pr(d_0) - 10n_{q,m} \log_{10}\left(\frac{d_{q,m}}{d_0}\right). \tag{4}$$

Search for the most matching RSS-F among the generated RSS-F:

$$\inf(gap_{is}) = \min\left(\frac{1}{m} \sum_{n=1}^{m} \left|RSS_{q,ms} - RSS_m\right|\right). \tag{5}$$

Where the physical coordinates of the chosen auxiliary RSS-F is the localization result of the initial point.

- Localization method for test points:

Find the most matching VD AP_q through the matching Eq. 1:

$$\inf(gap_i) = \min\left(\frac{1}{m} \sum_{n=1}^{m} \left|RSS_{i,m} - RSS_m\right|\right). \tag{6}$$

Where RSS_m denote the real-time RSS after the NLOS compensation above and the physical coordinates of the chosen RSS-F is the localization result of the proposed VBFL.

3   Experimental Result and Analysis

3.1   Experimental Environment Evaluation

In order to verify the performance of VBFL proposed in this thesis, a wireless indoor positioning system is developed. The environment to be positioned is shown in Fig. 3:

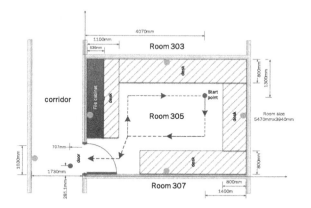

Fig. 3.   Indoor localization environment

As shown in the Fig. 3, the indoor localization environment of this experiment is a 5470 mm * 3940 mm laboratory. The localization system includes five beacon nodes, a wireless measurement node and a processor for centralized data processing. The CC2530 chip of TI with wireless communication is adopted as beacon node. The beacon node is fixed in the indoor positioning environment shown in Fig. 3, move the mobile intelligent terminal to the pre-set distances shown in the following table away from BP and receive the continuous RSS for 3 min time to test the transfer stability of the chip in the real positioning environment.

As shown in Table 1, the transmission of signals is very stable within 2.5 m. Due to the impact of multi-path propagation of the measured environment, the packet loss happens mainly at 3 m. Overall, in this environment, the confidence of RSS propagation distant is 2.5 m. So five BPs are deployed in Fig. 3, the deployment topology between BP and the mobile intelligent terminal is star topology, the BP is fixed 1.2 m away from the ground on the four wall of the localization space. The processor is set to be centralized process mode. The mobile intelligent terminal is connected with the processor via Ethernet. Both the measurement node and the beacon node work in the 2.4 GHz frequency band. The transmit power is configured by default to +1 dBm. The mobile intelligent terminal communicates with the BP in each gait, transfer the signal strength value RSS corresponding to the BP and the BP label to the processor via the Ethernet, then the estimated position is obtained.

Table 1. The variance of RSS in the location environment

Distance	0.5 m	1.0 m	1.5 m	2.0 m	2.5 m	3.0 m	3.5 m	4.0 m
Variance	0.376	0.050	0.142	0.282	0.862	0.456	0.469	0.448

The AP point is randomly distributed in every 0.5 m^2. In the offline acquisition stage, the wireless measurement nodes (the mobile intelligent terminal) are deployed at each AP to collect 200 groups of RSS data to establish the RSS-F map database through the proposed eigenvalue extraction method.

3.2   Improved Indoor RSS-F Localization Algorithm Performance Evaluation

The Voronoi Based Fingerprint Localization (VBFL) is proposed for self-adaptive NLOS-detection and compensation. The evaluation of the proposed VBFL is performed in the localization environment established as Fig. 3.

First of all, the RSS-F database is collected in the offline phase can be regionalization according to the method described in Sect. 2.1, the result of the regionalization is shown in Fig. 4.

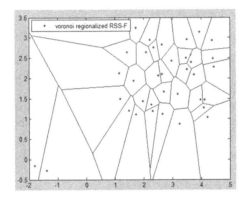

Fig. 4. Fingerprint Database Regionalization

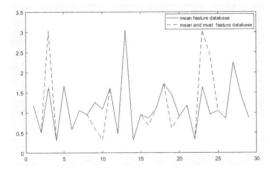

Fig. 5. Localization Error Based on Different RSS-F

Figure 4 shows that the localization area is divided into 37 VDs according to the preset APs. The tester carries wireless measurement node clockwise move from the upper right corner to the lower left corner then go through the door at a non-uniform velocity, meanwhile, 29 gaits are selected to test the localization performance. Firstly, the traditional fingerprint localization algorithm is used to search the difference of using different RSS-F database establish methods. The localization result are shown in the Fig. 5.

In the Fig. 5, the solid line represents the localization error using the database of mean feature, and the dashed line represents the result of using the database of mean and most value eigenvalue. Figure 5 shows that there exists small difference on the localization error based on different RSS-F database. But Fig. 6 shows the chosen VD is entirely different in the similar localization error. Take the 10th gait localization performance as an example, the true location is represented by the blue star, the red cycle represents the localization result using the database of mean and most value eigenvalue, while the red cross represents the localization result using the database of mean feature. Obviously, the localization performance using the database of mean and most value eigenvalue is better.

Fig. 6. Localization performance based on different RSS-F at the 10th gait

Furthermore, the Fig. 5 shows that the localization reliability is relatively high under the line-of-sight circumstance, the general localization error is about 1 m. However, the localization error rises to 3 m in the existence of NLOS signal propagation. Aiming at this problem, the Voronoi based fingerprint localization (VBFL) is proposed, and the localization error is shown in the Fig. 7.

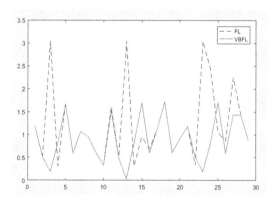

Fig. 7. Localization error contrast of VBFL and FL

It can be seen from Fig. 7 that the VBFL algorithm proposed in this thesis effectively reduces the influence of the localization error caused by non-line-of-sight signal propagation, and adaptively identifies and compensates the NLOS propagation. So that the maximum error reduced from 3.04 m to 1.71 m, and the mean error reduced from 1.21 m to 0.90 m, the comparison with other localization methods under the NLOS environment shows that the proposed VBFL can effectively optimize the localization of RSS based fingerprint localization algorithm in dynamic environment.

4   Conclusion

In this thesis, aiming to reduce the localization error caused by the NLOS interference exist in the complex environment. Firstly, the Voronoi diagram is used to regionalize the localization area adaptively. Then, the signal propagation environment similarity between the adjacent VDs is utilized to recognize NLOS and realize the compensation. The verification experiment shows the maximum localization error can be reduced from 3.04 m to 1.71 m, the mean localization error decreased to 0.90 m, and the localization time (contain 10 test points) consumption lowered to 2.113 s. Experiments show that the proposed algorithm can achieve a good localization performance in complex NLOS localization environment (Table 2).

Table 2. Localization error of different methods

Methods	Maximum error (m)	Average error (m)	Time consumption
RSS-F	3.08	1.21	Few minutes
CRLB [17]	3.43	1.44	Few minutes
PF	4.98	2.42	N/A
IMM [18]	4.50	2.42	N/A
VBFL	1.71	0.90	2.113 s

Acknowledgment. This work was financially supported by the Science and Technology Commission of Shanghai Municipality of China under Grant (No. 17511107002).

References

1. Xiao, J., Zhou, Z., Yi, Y.: A survey on wireless indoor localization from the device perspective. ACM Comput. Surv. **49**(2), 1–31 (2016)
2. Chen, L., Yang, K., Wang, X.: Robust cooperative Wi-Fi fingerprint-based indoor localization. IEEE Internet Things J. **3**(6), 1406–1417 (2016)
3. Kumar, S., Hegde, R.M., Trigoni, N.: Gaussian process regression for fingerprinting based localization. Ad Hoc Netw. **51**(1), 1–10 (2016)
4. Liu, W., Fu, X., Deng, Z.: Coordinate-based clustering method for indoor fingerprinting localization in dense cluttered environments. Sensors **16**(12), 1–26 (2016)
5. Yiu, S., Dashti, M., Claussen, H.: Wireless RSSI fingerprint localization. Sig. Process. **131**(1), 235–244 (2017)

6. Yiu, S., Yang, K.: Gaussian process assisted fingerprinting localization. IEEE Internet Things J. **3**(5), 683–690 (2016)
7. Huang, Y., Zheng, J., Xiao, Y., Peng, M.: Robust localization algorithm based on the RSSI ranging scope. Int. J. Distrib. Sens. Netw. **10**(1155), 1–8 (2015)
8. Chen, Q.Y., Wang, B.: FinCCM: fingerprint crowdsourcing, clustering and matching for indoor subarea localization. IEEE Wirel. Commun. Lett. **4**(6), 677–680 (2015)
9. Hernandez, N., Alonso, J.M., Ocana, M.: Hierarchical approach to enhancing topology-based WiFi indoor localization in large environment. J. Multiple-Valued Logic Soft Comput. **26**(3), 221–241 (2016)
10. Li, J., Tian, J., Fei, R.: Indoor localization based on subarea division with fuzzy C-means. Int. J. Distrib. Sens. Netw. **12**(8), 1–16 (2016)
11. He, C., Guo, S., Wu, Y.: A novel radio map construction method to reduce collection effort for indoor localization. Measurement **94**(1), 423–431 (2016)
12. Pak, J., Ahn, C.K., Shmaliy, Y.S., Lim, M.T.: Improving reliability of particle filter-based localization in wireless sensor networks via hybrid particle/FIR filtering. IEEE Trans. Industr. Inf. **11**(5), 1089–1098 (2015)
13. Chen, G.L., Meng, X.L., Wang, Y.J., Zhang, Y.Z.: Integrated WiFi/PDR/Smartphone using an unscented Kalman filter algorithm for 3D Indoor localization. Sensors **15**(9), 24595–24614 (2015)
14. Tian, Z., Jin, Y., Zhou, M.: Wi-Fi/MARG intergration for indoor pedestrian localization. Sensors **16**(12), 1–24 (2016)
15. Li, L., Yang, W., Bhuiyan, M., Zakirul, A.: Unsupervised learning of indoor localization based on received signal strength. Wirel. Commun. Mobile Comput. **16**(15), 2225–2237 (2016)
16. Zhou, M., Qiu, F., Xu, K.J., Tian, Z.S., Wu, H.B.: Error bound analysis of indoor Wi-Fi location fingerprint based positioning for intelligent Access Point optimization via Fisher information. Comput. Commun. **86**(1), 57–74 (2016)
17. Zhao, Y.B., Fan, X.P., Xu, C.Z., Li, X.F.: ER-CRLB: an extended recursive cramer-rao lower bound fundamental analysis method for indoor localization systems. IEEE Trans. Veh. Technol. **66**(2), 1605–1618 (2017)
18. Ru, J., Wu, C., Jia, Z.: An indoor mobile location estimator in mixed line of sight/non-line of sight environments using replacement modified hidden markov models and an interacting multiple model. Sensors **15**(6), 14298–14327 (2015)

Modeling and Simulation of Life Systems

Co-simulation Using ADAMS and MATLAB for Active Vibration Control of Flexible Beam with Piezoelectric Stack Actuator

Haotian Liu, Yubin Fang, Bing Bai, and Xiaojin Zhu[✉]

School of Mechatronic Engineering and Automation, Shanghai University,
Shanghai 200072, People's Republic of China
mgzhuxj@shu.edu.cn

Abstract. Co-simulation using ADAMS and MATLAB is implemented for active vibration control of flexible beam with piezoelectric stack actuator. The virtual prototype of flexible beam with piezoelectric actuator is created in ADAMS, and the implement of prototype provides an approach for acquiring the information of dynamic and kinematic properties. When the properties analysis is finished, the controller based on FXLMS algorithm is established in MATLAB. The controller calculates the signals of acceleration which are measured from virtual prototype, then the force is generated to suppress the vibration of flexible beam. The results and analysis prove that active vibration control for flexible beam has a great suppression performance.

Keywords: Co-simulation · Flexible beam · Piezoelectric stack actuator · Vibration control

1 Introduction

Flexible body is widely used because of its flexibility. Especially in aerospace field, flexible structure is also essential, such as solar panels, satellite antenna, and the space manipulator, all of them tend to be flexible [1]. However, flexible structure also has shortcomings, there will be elastic vibration caused by external disturbance. The vibration affects the accuracy and stability of space instrument seriously, so it is important to suppress the vibration of flexible structure. In general, many flexible structures as research objects can be replaced by flexible beams [1]. In theory, the analysis process of flexible structure is complex because of its infinite dimension, so the flexible structure is usually modeled and analysis with the discretization [2]. ADAMS is implemented for kinematics and dynamics analysis of mechanical system. The virtual prototype created by ADAMS is significant in engineering. And co-simulation using ADAMS and

This work is supported by National Natural Science Foundation (NNSF) of China under Grant 51575328. Mechatronics Engineering Innovation Group project from Shanghai Education Commission and Shanghai Key Laboratory of Power Station Automation Technology.

© Springer Nature Singapore Pte Ltd. 2017
M. Fei et al. (Eds.): LSMS/ICSEE 2017, Part I, CCIS 761, pp. 315–323, 2017.
DOI: 10.1007/978-981-10-6370-1_31

MATLAB provides the theoretical basis for practical system. In these years there are some researches on co-simulation using ADAMS and MATLAB for flexible structure. Malcolm D.J. did research on modeling of blades as equivalent beams for aeroelastic analysis [3]. Boscariol P. did the research on design and implementation of a simulator for 3D flexible-link serial robots [4]. Brannan J.C. did research on modeling flexible-body Dynamics in Real-Time Robotic Systems used in Satellite Servicing Simulations [5]. On one hand, the virtual prototype breaks through the limitations of the use of the test equipment and avoids the possibility that the experimental method may cause damage to the equipment, on the other hand, it avoids the complex mechanical derivation process, the object parameters are only needed to consider. In this paper, the model of flexible beam with piezoelectric stack is created in ADAMS, the vibration controller based on FXLMS algorithm is established in MATLAB. The result of co-simulation shows that the vibration of flexible can be suppressed by active vibration control system, which consists of piezoelectric and controller based on FXLMS algorithm.

2   Design for Flexible Beam with Piezoelectric Stack Actuator

The structure model of flexible beam with system is shown in Fig. 1, including the ground base, one beam, and one piezoelectric stack actuator. The beam is fixed on base station. The piezoelectric stack actuator exerts force to control the vibration of beam, and the piezoelectric stack actuator is regarded as stiffness-damper for mechanical analysis. The piezoelectric stack actuator consists of a piezoelectric stack and a rubber block. And the piezoelectric stack is fixed above the rubber block. In this model, the vibration of Y direction will only be studied [6, 7].

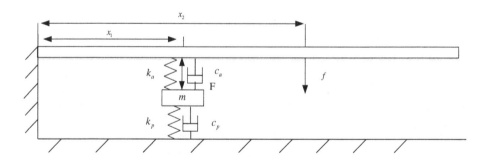

Fig. 1. The structure model of the beam with piezoelectric stack actuator

The force and motion equation of piezoelectric actuator is expressed:

$$m\ddot{x}_m - c_c\dot{x}_M + (c_c + c_b)\dot{x}_m - k_c x_M + (k_c + k_r)x_m = F \tag{1}$$

$$M\ddot{x}_M + c_a\dot{x}_M - c_a\dot{x}_M + k_a x_M - k_a x_m = -F \tag{2}$$

In these equation, F is control force, M is the mass which the force affect, m is the middle mass which connects the piezoelectric stack with rubber block, k_c is the piezoelectric stack stiffness coefficient, k_r is the rubber block stiffness coefficient, c_c is the active isolation's damping coefficient, c_r is the rubber block damping coefficient, x_m is the displacement of the middle mass, x_M is the displacement of the object.

In generalized coordinate system, the motion equation of the flexible beam which is based on the analytical mechanics theory is expressed as:

$$\frac{d}{dt}\left(\frac{\partial L}{\partial \dot{\phi}}\right) - \frac{\partial L}{\partial \phi} + \frac{\partial B}{\partial \dot{\phi}} + \left[\frac{\partial \psi}{\partial \phi}\right]^T \lambda = Q \tag{3}$$

$$\psi = 0 \tag{4}$$

$$L = T - V \tag{5}$$

L is Lagrange variable, B is energy loss function, ψ is constraint function, λ is Lagrange coefficient, Q is generalized coordinate is generalized force, T is kinetic energy, V is potential energy. According to the analysis of the system, there are some parameters should be designed:

X: The length of the beam
H: The width of the beam
H: The thickness of the beam
K_r: The stiffness coefficient of the rubber block
C_r: The damping coefficient of the rubber block
K_a: The stiffness coefficient of the piezoelectric stack
C_a: The damping coefficient of the piezoelectric stack
x_1: The location where the piezoelectric stack used to suppress vibration fixed on the beam
x_2: The location where the piezoelectric stack used to produce vibration fixed on the beam.

3 Build the Model of Flexible Model with Piezoelectric Stack Actuator

ADAMS provides a module for flexible body creation. There are two kinds of approach for modeling flexible body in ADAMS. One is discretization, and the other is extension. The geometry model is divided into multiple rigid elements, and the mechanical properties of each element will be defined. Both of them create the modal neutral file. Extension method is implemented to create the flexible beam in this model.

According to the analysis above, the geometry model consists of three parts is built. In this step, the flexible beam is created with extension method. At first, flexible beam endpoints and attachments are defined. The endpoint determines the stretch path for flexible beam. Then Cross section is drawn based on the center line. At this point, the geometry shape of the flexible beam is determined. Next, the properties of the element units can be setup. In this way, the element shape is hexahedra. This method can better represents the elasticity of the beam. Then the attachments are chosen to establish the constraints between beam and other parts. After that, the properties of these attachments should be determined. At last, aluminum is chosen as the material of beam. Meanwhile, the number of modes used to calculate should be determined [10].

When the modeling of flexible beam is finished, the other parts are created, and the connection between the flexible beam and other parts is established with attachments. The holder is fixed on ground with fixed joint, then the beam is fixed on holder, and the piezoelectric stack actuator is fixed on beam. Then attachments exist between beam and holder, and they also exist between beam and piezoelectric stack actuator. The piezoelectric is modeling as a translational spring-damper [8, 9] (Table 1).

Table 1. The parameters of the rigid system

Parameter	Size	Unit
L	1500	mm
W	60	mm
T	15	mm
K_a	20702	N/mm
C_a	28.681	N/mm
K_r	118.684	N/mm
C_r	0.91602	N/mm
x_1	450	mm
x_2	1050	mm

After all of operations have been finished, the computer started the calculation process of a flexible beam. It shows that the flexible beam with meshes in Fig. 2.

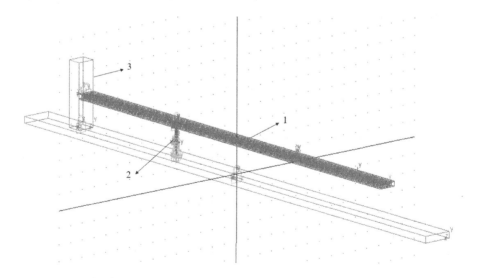

Fig. 2. The prototype of flexible beam with piezoelectric stack actuator

Explanation of annotations:

1: The flexible beam
2: The piezoelectric stack actuator
3: The holder.

4   Establish the Co-simulation Module

In mechanical system dynamics and kinematics analysis, ADAMS possesses powerful functions, but it is not good at control system design. Simulink possesses powerful function of control simulation, although mechanics module in MATLAB is able to establish mechanical systems, in terms of programming for dynamics and kinematics analysis, the work is complex. Co-simulation combines their advantages effectively, and it establishes a connection between ADAMS and MATLAB. The coupled data relation is defined as state variable before the softwares begin to work. ADAMS/Controls provides the interface for data interaction. State variables are divided into two categories: input signal and output signal. Input signals presents the state variables which come from MATLAB controller, they will affect the system in ADAMS, and output signals presents the state variables which calculated by ADAMS, they will work as the input of controller. In Table 2, four variables from system in ADAMS are designed for the controller. Source force is used to produce the vibration, and control force is used to suppress the vibration. The accelerations work as reference signal and error signal for controller. The reference signal is measured near the disturbance signal, and the error signal is measured near the end of flexible beam.

Table 2. The state variables of input and output

Input	Output
SF	A_1
CF	A_2

At last, run interaction file produced by ADAMS/controls, and use the command to open the interaction module in Simulink. In this way, the variables in ADAMS are verified same as the variables in MATLAB.

5   Controller Based on FXLMS Algorithm

The principle of active vibration control is following the original signal to generate the action signal for vibration suppression. FXLMS algorithm is shown in Fig. 3.

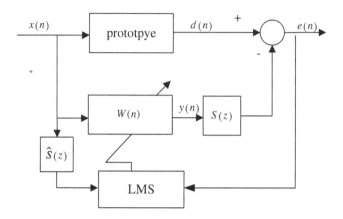

Fig. 3. The diagram of FXLMS algorithm

FXLMS algorithm relies on the reference signal to produce the desired signal which is used to control vibration [11]. FXLMS algorithm constantly adjusts the weight of filter until the error is eliminated to convergence. LMS algorithm is usually used as an update algorithm of weight, its objective function is the square of instantaneous error:

$$e^2(n) = (d(n) - y(n))^2 \tag{6}$$

In this equation, $y(n)$ is the vibration of object, $d(n)$ is the desirable signal.

The gradient descent method is proposed to minimize the objective, the update function of weight is that:

$$W(n + 1) = W(n) + \mu e(n)x(n) \tag{7}$$

In this equation, μ is the step size, $x(n)$ is the reference signal. $W(n)$ is the weight of filter.

In order that the convergence of the algorithm is not destroyed, it is usually necessary to estimate the secondary path of the system. $S(z)$ is estimation of secondary path.

6   Results and Analysis of Co-simulation

The connection module of co-simulation is imported into SIMULINK to establish the controller. In Fig. 4, the acceleration, which is error signal, is input to the controller. After the controller has processed date, the force signal is output from interface y. The voltage signal drives the piezoelectric stack to suppress vibration of flexible beam until the error signal tend to converge.

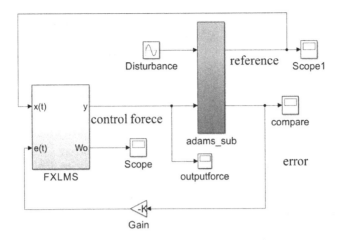

Fig. 4. The control system with FXLMS algorithm in MATLAB

In order to make the experimental results more intuitive, the error signal without control force will be compared with the error signal which is suppressed by FXLMS controller. At first, the signal which only contains the disturbance signal is input to make vibration on the flexible beam in this simulation. The frequency of the disturbance signal is 99 Hz, and the amplitude of the signal is 0.1 N. The amplitude of acceleration error signal produce by disturbance force is approximately 15 mm/s². In Fig. 5, the controller is set to start working after 1 s, and it tends to outputs a stable signal to drive the piezoelectric stack. The result of co-simulation is shown in Fig. 6, the error signal with FXLMS algorithm is compared with error signal without FXLMS algorithm. Obviously, when the controller starts working, the error signal converges quickly, and it attenuates to zero in 0.15 s.

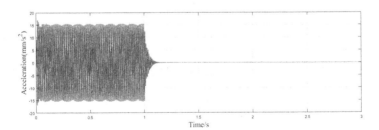

Fig. 5. The result of co-simulation for active vibration

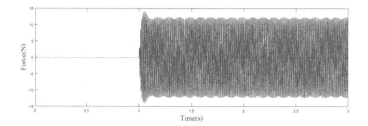

Fig. 6. The control force calculated by controller

Figure 6 is related to Fig. 5. In Fig. 6 it shows the variation of the control force. In practice, the force is produced by stack actuator, and the force is changed by FXLMS controller. The control force changes according to the error signal, and it tends to stabilization when the error signal convergence.

In this paper, active vibration control of flexible beam with piezoelectric stack actuator is investigated. The virtual prototype is the method which is useful in engineer analysis. In this way, the virtual prototype of flexible beam vibration system with piezoelectric stack actuator is designed in ADAMS, and the stack actuator can be analyzed as the stiffness-damping system which outputs the driving force, meanwhile, the vibration is suppressed by piezoelectric stack actuator effectively. It illustrates that co-simulation using ADAMS and MATLAB for active vibration control of flexible beam is feasible. In the other word, the virtual prototype of flexible beam system can be implemented in active vibration control. In addition, the parameter of prototype can be modified for different experiment.

7   Conclusion

Co-simulation using ADAMS and MATLAB is implemented for active vibration control of flexible beam with piezoelectric stack actuator. The simulation result shows that the active vibration algorithm has a good effect on the vibration of the flexible beam. The technology of virtual prototype provides an effective mean for the verification of different experimental methods, at the same time, it provides the theoretical basis for verification of engineering applications. Of course, the optimization design of virtual prototype is also a meaningful research content. In the future, the vibration control of flexible structure and mechanical multi-rigid body structure will be further studied.

References

1. Li, L.: Study on active vibration control of flexible beam using piezoelectric actuators. Zhejiang University (2008)
2. Chen, R.J.: A dynamic model of large flexible multi-body spacecraft structure and simulation platform for control. National University of Defense Technology (2002)
3. Malcolm, D., Laird, D.: Modeling of blades as equivalent beams for aeroelastic analysis. In: ASME Wind Energy Symposium, pp. 293–303 (2011)

4. Boscariol, P., Gasparetto, A., Giovagnoni, M., Moosavi, A.K., Vidoni, R.: Design and implementation of a simulator for 3D flexible-link serial robots. In: Biennial Conference on Engineering Systems Design and Analysis, ASME 2012, vol. 3, pp. 155–164 (2012)
5. Brannan, J.C., Carignan, C.: Modeling flexible-body dynamics in real-time robotic systems used in satellite servicing simulations. In: AIAA Modeling and Simulation Technologies, AIAA 2013-5157 (2013)
6. Pu, A.T., Harrison, A.S., Robertson, J.M., et al.: Vibration control of beams by beam-type dynamic vibration absorbers. J. Eng. Mech. **118**(2), 248–258 (1992)
7. Samani, F.S., Pellicano, F., Masoumi, A.: Performances of dynamic vibration absorbers for beams subjected to moving loads. Nonlinear Dyn. **73**(72), 1065–1079 (2013)
8. Kim, S.H., Choi, S.B., Hong, S.R., et al.: Vibration control of a flexible structure using a hybrid mount. Int. J. Mech. Sci. **46**(1), 143–157 (2004)
9. Li, P., Fu, J., Wang, Y., et al.: Dynamic model and parameters identification of piezoelectric stack actuators. In: The 26th Chinese Control and Decision Conference, pp. 1918–1923. IEEE (2014)
10. Xing, J.W.: ADAMS_Flex and AutoFlex Training Tutorial. Science Press, Beijing (2006)
11. Kuo, S.M., Morgan, D.R.: Active noise control: a tutorial review. Proc. IEEE **87**(6), 943–973 (1999)

Review of Research on Simulation Platform Based on the Crowd Evacuation

Pei-juan Xu[1(✉)] and Ke-cai Cao[1,2]

[1] Nanjing University of Posts and Telecommunications, Nanjing 210023, China
1282845115@qq.com
[2] Nanjing University of Aeronautics and Astronautics, Nanjing 210016, China

Abstract. As the security accidents in public places frequently emerge, the research based on crowd evacuation gets more and more people's attention. Now, the crowd evacuation research has shifted from the traditional live exercise to computer simulation. This paper chose five kinds of crowd evacuation simulation platform and summarized Cellular Automata, Agent-based model, network model they involved. Then, the thesis introduced the software, analyzed the performance of them and stated the respective advantages and disadvantage in order to help user choose proper platform to achieve fast and efficient results of crowd evacuation simulation.

Keywords: Crowd evacuation simulation · Cellular Automata · Agent-based model · Network model

1 Introduction

In recent years, with the growing pace of process of urbanization, large quantities of people inrush into the city, therefore, the risk of public security may exist in airport, stadiums and other crowded gathering places. Domestic and foreign media has reported that many public places have occurred congestion and people could not escape, which led to a serious loss of life and property. Thus, the research of how to make effective management of large public facilities and complex and huge crowd in emergencies such as fires, earthquakes and other hazardous conditions is a vital and valuable research topic.

It usually takes three steps to make scientific and effective guidance for people in an emergency situation. Firstly, design relevant emergency plans according to the environment in public and security crises which could happen. Secondly, verify the rationality and effectiveness of the emergency plan through the scientific method. Thirdly, rehearse the plan together with fire departments, public security departments, medical care department, etc. However, the complexity of the security problems that may appear far beyond a calculation analysis ability of one expert or even a group of experts as continuous appearance of large-scale and even super-large scale public facilities. Moreover, due to staff size, space limitation, more safety factors, the traditional way of practice cannot be achieved and the effectiveness of emergency plans is difficult to verify [1].

© Springer Nature Singapore Pte Ltd. 2017
M. Fei et al. (Eds.): LSMS/ICSEE 2017, Part I, CCIS 761, pp. 324–333, 2017.
DOI: 10.1007/978-981-10-6370-1_32

In a gesture to solve difficulties above, the domestic and foreign researchers have tried to apply computer simulation to the field of public safety. Over the years, a large number of simulation platforms based on the crowd evacuation have been available, commonly used STEPS, Building EXODUS, Legion based on Cellular Automata [2, 3], Pathfinder based on Agent-based model [4], EVACNET4 based on network model [5], etc.

This article first summarizes the models involved in crowd evacuation simulation platforms mentioned above, then, introduces these five software, analyzes the performance of each simulation platform and describes their advantages and disadvantages, which can provide the basis for the selection of the simulation platform.

2   Simulation Model Based on Crowd Evacuation

Study on models is the basis of the research of crowd evacuation. The models in the achievements that researchers both at home and abroad have made in the field of crowd evacuation can be divided into macro model and micro model from different angles. Macro models include network model and fluid dynamics model, etc. Micro models include Cellular Automata model, Agent model, Social force model, etc. [3]. Here mainly introduces Cellular Automata, Agent-based model and network model.

2.1   Cellular Automata

In the late 1940s, John von Neumann put forward Cellular Automata [6]. In this model, cell, neighborhood, state, rule and other factors can realize the discretization of time and space dimension and control specific state of each cell at any time [7]. In the process of simulation based on crowd evacuation, Cellular Automata discretizes the scene into grids. Each person occupies a grid and interacts with others and walls by adjacent cells. Each cell has only two elements $\{0, 1\}$ and they represent two states of cell and each state has its own corresponding integral value. Neighborhood of Cellular Automata usually can be divided into three forms including Von Neumann, Moore and expand Moore type, as shown in Fig. 1.

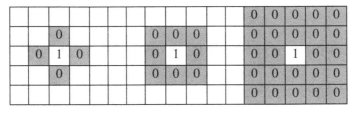

(a) Von Neumann   (b) Moore   (c) expand Moore type

Fig. 1. Neighborhood of Cellular Automata

According to the current state, Cellular Automata decides next behavior and synchronously updates all the states through neighborhood. The updated rule is:

$$f:s_i^{t+1} = f(s_i^t, s_{i+1}^t, s_{i+2}^t \cdots s_{i+n}^t) \tag{1}$$

Where s_i^t is the state of Cellular Automata at time t, f is mapping rules.

2.2 Agent-Based Model

In 1986, Minsky first proposed the concept of Agent. There are many definitions of Agents. The well recognized definition is proposed by wooldridge [8]. Agent-based model has autonomy, reactivity, initiative and sociality.

Agent-based model is a model based on Agent. It considers each Agent in the system as an individual and uses different attributes and interactions with each other to describe the behavior of the system. The action decision set of an agent system is a rule library into which the information is entered according to the environment. Then, the rule library makes decision and the output acts on the environment later. Individual behaviors among Agent-based models are decided by interactions between agents, which is not well consistent with real the situation in the large-scale crowd evacuation process [4].

2.3 Network Model

A network model consists of a series of nodes and paths. The nodes represent different rooms, passages, stairwells, safety exits, and other spaces, outdoor or other safe locations in the building are defined as target nodes. Adjacent nodes are connected through a virtual path. In addition, direction of the path needs to be set up to reflect the direction of personnel flow within the network [9].

Two parameters need to be defined respectively for each node: the initial number of people at the beginning of the simulation which usually can be set to 0 and the maximum number of people that the node can hold, the default value is set to infinity.

Each path also needs to define two parameters:

(1) Time the person takes to pass through the path:

$$T = L/V \tag{2}$$

Where, L is the length of the path, V is the speed of walking through the path for people.

(2) Number of people passing through the path per unit time:

$$N = W \times q \tag{3}$$

Where, W is the effective width of the path, q is the human traffic on the unit width of the path.

3   Simulation Platform Based on Crowd Evacuation

At present, more than 50 kinds of evacuation simulation platform have been put forward both at home and abroad and commonly used STEPS, Building EXODUS, Legion, Pathfinder, EVACNET4, SIMULEX, SimWalk, etc. Here, the first five platforms are chosen to introduce. Their performance is analyzed and advantages and disadvantages are summarized respectively.

3.1   STEPS

STEPS is a software based on entity and Cellular Automatic model and designed by the British company Mott MacDonald [2]. Currently, this platform is diffusely used in crowd evacuation happen in airports, subways, office buildings and large shopping centers, etc.

STEPS offers abundant modeling tools such as Shapes, Planes, Exits, Stairs, Lifts, etc. The tool called Vehicles is also provided since STEPS need to simulate the evacuation among crowds in subway station. The platform describes the geometric characteristics of the person with height, shoulder width and chest thickness and eight different types of the crowd, for example, people who carry handbags. It also provides two methods to define the person's walking speed, one is setting by users and the other is random normal distribution.

STEPS supplies output of data files in the form of CSV, covering human traffic, crowd density, space utilization, etc. Besides these data, interactive two-dimensional and three-dimensional visual graphics, images in the form of JPG, TIFF, PNG and BMP and animations in the form of AVI are also supported. Through connecting with ANSYS CFX and importing simulation data of smoke and gas, STEPS can be exploited to study the effect of crowd's speed caused by toxic substance in evacuation mode.

3.2   Building EXODUS

Building EXODUS is one of the most successful evacuation platforms developed by the University of Greenwich and the Fire Safety Engineering Group [3]. The simulation of crowd evacuation takes Cellular Automatic as a model and the scene is viewed as network nodes on which pedestrians can only move. Now, in the evacuation simulation in supermarkets, hospitals, stations, schools, airports and other buildings, Building EXODUS is proverbially utilized.

Floor plan in Building EXODUS can be generated by DXF file while the 3D model is completed by Revit Architecture and then stored in the graphics library. The biggest difference between this platform and others is that it takes into account the interplay among evacuees, evacuees and fires, evacuees and buildings. The software relatively true simulates several attributes of scene and behaviors of people and gives a more comprehensive and detailed prediction results owing to combination of social factors which covers 22 individual characteristics and social features such as age, sex, walking speed, death, etc.

Besides the total time of evacuation, a variety of other results containing the prediction of bottleneck position, speed, starting time, terminal time and evacuation process are also produced by Building EXODUS. Connection between Building EXODUS and CFAST allows historical data of CFAST to transfer to the former automatically. Searching for huge data output and extracting specific data selectively and efficiently with software mentioned above can help create a virtual reality graphical environment, and provide 3D animation presentations on evacuation.

3.3 Legion

Legion is a kind of simulation software developed by Legion in Britain [1] and be based on the Cellular Automata model. It is applied to make and evaluate plans on emergent evacuation happen in places such as airport, railway stations, subway stations where people is highly gathered.

Legion provides a simplified and automated modeling process to help simulate complex operational processes more easily and faster, at the same time, simulate and analysis pedestrian behavior. With interaction between pedestrians and mutual effect between pedestrians and obstacles in the surrounding environment taken into consideration, Legion is able to emulate pedestrian walking, where each person is modeled as a two-dimensional entity who moves to destination by looking for the minimized target cost function. The cost is a weighted average of three parts, namely, inconvenience, frustration and discomfort [2]. Entities can not only learn and adjust the weight of these three parts to adapt to the surrounding environment, but also communicate with adjacent entities. The parameters of the entity need to be set and calibrated according to local conditions, including the physical radius, the free speed depending on preference, space factor, etc.

Legion supports the output of graphics including distribution of flow density, space utilization, duration distribution of maximum density in active area and data in terms of walking time, walking speed, evacuation time, etc. Provided that pedestrian location data is saved as a XML file, a hybrid simulation of people and cars can be achieved together with the traffic library.

3.4 Pathfinder

Pathfinder based on Agent-based model is invented by the United States Thunderhead Engineering. It is proper for evacuation in small-scale buildings.

Movement mode of the crowd is split into SFPE and Steering. Under the SFPE, the pedestrians automatically move to the nearest export, regardless of the interaction between themselves. Under the Steering, the movement of the crowd is controlled through combination of path planning, guidance mechanism and collision processing. When the distance between the staff or the path between the nearest points exceeds a certain threshold, a new path will regenerate in a gesture to adapt to the new form [4]. Pathfinder provides Polygonal Room Rectangular, stairs and other modeling tools, and describes the height and shoulder width of a person. The user can set some parameters including updated step length, degree of attraction to pedestrians comes from exits, the minimum width of exit pedestrians can pass, acceleration time, export flow, etc.

Pathfinder not only outputs the distribution of flow density, the direction of personnel flow and so on, but also furnishes users with two-dimensional and three-dimensional visual interface and chart.

3.5 EVACNET4

EVACNET4 is a platform that is applied to evacuation simulation in multistoried buildings and indoor environment and it is developed by the University of Florida. The software describes the scene as a network and simulates the flow of people within the network. When all the pedestrians finally reach the safe location, the emulation process stops.

In EVACNET4, movement of crowd in the network is viewed as flow of water in pipe without regard to individual characteristics of pedestrians. Assuming that evacuation is an ideal process, that is to say, evacuated personnel have the same characteristics and enough physical condition which helps them to reach a safe place; staffs are sober, always keep orderly and not halfway return; personnel are always evacuated from the available exits at the same speed; the human traffic is always proportional to the width of the evacuation channel. To make up for some of the uncertainties in the evacuation process, the actual evacuation time is usually multiplied by a safety factor which takes 1.5 to 2.0. Initial number in nodes and the maximum number of people that nodes can accommodate are set by users in simulation evacuation.

The evacuation results include the total evacuation time respectively in the entire scenario, in each floor and within each node; the personnel information at each exit; and the bottleneck in the evacuation process. EVACNET4 provides users with data output, two-dimensional visual interface which reflects the geometry of the building and the location of the exit.

3.6 Descriptions of Performance, Advantages and Disadvantages

These 5 platforms introduced above all have their own features and descriptions about their performance, advantages and disadvantages are described in Tables 1 and 2. Here, advantages and disadvantages are briefly analysis mainly from three aspects including the setting of behavior and scene, capability of simulating large crowds and price. Then, an assessment is made.

Table 1. Technical performances

Name of platforms	Methods of modeling	Input	Output
STEPS	Cellular Automata	Arrangement of architectural space, three-dimensional size of people, walking speed, patience, etc.	Data files(pedestrian volume, crowd density, space utilization, etc.); supporting the output of two-dimensional and three-dimensional visual graphics, animation
Building EXODUS	Cellular Automata	Arrangement of architectural space, personnel characteristics, (number of people, gender, patience, etc.), social factors	Data files (total evacuation time, bottleneck position, evacuation speed, evacuation start time and termination time, evacuation process, etc.); supporting three-dimensional animation presentations
Legion	Cellular Automata	Arrangement of the architectural space, physical radius, the free speed of the preference, space factor, etc.	Data files(walking time, walking speed, evacuation time, length of the queue, etc.); supporting the output of graphics, data, and chart
Pathfinder	Agent-based model	Arrangement of architectural space, two-dimensional size of people, maximum walking speed, acceleration time, export flow, etc.	Data files(distribution of flow density, direction of personnel flow in active area, etc.); supporting two-dimensional and three-dimensional visual interface and chart
EVACNET4	Network model	Network arrangement of architectural space, initial number in nodes, the maximum number of people nodes can accommodate	Data files(evacuation time, personnel information at each exit, the bottleneck in the evacuation process); supporting two-dimensional visual interface

STEPS can be used for evacuation (emergency) and normal (non-emergency) conditions. It can not only simulate the behavior of pedestrians in emergency situations, but also in non-emergency situation find ways to improve the environment for more comfortable and faster movement of the crowd. Users can give personnel specific actions including staying for a specific time or queuing in a place and specific evacuation route by using STEPS. The path decision system allows the staff to choose the exit, which is closer to the reality; large-scale crowd evacuation can show some characteristics of the crowd, which is more corresponding with the real scene. The software is able to simulate a model with more than 20000 pedestrians [10]. Users can get a one year license through spending a few thousand pounds, which is perhaps possible within a limited budget. STEPS is a platform based on Cellular Automata and the physical space is discrete, therefore, the unconscious individual in the model moves in the direction that the integral value is high to low which makes the movement rule be simple and the simulation be less closer to the reality.

Table 2. Advantages and disadvantages

Name of platforms	Advantages	Disadvantages
STEPS	Setting of behavior and scene: specify part of the action; Export choice is closer to the reality; large-scale crowd evacuation is corresponding with the real scene Price: relatively low Capability of simulating large crowds: tens of thousands of pedestrians	Setting of behavior and scene: the individual movement rule is simple which makes simulation be less closer to the reality
Building EXODUS	Setting of behavior and scene: more truly simulate behaviors of evacuation personnel; the setting of the evacuation scene is the most realistic Price: relatively low	Setting of behavior and scene: it is difficult to simulate scene with high density crowd Capability of simulating large crowds: thousands of pedestrians
Legion	Setting of behavior and scene: simulate scene with high density crowd; evaluate passenger flow service level in a place	Setting of behavior and scene: the mechanism of human action is not perfect Price: very high Capability of simulating large crowds: thousands of pedestrians
Pathfinder	Setting of behavior and scene: the individuals have rich attributes; the complete collision handling mechanism is close to the reality Price: relatively low Capability of simulating large crowds: tens of thousands of pedestrians	Setting of behavior and scene: the setting of geometric model and moving position of staff are not very reasonable; ignores the group psychology of individuals; large-scale evacuation results are relatively poor
EVACNET4	Setting of behavior and scene: belong to an optimization model and runs faster in the calculation; it is proper for the occasions with higher crowd density Price: relatively low Capability of simulating large crowds: tens of thousands of pedestrians	Setting of behavior and scene: the hypothesis of the model is too ideal

Building EXODUS is suitable for simulating the escape and action of individuals in complex buildings. This software more truly simulates a number of attributes of scenes and behaviors of evacuation personnel and tracks many details in evacuation process, thus gives more comprehensive simulation results and makes the evacuation scene be closest to reality [11]. In 2011, the procurement item of software licensing organized by China Academy of Safety Sciences and Technology indicated that the transaction price was almost 5,000 pounds [12]. The scale of visible crowd is limited because the largest number of pedestrians in current case is less than 10000. It is also difficult to simulate scene with high density crowd.

Legion can also be used for evacuation and normal situations. It can simulate scene with high density crowd. Through simulating the passenger flow in a place, crowded point in this area will be determined and the passenger flow service level can be evaluated. Legion is the most expensive of these five platforms (more than 20000 pounds every year). A model with thousands of pedestrians can be simulated, that is to say, the capability of simulating large crowds is limited. Legion assumes that pedestrians only move at 45°, which results in limitation of walking direction [13]. If the pedestrians want to move at other angles, they must move many times. Compared with the actual moving distance, moving in this way makes distance increase and affects the evaluation result.

Pathfinder can only simulate emergency evacuation. The individuals in the model have rich attributes and their abilities are close to the reality. A complete collision handling between pedestrians, pedestrians and walls lets the situation be close to the reality. The price of Pathfinder is 19700 yuan for a year, nearly 2000 pounds. Pathfinder can at least simulate evacuation scenarios with 50,000 people, but the result is relatively poor. Pathfinder only supports some dynamic geometric models such as elevators, escalators and doors while not including trains and other moving objects. Personnel can only move from the current location to the export and without group psychology of individuals into consideration.

EVACNET4 is suitable for evacuation in public buildings. It belongs to an optimization model and runs faster in the calculation. The node capacity is related to itself and the number of square feet associated with the maximum assumed density of the person during the evacuation process. The software is proper for the occasions that have higher crowd density [14]. The copy of this platform costs less than 200 pounds. The price may be low but the capability of simulating large crowds is strong which reaches about 50000 [15]. Because of the ideal hypothesis of the model, a large deficiency exists in the process of evacuation. It is difficult to describe the detailed characteristics of the local evacuation and the influence among the pedestrians.

From the analysis of the advantages and disadvantages, we know legion is very expensive and the choice of it will depend on the budget assigned to the project while choosing other four platforms is perhaps possible within a limited budget. STEPS, pathfinder and EVACNET4 can be used to simulate large-scale occasion since the size of the crowd reaches tens of thousands. Besides these two factors taken into account, users finally select the software should also depend on the model and the behavior of the crowd since it will let the simulation results be closer to reality.

4   Conclusion

The simulation based on crowd evacuation has important research significance in those fields such as evaluation and validity verification of emergency evacuation plan, interior design of building and design of transportation hub. The combined effects of pedestrians and environment need to be taken into account due to the complexity of the research on crowd evacuation. The existing simulation platforms not only help achieve rapid and effective modeling of crowd evacuation, but also realize scene simulation and motion analysis. Users should consider the price, crowd size, behaviors of personnel and other factors to choose the platform. In the future, the work of designing simulation software includes detailed assessment of the existing simulation methods, comprehensive consideration of the influencing factors in the evacuation process and improvement of the existing problems in setting of behavior and scene.

References

1. Wang, Z., Mao, T., Jiang, H., Xia, S.: Guarder: virtual drilling system for crowd evacuation under emergency scheme. J. Comput. Res. Dev. **47**(6), 969–978 (2010)
2. Hu, M.W., Shi, Q.X.: Comparative study of pedestrian simulation model and related software. J. Transp. Inf. Saf. **27**(4), 122–127 (2009)
3. Zhu, J.L.: Research on Several Key Technologies of the Crowd. Jilin University (2016)
4. Du, C.B., Zhu, G.Q., Li, J.Y.: Comparative study on evacuation simulation software STEPS and pathfinder. J. Fire Sci. Technol. **34**(4), 456–460 (2015)
5. Lv, A.B., Wang, C., Ding, S.J., Liu, J.P.: Study on personnel evacuation simulation of subway station based on EVACNET4. J. Value Eng. **33**(14), 230–232 (2015)
6. Codd, E.F.: Cellular Automata. Academic Press, New York (1968)
7. Liu, Z.Y., Huang, P., Xu, Y.B., Guo, C., Li, J.: The quantitative risk assessment of domino effect caused by heat radiation. In: 5th International Conference on Systems, pp. 120–124. IEEE Press, Jiang Su (2010)
8. Wooldridge, M., Jennings, N.R.: Intelligent agents: theory and practice. J. Knowl. Eng. Rev. **10**(2), 115–152 (1995)
9. Li, Q., Wang, Z.G., Tian, Y.: Preliminary discussion on establishing the model of building network by EVACNET4. J. Fire Sci. Technol. **23**(5), 442–445 (2004)
10. STEPS Website. https://www.mottmac.com/skillsandservices/software/stepssoftware. Accessed 11 May 2007
11. Wang, G.F., Zhang, X.L., Yan, W.D.: Building fire behavior modeling of EXODUS. J. Saf. Sci. Technol. **7**(8), 67–72 (2011)
12. China mining for networking. http://www.bidcenter.com.cn/newscontent-7457467-4.html. Accessed 4 Mar 2011
13. Gan, Y.: Pedestrian simulation in large-scale activities based on legion——a case study of pedestrian simulation of the opening ceremony at the 16th Asian games. J. Transp. Inf. Saf. **30**(1), 76–81 (2012)
14. Tao, Y., Zhang, Y.F., Hou, Z.Z., Xiao, X.F.: Comparative study of evacuation time prediction methods. J. Fire Sci. Technol. **28**(3), 181–186 (2009)
15. Li, J.M., Wang, N.: Performance-based research of evacuation of Guiyang Olympic Center. J. Fire Sci. Technol. **28**(9), 642–645 (2009)

A TopicRank Based Document Priors Model for Expert Finding

Jian Liu[1,2,3], Bei Jia[1,3], Hao Xu[1,3], Baohong Liu[2,3], Donghuai Gao[1,3], and Baojuan Li[4,3(✉)]

[1] Network Center, Fourth Military Medical University, Xi'an 710032, Shaanxi, China
{liu-jian,xuhao,donghuai}@fmmu.edu.cn, jiabeihn@163.com
[2] College of Information System and Management, National University of Defense Technology, Changsha 410073, Hunan, China
lbh_nudt@126.com
[3] Xi'an Communication Institute, Xi'an 710106, Shaanxi, China
[4] School of Biomedical Engineering, Fourth Military Medical University, Xi'an 710032, Shaanxi, China
libjuan@163.com

Abstract. Document priors that encode our prior knowledge about the importance of different documents are essential to an expert finding system. This study proposed a TopicRank-based document priors model for expert finding. TopicRank algorithm is an extension of the DocRank algorithm. Latent dirichlet allocation was used to extract topics of the documents. We assumed there was a link between two documents that share common topics. Link analysis techniques were then used to obtain document priors. The proposed model was evaluated using the CSIRO Enterprise Research Collection and the results showed that the performance of the expert finding system was dramatically improved by introducing TopicRank-based document priors. In particular, Mean Average Precision increased 19.9% while Mean Reciprocal Rank rose as much as 23.4%.

Keywords: Expert finding · Document priors · TopicRank

1 Introduction

Experts are valuable for an organization for that they have specific knowledge, expertise and experiences that can benefit the organization. Interaction with experts directly may be the most effective way in many systems. However, to ensure effective access and utilization of the experts' expertise is difficult [1]. Recently expert finding systems have tried to solve this problem using information retrieval techniques. Expert finding is also referred to as expert recommendation [2], expert search [3, 4], expert identification [5] in the literatures. Given a query, the goal of the expert finding system is to find experts who are knowledgeable about the query and subsequently rank these experts according to their level of expertise.

One of the core elements in classical expert finding models is the document priors model. Document priors that encode our prior knowledge about the importance of

© Springer Nature Singapore Pte Ltd. 2017
M. Fei et al. (Eds.): LSMS/ICSEE 2017, Part I, CCIS 761, pp. 334–341, 2017.
DOI: 10.1007/978-981-10-6370-1_33

different documents are essential to an expert finding system. Different types of documents may be of different importance. For example, journal articles may provide more reliable information about who is an expert than code files. Even if all the documents are of the same type, we may still need to treat the documents differently. A web page with shorter url may be more important than another one with longer url. However, for simplicity, document priors are usually assumed to be uniform in most studies, which generally results in a performance degradation of the expert finding system. Some research recently have attempted to improve the results of expert search by extending the classical models. In literature [6], Balog et al. proposed a method which used the length of the urls to obtain document priors. Another model which used the retrieval performance of different document types to get the document priors was proposed in [7]. For one model, the authors achieved a raise of 7% in Mean Average Precision (MAP) and 10% in Mean Reciprocal Rank (MRR). In addition, Deng et al. proposed a weighted language model which used the citation number to evaluate the importance of the document. This model was evaluated on the Digital Bibliography & Library Project (DBLP) dataset and the results showed that the MAP was increased by 6.6% [5]. Although it has been demonstrated that incorporate a document's importance to classical expert finding models can improve the results, uniformed document priors are still used in most cases when specific information such as the retrieval performance of each document type, the citation numbers are not available in the data corpus. In the current study, we proposed a TopicRank-based document priors (TBDP) model for expert finding. The TBDP model was evaluated using the CSIRO Enterprise Research Collection (CERC). The results showed that the performance of the expert finding system was largely improved by using TopicRank-based document priors.

2 Methods

2.1 The Baseline Model

The baseline model in this paper is the same with model 2 in [7]. Given a query q, the probability of the co-occurrence of candidate ca and query q can be written as:

$$p(ca, q) = p(q \mid ca)p(ca) \tag{1}$$

where $p(ca)$ is the prior probability that candidate ca is an expert and $p(q|ca)$ represents the probability of query q given candidate ca. Extending Eq. 1 by using documents in the collection D to connect candidate ca and query q, we can obtain:

$$p(q|ca) = \sum_{d \in D_{ca}} p(q \mid d, ca)p(d \mid ca) \tag{2}$$

where D_{ca} is the collection of the documents associated with candidate ca. $p(d|ca)$ in Eq. 2 can be calculated in the following way:

$$p(d \mid ca) = \frac{p(ca \mid d)p(d)}{p(ca)} \tag{3}$$

Finally, absorbing Eqs. 2 and 3 into Eq. 1, the occurrence of candidate ca and query q can be expressed as:

$$p(ca, q) = \sum_{d \in D_{ca}} p(q \mid d, ca)p(ca \mid d)p(d) \tag{4}$$

In traditional studies, $p(d)$ is usually assumed to be uniform [8–11].

2.2 TopicRank-Based Document Priors Model

In this section, we proposed a new document priors model for expert finding based on probabilistic topic models.

Probabilistic Topic Models

Probabilistic topic models were first proposed to analyze the content of documents and the meaning of words [12–15]. There have been a variety of topic models developed in the field of machine learning and natural language processing such as probabilistic latent semantic analysis (PLSA) [13], Latent dirichlet allocation (LDA) [14], etc. Different topic models make slightly different statistical assumption but share the same basic idea [15]. Topic models treat the documents as mixtures of topics, while each topic has a probability distribution over words. In this paper, we used the LDA model to get the topic distribution of each document in the corpus. As one of the most commonly used topic models, LDA has many advantages compared to classical topic models. The number of model parameters in LDA does not grow with the size of the corpus. A graphical model representation of the LDA model used for expert finding is shown in Fig. 1. In principle, a topic model is a generative model. In order to generate a document, one first chooses a topic distribution θ_d for this document from a dirichlet distribution parameterized by α. For each term t in the document, a topic z is then chosen from θ_d. Finally, a term is generated using the topic-term distribution φ parameterized by γ. Gibbs sampling was used to invert the LDA model and estimate the topic distribution of each document.

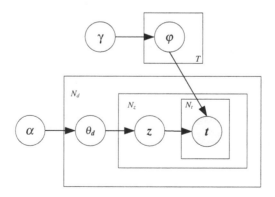

Fig. 1. Graphical model representation of the LDA model

TopicRank-based Document Priors

The TopicRank-based model we proposed here is an extension of the PageRank algorithm and the DocRank algorithm [16]. Since been introduced in 1998, the PageRank algorithm has become one of the most successful link analysis models [16]. Later, the DocRank algorithm which can deal with documents without links was proposed by Marmanis and Babenko [16]. The DocRank algorithm assumes that there is a link between two documents that share some high frequency words. Document priors were then calculated similarly as the PageRank algorithm. However, due to the inability to deal with the synonyms problem, this algorithm is unable to furnish reasonable document priors in some cases.

Generally speaking, web pages connected to each other with hyperlinks may share common topics. Documents that share common words may also discuss about common topics. A topic derived by analyzing the content of the whole document usually represents more precise information about the document rather than a word. Taken these factors into consideration and inspired by the DocRank algorithm, we proposed the TopicRank-based document priors model and assume there is a link between two documents that share common topics. An adjacent matrix H is constructed as follows:

$$H = \left\{ \begin{array}{cccc} w_{11} & w_{12} & \cdots & w_{1n} \\ w_{21} & w_{22} & \cdots & w_{2n} \\ & & \vdots & \\ w_{n1} & w_{n2} & \cdots & w_{nn} \end{array} \right\} \tag{5}$$

where w_{ij} represents the importance of document d_j over document d_i and is calculated over the common topics $z \in Z_c$ shared by document i and j:

$$w_{ij} = \sum_{z \in Z_c} [\tanh(\frac{p(z \mid d_j)}{p(z \mid d_i)})] \tag{6}$$

where $p(z \mid d_j)$ is the probability of topic z given document d_j. A transition matrix P can then be constructed in the following way:

if $w_{ij} \neq 0$, then

$$P(i,j) = \frac{(1 - \alpha)H(i,j)}{\sum\limits_{j=1}^{N} H(i,j)} + \alpha/N \tag{7}$$

if $w_{ij} = 0$ but $w_{ik} \neq 0 (k \neq j)$, then

$$P(i,j) = \alpha/N \tag{8}$$

if $w_{ik} = 0 \ (k = 1, \ldots, N)$, then

$$P(i,j) = 1/N \tag{9}$$

Finally, document priors can be obtained by solving the eigenvector of matrix H:

$$p(D) = \begin{bmatrix} p(d_1) & p(d_2) & \cdots & p(d_{N_d}) \end{bmatrix}^T = P \times p(D) \tag{10}$$

where N_d is the number of the documents associated with query q.

3   Experimental Setup

The CERC corpus (http://es.csiro.au/cerc/) provided by the 2007 edition of the TREC Enterprise track was used to evaluate the TBDP model. This corpus contains 370,715 documents crawled from the CSIRO web site, adding up to 4.2 GB. In addition, this data set is made up of different types of documents including web pages, sources codes, pdf documents, etc. Three kinds of evaluation metrics including MAP, MRR, and precision at rank n (P@5, P@10) were used to evaluate the performance of the expert finding system.

4   Results and Discussion

The results showed that the performance of the expert finding system was dramatically improved by introducing TopicRank-based document priors (Table 1). The TBDP model outperformed the baseline model on all the metrics. In particular, MAP increased 19.9% while MRR rose as much as 23.4%.

Table 1. Performances of the baseline model and the TBDP model

Model	MAP	MRR	P@5	p@10
Baseline	0.3347	0.4477	0.2040	0.1360
TBDP	**0.4014**	**0.5524**	**0.2320**	**0.1460**
Increased (%)	19.9	23.4	13.7	7.4

As for the TBDP model, we have two new parameters: N_z which denotes the number of the topics extracted from the documents and N_c which represents how many common topics were used to calculate the TopicRank value. We then investigated the effects of these parameters on the performance of the TBDP model. We show changes of all the metrics while N_z varying from 15 to 50 in Fig. 2. The range of the parameter was chosen according to the number of the documents associated with each query. According to Fig. 2, the ideal N_z for MAP and MRR is 20. Specifically, MAP and MRR increase dramatically while N_z increases at the early stage and reached their peak value. These metrics then decrease while more topics are extracted from the documents. In contrast, changes in P@5 and P@10 show very different pattern.

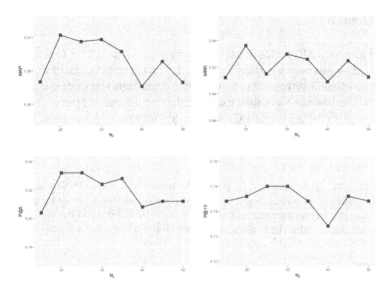

Fig. 2. Changes of MAP, MRR, P@5 and P@10 while N_z varying from 20 to 50

Figure 3 illustrates the effects of N_c on all the metrics while N_z is set to its ideal value 20. It is quite interesting to note that MAP, MRR, and P@5 normally decrease while N_c increases. As shown in Fig. 3, the ideal value for N_c lies between 2 and 4. This may be because most of the documents generally focus on limited topics.

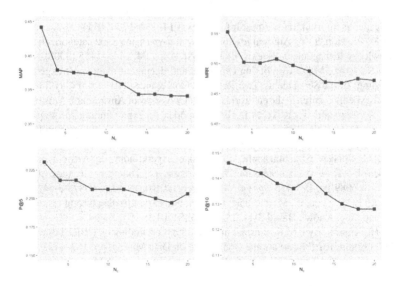

Fig. 3. Changes of MAP, MRR, P@5 and P@10 while N_c varying from 2 to 20

5   Conclusion

This paper proposed a TopicRank-based document priors model for expert finding. In this model, LDA was used to extract topics in the documents associated with the given query. We assumed there was a link between two documents that share common topics. Link analysis techniques were then used to calculate document priors. The proposed model was evaluated using the CERC collection and the results showed that the performance of the expert finding system was largely improved by using TopicRank-based document priors.

Acknowledgement. This work is supported by the Industrial Research Project of Science and Technology in Shaanxi Province (No. 2016GY-094), the Social Development Research Project of Science and Technology in Shaanxi Province (No. 2016SF-255), and the National Natural Science Foundation of China (Li Baojuan for No. 81301199, Liu Baohong for No. 61374185).

References

1. Krisztian, B., Leif, A., Maarten, D.R.: A language modeling framework for expert finding. Inf. Process. Manage. **45**, 1–19 (2009)
2. Li, M., Liu, L., Li, C.-B.: An approach to expert recommendation based on fuzzy linguistic method and fuzzy text classification in knowledge management systems. Expert Syst. Appl. **38**, 8586–8596 (2011)
3. Guan, Z., Miao, G., McLoughlin, R., Yan, X., Cai, D.: Co-occurrence based diffusion for expert search on the web. IEEE Trans. Knowl. Data Eng. **25**(5), 1–16 (2012)
4. Fang, Y., Si, L., Mathur, A.: Discriminative probabilistic models for expert search in heterogeneous information sources. Inf. Retrieval **14**, 158–177 (2011)
5. Yang, K.-W., Huh, S.-Y.: Automatic expert identification using a text categorization technique in knowledge management systems. Expert Syst. Appl. **34**, 1445–1455 (2008)
6. Balog, K., Rijke, M.D.: Combining candidate and document models for expert search. In: Proceedings of the Seventeenth Text Retrieval Conference (TREC 2008), NIST (2008)
7. Balog, K.: People search in the enterprise. Ph.D. University of Amsterdam, Amsterdam (2008)
8. Balog, K., Azzopardi, L., Rijke, M.D.: Formal models for expert finding in enterprise corpora. In: Proceedings of the 29th Annual International ACM SIGIR Conference on Research and Development in Information Retrieval, pp. 43–50. ACM, Seattle (2006)
9. Bordea, G.: Concept extraction applied to the task of expert finding. In: Aroyo, L., Antoniou, G., Hyvönen, E., Teije, A., Stuckenschmidt, H., Cabral, L., Tudorache, T. (eds.) ESWC 2010. LNCS, vol. 6089, pp. 451–456. Springer, Heidelberg (2010). doi:10.1007/978-3-642-13489-0_42
10. Daud, A., Li, J., Zhou, L., Muhammad, F.: Temporal expert finding through generalized time topic modeling. Knowl. Based Syst. **23**, 615–625 (2010)
11. Deng, H., King, I., Lyu, M.R.: Formal models for expert finding on DBLP bibliography data. In: 2008 Eighth IEEE International Conference on Data Mining, pp. 163–172 (2008)
12. Jiang, P., Yang, Q., Zhang, C., Niu, Z., Fu, H.: A probability model for related entity retrieval using relation pattern. In: Xiong, H., Lee, W.B. (eds.) KSEM 2011. LNCS, vol. 7091, pp. 318–330. Springer, Heidelberg (2011). doi:10.1007/978-3-642-25975-3_28
13. Agerri, R., Granados, R., García Serrano, A.: Enrichment of named entities for image photo retrieval. In: Detyniecki, M., García-Serrano, A., Nürnberger, A. (eds.) AMR 2009. LNCS, vol. 6535, pp. 101–110. Springer, Heidelberg (2011). doi:10.1007/978-3-642-18449-9_9

14. Blei, D.M., Ng, A.Y., Jordan, M.I.: Latent dirichlet allocation. J. Mach. Learn. Res. **3**, 993–1022 (2003)
15. Steyvers, M., Griffiths, T.: Probabilistic topic models. In: Landauer, T., McNamara, D.S., Dennis, S., Kintsch, W. (eds.) Handbook of Latent Semantic Analysis, vol. 427, pp. 424–440. Erlbaum, Hillsdale (2007)
16. Marmanis, H., Babenko, D.: Algorithms of the Intelligent Web. Manning, Greenwich (2009)

Algorithm Design for Automatic Modeling of the First and the Second Level of Airway Tree

Yue Lou[(✉)] and Xin Sun

School of Mechatronic Engineering and Automation, Shanghai University,
Shanghai 200072, China
ly603341486@gmail.com

Abstract. The models of airway tree designed in this paper are different from the general visual models. The model preserves all the information of the space data when it is created. The space data is mainly consisted by the coordinates of the boundary pixels and the spatial functions of the model surfaces. The algorithm consists of three main steps. Firstly, the boundaries of the airway tree are extracted by Sobel operator. Then, the boundary pixels are ring-likely sorted according to the distance between each other. Finally, each three pixels belong to the adjacent layers form a surface. An airway tree model can be eventually created by iterating the main steps. What's more, the algorithm has also been optimized, we can mostly get a model in 50 s.

Keywords: CT image · Airway tree · Edge detection · 3D modeling

1 Introduction

Nowadays, doctors can utilize the MDCT to obtain the high resolution CT thin scan images of the patients by one-off breath holding scanning. However, the data of the airway tree structure from the MDCT cannot demonstrate the pulmonary ventilating function explicitly, an algorithm is demanded to separate the air tree from the lung and remodel it in order to improve the accuracy of the diagnosis [1, 2].

The adjacent layers of the CT images are similar and successive, the air tree model can be built firstly in MATLAB. Because the spatial structure and data of the model have been preserved, the model can be input into the model analysis software such as ANSYS, FLUENT to get a further gas dynamics analysis [3].

2 Material

This model contains 255 images, scanned by the Discovery CT750 HD of GE company. The scanner works with the AW (advantage workstation) 4.4 that can observe the CT thin scan images. The scale of the matrix is 512.

© Springer Nature Singapore Pte Ltd. 2017
M. Fei et al. (Eds.): LSMS/ICSEE 2017, Part I, CCIS 761, pp. 342–348, 2017.
DOI: 10.1007/978-981-10-6370-1_34

3 Procedures and Results

3.1 The DICOM Format and the Image Preprocessing

A CT image is shown in MATLAB by the DICOM format. A DICOM image contains a large sum of information. A DICOM file consists of the file header and the dataset. DICOM dataset is formed by DICOM data elements which follow a typical sequence [4]. The DICM data elements is the basic units of the DICOM files, consists of tag, data type, date length and data field.

The median filtering is processed firstly in order to get rid of some irrelevant information such as tiny vessels. In that case, the airway boundaries can be detected and extracted with a high efficiency.

3.2 Detection and Extraction of the Airway Tree Boundary

More than one closed boundaries can be detected in a CT images, an algorithm is demanded to find the airway boundary automatically and accurately among all the detected boundaries. The utilization of the centric coordinates performs well.

The adjacent layers between the two CT images is similar so their airway boundaries are also alike. Once the centric coordinate of the airway boundary in the previous CT image is fixed, the airway boundary in the next CT image can be also confirmed. Because the centric coordinate of the correct boundary should be the nearest from the previous one among all the detected boundaries. As long as the initial centric coordinate in the

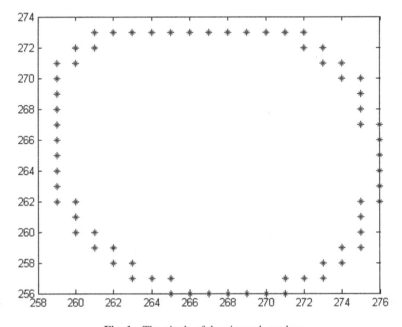

Fig. 1. The pixels of the airway boundary

first CT image is defined, all the airway boundaries in each CT image will be automatically obtained in this way.

The airway boundary consists of a set of pixels actually. When the airway boundary is chosen successfully, the pixels of the boundary are put into the space rectangular coordinate system.

The pixels of the extracted boundary are shown in Fig. 1. The shape of the pixels should be ring-like because of the shape of the airway. The following procedures are based on these pixels.

3.3   The Sort of the Airway Boundary Pixels

MATLAB has an inherent algorithm to sort a point set automatically. The subset whose x-coordinate value is minimum is chosen firstly, then, sort the subset by the y-coordinate form small to large. Another subset is chosen whose x-coordinate is the second smallest value, iterate the procedure, a certain point sequence is formed eventually. The Fig. 2 shows the automatic sequence algorithm of the MATLAB.

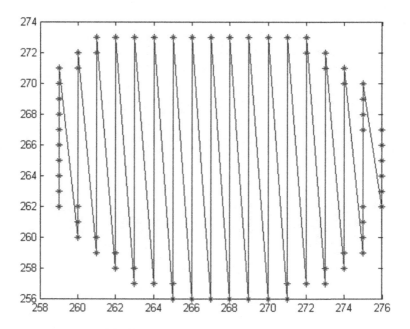

Fig. 2. The default sequence of the pixels

This sequence cannot be used for our modeling, the point set is expected to have an automatic self-ordering to a ring-like shape. So, a further procedure is demanded to sort these points.

An accurate algorithm is proposed in the paper. The point set can be sorted annularly according to their distances between each other. We define a initial point whose y-coordinate is the smallest in the subset whose x-coordinate is the smallest among all the x-coordinates. Finding the next point which is closest to the initial one. Then, replace the initial point by the next one, finding the next point in the same way until all the points are used.

An eligible sequence is formed and the consequence is shown in Fig. 3.

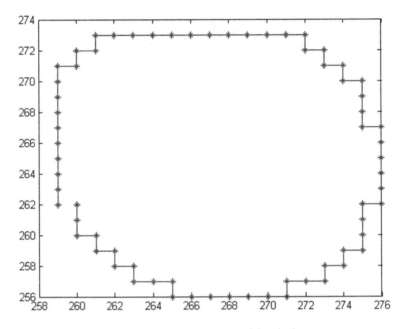

Fig. 3. The ideal sequence of the pixels

3.4 Modeling According to the Scattered Points

Because the adjacent lays are similar, the side surface can be generated directly according to these scattered points which are extracted from two adjacent CT images. The patch function in MATLAB is used to generate a surface through each three points. The consequence is shown in Fig. 4.

The final model is created by a series of CT that contains 255 images. The integral model is shown in Fig. 5.

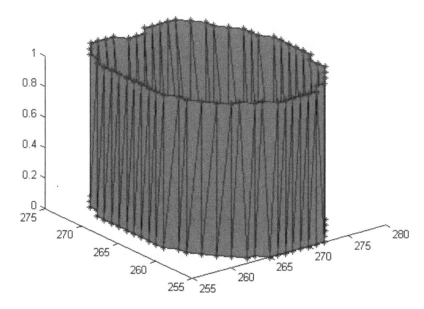

Fig. 4. The side surface generated by two adjacent layers

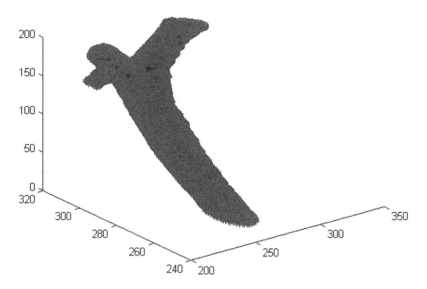

Fig. 5. The model of airway tree

4   Discussion

4.1   The Bwlabel Function Is Disabled in Several Images

The CT image preprocessing by median filtering is the trade-off between the reconstruction speed and the image accuracy. For most of the CT image, a relatively accurate boundary can be extracted in this way and the efficiency of the algorithm is also improved significantly.

However, it is inevitable that some information will be missed when a CT image is processed by median filter. So, not all of the airway boundaries in any CT images can be extracted correctly. When this happens, the whole modeling project is interrupted. So, a feedback loop is designed to deal with this individual situation.

Known from Sect. 3.2, we pick up the airway boundary according to the coordination of the previous airway boundary. A constraint is added to the algorithm that if the shortest distance from the previous centric coordinate is still lager than the default value, it is reasonable that the boundary information dropout occurs, result from the median filter processing. In that case, the information dropout layer copies the airway boundary from the previous layer. Although local information is distorted in this way, the experimental result shows that only 7 CT images drop their boundary information among all the 255 CT images. Considering that the system efficiency of image processing can be significantly improved by median filter, the information distortion of several images is acceptable.

4.2   The Number of the Pixels Extracted from the Adjacent CT Images Do not Match

Known from Sect. 3.4, the rudimentary way of the airway modeling is extracting the pixels from the two adjacent CT images, generating a side surface by every three points and the whole airway tree model is connected by all the CT images. But the number of the pixels extracted from the two adjacent layers is usually different, the difference values ranges from 2 to 5. The whole side surface can be only generated when the number of the pixels extracted from the two layers are equal. If the difference value is N, the lay whose pixels is more than the other one is divided evenly into N parts, delete one pixel from each part. There are generally at least 70 pixels in an airway boundary and these superfluous point is temporarily deleted, still be used in the next layer, so the model is built and the accuracy can be guaranteed at the same time.

5   Conclusion

More than hundreds of CT thin scan images are used to create the airway tree modeling and preserves its spatial data at the same time. So, it takes a long time for a general image traversal algorithm to accomplish the model. The key for the modeling is a suitable image processing algorithm. The algorithm proposed in the paper not only keep the model accuracy but also preserve the spatial data of the model. The model can be imported into model analysis software such as ANASYS for a further analysis, and it is available for a further analysis of the gas dynamics characteristic of human airway trees.

Acknowledgement. The study and the material were supported by Beijing Cancer Hospital.

References

1. Sluimer, I., Schilham, A., Prokop, M., et al.: Computer analysis of computer tomography scans of the lung: a survey. IEEE Trans. Med. Imaging **25**(4), 380–402 (2006)
2. Hoffman, E.A., Simon, B.A., McLennan, G.: A structural and functional assessment of the lung via multidetector-row computed tomography. In: Proceedings of the American Thoracic Society 2006, pp. 523–530. American Thoracic Society, New York (2006)
3. Li, Z H.: Airflows in human airway based on computational fluid dynamics. Northeastern Univ., 44–55 (2014)
4. Lv, X.Q., Liu, J.X.: Technology research based on the windows platform of DICOM medical image display. Chin. J. Biomed. Eng. **25**(4), 410–425 (2006)

Light-Weight Mg/Al Dissimilar Structures Welded by CW Laser for Weight Saving Applications

Qiong Gao[1], Sonia Meco[2], Kehong Wang[1(✉)], and Shun Guo[1]

[1] School of Material Science and Engineering, Nanjing University of Science and Technology,
Nanjing 210094, People's Republic of China
wkh1602@126.com
[2] Welding Engineering and Laser Processing Centre, Cranfield University, University Way,
building 46, Cranfield, Bedfordshire MK43 0AL, England, UK

Abstract. With the increasing demand of light-weight alloys, such as magnesium (Mg) and aluminum (Al), the need for joining these two alloys is unavoidable. In this study, AZ31B Mg and 1060 Al alloys were joined by continuous wave laser micro-welding using a 0.05 mm thick Cu/Zn interlayer. The microstructure and phases constituent of the weld seam were examined by optical microscope, SEM and EDS. The formation and distribution of the intermetallic compounds (IMCs) and the relationship between these structures and the microhardness of the weld were discussed in detail. The effect of Cu/Zn interlayer on the performance of Mg/Al joint was also analyzed. The results showed that Mg/Al IMCs were formed in the weld, which indicates that the Cu/Zn foil could not prevent the reaction between Mg and Al. However, the addition of Cu and Zn into the weld pool refined the microstructure by improving the number of eutectic structures. The micro-hardness of Mg/Al IMCs in the middle of the weld was very high which can be detrimental to the toughness of the Mg/Al joints.

Keywords: Magnesium · Aluminum · Laser · Micro-welding · Microstructure · Intermetallic compounds

1 Introduction

Light-weight components are of high interest for aerospace and automotive industries because such components permit reducing fuel consumption and ease pressure on the fuel economy. This has led to the development of novel materials and advanced processing techniques. Magnesium and aluminum are preferred for their outstanding performance, low density (light weight) and high strength to weight ratio which are the most important characteristics in the manufacturing industries within the transportation sector. The continuous exploitation of applications for these alloys will lead to a broader use of Mg/Al composite structures in order to reduce the total weight of the structures without compromising the required strength [1–3]. However, it is still a challenge to produce reliable Mg/Al joints due to their different physical properties and the formation of excessive Mg-Al intermetallic compounds (IMCs) when both metals are mixed together. Hence, the way to produce high quality and high strength Mg/Al joints is still

© Springer Nature Singapore Pte Ltd. 2017
M. Fei et al. (Eds.): LSMS/ICSEE 2017, Part I, CCIS 761, pp. 349–357, 2017.
DOI: 10.1007/978-981-10-6370-1_35

one of the main research topics worldwide. Previous researchers have focused their work on controlling the reaction between Mg and Al to a minimum level to minimize the formation of Mg-Al IMCs. The authors have used the following welding processes: TIG [4], spot welding [5], MIG [6], laser [7, 8], laser-MIG hybrid [9], friction stir welding [10, 11] and diffusion bonding [12] between others. However, the formation of Mg/Al IMCs is still.

Laser processing uses a high power density heat source and offers significant advantages when compared to arc welding low heat input is the main advantage because it not only reduces the distortion [13] but also minimizes the dilution of Mg into Al and vice-versa. Therefore, laser welding is a promising process to join Mg to Al. An alternative approach to reduce the formation of Mg/Al IMCs is by changing the chemical composition of the melt pool by using an interlayer and alloying the weld. No research about Mg/Al dissimilar welding with Cu/Zn interlayers has been carried out. Hence, in this study, 1060 Al alloy and AZ31B Mg alloy were welded by continuous wave (CW) laser welding. The microstructure and composition of the welds were investigated as well as the mechanical properties of the joints.

2   Experimental Details

Sheets of 1050 aluminum ($0.5 \times 30 \times 105$ mm) and AZ31B magnesium ($1 \times 30 \times 105$ mm) were used as base materials in this work. The typical chemical composition of the base metals is shown in Table 1. Interlayers of 0.05 mm thick copper brass (63 wt% Cu and 37 wt% Zn) were added between the Mg and Al sheets. The schematic diagram of the laser welding process set-up is shown in Fig. 1. The Mg sheet was placed on the top of Al sheet because the absorption of the laser radiation (1070 nm of wavelength) by Al is lower than that of Mg. Thus, the laser process becomes more efficient with Mg on the top because more energy is absorbed by the material. Mg and Al alloys are both active. Thus, shielding gas is generally required during the welding process. In this work pure shield argon was used with 15 l/min flow rate. Before welding, the surfaces of the base materials were ground with a grinder and degreased with acetone to remove the oxides and oil from the material surface. A CW fibre laser with 3 kW of maximum power was used. The characteristics of laser is listed in Table 2. The laser beam was characterized using a Primes GmbH focus monitor system and the D4σ method. The optimal parameters were power 525 W, travel speed 96 mm/s. The characteristics of laser used in this study is listed in Table 2.

Table 1. Chemical compositions [wt%] of AZ31B and 1050 alloys.

Material	Al	Zn	Mn	Si	Fe	Cu	Ca	Cr	Mg	Other
AZ31B	2.5–3.5	0.6–1.4	0.2	0.1	0.005	0.05	0.04	–	Bal	0.3
1050	Bal	0.05	0.05	0.25	0.4	0.05	–	–	0.05	0.10

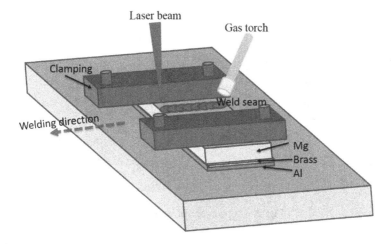

Fig. 1. Schematic of Mg/Al laser welding process

Table 2. Laser characteristics

Laser	Max. Power (KW)	Wavelength (nm)	Focusing lens	Delivery fiber (μm)	Beam diameter (μm)
Fiber laser	3	1070	FL 300	100	192

After welding, the cross sections of the Mg/Al joints were prepared for microstructural analysis. The samples were ground and polished by abrasive papers and diamond polishing paste, following standard metallographic procedure. The micrographs of the full section of the welds were taken by a camera installed in the optical microscope. The microstructure of the welds was observed under a scanning electron microscope (SEM) (FEI Quanta 250F environment scanning electronic telescope) equipped with an energy dispersive x-ray spectrometer (EDS). The composition and distribution of the IMCs present in the joints were determined by EDS. Micro-hardness tests were carried out by a Vickers hardness tester (HVS-1000Z) with 0.98 N.

3   Results and Discussion

3.1   Macrostructure of Mg/Al Joint

The initial observation of the results indicates that Mg and Al alloys can be successfully welded by CW laser welding with brass foil. The appearance of the Mg/Al joint with brass interlayer can be seen in Fig. 2. The micrograph shows a smooth weld seam with no visible defects. Due to the small laser beam diameter (192 μm) the width of weld seam is extraordinary small, nearly 1 mm wide, which is much smaller than the width of the weld seams produced by the arc welding processes, normally (10 mm wide).

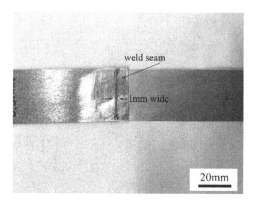

Fig. 2. Appearance of weld surface

The cross section of the weld is shown in Fig. 3. The weld seam is 943.3 μm wide and 1450.6 μm deep, which means that the weld is not fully penetrating. It is also clear that the microstructure within the melt pool is not uniform because different zones with different colours can be observed under the optical microscope. The dark, light black and white zones are set as zone A, B, C, respectively. The EDS elemental mapping is presented in Fig. 4(b), (c). It is clearly seen that every element is unevenly distributed within the weld pool. Mg diffused into Al substrate and was mixed with Al. The elements from the interlayer, Cu and Zn, are evenly dispersed in the weld. The uneven distribution of each element explains the distinct zones with different colours visible in the micrograph. The chemical composition of each zone was investigated by SEM and EDS analysis. The results are discussed as follows.

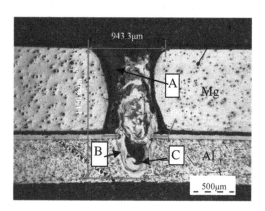

Fig. 3. Morphology of weld cross section

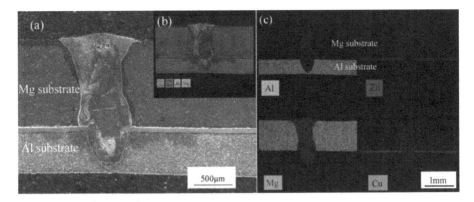

Fig. 4. (a) SEM morphology of weld, (b) EDS mapping picture, (c) EDS mapping pictures for Al, Mg, Zn, Cu, respectively.

3.2   Microstructure of Mg/Al Joint

Figure 5(a) shows the SEM micrographs of zone A in Fig. 3. Figure 5(b) presents the microstructure of the interface and bottom of the weld. The micrographs with higher magnification of zones I, II, III in Fig. 5(b) are shown in Fig. 5(c), (d), (e), respectively. The chemical composition of the points 1 to 6 in Fig. 6 is listed in Table 3. The grey dendritic crystal shown in Fig. 5(a) (point 1) is composed by 76.8 at % Mg, 20 at % Al, 1.6 at % Zn and 1.6 at % Cu. This phase has been identified as Mg-based solid solution and Mg-Al eutectic structure, according to the Mg-Al binary phase diagram (see Fig. 6). According to Fig. 5(c), the microstructure of zone I (zone B in Fig. 3) is mainly composed by a continuous IMC with dendritic structure, identified as Al12Mg17. However, eutectic structures were found to segregate the continuous dendrite to a dispersed and small size dendrite matrix. It can be indicative that the Cu and Zn has a high tendency to form the eutectic dendritic structures. Figure 5(d) shows the microstructure of the bottom of weld which turned to be eutectic dendritic corresponding to zone C in Fig. 3. Point 3 is composed by 19.4 at % Mg, 75.7 at % Al, 1.5 at % Zn, and 3.4 at % Cu, which is considered as Al-based solution and Al_3Mg_2 IMCs. The content of Cu in point 3 is higher than that in point 1. That could be explained by the solubility of Cu in Al to be higher than that in Mg. More Cu has been dissolved in Al solution. The composition in point 4 is 36.4 at % Mg, 59.4 at % Al, 1.5 at % Zn and 2.7 at % Cu corresponding to Mg based solution and $Al_{12}Mg_{17}$ eutectic structure. The magnified morphological features of zone III is presented in Fig. 5(e), which shows two parallel structures with different morphologies. In point 5, the microstructure has lamellar eutectic structures regarded as Mg-Al eutectic (Al_3Mg_2, $Al_{12}Mg_{17}$ and Mg based solution). In addition, the microstructure in Point 6 is a cellular crystal with composition of 54.7 at % Mg, 41.6 at % Al, 2.1 at % Zn and 1.6 at % Cu, which can be considered as Mg based solution and $Al_{17}Mg_{12}$.

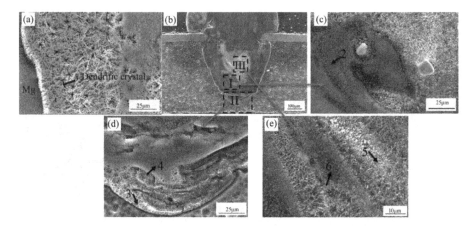

Fig. 5. (a) Microstructure of the top of weld seam, (b) microstructure of interface and bottom of the weld seam, (c) magnified picture of zone I, (d) magnified picture of zone II, (e) magnified picture of zone III.

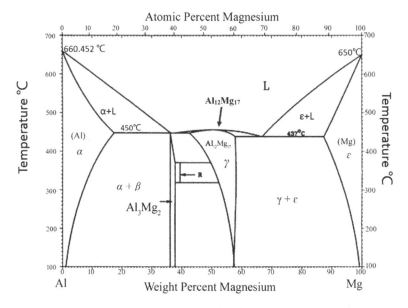

Fig. 6. Mg and Al binary phase diagram

It can be concluded that the addition of 0.5 mm brass interlayer could not prevent the diffusion of Mg and Al due to the full fusion of both alloys, Mg and Al. The addition of Cu and Zn element promoted the formation of branch crystals. In Fig. 3, the darker phase corresponded to higher content of Mg. Hence, the darkness feature in Fig. 4 (zone A) is Mg based solution and eutectic structure. The light dark feature in zone B in Fig. 3 is a mixture of Mg and Al which corresponds to Mg-Al IMCs. The white structures in zone C of Fig. 3 were Al based solution and eutectic structures.

Table 3. Phases Composition at.%

Point	Mg	Al	Zn	Cu	Possible Phase
1	76.8	20.0	1.6	1.6	Mg solid solution and Mg-Al eutectic structure
2	51.2	43.4	3.3	2.1	$Al_{12}Mg_{17}$
3	19.4	75.7	1.5	3.4	Al solid solution and Mg-Al eutectic structure
4	36.4	59.4	1.5	2.66	Mg_2Al_3
5	60.1	31.6	1.7	5.7	Al_3Mg_2, $Al_{12}Mg_{17}$ and Mg based solution
6	54.7	41.6	2.1	1.6	Mg based solution and $Al_{17}Mg_{12}$

3.3 Micro-hardness Analysis

The micro-hardness distribution of the weld seam has an important effect on the performance of the joint. Based on the micro-hardness of the weld it is possible to infer the strength and identify the weakest points on the joints. Figure 7(c) shows the direction and relative position of the indentations to the weld pool. The micro-hardness distribution along the vertical path, represented by line 1, is shown in Fig. 7(a), while the micro-hardness distribution along the horizontal path, represented by line 2 is drawn in Fig. 7(b). In Fig. 7(a), the micro-hardness in the middle of the weld seam is higher, above 300 HV, than that in the top and bottom of the weld seam. According to the previous analysis of the microstructure, the phase composition in the top of the weld seam is Mg-based solution and eutectic structure. Therefore, the micro-hardness in this zone is lower than others but higher than that of Mg substrate. The feature in the middle of the weld is a mixture of Mg and Al as a form of Mg and Al IMCs which explains the higher micro-hardness. However, the value of the micro-hardness in the middle of the weld is not constant. The drop-down values confirms that the phase composition is not continuous. In the Al side, the average micro-hardness of the weld is lower due to the formation of Al based solution and eutectic structures.

In Fig. 7(b), the micro-hardness of weld seam near fusion line is about 150 HV, which is lower than the value in middle of weld seam but higher than that of the Al alloy. One possible reason is that in that zone the branch crystal is an Al based solution which has a relatively low value of micro-hardness. This can have a positive impact on the toughness of Mg/Al joint. In the middle of the weld, the micro-hardness reaches just above 300 HV due to the formation of Mg/Al IMCs.

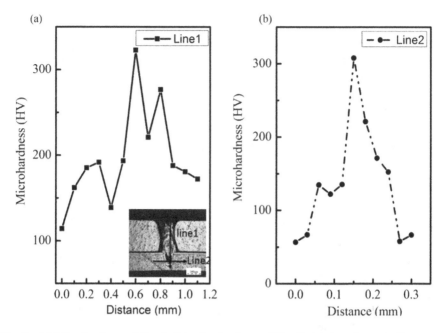

Fig. 7. (a) Microhardness of line 1, (b) microhardness of line 2, (c) the schematic diagram of microhardness test

4 Conclusion

AZ31B Mg and 1060 Al thin sheets were successfully joined by CW laser micro-welding using Cu/Zn composite interlayer. The addition of 0.05 mm thick Cu/Zn interlayer could not avoid the formation of Mg/Al IMCs but refined the microstructure of weld seam (improve the formation of dendritic structure), which can improve the toughness of Mg/Al joints.

The microstructure of weld seam was uneven and formed by different phases. In the top of weld, zone A was composed of Mg-based solution, zone B was composed of Mg/Al IMCs and zone C composed of Al-based solution.

The distribution of IMCs in the weld seam affects the micro-hardness. The micro-hardness in zone A, top of the weld pool, was lower than that of the Mg and Al mixture region composed of Mg/Al IMCs. The micro-hardness on the sides of the weld seam was lower than that in the middle of weld due to the formation of solid solution, which indicates that the increasing number of solid solution can improve the toughness of Mg/Al joint.

References

1. Schubert, E., Klassen, M., Zerner, I., Walz, C., Sepold, G.: Light-weight structures produced by laser beam joining for future applications in automobile and aerospace industry. J. Mater. Proc. Tech. **115**, 2–8 (2001)
2. Miller, W.S., Zhuang, L., Bottema, J., Wittebrood, A.J., De Smet, P., Haszler, A., Vieregge, A.: Recent development in aluminum alloys for the automotive industry. Mater. Sci. Eng., A **280**, 37–49 (2000)
3. Mordike, B.L., Ebert, T.: Magnesium: properties-applications-potential. Mater. Sci. Eng., A **302**, 37–45 (2001)
4. Zhang, H.T., Dai, X.Y., Feng, J.C.: Joining of aluminum and magnesium via pre-roll-assisted A-TIG welding with Zn interlayer. Mater. Lett. **122**, 49–51 (2014)
5. Macwan, A.: D.L. Chen.: Ultrasonic spot welding of rare-earth containing ZEK100 magnesium alloy to 5754 aluminum alloy. Mater. Sci. Eng., A **666**, 139–148 (2016)
6. Wang, J., Feng, J.C., Wang, Y.X.: Microstructure of Al–Mg dissimilar weld made by cold metal transfer MIG welding. Mater. Sci. Tech. **24**, 827–831 (2008)
7. Liu, L., Wang, H.: Microstructure and Properties Analysis of Laser Welding and Laser Weld Bonding Mg to Al Joints. Meta. Mater. Trans. A **42**, 1044–1050 (2011)
8. Scherm, F., Bezold, J., Glatzel, U.: Laser welding of Mg alloy MgAl3Zn1 (AZ31) to Al alloy AlMg3 (AA5754) using ZnAl filler material. Sci. Tech. Weld. Join. **17**, 364–367 (2012)
9. Liu, L.M., Liu, X.J., Liu, S.H.: Microstructure of laser-TIG hybrid welds of dissimilar Mg alloy and Al alloy with Ce as interlayer. Scrip. Mater. **55**, 383–386 (2006)
10. Liang, Z., Qin, G., Wang, L., Meng, X., Li, F.: Microstructural characterization and mechanical properties of dissimilar friction welding of 1060 aluminum to AZ31B magnesium alloy. Mater. Sci. Eng., A **645**, 170–180 (2015)
11. Champagne III, V.K., West, M.K., Rokni, M.R., Curtis, T., Champagne Jr., V., McNally, B.: Joining of Cast ZE41A Mg to wrought 6061 al by the cold spray process and friction stir welding. J. Therm. Spray. Tech. **25**, 143–159 (2016)
12. Shang, J., Wang, K.H., Zhou, Q., Zhang, D.K., Huang, J., Ge, J.Q.: Effect of joining temperature on microstructure and properties of diffusion bonded Mg/Al joints. Trans. Non. Metal. Soc. China **22**, 1961–1966 (2012)
13. Meco, S., Ganguly, S., Williams, S., Mcpherson, N.: Effect of laser processing parameters on the formation of intermetallic compounds in Fe-Al dissimilar welding. J. Mater. Per. Eng. **23**, 3361–3370 (2014)

Modeling and Simulation of Intelligent Substation Network Under Intrusion Attack

Xiaojuan Huang[1(✉)], Rong Fu[1], Yi Tang[2], Mengya Li[2], and Dong Yue[1]

[1] College of Automation, Nanjing University of Posts and Telecommunications, Nanjing 210023, China
huangxiaojuannj@163.com
[2] College of Electrical Engineering, Southeast University, Nanjing 210023, China

Abstract. Recent advancement in the integration of power systems and information communication technology has brought the key concerns towards security operation of cyber physical power system. This paper focuses on realizing the unified system modeling under intrusion attacks and refining the attack effects on communication network by simulation research. We start this survey with an overview of the system operation and crucial intrusion attacks associated with operational security from fusion system perspective. A novel limited stochastic Petri net (LSPN) graph theory is introduced to establish the unified firewall protection system model of intelligent substation network. By proposing quantitative computational methodology of communication throughput variation, the potential consequence on the communication network is determined with information transmission constraints. The final test on IEEE-30 node power system illustrates the usefulness of the proposed model analysis. The research work would raise awareness of the cyber intrusion threats and provide the basis for security defense.

Keywords: Cyber physical system · Communication throughput variation · Intelligent substation network · Petri net theory

1 Introduction

The modern power systems are evolving to accommodate increased renewable sources of energy and active distribution systems with the integration of communication network overlay by information communication technology (ICT). This integration, however, makes facilities in open network [1] be more vulnerable for cyber attackers to invade system safety operation. Ukraine blackout accident [2] was a typical case of power outage with its secondary network suffering from network intrusion attack.

Significant research exists in modeling network intrusion attacks and assessing their impact on cyber physical system (CPS). Literature [3] attempted to characterize attacks on an Industrial Control System, where does not include several aspects of attacks such as start states, intents, and attack points. Generally, many approaches are to base attacks on traditional models derived from information and network security. Such as

© Springer Nature Singapore Pte Ltd. 2017
M. Fei et al. (Eds.): LSMS/ICSEE 2017, Part I, CCIS 761, pp. 358–367, 2017.
DOI: 10.1007/978-981-10-6370-1_36

influences on the system stability with error data injection and various cryptographic attacks. Graph based modeling techniques are derived from research in network security [4, 5]. The attack graph model, which fully taking into account the network topology information, not only can visual the attack behavior of the infiltration process, but also make it more suitable for complex network attack modeling with the support of automated tools [6]. As a typical method, Petri net graph theory shows a great deal of flexibility and a limited stochastic Petri net (LSPN) can specific attack source propagation path to quantify the impact on the target system. Literature [7] proposed a concept about Petri-net-based CPS fusion modelling by implementing service specification into CPS service control flow. On the basis of studying the demand of active distribution system, a CPS control model and control method base on hybrid system were presented in literature [8].

In the evaluation of the security performance of CPS, development and implementation of SCADA cyber security testbeds are reported in [9–11]. Although the testbeds enable accurate simulations, the cost is high when to incorporate the entire cyber and power system models. In literature [12], the risk of attacking the system was obtained to evaluate successful attack probability by modeling the communication network and the power network respectively. The authors in [13] proposed the method that suggests the use of public key infrastructure technologies along with trusted computing elements, supported by firewalls, strong user and device authentication. Literature [14–16] analyzed the importance of cyber infrastructure security together with power grid security and the need for intrusion attack prevention. However, there is still a lack of methodologies to model the integration of CPDS for security assessment under cyber intrusion scenes.

In view of theoretical study, limited stochastic Petri nets are used in this paper to describe the attacked network states and attack information transmission process by using state transition graph. A unified firewall protection system model is established with highly abstraction of intrusion process and refinement of component constructions. The communication throughput variation is proposed based on the steady-state probability to quantify the impact on substation network when the corresponding power node loses efficacy due to intrusion attacks.

2   System Model and Network Attacks

Cyber Physical System (CPS) is built from, and depends upon the integration of computational algorithms and physical components [17, 18]. Figure 1 illustrates the communication architecture of the cyber-physical power system model, corresponding to the simulation environment. For the power system, the usual power model is showed. With regard to general communication infrastructures, they include wireless transmission networks, SCADA, control center, etc. Typical hardware is known as components in the control center, such as isolated firewall, engineering workstations, and various servers which can store and process the data. And the IEDs reside usually consist of the remote terminal units (RTU), advanced metering infrastructure and the programmable logic controller (PLC). The servers store and process the information

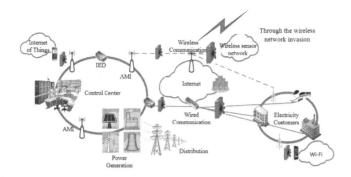

Fig. 1. Cyber physical power system communication infrastructures

sent from and to the RTUs, and the RTU or PLC controls the process of the field devices.

The Report No. 7628 《American Institute of Standards and Technology》 [19] points out that the three elements of cybersecurity are confidentiality, integrity and availability, commonly referred to as "CIA" security objectives [20]. Concerning CIA, private networks in power system are becoming more vulnerable to IP-based intrusion attacks with TCP/IP and Ethernet technologies.

At the beginning of the network operation, the initial measurements of state variables are collected by IEDs and then transmitted to the communication network at a certain interval time. After crossing the firewall detection process, the converted digital signals, such as IP address, then are transmitted to the control centre as the input values with TCP/IP communication protocol. Whenever control commands are sent by control centre after being processed by the state estimator to the actuators, the control variables are modified accordingly and the substation network shows a new operating state. For an IP-based intrusion attack, the proposed graph-theoretic system model in Fig. 2 enables the detailed firewall and password protection model analysis for information transfer process in the communication part to substation network. Such behaviors are studied based on the methodological modelling that provides the boundary inspection of malicious packets and intrusion attempts on each computer system.

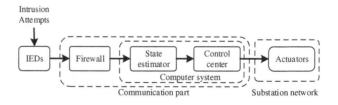

Fig. 2. Proposed intrusion model

3   Unified System Model Setup Under Intrusion Attacks

A. Limited Stochastic Petri Net (LSPN) theory

A basic LSPN consists of the following marks: places, tokens, black spots, transitions and the arc. The relationship between the local states is determined by arc. The directed arcs formed by the direct reachability relation are called the reachable marking states. In this paper, the performance analysis of the LSPN model is based on the isomorphism of its state space and Markov Chain (MC). Each place of LSPN is mapped to state space in MC and the rate of change in the reachability graph corresponds to the transition rate between MC states.

B. Component construction extensions

According to the cyber intrusion activity model introduced in Fig. 2, the modelling method is a high level of abstraction for the intelligent substation network. For the network infrastructures with point-to-point transmission that involve no monitoring capabilities, such as state estimate, IEDs and actuators in this paper, the probability of successful information transmission is considered to be 100%.

A successful intrusion attack means that necessary information needs to be acquired from different tools and resources to determine IP addresses in the network firewalls. So, an exact rule set for a secure firewall is very necessary. A mixed firewall filtering rules combining transport layer protocol type and packet IP address are shown in the following Table 1. The computer control center is generally the ultimate goal to realize data tampering. The password model is used to show the monitoring ability of the computer, which includes two parts: failed logon probability and the response rate.

Table 1. The specific firewall filtering rules

Rule	Protocol	S_IP	Action
1	TCP/IP	123.45.67.89	Accept I
2	TCP/IP	123.45.67.88	Accept II
3	TCP/IP	123.45.67.87	Accept III
0	UDP/IP	123.45.67.86	Deny

C. Unified model for substation level network

Based on properties of the transient and time delay transitions in a LSPN network, the system dynamic equilibrium can be achieved. With regard to unified network modelling, system elements can be grouped into substation nodes considering that a transmission line connects two substations. Such as the generators, DERs, transmission lines and load nodes, can be grouped into substation nodes. Based on the LSPN theory, the unified model is concerned with refined communication network components in B part. The state of the random process reflects the cyber intrusion abnormal activity which includes the malicious packet flow in firewall and the password login failure. As shown in the Fig. 3, the firewall protection model includes the firewall based on information detection technology and the realization of authentication login protection. The time transition is consistent with the time required by the attacker to obtain the

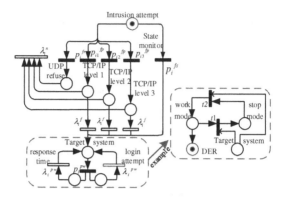

Fig. 3. Unified firewall protect model for substation network

system response. The tokens are used to represent the intrusion attempts after the attack begins. Especially for distributed energy resource (DER) node, we use LSPN theory to describe and model the switching process of the operating mode effectively that includes work and stop mode.

In the model, the places 'TCP/IP level 1, 2 and 3' represent the specific firewall filtering rules and the 'state monitor' means the information state monitoring function for malicious packets. Also, 'UDP refuse' means that the intrusion attempt is invalid with certain penetration probability. Each transient transition of the firewall can be calculated based on the firewall log [12]. The probability of passing the firewall through each rule is Eq. (1):

$$P_{i,j}^{fp} = \frac{f_{i,j}^{fp}}{N_{i,j}^{fp}}, \; P_i^{fr} = \frac{f_i^{fr}}{N_i^{fr}}, \; P_i^{fs} = \frac{f_i^{fs}}{N_i^{fs}}, \; P_i^{pw} = \frac{f_i^{pw}}{N_i^{pw}} \tag{1}$$

Where $f_{i,j}^{fp}$ represents the frequency through the firewall. $N_{i,j}^{fp}$ is the total recorded number of the firewall rule j. f_i^{fr} is the number of rejected packets. N_i^{fr} and N_i^{fs} both are the total number of firewall records. f_i^{fs} is the number of packets directly through state monitoring. The firewall execution speed λ_i^f is the number of instructions executed per second. The average response speed λ_i^{nr} depends on the network transmission status which is estimated using the ping command. For computer system i, the transition probability which respectively represents the computer response time, login attempt and the final target system. The f_i^{pw} is the number of intrusion attempts. N_i^{pw} is the total number of records except the login attempt within a specific time interval which is regarded as an common user input error. The response speed λ_i^{pw} is the delay of the repeated login.

4   Quantitative Analysis on Substation Network

The proposed security assessment methodology can be summarized as a two-step approach. (1) The first step is to analyse the computer network topology in the system for deriving possible intrusion attack paths to the control centre. The net modelling with LSPN defines the intrusion scenarios and quantifies the steady invasion state probability. (2) In the second step, the consequence severity of the communication malfunctions of the substation nodes is determined with communication throughput variation. The integration of these two steps makes it possible to quantify the impacts caused by a potential cyber intrusion attack.

A. Quantitative computational theory

According to the analysis of communication network in cyber side, the intelligent devices integrated on the computer can be mapped to the communication data point. The steps a successful network intrusion attack must complete are: (1) obtain the availability of computer systems in the network; (2) attempt to invade the computer; (3) understand how to attack through the communication network with appropriate attack access point.

Generally, the state transition matrix W can be obtained through determining the instantaneous transition and the time delay transitions to describe the intrusion attack behaviours [12]. The Markov equilibrium equation is solved with corresponding Markov chain state. The specific steady-state probability equation is:

$$\begin{cases} \tilde{\pi}W = \tilde{\pi} \\ \sum_{M \in T \cup V} \tilde{\pi} = 1 \end{cases} \tag{2}$$

Where T and V are the set of identities that transient changes and latency changes respectively. W represents a transfer matrix formed under different attacks. The vector π represents the embedded MC states.

B. Attack influence on substation communication network

The focused network intrusion attack refers to the behavior that across the firewall to reach the control center of the computer and make the corresponding target node fail to receive or transmit the correct information by the means of blocking the communication channel. Specifically, the attacker can control the packet size of the channel information transmission directly by changing the packet loss rate to congest network channel. To quantify the communication volume, the change of the channel throughput variation T is analyzed. While the data packets affected by the cyber intrusion attack become malicious, its packet loss rate is seen as the steady state probability value π. Thus, the communication throughput variation under the intrusion attack can be expressed as:

$$T = (1 - \frac{\pi_i N_2}{N_2 + N_1}) \times \frac{LR}{L+H} = (1 - \frac{\pi_i N_2}{1 + N_1/N_2}) \times \frac{LR}{L+H} \tag{3}$$

Where π_i is the steady-state probability after intrusion attack, which can be obtained by Eq. (2). Attack ratio N_1/N_2 depicts the characteristics of the interference attack, which means how well the attacker knows about the substation network. L, H are respectively the length of the original data message and the preamble. The transmission rate is R. The smaller throughput variation, the stronger the operation robustness of the corresponding communication network in the cyber side.

5   Case Simulations and Implementation

In this section, we evaluate the throughout variation caused by the intrusion attacks on communication networks of the IEEE30 power system. Based on the system wiring diagram, the three-winding transformer bus lines 4, 12, 13 and 6, 9, 10, 11, and the double-winding transformer bus lines 27 and 28 are used as one substation. The system has a total of 24 communication network models. We define the intrusion attack process of each substation into three models: (1) Directly attack the substation network and attempt to reach the control center (as shown ⊕); (2) Attack through the distribution network or substation network (as shown ◉); (3) Jointly attack through the power plant process control network, distribution network and substation network (as shown ◯). Assume that there is two-way firewall isolation among the three networks, and only substation network can directly connect to the control center (Fig. 4).

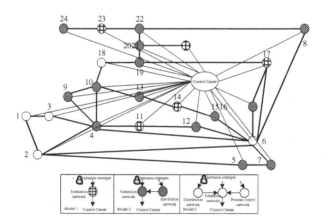

Fig. 4. The communication diagram of substation

According to the above Limited Stochastic Petri net (LSPN) theory, the transfer of tokens among states is similar to the Markov process. The penetration probabilities of firewall rules and packet information state monitors are assumed to be $P_i^{fp} = (0.0095324, 0.0181514, 0.0019415)$, $P_i^{fs} = (0.0083154)$ respectively. Also, the rejection probability of malicious packets is $P_i^{fr} = (0.71457)$. The logon failure probability for each machine is designed to be 10%. The computer response rates of computer and firewall are set to be $\lambda_1^{pw} = \lambda_2^{pw} = 12 \times 10^{-10}$ and $\lambda_1^f = \lambda_1^{nr} = 63 \times 10^{-7}$. The above values are generated by a random number generator.

To support the intuitive judgment, the Table 2 shows the steady-state probability corresponds to the 3 kinds power communication network topologies with different intrusion access points. The values in lines 1 to 4 of the table represent the steady-state values of intrusion behaviors at different locations with different access points A and B. The steady state value of the external attack point is generally less than the internal intrusion scene. Because that the internal firewall is the first firewall from outside to the internal network, which makes the probability of successfully penetrating into the control center increases. The main factor affecting the internal attack vulnerability is the configuration of the communication network and more complex structure (model 3). Therefore, the network configuration and protection of the key nodes in the actual power grid can refer to the above situation.

Table 2. Transmission probability with different intrusion access points

	A (inside the firewall)			B (outside the network)		
	Node17 (model 1)	Node4 (model 2)	Node11 (model 3)	Node17 (model 1)	Node4 (model 2)	Node11 (model 3)
Substation intrusion	0.0000076	0.000796	0.00106	0.00019	0.00019	0.0002
Distribution network intrusion	–	0.00004	0.00442	–	0.0199	0.334
Process control network intrusion	–	–	0.00018	–	–	0.0133
Control center arrival	0.000792	0.000396	0.000176	0.0398	0.0199	0.0133

For the network structure model 3, the change trend of the communication throughput variation with the different attack access points is shown in the Fig. 5. The network intrusion attacks from outside the firewall have a strong effect on the throughput of data communications (strongly opposed to the other two models), and from within the firewall. When the node is attacked, its throughput is changed by the communication topology of the larger impact. Because the low probability attack inside the network makes the throughput of data transmission less affected by the network communication structure. The range of communication throughput changes in model 3 are more affected by the data transmission path and network topology model which indicates the more robustness of the corresponding communication network model.

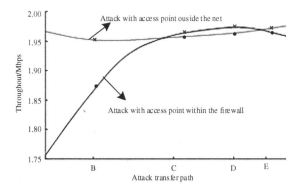

Fig. 5. Throughput variations in model 3

6   Conclusion

In this paper, the intelligent substation network modeling problem has been investigated considering the information transmission process in a cyber-physical system subject to effects of the intrusion attacks. Towards the fine modeling for network components equipped with different protection methods, here the firewall and computer login system protection methods were studied and a unified firewall protection communication model of substation network was established. In the case of distinguishing the three different levels of attack paths with two access points, the proposed analytical framework evaluated the steady-state attack probability of a successful attack based on LSPN graph theory. Then the communication throughput variation is proposed to quantify the impact on intelligent substation network. Moreover, the network intrusion attack definition, intrusion attack scenarios, specific throughput variation formula and its simulation results mentioned in this paper can provide a reference for identifying the vulnerability of substation network in the face of network intrusion attacks.

Acknowledgment. This work was supported by Key project of National Natural Science Foundation of China (Research on complex networked system architecture and system operating state security assessment with fault dynamic evolution mechanism) (Grant No. 61633016), a Communication and Network Technology National Engineering Research Center Open Fund project and an open project of National Key Laboratory (Research on operation security analysis and defense strategy of cyber physical system). We gratefully acknowledge the support from the Professor Fei Minrui of Shanghai University and Professor Hu Songlin.

References

1. Liu, Z.: Smart Grid Technology, 4th edn. China Electric Power Press, Beijing (2010). ISBN 9787512302235
2. Liang, G., Zhao, J., Weller, S.R.: The 2015 Ukraine blackout: implications for false data injection attacks. IEEE Trans. Power Syst. **40**, 149–151 (2016)

3. Berman, D., Butts, J.: Towards characterization of cyber attacks on industrial control systems: emulating field devices using Gumstix technology. IEEE Xplore, Resilient Control Systems (ISRCS), pp. 63–68 (2012)
4. Buford, J., Sanchez-Aarnoutse, J.C., Chen, T.M.: Petri net modeling of cyber-physical attacks on smart grid. IEEE Trans. Smart Grid **2**, 741–749 (2011)
5. Tran, B.N., Lakshminarayana, S.: SecureRails: towards an open simulation platform for analyzing cyber-physical attacks in railways. IEEE Xplore, pp. 95–98 (2016)
6. Hong, J., Panciatici, P., Liu, C.-C., Stefanov, A.: Intruders in the grid. IEEE Power Energy Mag. **10**, 58–66 (2012)
7. Ahmadon, M.A.B., Yamaguchi, S.: On service orchestration of cyber physical system and its verification based on petri net. In: 2016 IEEE 5th Global Conference on Consumer Electronics, pp. 1–4 (2016)
8. Liu, D., Li, Q-S., Wang, Y.: A hybrid system based CPS model and control of loads in active distribution network. In: 2016 IEEE International Conference on Power System Technology, pp. 1–8 (2016)
9. Adnan, R., Hahn, A., Higdon, M., Manimaran, G., Sridhar, S.: Development of the power cyber SCADA cyber security testbed cyber security and information intelligence research workshop. Oak Ridge National Lab, vol. 21, pp. 1–4 (2010)
10. Davis, C.M., Tate, J.E., Okhravi, H.: SCADA cyber security testbed development. In: 38th North American Power Symposium NAPS, pp. 483–488 (2006)
11. Xie, X., Hu, Y., Xin, Y.: Power information systems security: modeling and quantitative evaluation. In: Proceedings of IEEE Power Engineering Society General Meeting, vol. 1, pp. 905–910 (2004)
12. Liu, C.-C., Manimaran, G., Ten, C.W.: Vulnerability assessment of cybersecurity for SCADA systems. IEEE Trans. Power Syst. **23**, 1836–1846 (2008)
13. Metke, A.R., Ekl, R.L.: Security technology for smart grid networks. IEEE Trans. Smart Grid **1**, 99–107 (2010)
14. Jajodia, S., Noel, S.: Advanced cyber attack modeling, analysis, and visualization, Final Technical report AFRL-RI-RS-TR-2010-078, George Mason University (2010)
15. Hahn, A., Sridhar, S., Manimaran, G.: Cyber-physical system security for the electric power grid. Proc. IEEE **100**, 210–224 (2012)
16. Bompard, E., Napoli, R., Xue, F.: Vulnerability of interconnected power systems to malicious attacks under limited information. Eur. Trans. Electr. Power **18**, 820–834 (2008)
17. Ernster, T., Morris, T., Pan, S., Srivastava, A., Vellaithurai, C.: Modeling cyber-physical vulnerability of the smart grid with incomplete information. IEEE Trans. Smart Grid **4**, 235–244 (2013)
18. Liu, D., Lu, M.Y., Wang, X.S., Wang, Y.: Key technologies and trends of cyber physical system for power grid. In: Proceedings of the CSEE, vol. 35, pp. 3522–3531 (2015). (in Chinese)
19. National Institute for Standards and Technology (NIST): Guidelines for smart grid cyber security: vol. 3, supportive analyses and references: NISTIR 7628[S] (2010)
20. Li, M.Y., Rahman, S., Wang, Q., Tang, Y.: Framework for vulnerability assessment of communication systems for electric power grids. IET Gener. Transm. Distrib. **10**, 477–486 (2016)

Data Driven Analysis

Analysis of Temperature and Gas Flow Distribution Inside Safety Helmet Based on Numerical Simulation

Heng Ma[✉], Rui Li, Ke Qian, Yibo Gao, and Ling Chen

Zhejiang Key Laboratory for Protection Technology of High - Rise Operation,
Zhejiang Huadian Equipment Testing Institute, Hangzhou 310015, Zhejiang, China
maheng38923@163.com

Abstract. Aiming at the problem that the low heat dissipation of the safety helmet has an impact on the workers which can easily lead to safety accidents under high temperature and high strength, this paper analyzed the temperature field and gas flow field inside the safety helmet by using Finite Element Analysis, modeled and meshed the real safety helmet, obtained the distribution nephogram of temperature and heat flux by steady state and transient thermal analysis, calculated the results of the hot gas flow inside safety helmet and the gas flow nephogram by turbulence model. The results show that the temperature inside the safety helmet decreases gradually from bottom to top, the heat transfer rate of the contact part with the head is faster, only a small amount of hot gas can be vented from the vent while most of the hot gas is still concentrated inside the safety helmet difficult to vent. Therefore, it is necessary to improve the design of safety helmet and the comfort of the operation to ensure the safety of the workers.

Keywords: Thermodynamics · Temperature field · Fluid dynamics · Finite element · Safety helmet

1 Introduction

As the most basic personal protective equipment, safety helmet is mainly used to protect head from the impact of high impact [1, 2]. Because of the high tightness of the safety helmet, the workers' heads are hot and the sweat is difficult to evaporate, that will affect the normal work and have serious security risks. The gas flow and temperature distribution inside the safety helmet are very important but also very complex, because it is difficult to obtain accurate results by general test methods.

Finite element analysis is a method of simulating a real physical system (geometry and load conditions) using mathematical approximation [3]. By using simple and interacting elements, a finite number of unknowns can be used to approximate the infinite number of real systems. In recent years, finite element analysis is used more and more widely in engineering analysis because of its high computational accuracy and adaptability to complex shapes. In this paper, the finite element analysis method is used to simulate the temperature and gas flow inside the safety helmet, which provides a theoretical basis for the improved design of safety helmets.

© Springer Nature Singapore Pte Ltd. 2017
M. Fei et al. (Eds.): LSMS/ICSEE 2017, Part I, CCIS 761, pp. 371–379, 2017.
DOI: 10.1007/978-981-10-6370-1_37

2   Theoretical Model

2.1   Thermodynamic Model

Thermodynamic analysis is used to research the thermal response of structures under thermal loads, the theoretical basis of this paper is the heat conduction equation based on Fourier's law. In general, the key to the thermodynamic analysis is the temperature and the rate of heat flow of the structure, although the heat flux. According to the laws of physics, it is known that the general nonlinear thermal balance matrix equation is as follows [4]:

$$[C(T)]\{T'\} + [K(T)]\{T\} = [Q(T)] \tag{1}$$

where $\{T\}$ is the node temperature vector, $[C]$ is the specific heat matrix(heat capacity), considering the increase of internal energy, $[K]$ is the conduction matrix, including thermal conductivity, convection coefficient, emissivity and form factor, T' is the derivative of the temperature to time.

Transient heat transfer refers to the heating or cooling of a system. In this process, the temperature, heat flux, thermal boundary condition and internal energy of the system have obvious changes over time. According to the principle of energy conservation, the transient heat balance can be expressed as follows [5]:

$$[C]\{T'\} + [K]\{T\} = \{Q\} \tag{2}$$

where $\{Q\}$ is the node heat flux vector, including heat generation.

If the net heat flow rate of the system is 0, that is, the amount of heat flowing into the system and the heat generated by the system is equal to the heat of the system. $q_{inflow} + q_{generate} - q_{flowout} = 0$, the system is in steady state, the temperature of any node doesn't change with time in steady state thermal analysis. The energy balance equation of steady state thermal analysis is as follows:

$$[K]\{T\} = \{Q\} \tag{3}$$

In the steady state thermal analysis, all items related to time are not considered(certainly nonlinear phenomena is possible). The thermal analysis in this paper is based on the heat conduction equation, internal heat flow is the basis of $[K]$, the heat flux, rate of heat flow and convection are considered as boundary conditions in $\{Q\}$.

2.2   Heat Transfer Model

To solve the temperature field, the heat transfer model should be determined. The heat transfer inside the safety helmet is mainly heat conduction, heat convection, which mainly heat convection [6].

When there is a temperature difference inside the safety helmet, the heat is transferred from the high temperature part to the low temperature part. When the head is in contact with the helmet and there is a temperature difference, the heat is transferred from the higher temperature head to the lower temperature safety helmet, this way of heat transfer is heat conduction. The heat transfer follows the Fourier law as follows:

$$q'' = -k\frac{dT}{dx} \tag{4}$$

where q is the heat flux density and its unit is W/ m^2, k is the thermal conductivity and its unit is W/(m$^\circ$C).

Convection refers to the heat transfer caused by the relative motion of the fluid at different temperatures. The temperature of the head is higher, and the air near the surface is expanded by heat, upward flow. The cold air near the inner surface of the helmet will fall to replace the original heated air and cause convection. Thermal convection is divided into two types of natural convection and forced convection and satisfies the Newtonian cooling equation as follows:

$$q'' = h(T_s - T_b) \tag{5}$$

where h is the convective heat transfer coefficient (or membrane coefficient), T_s is the solid surface temperature, T_b is the ambient fluid temperature.

2.3   Gas Flow Model

The gas flow inside the safety helmet is low-speed flow, so the gas can be treated as an incompressible fluid, and the gas temperature does not change much, that is, the density does not change much, so it can be considered that the gas flow is in accordance with the Bonssinesq hypothesis [7]. This assumption includes: (1) The viscous dissipation in the fluid is negligible. (2) Other physical properties except density are constants. For density, only the terms related to the volume force in the momentum equation are considered, the density of the rest is also constant. Air viscosity cannot be ignored, that is, gas flow is usually turbulent flow, so the physical model of gas flow can be simplified as follows [8, 9]:

1) Normal temperature, low speed, incompressible fluid flow.
2) Consistent with the Boussinesq hypothesis.
3) Turbulence.
4) Steady and unsteady state.

The mathematical model of gas flow and heat transfer inside the safety helmets is as follows:

Continuity equation:

$$\frac{\partial u}{\partial x} + \frac{\partial v}{\partial y} + \frac{\partial w}{\partial z} = 0 \tag{6}$$

Momentum equation:

$$\rho \frac{dv}{dt} = \rho F - \text{grad}p + \mu \ \nabla^2 v \tag{7}$$

Energy equation:

$$\frac{\partial}{\partial t}(\rho E) + \frac{\partial}{\partial x_i}\left[u_i(\rho E + p)\right] = \frac{\partial}{\partial x_i}\left[k_{\text{eff}}\frac{\partial T}{\partial x_i} - \sum_{j'}h_{i'}J_{i'} + u_j(\tau_{ij})_{\text{eff}}\right] + S_h \tag{8}$$

where $E = h - \frac{p}{\rho} + \frac{u_i^2}{2}$; $k_{\text{eff}} = k + k_t$, k_t is the turbulence heat transfer coefficient which is defined according to the turbulence model.

Select the hot gas inlet from the top of the head into the safety helmet. The hot gas outlets are the vents in front of the safety helmet and on both sides of the top. Select the standard k-e turbulence model, turbulence kinetic energy equation k is as follows:

$$\frac{\partial}{\partial t}\rho k + \frac{\partial}{\partial x_i}\left(\rho k \mu_i\right) = \frac{\partial}{\partial x_j}\left[\left(\mu + \frac{\mu_t}{\sigma_k}\right)\frac{\partial k}{\partial x_j}\right] + G_k + G_b - \rho\epsilon - Y_M + S_k \tag{9}$$

Diffusion equation e is as follows:

$$\frac{\partial}{\partial t}\rho\epsilon + \frac{\partial}{\partial x_i}\left(\rho\epsilon\mu_i\right) = \frac{\partial}{\partial x_j}\left[\left(\mu + \frac{\mu_t}{\sigma_e}\right)\frac{\partial\epsilon}{\partial x_j}\right] + C_{1\epsilon}\frac{\epsilon}{k}\left(G_k + C_{3\epsilon}G_b\right) - C_{2\epsilon}\rho\frac{\epsilon^2}{k} + S_\epsilon \tag{10}$$

where G_k in the equation represents the turbulent kinetic energy generated by the laminar velocity gradient and G_b is the turbulent kinetic energy generated by buoyancy.

3   Results and Discussion

Establish a safety helmet model according to the shape and size of the real helmet in Fig. 1. As the safety helmet is curved and irregular shape, it is difficult to model in the ANSYS classic interface or ANSYS-workbench, so this paper chose to import ANSYS-workbench after modeling in SolidWorks, This will not only ensure the accuracy of the

Fig. 1. Real safety helmet

model, but also ensure the accuracy of the results of the analysis, to achieve the purpose of saving time and reducing workload, and the results are also very intuitive, the modeling results are shown in Fig. 2. The safety helmet shell is hemispherical, including top reinforcement, peak, brim and vent, due to the headband, chins trap and other accessories have little effect, it is not for analysis.

Fig. 2. Model of safety helmet

Importing the model into ANSYS-workbench, the boundary conditions and geometrical features of the safety helmet finite element should be separated from the grid meshing to make the model easier to generate. After the model is completed, it is not involved in the finite element analysis, the model must be meshed, all the temperature and gas flow inside the safety helmet must be transferred to the finite element model. The meshing result of the model is shown in Fig. 3, the boundary of the grid is 2 mm, the total number of units is 175553, the total number of nodes is 276023, the grid quality reaches 0.77 indicates the higher accuracy.

Fig. 3. Mesh of the model

Safety helmet shell material can be made of a wide range of synthetic resin materials. In this paper, high molecular polyethylene is chosen as the simulation material with a specific heat of 2300 J/ (kg °C). Set head temperature 45°C, wind speed 0.85 m/s, relative humidity 50%, continuous working time 30 min. The fluid material is gas with a density of 1.293 g/L, a viscosity of 1.7894e-5 kg/ m-s, the cooling factor is selected for hysteresis (11°C-31°C), convective heat transfer coefficient is 5.0 W/(m²°C).

Figure 4 is the result of temperature distribution of steady-state thermal analysis. According to the analysis results, the internal temperature of the safety helmet gradually decreases from bottom to top, the top temperature is about 37.6°C, close to the ambient temperature.

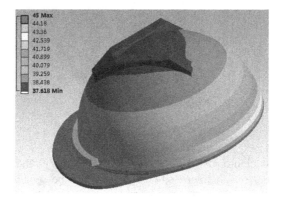

Fig. 4. Temperature distribution nephogram of steady state thermal analysis

Figure 5 is the heat flow distribution of the steady-state thermal analysis. Consistent with the temperature distribution, the heat flow of the safety helmet decreases gradually from bottom to top. The bottom heat flow is about 195 W/m², and the top heat flow is about 14 W/m², indicating that the heat transfer rate is faster where the safety helmet is in contact with the head.

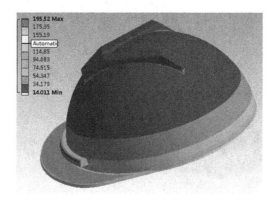

Fig. 5. Heat flow distribution nephogram of steady state thermal analysis

Transient thermal analysis of the safety helmet, as shown in Fig. 6. In 5 s the temperature of the top of the safety helmet is increased by about $0.06\,^{\circ}$C, indicating that the heat transfer speed is very fast inside the safety helmet. The temperature variation in 5 s is shown in Fig. 7.

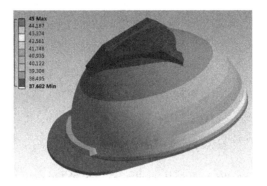

Fig. 6. Temperature distribution nephogram of transient state thermal analysis

Time [s]	✔ Minimum [°C]	✔ Maximum [°C]
0.25	37.627	45.
0.68373	37.637	45.
1.1837	37.646	45.
1.6837	37.653	45.
2.1837	37.659	45.
2.6837	37.664	45.
3.1837	37.668	45.
3.6837	37.672	45.
4.1837	37.676	45.
4.6837	37.68	45.
5.	37.682	45.

Fig. 7. Temperature change in 5 s of transient state thermal analysis

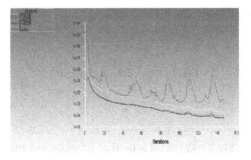

Fig. 8. Residual curve after convergence

The fluid is calculated under the number of iterations 500, its residual curve after convergence is shown in Fig. 8, the results are shown in Fig. 9.

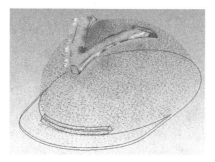

Fig. 9. Simulation results

According to the results in Fig. 9, as the hot gas of the head is rising gas, so the density of the gas near the vent on both sides of the top of the safety helmet is higher, the gas flow on both sides of the top and in front is faster while most of the gas is concentrated in the area between the safety helmet and head which is difficult to discharge.

According to the gas flow cloud diagram of Fig. 10, the hot gas flow in the center of the vent is faster while in the boundary position the flow boundary has a certain flow resistance to the hot gas, so the hot gas flow slows down. The resistance is determined by the shape and length of the object and the roughness of the surface of the object.

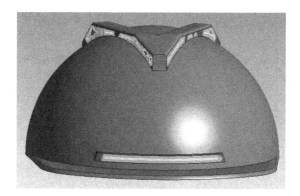

Fig. 10. Gas flow nephogram

4   Conclusion

In this paper, the finite element method is used to simulate and analyze the temperature field and gas flow field in the safety helmet. Analysis results show as below:

(1) The temperature filed and air flow simulation were successfully analyzed by ANSYS finite element analysis, which can simulate the usage of helmets.

(2) The finite element analysis results show that the temperature inside the safety helmet decreases gradually from bottom to top, the heat transfer rate of the contact part with the head is faster, only a small amount of hot gas can be vented from the vent while most of the hot gas is still concentrated inside the safety helmet difficult to vent.

(3) Although the vent of ordinary safety helmets can speed up the gas flow inside the safety helmet, it is still difficult to meet the needs of the timely release of hot gas from head. It makes the workers affected by region, temperature, place, working hours and working intensity in the course of work, form a lot of sweat in the head.

(4) At present, most of the sweatband in safety helmet is mostly made of sponge and cloth cover which have low sweat absorption rate and poor air permeability, it can't fully absorb the sweat on forehead and there are serious security risks. In this paper, the simulation of the temperature and gas flow inside the safety helmet provides a solid theoretical basis for the improved design of the safety helmets.

Acknowledgements. This work is financially supported by 2017 Zhejiang Electric Power Company Science and Technology Project " Research and Application of Conductive Sweat Type Safety Helmet".

References

1. Zang, L., Xiao, Y., Yang, W.:Development trend of helmet at home and abroad. China Pers. Protective Equipment (5), pp. 17–18 (2004)
2. Yu, Y., Chen, H.: Current situation and development trend of safety helmet industry in China. Mod. Occup. Saf. (109), pp. 31–33 (2010)
3. Chen, Q.: Current situation and future development trend of safety helmet industry in China[J].Occup. Saf. Health (5), pp. 47–49 (2013)
4. Ju, X.: Application of finite element analysis in the design of safety helmet. J. Saf. Sci. Technol. **8**(7), 143–147 (2012)
5. Wang, Q., Zuo, F., Hu, R.: ANSYS10.0 Advanced Application Examples of Mechanical Design. Machinery industry press, Beijing (2006)
6. Jebelli, H., Ahn, C., Stentz, T., et al.: Fall risk analysis of construction workers using inertial measurement units: validating the usefulness of the postural stability metrics in construction. Saf. Sci. **84**, 161–170 (2016)
7. Martin, J.B., Rivas, T., Matias, J.M., et al.: A Bayesian network analysis of workplace accidents caused by falls from a height. Saf. Sci. **47**(2), 206–214 (2009)
8. Sa, J., Seo, D., Choi, S.D., et al.: Comparison of risk factors for falls from height between commercial and residential roofers. J. Saf. Res. **40**(1), 1–6 (2009)
9. Forward, S.: The theory of planned behavior: the role of descriptive norms and past behavior in the prediction of drivers' intentions to violate. Transp. Res. Part F-Traffic Psychol. Behav. **12**, 198–207 (2009)

Analysis of Influence of Moving Axial Load on Elevated Box Bridge of Slab Track

Xiaoyun Zhang, Guangtian Shi[⊠], Xiaoan Zhang, and Yanliang Cui

School of Mechanical Engineering, Lanzhou Jiaotong University,
Lanzhou 730070, China
zxa_lzjtu@163.com

Abstract. Due to the increasingly usage of the heavy-load train, the effect of the moving axial load on the noise radiation of elevated bridge structure should be concerned. In this paper, high speed train-track-bridge coupled model is established, and the train axle load, which is taken as the boundary condition of the load, is applied to the finite element model of elevated box bridge structure to calculate the vibration response of the surface of a box bridge. In this model, the vibration response is taken as the acoustic boundary conditions and is added to the boundary element model of elevated box bridge structure to study its sound radiation. The results show that the plate-shell unit can well reflect the overall and local vibration characteristics of the bridge structure, and under the effect of moving axial load, the vibration frequencies of box bridge structure concentrate in 0–300 Hz and the main peaks are in 10–160 Hz; the finite element-boundary element method can effectively analyze the low frequency noise radiation of box bridge caused by the moving axial load; the most of structure noise induced by moving axial load is below the audible range of which the noise is greatly harmful to human body, thus it must be taken seriously.

Keywords: High speed railway · Elevated box bridge · Moving axial load · Finite element-boundary element method · Sound radiation

1 Introduction

For dynamic interaction of railway system and structure noise, in early researches, Zhai W.M. et al. analyzed the dynamic mechanism between the locomotive and the track structure of the high-speed trains through the bridge. On this basis, train-track-bridge coupled model was established, and then being validated and analyzed based on the field measurements of Qinhuangdao-Shenyang Passenger Dedicated Line. The dynamic characteristics of the high-speed trains passing through the bridge at different speeds are characterized by the compiled program TTBSIM [1–3]. He Z.X. et al. studied the ground vibration under the simple harmonic loads by semi-analytical method and analyzed the energy transfer characteristics between the track, the foundation and the effect of the simple harmonic load on the vibration attenuation. The ground vibration caused by the train axial load and its propagation characteristics and the vibration damping effect of the isolation groove are simulated and analyzed [4, 5].

© Springer Nature Singapore Pte Ltd. 2017
M. Fei et al. (Eds.): LSMS/ICSEE 2017, Part I, CCIS 761, pp. 380–389, 2017.
DOI: 10.1007/978-981-10-6370-1_38

Thompson et al. made extensive researches on the vibration and control of railway noise and developed the corresponding software to guide the low noise and vibration design of train-track systems and noise reduction of the existing tracks [6–8]. Li X.Z. proposed a method for full-band noise prediction of high-speed railway bridge structures, a span of 32 m double-line simply supported concrete box bridge are regarded as the case study, and the field test and simulation analysis [9–11]. Augusztinovicz F. et al. studied the structural noise of orthotropic box-bridge [12]. Ouelaa N. et al. calculated the acoustic pressure of the field, in which the three-span continuous bridge transient acceleration as the monopole excitation source is used by considering the coupling effect of vehicle-bridge and track irregularities and, taking only two degrees of freedom into account for each train [13]. Song et al. established the vertical coupling dynamic model of vehicle-track-elevated structure, and obtaining the force between the wheel-rails as the input of the statistical energy analysis to research on the vibration and acoustic radiation of elevated structure [14]. Ngai K.W. et al. proposed a vibration and noise test on a concrete box bridge of an elevated railway in Hong Kong. The results show that the noise and vibration frequency range of the elevated bridge is from 20 Hz to 157 Hz when the train runs at 140 km/h [15]. Li Qi et al. used the modal superposition method to calculate the train-track-bridge coupling dynamic of the urban rail transit trough girders, and calculating the structural noise of trough beam using the modal acoustic transmission vector and the modal coordinate response of the beam [16].

In this paper, the train axle load, taken as the load boundary conditions, is applied to the finite element model of elevated box bridge structure to calculate the vibration response of the surface of elevated box bridge. The vibration response of the box bridge structure is used as the acoustic boundary condition, and the noise radiation of elevated box bridge structure is analyzed by the Helmholtz boundary integral equation.

2   Noise Prediction Model of Elevated Box Bridge Under Axial Load

2.1   Train Axle Load

Generally, to the vehicle-track-bridge coupled system, the excitation sources of vibration mainly consist of two parts: the axle load and the dynamic loads induced by the track irregularities. When a train moves on the elevated bridge, the effects of the axle load on the vibrations of the bridge cannot be ignored because of the large size of the axle loads. It is therefore necessary to research the structural noise of the bridge originated from the axle loads. In this paper, the wheel-rail force is simplified as a series of axle loads while ignoring the influence of track irregularities, as shown in Fig. 1.

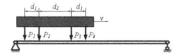

Fig. 1. The model of moving axle load on train

Assume the velocity of train is v and the wheel/rail vertical forces are p_1, $p_2...p_n$, (n is the number of the wheels). The coordinate a_i ($i = 1, 2, ...n$) of each wheel is determined by d_1 and d_2 in Fig. 1. In the train operation, the force on track structure can be expressed as using Dirac-delta function (1).

$$P(t) = \sum_{j=1}^{N_w} \delta(x - vt - \alpha_j)p_j e^{\frac{2\pi vt_j}{\lambda}} \tag{1}$$

2.2   The Equations of Motion and Solution

Because the axle loads directly act on the rail beams, the differential equation of motion of the rail can be described as

$$E_r I_r \frac{\partial^4 W_r(x,t)}{\partial x^4} + M_r \frac{\partial^2 W_r(x,t)}{\partial t^2} + C_{rs}[\frac{\partial W_r(x,t)}{\partial t} - \frac{\partial W_s(x,t)}{\partial t}] + K_{rs}[W_r(x,t) - W_s(x,t)]$$
$$= -\sum_{i=1}^{N_s} F_{rVi}(t)\delta(x - x_{Fi}) + \sum_{j=1}^{N_w} \delta(x - ct - \alpha_j)p_j e^{\frac{2\pi ct_j}{\lambda}} \tag{2}$$

Where, E_r and I_r denote the elasticity modulus and inertia moment of rail, respectively. The bending rigidity can be expressed as the product of elasticity modulus and inertia moment. M_r is the quality of each unit length of the rail. W_r, W_s denote the rail vertical displacement and the track vertical displacement. C_{rs}, K_{rs} denote the damping coefficient and stiffness coefficient connecting the rail and the track. N_s and N_w are the total number of the fastener and the wheel axle in the rails, respectively. F_{rVi} is the vertical reaction force of the ith supporting point of the rail.

The differential equation of motion of the track slab can be described as:

$$\frac{\partial^4 W_s(x,y,t)}{\partial x^4} + 2\frac{\partial^4 W_s(x,y,t)}{\partial x^2 \partial y^2} + \frac{\partial^2 W_s(x,y,t)}{\partial y^4} + \frac{C_s}{D_s}\frac{\partial W_s(x,y,t)}{\partial t} + \frac{\rho_s h_s}{D_s}\frac{\partial^2 W_s(x,y,t)}{\partial t^2}$$
$$= \frac{1}{D_s}[\sum_{i=1}^{N_p} P_{rVi}(t)\delta(x - x_{Pi})\delta(y - y_{Pi}) - \sum_{j=1}^{N_F} F_{sVj}(t)\delta(x - x_{Fj})\delta(y - y_{Fj})] \tag{3}$$

In the dynamic analysis of the bridge structures, the bridge system is divided as an assembly of shell element, and then selecting the shape functions to calculate the kinetic and potential energy of each unit, and consequently the entire structure. Finally, the dynamic equations of motion of the whole system can be derived by the Hamilton principle [1], which is

$$[M_b]\{\ddot{u}_b\} + [C_b]\{\dot{u}_b\} + [K_b]\{u_b\} = \sum_{j=1}^{N_F} F_{sVj}(t)\delta(x - x_{Fj})\delta(y - y_{Fj}) \tag{4}$$

Where, $[M_b]$, $[C_b]$ and $[K_b]$ denote the mass matrix, damping matrix and stiffness matrix of the bridge; $\{u_b\}$, $\{\dot{u}_b\}$ and $\{\ddot{u}_b\}$ denote the generalized vectors of displacement, velocity and acceleration of the bridge, respectively.

The Newmark method is used to calculate the dynamic response of the bridge by satisfying the following integral scheme.

$$\begin{cases} \{u_{n+1}\} = \{u_n\} + \{\dot{u}_n\}\Delta t + [(1/2 - \alpha)\{\ddot{u}_n\} + \alpha\{\ddot{u}_{n+1}\}]\Delta t^2 \\ \{\dot{u}_{n+1}\} = \{\dot{u}_n\} + [(1 - \beta)\{\ddot{u}_n\} + \beta\{\ddot{u}_{n+1}\}]\Delta t \end{cases} \quad (5)$$

The integral form is unconditionally stable if taking $\alpha = 1/4$, $\beta = 1/2$. The time interval of the integration is 0.01 s. Use u_{n+1} to express $\dot{u}_{n+1}$ and $\ddot{u}_{n+1}$ in above formulas, and substitute them into the dynamic equations of the bridge system, that is,

$$[M_b]\{\ddot{u}_{b,n+1}\} + [C_b]\{\dot{u}_{b,n+1}\} + [K_b]\{u_{b,n+1}\} = \{F_{b,n+1}(t)\} \quad (6)$$

Therefore, the following equation can be obtained by resorting the above dynamic equation.

$$(a_0[M_b] + a_1[C_b] + [K_b])\{u_{b,n+1}\} = \\ \{F_{b,n+1}(t)\} + [M_b](a_0\{u_{b,n}\} + a_2\{\dot{u}_{b,n}\} + a_3\{\ddot{u}_{b,n}\}) + [C_b](a_1\{u_{b,n}\} + a_4\{\dot{u}_{b,n}\} + a_5\{\ddot{u}_{b,n}\}) \quad (7)$$

With, $a_0 = \frac{1}{\alpha\Delta t^2}$, $a_1 = \frac{\beta}{\alpha\Delta t}$, $a_2 = \frac{1}{\alpha\Delta t}$, $a_3 = \frac{1}{2\alpha} - 1$, $a_4 = \frac{\beta}{\alpha} - 1$, $a_5 = \frac{\Delta t}{2}\left(\frac{\beta}{\alpha} - 2\right)$.

The following formulas can be obtained by (5) and (7).

$$\begin{cases} \{\dot{u}_{b,n+1}\} = \{\dot{u}_{b,n}\} + a_6\{\ddot{u}_{b,n}\} + a_7\{\ddot{u}_{b,n+1}\} \\ \{\ddot{u}_{b,n+1}\} = a_0(\{u_{b,n+1}\} - \{u_{b,n}\}) - a_2\{\dot{u}_{b,n}\} - a_3\{\ddot{u}_{b,n}\} \end{cases} \quad (8)$$

With, $a_6 = \Delta t(1 - \beta)$, $a_7 = \beta\Delta t$.

Therefore, the displacement, velocity and acceleration of the bridge system can be obtained.

2.3   Sound Field Solution

Because the dynamic responses of the box bridge are obtained in time domain, therefore, it is a necessity to transform the time-domain results into the frequency domain. Using the discrete Fourier transform, the responses of continuous frequency points can be obtained by the vibration responses of the box bridge structure in time domain. Then, we perform filtering on the results, that is, the time histories of the response can be obtained by the inversely discrete Fourier transform, and then calculating the root mean square (RMS) in every 1/3 octave band range as the outputs of the vibration response at 1/3 octave center frequency. The dynamic response of a node in directions toward different degrees of freedom can be projected to the normal direction of the structure, which is treated as the vibration boundary conditions of the structural element.

In the acoustic radiation of the closed structure in the fluid medium, the acoustic wave equation: $\nabla^2 \vec{p} + k^2 \vec{p} = 0$, the Neumann boundary condition of the fluid-solid interface: $\frac{\partial \vec{p}}{\partial \bar{n}} = -i\rho_0 \omega \vec{v}_n$,   and   the   Sommerfeld   radiation   condition: $\lim\limits_{r \to \infty} \left[r \left(\frac{\partial \vec{p}}{\partial r} - ik\vec{p} \right) \right] = 0$ must be satisfied. ∇, $\vec{p}$, ω, c, and ρ_0 represent the Laplacian operator, sound pressure, circular frequency, sound velocity and fluid density, respectively. i is the imaginary unit. k stands for wave number, $k = \omega/c$. $\vec{v}_n$ is the normal velocity vector on the interface between the fluid and structure. r is distance between any point X on the structure surface Y, $r = |\vec{X} - \vec{Y}|$. A is the surface area of the structure.

The Helmholtz integral equation can be obtained by the free space Green's formula,

$$C(Y)\vec{p}(Y) = \int_A \left(-i\omega\rho_0 \frac{e^{-ikr}}{4\pi r} \vec{v}_n(X) + \frac{e^{-ikr}}{4\pi r} \left(ik + \frac{1}{r} \right) \cos \beta \vec{p}(X) \right) dA(X) \quad (9)$$

With, $C(Y) = \begin{cases} 1, Y \in D \\ 1 - \int\limits_A \frac{\cos\beta}{4\pi r^2} dA, Y \in A \\ 0, Y \notin (D \cup A) \end{cases}$

Where, β is the angle between normal vector of X point on the structure surface and radius vector r, D is the fluid domain.

There are M elements and N nodes on the boundary by dividing the surface A of the vibrating structure. L is the nodes of each element. Suppose that the local coordinates of any point on the unit (x, y, z) is (ξ, η). We get,

$$\begin{cases} \vec{p}(x, y, z) = \sum\limits_{i=1}^{L} N_l(\xi, \eta)\vec{p}_l \\ \vec{v}_n(x, y, z) = \sum\limits_{i=1}^{L} N_l(\xi, \eta)\vec{v}_{nl} \end{cases} \quad (10)$$

Where, $N_l(\xi, \eta)$ is interpolating shape function. Each node on the boundary is used as the source in turn, the formula of (11) can be obtained by discretizing the Helmholtz integral equation.

$$Q\vec{p} = P\vec{v}_n \quad (11)$$

Q and P are all symmetric plural full rank matrix, which are functions of excitation frequency and are related to the surface shape, size and interpolating shape function of the structure. $\vec{p}$ and $\vec{v}_n$ are column vectors at N dimensions.

$$\vec{p} = Z\vec{v}_n \quad (12)$$

Where, $Z = Q^{-1}P$, Z is the vibration structure acoustic impedance matrix, any unit z_{ij} represents the contribution of the unit velocity at the node j to the sound pressure at the node i.

With the acquisition of $\vec{p}$ and $\vec{v}_n$, the radiated sound pressure at any point in the sound field can be obtained by the Helmholtz integral equation $(Y \in D)$,

$$\vec{p}(Y) = q^T \vec{p} + k^T \vec{v}_n \tag{13}$$

Where, q and k are interpolation shape function column vectors related to the shape of structure surface as well as the location of Y, determined by Eq. (9).

For the non-closed elevated box bridge structure, the sound pressure at any point in the sound field can be obtained by calculating the difference between Helmholtz integral equations on both sides of the boundary surface,

$$\vec{p}(Y) = \int_A \left(\frac{e^{-ikr}}{4\pi r} \Delta \vec{v}(X) + \frac{e^{-ikr}}{4\pi r} \left(ik + \frac{1}{r} \right) \cos \beta \Delta \vec{p}(X) \right) dA(X) \tag{14}$$

Where, $\Delta \vec{v}(X)$ is the speed difference of X point on both sides of the surface, $\Delta \vec{p}(X)$ is the sound pressure difference of X point on both sides of the surface.

Being respectively expressed as

$$\begin{cases} \Delta \vec{v}(X) = -i\rho_0 \omega (\vec{v}_{n1}(X) - \vec{v}_{n2}(X)) \\ \Delta \vec{p}(X) = \vec{p}_1(X) - \vec{p}_2(X) \end{cases} \tag{15}$$

Where, $\vec{p}_1(X)$ and $\vec{p}_2(X)$ are pressure of X point on both sides of the structural surface; $\vec{v}_{n1}(X)$ and $\vec{v}_{n2}(X)$ indicate normal vibration speed of X point on both sides of the structure surface.

When the structural surface is separated by the boundary elements, the difference of the velocity and the sound pressure of the nodes can be expressed by

$$\begin{Bmatrix} \Delta \vec{v}(X) \\ \Delta \vec{p}(X) \end{Bmatrix} = C^{-1} F \tag{16}$$

Where, C is a symmetric-complex full rank matrix which is related to the shape, size and interpolated shape of the structural surface and is a function of the exciting frequency. F is external excitation vector depending on the structural surface vibration velocity.

Sound pressure of any point outside the structural surface $\vec{p}$ can be obtained by (14).

$$\vec{p}(Y) = B \begin{Bmatrix} \Delta \vec{v}(X) \\ \Delta \vec{p}(X) \end{Bmatrix} = BC^{-1} F \tag{17}$$

Where, $\vec{p}(Y)$ is the sound pressure of any observation point, B is the interpolation matrix determined by Eq. (14).

3 Dynamic Analysis of the Box Girder Bridge

When a train moves through the elevated box girder bridge at a velocity of 200 km/h, the train axial load is 191.1kN, corresponding to p_1–p_4 in Fig. 1, the parameters used to confirm the wheel location are d_1 = 2.5 m, d_2 = 17.37 m, respectively, as shown in Fig. 1. The simply supported concrete single box single cell elevated box bridge, which is commonly used in high-speed railway, is adopted for the case studies. The total length of the track slab is 32 m and the thickness of the slab is 3 m, the thickness of the deck plate, the web and the bottom plate is 0.315 m, 0.480 m and 0.300 m, respectively. The bridge uses CRTS type I double-track slab ballastless structure; the length, width and thickness of track plate is 4.93 m, 2.4 m and 0.2 m, respectively, the rail type is 60 kg/m. other specific parameters are shown in Table 1.

Table 1. Track and bridge dynamics parameters

Dynamic parameters	Parameter values	Dynamic parameters	Parameter values
Elastic modulus of rail/(N·m^{-2})	2.1×10^{11}	Adjust the layer stiffness under the track plate/(N·m^{-1})	9.375×10^9
Inverse moment of rail/(m^4)	3.215×10^{-5}	Adjust the layer damping under the track plate/(N·s·m^{-1})	7.5×10^5
Poisson's ratio of rail	0.3	Elastic modulus of base/(N·m^{-2})	3.3×10^{10}
Rail line density/(kg·m^{-1})	60.64	Poisson ratio of Base	0.2
Rigidity of cushion under the track/(N·m^{-1})	4×10^7	Density of Base/(kg·m^{-3})	2500
Damping of cushion under the track/(N·s·m^{-1})	2.2656×10^4	Elastic modulus of bridge/ (N·m^{-2})	3.8×10^{10}
Elastic modulus of track plate/(N·m^{-2})	3.6×10^{10}	Poisson ratio of bridge	0.25
Poisson ratio of track plate	0.2	Density of bridge/(kg·m^{-3})	2500
Track plate density/(kg·m^{-3})	2500		

Figure 2 shows the time-varying acceleration and spectrum of the various parts of the elevated box bridge across the center.

It can be observed from Fig. 2 that the vibration of the deck plate is the largest compared to the web and the bottom plate. From the spectrum shown in Fig. 2(b), the vibration frequency is mainly concentrated at the frequency range of 0–300 Hz, most of the peaks appear at 10–160 Hz in the low frequency band.

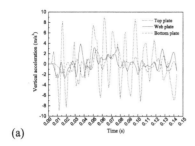
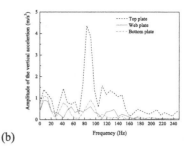

(a)   (b)

Fig. 2. Acceleration response of bridge structure vibration at mid span (a. Vertical acceleration time histogram, b. Vertical acceleration spectrum)

4 Sound Radiation Analysis

The dynamic responses of the box bridge structures are used as the acoustic boundary condition to calculate the influence of the box bridge structure on the sound field of the surrounding space. Assuming that the height of bridge pier is 20 m, the field points are selected, as shown in Fig. 4(a). This paper uses linear weighted sound pressure levels in the analytical process.

Both Figs. 3 and 4(b)(c)(d) show the two-dimensional sound field distribution and the sound pressure level (SPL) of the selected field points. These figures show that the elevated box bridge under the train moving axial load can produce large structural noise, which is more intensive at the upper and bottom of the structures. As the train runs on the single line, the sound field distribution on both sides of the bridge is not symmetrical. When the frequency is low, the directivity of the sound field is strong, and the radiation noise decreases with the increase of the frequency. The SPLs of the field points of the bottom of the bridge is the largest, which increases with the increase of the ground distance; when the distance is closer to the box bridge structure at the center frequency of 80–100 Hz, the SPLs is obviously higher than other field points; The SPLs of the transverse field points decreases with the increase of the distance, and the SPL of the near field points of the box bridge is slightly higher than that of the bottom. For the far field region in the vertical direction, the structural noise changes are more complicated.

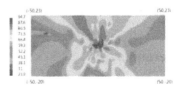

Fig. 3. Two-dimensional sound field distribution cloud at mid span (10 Hz-left, 118 Hz-right)

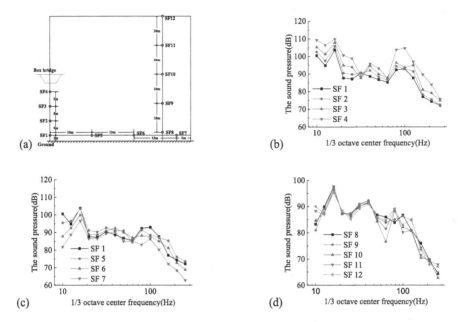

Fig. 4. Field points velocity SPL spectrum curve at 200 km/h (a. Field points distribution, b. SPL at the bottom, c. SPL of transverse, d. SPL in far-field)

5   Conclusions

(1)   The plate-shell element can reflect the whole and local vibration characteristics of the box bridge well, and under the effect of moving axial load, the vibration frequencies of box bridge structure concentrate in 0–300 Hz and the main peaks are in 10–160 Hz.

(2)   The finite element-boundary element method can effectively calculate the low frequency noise radiation of box bridge induced by the moving axial load; most of the structural noise induced by moving axial load is below the audible range that is greatly harmful to human body, therefore, it must be dealt with seriously;

Acknowledgment.   This research was supported by the Open Project of State Key Laboratory of Traction Power, Southwest Jiaotong University (TPL1604); Gansu Province Natural Science Foundation of China (1606RJZA014); Young Fund of Lanzhou Jiaotong University (2015025).

References

1.   Zhai, W.M., Xia, H.: Train-Track-Bridge Dynamic Interaction: Theory and Engineering Application. Science Press, Beijing (2011)
2.   Zhai, W.M., Cai, C.B., et al.: Mechanism and model of high-speed train-track-bridge dynamic interaction. China Civil Eng. J. **38**(11), 132–137 (2005)

3. Cai, C.B., Zhai, W.M., et al.: Dynamics simulation of interactions between high-speed train and slab track laid on bridge. China Railway Sci. **25**(5), 57–60 (2004)
4. He, Z.X., Zhai, W.M., Yang, X.W., et al.: Moving train axle-load induced ground vibration and mitigation. J. Rail Way Sci. Eng. **4**(5), 73–77 (2007)
5. He, Z.X., Zhai, W.M., Wang, S.Z., et al.: Semi-analytical study on ground vibration induced by railway traffics with axle loads. J. Vib. Shock **26**(12), 1–5 (2007)
6. Janssens, M.H.A., Thompson, D.J.: A calculation model for the noise from steel railway bridge. J. Sound Vib. **193**(1), 295–305 (1996)
7. Bewes, O.G., Thompson, D.J., Jones, C.J.C., et al.: Calculation of noise from railway bridge and viaducts: experimental validation of a rapid calculation mode. J. Sound Vib. **293**(3–5), 933–943 (2006)
8. Thompson D.J., Jones C.J.C., Bewes O.G.: Software for Predicting the Noise of Railway Bridge and Elevated Structures-Version 2.0. Southampton: Institute of Sound and Vibration Research (2005)
9. Li, X.Z., Zhang, X., Liu, Q.M., et al.: Prediction of structure-borne noise of high-speed railway bridges in whole frequency bands (Part I): theoretical model. J. China Railway Soc. **35**(1), 101–107 (2013)
10. Zhang, X., Li, X.Z., Liu, Q.M.: Prediction of structure-borne noise of high-speed railway bridges in whole frequency bands (Parts II): field test verification. J. China Railway Soc. **35**(2), 87–92 (2013)
11. Zhang, X., Li, X.Z., et al.: Sound radiation characteristics of 32 m simply supported concrete box girder applied in high-speed railway. J. China Railway Soc. **34**(7), 96–102 (2012)
12. Augusztinovicz, F., Marki, F., Nagy, A.B., et al.: Derivation of train track isolation requirement for a steel road bridge based on vibro-acoustic analyses. J. Sound Vib. **293**(3/4/5), 953–964 (2006)
13. Ouelaa, N., Rezaiguia, A., Laulagnet, B.: Vibro-acoustic modeling of a railway bridge crossed by a train. Appl. Acoust. **67**(5), 461–475 (2006)
14. Song, L.M., Sun, S.G.: Noise predicting based on viaduct structure of railway. J. Beijing Jiaotong Univ. **33**(4), 42–45 (2009)
15. Ngai, K.W., Ng, C.F.: Structure-borne noise and vibration of concrete box structure and rail viaduct. J. Sound Vib. **255**(2), 281–297 (2002)
16. Li, Q., Xu, Y.L., Wu, D.J.: Concrete bridge-borne low-frequency noise simulation based on train-track-bridge dynamic interaction. J. Sound Vib. **331**(10), 2457–2470 (2012)

Low-Carbon Architectural Design and Data Analysis Based on BIM

Xiaoxing Ou[✉], Qiming Li, and Dezhi Li

Department of Construction and Real Estate, Southeast University,
Nanjing 210096, People's Republic of China
{ouxiaoxing,njldz}@seu.edu.cn, njlqming@163.com

Abstract. Through an analysis of the design contents of different architectural disciplines under the requirements of sustainable and low-carbon development, this study analyzes the low-carbon architectural design process and the makeup of relevant information from the various disciplines using a building information modeling (BIM) system. Based on BIM, we have constructed a carbon emission budgeting platform that captures the whole building life cycle, and have set forth evaluation criteria for the quantitative analysis of low-carbon buildings. In light of the above research, evaluation and optimization of low-carbon building designs, as well as the subsequent reduction of carbon emissions during the life-cycle of newly constructed buildings, can be achieved using BIM.

Keywords: BIM · Low-Carbon building · Architectural design · Data analysis

1 Introduction

The reduction of greenhouse gas emissions (also commonly referred to as carbon emissions) to mitigate global warming has become an urgent necessity for human development. According to the previous research statistics, 45% of the world's energy is used for building heating, lighting, ventilation at the operation stage, while 16% of the water used for building construction [1]. Control of greenhouse gas emissions from buildings thus presents an effective means of curbing increasing emissions. As architectural design requirements determine virtually all the performance indicators of a building, including life cycle carbon emissions, there is an urgent need to implement building carbon emission regulations at the design stage.

Building information modeling (BIM) utilizes 3D architectural information modeling to manage tasks and information required throughout the lifecycle of a building including design, construction, etc. A BIM system is capable of computerized virtual construction, as well as rapid building analysis using the powerful data processing ability of computers. With the growing application of BIM and the increasing demand for low-carbon buildings, there has been relevant research on the joint usage of the two. Wong et al. (2012) proposed the use of BIM visual modeling to analyze carbon emissions and develop low-carbon buildings [2]. Li et al. (2011) used the carbon footprint evaluation criteria to classify carbon sources in the construction process, and proposed a carbon emission calculation scheme for the construction process based on BIM.

© Springer Nature Singapore Pte Ltd. 2017
M. Fei et al. (Eds.): LSMS/ICSEE 2017, Part I, CCIS 761, pp. 390–399, 2017.
DOI: 10.1007/978-981-10-6370-1_39

Wang et al. (2016) established an information management system of building materials with BIM as technical core [3].

Against the backdrop of a low-carbon economy, both government agencies and academia are beginning to investigate and explore the possibilities of low-carbon construction. Much research has been conducted on the design of green buildings and energy-efficient buildings, but the quantitative research, design analysis, and optimization of low-carbon building systems are still lacking. Green building and low-carbon building approaches each have their own areas of emphasis. A fast and smooth transition from the former approach toward the latter requires an in-depth study of the optimization of low-carbon building design, which could, in turn, provide a more effective guide for the architect's work. Within the BIM approach, and considering the diversity of needs among the different disciplines of architectural design, this study explored the key factors for successfully controlling the life cycle carbon emissions in buildings, from both the perspective of information requirements, as well as from an approach for generating carbon emissions calculation and analysis. In this way, the authors sought to arrive at a method for the production of successful, complete design schemes while reducing the carbon emissions of future buildings.

2   The Main Content of Sustainable Architectural Design

In response to changes in the Earth's environment and the depletion of resources, the concept of sustainable development was proposed in the 1970 s, and has since impacted various aspects of human life. In the field of architecture, energy saving buildings, ecological buildings, green buildings, low carbon buildings were put forward, for reducing the energy consumption and resource consumption of the building life cycle. Architectural design should not only consider the composition of space and external image, but also need to consider the requirements of energy conservation and emission reduction.

To guide and evaluate the construction of buildings under the requirements of sustainable development, the nations of the world have established rating systems for green buildings or energy-efficient buildings. The earlier systems include the Building Research Establishment Environmental Assessment Method (BREEAM), which takes into account both the energy usage of the main building as well as its ecological impact on the surrounding site [4]. and the Leadership in Energy and Environmental Design (LEED) standard of United States, which inspects the building comprehensively, in terms of site impacts, water usage, building energy conservation and atmosphere, resource and materials usage, and even indoor air quality to evaluate the impact of the building on the environment [5]. The Ministry of Housing and Urban-Rural Development of China issued the "Green Building Evaluation Standard" in 2006, which assesses building design and new construction in five impact areas: energy, land, water, materials, and environment. Through an analysis of the evaluation criteria for energy-efficient building design and green building design, we can see the main design principles of building sustainability for the various disciplines of architectural design, as shown in Table 1.

Table 1. Main tasks of energy-efficient building and green building design

Discipline	Green building design principles
Architecture	Use enclosure structure with low energy consumption, reduce decorative material consumption; integrate civil and decorative works; and incorporate removable partitions (walls) and their re-use
Structure	Optimize structural systems design; select prefabricated components from factories; undertake rational use of high-performance materials for the main structure; use building materials produced from waste materials
Drainage	Allowing comprehensive utilization of water resources; install water-metering devices; recycle cooling water; and use non-conventional water sources
Electricity and gas	Independently meter different sources of energy consumption; partition, time, and regulate illumination systems; employ a system for dynamic reactive power compensation and harmonic control of the electrical transmission and distribution system; and use renewable sources of energy
Intelligent design	Use monitoring system for indoor air quality, and equip spaces with carbon monitoring devices linked to the exhaust system
HVAC	Provide exhaust energy recovery; divide regulation and control of temperature into rooms (households); meter heat usage by household device (household heat sharing); utilize waste heat; provide cold and heat storage systems; use thermoelectric cooling; and use renewable sources of energy

3 BIM Design Process Analysis for Low-Carbon Buildings

3.1 Main Contents of Architectural Design

In every stage of a building's life, relevant requirements and work guides are set forth for its subsequent development. The tasks during the early stages of a building's life include planning, project establishment, feasibility assessment and architectural design—the most practical stage. Without BIM, messages are transmitted between stages by manually interpreting and translating information from and into blueprints or reports. In contrast, the practice of architecture in the information age uses BIM, which relies heavily on the construction of a building information model during the design phase. In addition, the initial building information model serves as the core model in BIM. Corresponding to the division of architectural design stages, the modeling of building information can be divided into three stages: scheme design, preliminary design, and architectural drawing design.

Scheme design is concerned with the planning of space, plane and vertical formations, as well as facade and environmental design for building groups and single units. It determines the building layout and function including the various entrances, traffic organization, fire prevention, and safety evacuation. Moreover, at this point in the design, the energy-saving measures are selected, and structure and equipment models for the building envelope are established. The scheme design model is mainly completed by architectural professionals, using BIM software such as Revit. During the modeling process, coordination between the architects and the structure and equipment professionals is needed. Considering the requirements of building space from architectural, structural, and construction equipment needs, the story height, as well as wall, door, and window placement within the overall structural plan are determined and used to complete a 3D model displaying the interior and exterior spaces of the building.

The preliminary design stage makes further advancements on the design scheme. This includes the analysis and calculation of building loads and required structural forms by the structural discipline in accordance with the architectural space established during the scheme design. This process determines of the size of the main structural components and the final structural layout. The equipment disciplines then compute the relevant equipment requirements such as water, heating and air-conditioning loads according to the requirements of building use, select the models for main equipment and pipelines, and determine their plane layout. Construction of the design model at this stage requires the participation of every discipline. The architectural discipline refines the model in Revit. Structural and equipment disciplines then use the same Revit model to add their own design components. The structural discipline can export the model to run structural and expect timely information has been included. Timely communication between the architect and the designers in each discipline facilitates the adjustment of the construction model to fulfill the requirements of structure and equipment layout. An adjustment of the 3D model is applied simultaneously to the plan, elevation, and section views of the building, and the adjusted building model can be made available on a public platform to provide updated relevant information for the design professionals of all other disciplines.

Using the traditional design method, the construction drawing design stage is in essence a complete refinement of the architectural design. The arrangement of main components and equipment completed in the preliminary design are assessed and included as a part of a detailed refinement of the plans. These construction drawings are used to produce blueprints for use during construction. However, using a BIM design method allows specific information on structural components and equipment to be added during the building system design and layout in the preliminary design stage. The structural designs for different parts of the building are also completed on the model during the preliminary design stage of BIM, and therefore, the preparation of construction drawings mainly requires the design analysis and adjustment of each discipline's work for optimization, rather than complete refinement for construction. The increased application of analysis software and the capability to deftly handle large amounts of information using computers have enabled BIM systems to compare different design schemes with increasing effectiveness, and to assist in the selection of the most desirable scheme. This in turn reduces the cost of construction and speeds up the entire design and construction process, while meeting the requirements of energy-savings and emissions reduction, and allowing better space usage both indoors and outdoors.

3.2   Information Structure of Low-Carbon Building Design

The design of a building is completed by the various different disciplines of architectural, structural and equipment design, each responsible for their own tasks as well as coordination with each other. Studies have shown that the carbon emissions of buildings can be divided into material consumption, construction activity consumption, building usage consumption, site afforestation, water consumption, etc., according to the source activities of the emissions. Low-carbon design for reduction of a specific class of carbon emissions is often achieved jointly by several disciplines. As labor is divided among the independent disciplines, it is necessary to study their individual information input into the low-carbon building design process. This input defines their design requirements in the cooperative BIM platform. Building energy consumption depends on various factors such as the type of building, construction standard, the environmental conditions, anticipated economic development, and lifestyle of the people using the building. An inspection of the contents of the design process yields the main tasks and related information provided by each discipline for a low-carbon design project, as shown in Fig. 1.

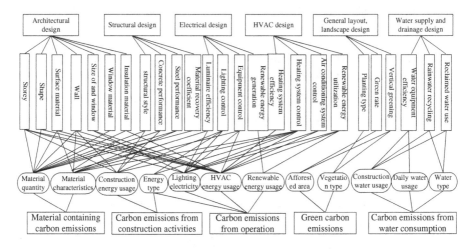

Fig. 1.   The contents of the design process for a low-carbon design project

4   Analysis of Low-Carbon Building Design in BIM

In traditional design and construction, calculation and analysis can be time-consuming because of the isolation of information and restriction of manual interpretation between disciplines, so less comparison, iteration and design optimization is conducted. The multi-dimensional information contained in BIM, coupled with the system's high data processing efficiency, overcomes the problem of time-consuming calculations. During the establishment of the BIM model, analysis and optimization can be performed to reduce cost and meet the requirements of energy conservation and green building before every discipline has refined its designs [6]. In this way, virtual building construction can be realized.

4.1   Using Analysis Software to Build a Carbon Emissions Budget Platform

The carbon emissions of buildings are usually computed with the carbon emission coefficient [7], which is expressed in Eq. 1.

$$C = AK \tag{1}$$

In Eq. 1, C is the carbon emissions, A is the data for an activity occurred, and K is the carbon emission coefficient corresponding to the activity, i.e., the carbon emissions of unit activity. This coefficient is the key factor describing the amount of carbon emissions that result from a given activity, and thus, it is designated as the "Emission Factor" on the IPCC greenhouse gas inventory. By knowing the number of activities generating carbon emissions and their corresponding carbon emission coefficients, the carbon emissions from these activities can be determined.

The carbon emission prediction for the whole lifecycle of the building, calculated with the design drawings, is an important basis for evaluating the low-carbon aspects of the building. Upon its completion, the design model is imported into other software and converted into a model for calculating specific system analysis functions such as building energy consumption, material consumption, and energy consumption of construction machinery and equipment. The Glodon GCL, GGJ, and GBQ software, for example, facilitates the calculation of the total quantity of all building materials, and the quantity of machinery needed during the construction and demolition stages. Through the use of analysis software such as Energy-plus, Ecotect, or Dest, the energy consumption during building usage can be calculated. In addition, any updates in materials required or used during the life of the building can be computed from the total quantity of materials and service life set in the measurement system. Figure 2 is a flow chart of building carbon emission calculations in BIM.

Fig. 2.   Flow chart of building carbon emission calculations in BIM

To aid the carbon emission budget calculations illustrated in Fig. 2, a user interface and database system could be established using the Microsoft Visual Studio software MFC (Microsoft Foundation Classes), on the Microsoft SQL Server database platform. Within the requirements of the carbon emission budget framework, carbon emission coefficients, material transport distances, material lifespans, and recovery rates could be incorporated to broaden the available data, allowing for the design of a budget platform accounting for building lifecycle carbon emissions that could make more complete predictions on future projects. Classification of carbon emissions on the basis of source can be used to derive the total carbon emissions for the whole building lifecycle, as well as in its different categories. These classification statistics serve as the foundation for the analysis and optimization of low-carbon building design.

4.2 Low-Carbon Design Evaluation and Relevant Standards

As shown in Fig. 3, the analysis software used during the design phase to optimize the low-carbon project requirements should also address continuing carbon emissions over the lifetime of the building. Evaluation criteria can be set to compare projected carbon emission values against the calculated values, so that the low-carbon properties of the building can be assessed according to the original carbon emission budget, and thus, the success of the low-carbon building design could be audited.

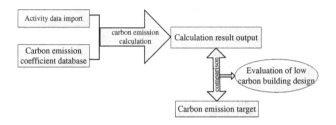

Fig. 3. Evaluation process of a low-carbon design based on IBM system

The lifecycle carbon emission budget of a building is an overall reflection of the low-carbon properties of the building. As shown in Fig. 1, the sources of carbon emissions are broken down among the products of various disciplines of architectural design to build a well-rounded evaluation system that provides an itemized and overall evaluation of the building design according to different low-carbon criteria and their corresponding discharge standards. Take, for example, the material content carbon emissions component of the total building carbon emissions: the building materials mainly undergo changes in physical arrangement at the construction site, and thus do not directly result in carbon emissions from the building project. This carbon emissions value instead refers to the energy consumption during material manufacture and the carbon emissions produced by the transformation of the materials. The material content carbon emissions during construction are influenced by the activities of industries outside of the building project [8]. The pertinent carbon emissions of each industry's manufacturing activities could be obtained using their correlated macroeconomic data, from which the portions corresponding to materials used in new building construction could be isolated to calculate the material content carbon emissions. The low-carbon design criteria for buildings can be derived from the low-carbon development standards of the industry, as shown in Eq. 2:

$$C_b = \sum_{i=1}^{n} C_i V_i K_b / S \qquad (2)$$

In Eq. 2, C_b is the material content carbon emissions of a unit construction area, and C_i is the carbon emissions of a unit output of class i material in its production industry. V_i is the input of that industry to construction, K_b is the proportion taken by the new building of the total output value of the construction industry, and S is the total area of the new building. The above relevant data can be obtained through state or local statistics.

Using Eq. 2, the present material content carbon emissions can be obtained for buildings in different locations. The development goals set by low-carbon economic development initiatives on the building design industry impose standard limits on the material content carbon emissions of buildings. Furthermore, as carbon emissions from construction activities contribute to the overall emissions of the construction industry, a standard can be developed to address the low-carbon development goals of the industry as well as green construction requirements.

The establishment of low-carbon standards for energy consumption, water consumption, and site utilization during building operation begins with the analysis of existing buildings, determining the resource consumption and carbon emissions during their the operation. These standards for low-carbon building design can be established through the analysis of existing evaluation standards for green buildings such as the green building evaluation index, as well as other standard practices.

4.3 Analysis of Calculation Results and Design Optimization

Building lifecycle carbon emissions were calculated and classified according to their sources using the design phase model. Substandard or poorly performing building emissions categories were identified by referring to corresponding evaluation criteria. This process forms the core of design and analysis, and provides the basis for design optimization. A carbon emissions budget platform is also built into the BIM system. The carbon emissions of various categories, as well as for the whole building lifecycle, are computed, and information from various software measurement systems are incorporated to establish the analysis program. The current low-carbon properties of the architectural design are then assessed by the analysis and necessary directions for optimization are determined from the results. The design data is subsequently adjusted and analyzed again using this high-efficiency information platform, and the final selection of optimization method is made. Figure 4 shows the BIM design model of a villa. Through calculation and analysis, it is revealed that the building carbon emissions in three areas of operation and use, material consumption, and water consumption are in

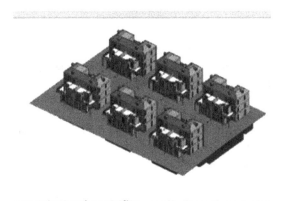

Fig. 4. The IBM design model of a villa

excess of the standard, and therefore, the buildings require design optimization and adjustment.

The designed building has a story height of 3.3 m. By reducing the story height to 3.2 m, the calculated carbon emissions can be reduced, mainly as a result of the reduction in wall material and insulation material consumption. Through calculation, it is found that the reduction in exterior wall materials and indoor space can reduce the power consumption of heating and air conditioning by 3%. The material carbon emission intensity decreases by 0.1628 $kgCO_2/m^2$.a, and heating and air conditioning carbon emissions decrease by 0.4998 $kgCO2/m^2$.a.

McLeod et al. (2013) studied passive housing and proposed the use of sunshades, as well as changing the window-wall ratio, to alter the energy consumption of buildings [9]. In this research, through the adjustment of BIM models, and by selecting the most optimal scheme via Green Building Studio (GBS) analysis, the width of each window on the east, west, and south walls were increased by 0.2–0.3 m. The window-wall ratios of the south, north, east and west walls increased from 20%, 38%, 10%, and 23% to 41%, 24%, 15%, and 26%, respectively, improving the passive solar heating effect. Sunshades were added to the windows of the second and third floors of the three buildings to the south, and the change in absorbed radiation during the summer was calculated for the windows before and after the installation of sunshades to illustrate the utility of this optimization. A south-facing window on the second-floor of a building was selected as the analysis object. The average solar radiation for each grid was 448521.56 Wh, which changed to 88492.95 Wh after sunshades were put up.

Green Building Studio software was used to calculate the annual energy consumption of the building model and the energy consumption of the whole lifecycle. The lifecycle energy consumption from heating and air conditioning was reduced by 461173 kWh, and the carbon emission intensity was reduced by 1.92 $kgCO2/m^2$.a.

In addition, through substituting the wall materials with Holiday Concrete, increasing the thickness of the insulation layer, and setting up the solar system, each index of optimized architectural design was adjusted, through BIM model changes and data analysis, to meet low-carbon building requirements.

5   Conclusions

Carbon emissions from construction-related activities constitute the main component of carbon emissions from human activities. With the development of BIM applications, it is possible to refine the carbon emissions information for an entire building lifecycle, allowing designers to evaluate, analyze, select, and optimize the architectural design, guaranteeing the low-carbon properties of buildings from the very beginning. On this basis, it is necessary to advance this study in the direction of evaluation criteria and sensing systems by further incorporating regional concerns and types of buildings, eventually realizing practical low-carbon design in architecture.

Acknowledgement.   This work was supported by the grants of the National Natural Science Foundation of China (No. 71301024).

References

1. Bao, J.Q., Ye, R.K.: The Study of Low Carbon Building (in Chinese), pp. 1–21. China Environmental Science Press, Beijing (2015)
2. Wong, J.K.W., Li, H., Wang, H.R., Huang, T., Luo, E., Li, V.: Toward low-carbon construction processes: the visualisation of predicted emission via virtual prototyping technology. Autom. Constr. **33**(8), 72–78 (2013)
3. Wang, Z., Zhao, Y., Ning, X.: Integrated information management system of building materials based on BIM technology in life cycle carbon emissions. In: Wu, Y., Zheng, S., Luo, J., Wang, W., Mo, Z., Shan, L. (eds.) Proceedings of the 20th International Symposium on Advancement of Construction Management and Real Estate, pp. 345–353. Springer, Singapore (2017). doi: 10.1007/978-981-10-0855-9_30
4. Seinre, E., Kurnitski, J., Voll, H.: Building sustainability objective assessment in estonian context and a comparative evaluation with LEED and BREEAM. Build. Environ. **82**, 110–120 (2014)
5. Heidarinejad, M., Dahlhausen, M., McMahon, S., Pyke, C., Srebric, J.: Cluster analysis of simulated energy use for LEED certified U.S. office buildings. Energy Build. **85**, 86–97 (2014)
6. Ding, L.Y., Zhou, Y., Akinci, B.: Building information modeling (BIM) application framework: the process of expanding from 3D to computable ND. Autom. Constr. **46**, 82–93 (2014)
7. Shang, C.J., Zhang, Z.H.: Building life cycle carbon emissions accounting. J. Eng. Manag. **24**(1), 7–12 (2010)
8. Lu, Y.J., Cui, P., Li, D.Z.: Carbon emissions and policies in China's building and construction industry (in Chinese). Evidence from 1994 to 2012. Build. Environ. **95**, 94–103 (2016)
9. McLeod, R.S., Hopfe, C.J., Kwan, A.: An investigation into future performance and overheating risks in passivhaus dwellings. Build. Environ. **70**, 189–209 (2013)

Data Reconciliation Based on an Improved Robust Estimator and NT-MT for Gross Error Detection

Shengxi Wu, Jinmeng Xu, Wei Liu, Xiaoying Wu,
and Xingsheng Gu[(✉)]

Key Laboratory of Advanced Control and Optimization for Chemical Processes,
East China University of Science and Technology, Ministry of Education,
No. 130, Meilong Road, Shanghai 200237, People's Republic of China
xsgu@ecust.edu.cn

Abstract. The quality of measurement data can be improved by data reconciliation. More accurate data will be provided for chemical process industry. However, the reconciliation results may be affected by gross errors. The influence of gross errors cannot be reduced effectively by classical method. Aimed at this problem, an improved robust NT-MT steady-state data reconciliation method is proposed in the paper. NT-MT method is used to detect suspicious nodes and variables with gross error. The suspicious variables are detected by critical value of adjustment detection. Robust estimator is used in data reconciliation. Finally, the measurement data is reconciled by the proposed robust estimator. The advantages of robust estimator and NT-MT method is combined together in this method. The simulation results show that the influence of gross error can be reduced effectively by the method proposed in the paper, thereby a better reconciliation results can be obtained.

Keywords: Data reconciliation · Gross error detection · Robust estimator · NT-MT · Measurement data

1 Introduction

Generally, data reconciliation technique is used to adjust the data generated from the chemical process to improve the quality of measurement data. The reconciled data is close to the true value and satisfies the rule of conservation of mass. However, gross errors may be contained in these measurements. The reconciliation results will be seriously affected by gross errors.

Statistical methods and robust estimation methods are very effective for the detection of gross errors and data reconciliation. Mah etc. [1] proposed measurement and detection method (MT, measurement test) to detect the gross error according to the deviation of the reconciled value and the measured value. However, in the traditional data reconciliation, the gross error will contaminate other reconciliation value, so that the type-1 error rate of this method is very high [2]. In order to overcome this shortcoming, many scholars put forward the improvement of MT method, such as Iterative measurement test (IMT) [3]. But when multiple gross errors in system simultaneously,

© Springer Nature Singapore Pte Ltd. 2017
M. Fei et al. (Eds.): LSMS/ICSEE 2017, Part I, CCIS 761, pp. 400–409, 2017.
DOI: 10.1007/978-981-10-6370-1_40

the IMT method may generate fault results. Node Test (NT) [4] method can confirm the measurement variables of suspicious node accurately, but it cannot identify the location of the suspicious measurement variables precisely. Yang [2] proposed the MT-NT detection method combined the measurement method with the node detection method, which including the advantages of MT and NT to overcome their disadvantages. Wang etc. [4] improved the traditional MT-NT method by applying a continuous compensation strategy. Mei Congli [5] proposed the NT-MT algorithm to research the problem that MT-NT algorithm cannot locate at the same node with two gross errors. Yan Xuefeng etc. [6] presented a novel method of NT-MT for data reconciliation based on the prior knowledge of domain experts. The simulation results showed that the influence of the gross error could be reduced by the robust objective function on the reconciliation results. Cauchy distribution was used to describe the distribution of measurement error by Shiyi Jin etc. [7]. Wu etc. [8] proposed a novel robust data reconciliation method based on GT distribution and historical data. A comparative performance analysis of robust data reconciliation strategies is presented by Llanos etc. [9].

Generally, NT-MT method can be used for the detection of gross errors. But after removing the gross error, the rank of parametric matrix may be insufficient. Aiming at this problem, a data reconciliation method for robust steady state NT-MT is proposed in the paper. An improved robust estimator is also presented, and the influence function illustrates that the robust estimator is robust to gross errors. At the same time, the robust estimator is also solved in the paper. Finally, the robust estimator proposed in the paper will allocate the corresponding weights according to the error of each variable in the iterative process of NT-MT algorithm. Therefore, good results can be obtained even in the presence of gross errors. The simulation results indicate that the proposed method is more effective.

2   Robust Data Reconciliation

Data reconciliation technique is commonly used to adjust the measured data to approach the true value and satisfies law of conservation of mass. When there is no gross error in the measured data, and the random errors are normally distributed, a good reconciliation result will be obtained by the traditional reconciliation method based on least squares estimation. However, when the measurement is mixed with outliers or when the process model is different from the actual system model, the reconciliation result will become inaccurate because the least squares estimation is sensitive to the gross error. Robust estimation is often used to reduce the effect of gross errors as the robust estimator assigns a small weight to these variables to ensure that the other measurements are not contaminated. The problem of robust data reconciliation can be described as follows:

$$\min_{x} \sum_{i=1}^{m} \rho(r_i)$$

$$s.t \; Ax = 0$$

(1)

Where ρ is a robust estimator. r_i is measurement error, and it can be calculated as $r_i = (\hat{x}_i - \tilde{x}_i)/\sigma_i$. σ_i^2 is the diagonal elements of Σ. The influence function is an important criterion to evaluate the robustness of robust objective function [10]. It can be described as follows:

$$IF \propto \frac{d\rho(r)}{dr} \tag{2}$$

The robust estimator is not affected by the error, so it can be demonstrated that the influence function of the robust estimator is bounded. The effect of the gross error on the objective function is expected to be reduced when the measurement error increases.

3   The Robust NT-MT Combined Method

Gross error detection is mainly to detect whether there is a gross error data, and whether the system contains leakage in the measurement. Compensation is operated to eliminate the effect of gross error on data reconciliation results. MT, NT and NT-MT are performed to detect gross error frequently. Robust estimation is often used to reduce the effect of gross errors as the robust estimator assigns a small weight to these variables to ensure that the other measurements are not contaminated.

3.1   Robust Estimator

An improved robust estimator can be defined as follows:

$$\rho(r) = \begin{cases} \frac{r^2}{2}, & |r| \le c \\ \frac{1}{\ln a} a^{r-c} + \frac{c^2}{2}, & |r| > c \end{cases} \tag{3}$$

Where a and c are scalars, $0 < a < 1$, adjusting under the necessity. Generally, Gross error ratio is between 1% to 10%. Therefore, c is expected to be set between 1 and 2. c is set to be 1.5, and a is set to be 0.5, in the paper. The influence function of the robust estimator is:

$$\psi(r) = \begin{cases} r, & |r| \le c \\ a^{r-c}, & |r| > c \end{cases} \tag{4}$$

From Eq. (4), when the measurement error increases, the influence function of the robust estimator is reduced. When the measurement error tends to infinity, the influence function approaches zero. But when the measurement error is small, the robust least squares estimator is equivalent to the least squares estimator. The robust estimator's influence function is bounded, so the effect of the error on the reconciliation results can be limited by the robust estimator.

3.2   Solution to Robust Data Reconciliation Problem

In order to solve the problem, $S = [s_1, \ldots, s_q]$ is defined. The column vectors of S spans the null space of A. Thus:

$$x = [s_1 \cdots s_q] \begin{bmatrix} \gamma_1 \\ \vdots \\ \gamma_q \end{bmatrix} = S\Gamma \tag{5}$$

Where $\gamma_1 \ldots \gamma_q$ are unknown vectors. Weight φ is defined as following:

$$\varphi(g) = \frac{\partial \rho}{\partial g} \frac{1}{g} \tag{6}$$

Where $g = \hat{x} - \tilde{x}$. Weight matrix Φ is defined as following:

$$\Phi = \underset{i=1,\ldots,v}{diag} (1/\varphi_i) \tag{7}$$

Where φ_i stands for $\varphi(\hat{x}_i - \tilde{x}_i)$. Thus:

$$\frac{\partial \rho}{\partial \Gamma} = S^T \Phi^{-1} g = 0 \tag{8}$$

$$\Gamma = (S^T \Phi^{-1} S)^{-1} S^T \Phi^{-1} \tilde{x} \tag{9}$$

$$\hat{x} = S(S^T \Phi^{-1} S)^{-1} S^T \Phi^{-1} \tilde{x} \tag{10}$$

According to Eqs. (3) and (6), the weight of the robust estimator is:

$$\varphi_i = \begin{cases} \frac{2}{\sigma_i^2}, & |r| \leq c \\ \frac{1}{g_i} a^{g_i - c}, & |r| > c \end{cases} \tag{11}$$

3.3   The Robust NT-MT Combined Method

To avoid the insufficient rank of coefficient matrix as NT-MT method removes gross error variables, not making full use of data information, an robust NT-MT method is proposed. The main idea of the algorithm is the potential error variable is obtained by the NT-MT algorithm primarily.

Then, the proposed robust estimator is performed to assign the corresponding weights to the variables depended on whether they contain gross errors, instead of remove the bad streams causing parameter matrix rank reducing. The variables without gross error will be assigned weight $2/\sigma_i^2$, while other variables contains gross error will be assigned weight $(1/g_i)a^{g_i - c}$, to reduce the credibility of these variables. The method scheme of this method is shown in Fig. 1. The main procedure is as following:

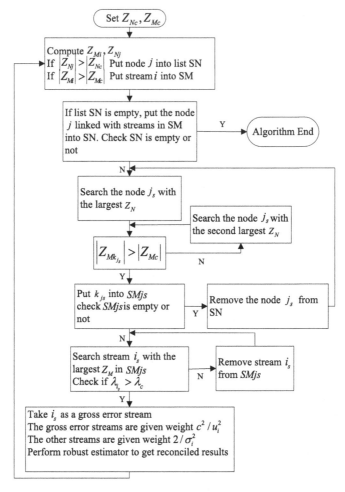

Fig. 1. Flow chart of the robust NT-MT method

Step 1. Set significance level α, and look up table to get Z_{Nc} and Z_{Mc}. Determine the critical value λ_c of adjustment detection based on expert prior knowledge.

Step 2. Calculate the reconciliation results of least squares objective function. Calculate all Z_{Mi} and Z_{Nj}.

Step 3. If $Z_{Nj} > Z_{Nc}(j = 1, 2, \ldots, m)$, add node j to a suspicious node set SN.

Step 4. If $Z_{Mi} > Z_{Mc}(i = 1, 2, \ldots, n)$, add stream i to a suspicious stream set SM.

Step 5. If set SN is empty, add node j connected to streams of set SM to the suspicious node set SN.

Step 6. If set SN is empty, the algorithm end. Or search node j_s with the largest Z_N in set SN. If $Z_{Nk_{j_s}} > Z_{Nc}(k_{j_s} = 1, 2, \ldots, n_{j_s})$, where n_{j_s} is the number of measurement variables connected to node j_s , add measurement variable k_{j_s} to a suspicious stream set SM_{j_s}.

Step 7. If SM_{js} is empty, remove node j_s, turn to step 6. Or search stream i_s with the largest Z_M in set SM_{js}. Computing adjustment $\lambda_{i_s} = \left|(\hat{x}_{i_s} - \tilde{x}_{i_s})/\tilde{x}_{i_s}\right|$, if $\lambda_{i_s} > \lambda_c$,turn to step 8. Or the measurement i_s is not containing gross error. Remove it from SM_{js}, and turn to step 7.

Step 8. Gross error exists in the measurement variable i_s. The corresponding weights are assigned to all processes by the proposed robust estimator according to Eq. (3). Bad variable i_s is assigned weight $(1/g_i)a^{g_i-c}$, while other variables are assigned weight $2/\sigma_i^2$. Get weight matrix by Eq. (7). Calculate the reconciliation results according to Eq. (10). Clear SN, SM and SM_{js}, then turn to step 2.

4 Case Study

A schematic diagram of crude oil distillation set is chosen for simulation in the paper as Fig. 2 (Zhou, 1999). The system is constituted by 7 nodes and 26 measured variables, no unmeasured variables included. In order to evaluate the performance of the new method more intuitively, two common performance criterions are adopted in this paper, Total Error Reduction (TER) and Sum of Squared Error (SSE), as evaluating indicator. The mathematical form is as follows:

$$TER = \frac{\sqrt{\sum_{i=1}^{n}(\tilde{x}_i - x_i^t)^2} - \sqrt{\sum_{i=1}^{n}(\hat{x}_i - x_i^t)^2}}{\sqrt{\sum_{i=1}^{n}(\tilde{x}_i - x_i^t)^2}} \tag{12}$$

$$SSE = \sum(\hat{x}_i - x_i^t)^2 + \sum(\hat{u} - u^t)^2 \tag{13}$$

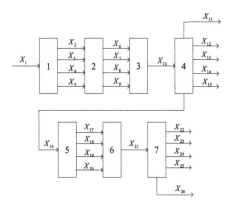

Fig. 2. Schematic diagram of crude oil distillation set

Where $\tilde{x}_i$ is value of measurement. $\hat{x}_i$ is value of reconciliation. x_i^t is true value of the variables. According to Eqs. (12) and (13), the performance of reconciliation is proportional to TER, but inversely proportional to SSE. So, TER is expected to be larger, and SSE to be smaller, leading to the value after reconciliation closer to the true value.

In the simulation analysis, the performance of the method proposed in the paper and the robust MT-NT method [10] is compared under the condition that there are one or more gross errors in the measurement network. c is set to be 1.5. λ_c is set to be 0.02.

Firstly, we assume that only stream 16 contains gross error. The results of the two methods are shown in Table 1. As can be seen from Table 1, robust MT-NT and robust NT-MT combined method perform well, as they can detect bad stream 16 and get a good reconciliation results. But, from the above analysis, TER is the larger the better,   while

Table 1. Reconciliation results of crude distillation set with one error

No.	True value	Measured value	Relative error %	Robust MT-NT	Relative error %	RobustNT-MT	Relative error %
1	308.0	300.4	2.47	305.0	0.97	305.3	0.88
2	77.0	74.0	3.90	75.0	2.60	75.1	2.47
3	77.0	75.7	1.69	76.7	0.39	76.8	0.26
4	77.0	75.3	2.21	76.3	0.91	76.4	0.78
5	77.0	76.0	1.30	77.0	0.00	77.1	0.13
6	62.0	62.4	0.65	61.1	1.45	61.2	1.29
7	82.0	84.1	2.56	82.8	0.98	82.9	1.10
8	82.0	81.4	0.73	80.3	2.07	80.4	1.95
9	82.0	82.2	0.24	80.9	1.34	81.0	1.22
10	308.0	309.4	0.45	305.0	0.97	305.3	0.88
11	8.5	8.3	2.35	8.4	1.18	7.6	10.59
12	5.5	5.7	3.64	1.3	76.36	5.0	9.09
13	30.0	29.7	1.00	29.8	0.67	29.0	3.33
14	20.0	20.7	3.50	20.8	4.00	20.0	0.00
15	14.0	14.3	2.14	14.4	2.86	13.6	2.86
16	230.0	**314.8**	36.87	230.5	0.22	230.0	0.00
17	57.5	58.3	1.39	57.5	0.00	57.4	0.17
18	57.5	58.3	1.39	57.5	0.00	57.4	0.17
19	57.5	59.9	4.17	59.1	2.78	59.0	2.61
20	57.5	59.9	4.17	56.4	1.91	56.3	2.09
21	230.0	230.0	0.00	230.5	0.22	230.0	0.00
22	12.0	11.9	0.83	12.3	2.50	12.2	1.67
23	49.0	48.5	1.02	48.9	0.20	48.8	0.41
24	38.0	36.6	3.68	37.0	2.63	36.9	2.89
25	21.0	21.4	1.90	21.8	3.81	21.7	3.33
26	110.0	110.0	0.00	110.4	0.36	110.3	0.27
	TER			0.91		0.93	
	SSE			52.41		31.39	

SSE is contrary. So the NT-MT method proposed in the paper performs better, as TER value is larger than MT-NT's, and SSE value is smaller.

After that, when measurements contain three gross errors, the reconciliation results are as shown in Table 2. By using the robust NT-MT method, in the first iteration, list SN is composed of 5 nodes: 2, 4, 5, 6, 7. List SM consists of 18 streams: 1, 10 ~ 26. Obviously, list SN is not empty. Then, node 5 is searched with the largest Z_N. Streams 16, 17, 18, 19, 20 are linked to node 5. Check out which stream's statistic of them exceed Z_{Mc}, and put them into list SM_{js}. List SM_{js} constitutes of streams 16, 17, 18, 19, 20. After that, stream 16 is searched with the largest Z_M. The corresponding relative adjustment of 16 exceed λ_c. Therefore, the first gross error is found.

Table 2. Reconciliation results of crude distillation set with three errors

No.	True value	Measured value	Relative Error %	Robust MT-NT	Relative Error %	Robust NT-MT	Relative Error %
1	308.0	300.4	2.47	309.9	0.62	305.4	0.84
2	77.0	74.0	3.90	76.2	1.04	75.1	2.47
3	77.0	75.7	1.69	78.0	1.30	76.8	0.26
4	77.0	75.3	2.21	77.5	0.65	76.4	0.78
5	77.0	76.0	1.30	78.3	1.69	77.1	0.13
6	62.0	62.4	0.65	62.4	0.65	61.2	1.29
7	82.0	84.1	2.56	84.0	2.44	82.9	1.10
8	82.0	81.4	0.73	81.4	0.73	80.4	1.95
9	82.0	82.2	0.24	82.2	0.24	81.0	1.22
10	308.0	309.4	0.45	309.9	0.62	305.4	0.84
11	8.5	8.3	2.35	8.3	2.35	7.5	11.76
12	5.5	5.7	3.64	5.7	3.64	4.9	10.91
13	30.0	29.7	1.00	29.3	2.33	28.9	3.67
14	20.0	20.7	3.50	20.5	2.50	19.9	0.50
15	14.0	14.3	2.14	14.2	1.43	13.5	3.57
16	230.0	**314.8**	36.87	232.0	0.87	230.6	0.26
17	57.5	58.3	1.39	57.9	0.70	57.5	0.00
18	57.5	58.3	1.39	57.9	0.70	57.5	0.00
19	57.5	59.9	4.17	59.5	3.48	59.1	2.78
20	57.5	59.9	4.17	56.8	1.22	56.4	1.91
21	230.0	**180.5**	21.52	232.0	0.87	230.6	0.26
22	12.0	11.9	0.83	11.7	2.50	11.9	0.83
23	49.0	48.5	1.02	45.9	6.33	48.5	1.02
24	38.0	36.6	3.68	35.1	7.63	36.6	3.68
25	21.0	21.4	1.90	20.9	0.48	21.4	1.90
26	110.0	**140.6**	27.82	118.4	7.64	112.2	2.00
	TER			0.89		0.94	
	SSE			117.8		37.42	

Then the proposed estimator is performed to assign weights to the variables according to whether it is potentially a gross error or not. The gross error variables are assigned weight $(1/g_i)a^{g_i-c}$. The variables that not contain a gross error provisionally are assigned weight $2/\sigma_i^2$. Then, the proposed robust estimator is used to get reconciled results. And start the second iteration. Finally, it turns out that the gross error variables are streams 16, 21, 26.

From the above analysis, TER is expected to be larger, and SSE to be smaller, leading to the value after reconciliation closer to the true value. As can be seen from Table 2, the TER and SSE of the method proposed in the paper is 0.94 and 37.42, while the TER and SSE of the robust MT-NT method are 0.89 and 117.83 respectively. It can be concluded that the method proposed in this paper can detect the multiple gross errors effectively, and obtain a better reconciled results.

If NT-MT method is applied in the case, streams 16, 21 and 26 will be removed after three times iterations. The rank of coefficient matrix will be reduced, resulting in NT-MT method accidental termination, unable to get correct results. However, the method proposed in the paper can solve this problem effectively, as the robust estimator assigns weights to all variables rather than removing the variables that contain the gross error.

5   Conclusion

A novel robust NT-MT combined method is presented in the paper. The NT-MT method and the proposed robust estimator are combined to avoid the insufficient rank of coefficient matrix. NT-MT is used to detect gross error by successive iteration, and the proposed robust estimator is performed to reconcile the measurements. The simulation results show that the proposed method in the paper can detect one or multiple gross errors efficiently, and get a better reconciliation results.

Acknowledgments. This work was supported by National Natural Science Foundation of China (61573144; 61673175) and Fundamental Research Funds for the Central Universities under Grant 222201717006.

References

1. Tamhane, A.C., Mah, R.S.: Data reconciliation and gross error detection in chemical process networks. Technometrics **27**(4), 409–422 (1985)
2. Yang, Y., Ten, R., Jao, L.: A study of gross error detection and data reconciliation in process industries. Comput. Chem. Eng. **19**, 217–222 (1995)
3. Serth, R.W., Heenan, W.A.: Gross error detection and data reconciliation in steam metering systems. AIChE J. **32**(5), 733–742 (1986)
4. Wang, F., Jia, X.P., Zheng, S.Q., Yue, J.C.: An improved MT-NT method for gross error detection and data reconciliation. Comput. Chem. Eng. **28**(11), 2189–2192 (2004)
5. Congli, M.E.I., Hongye, S.U., Jian, C.H.U.: An NT-MT combined method for gross error detection and data reconciliation. Chin. J. Chem. Eng. **14**(5), 592–596 (2006)

6. Yan, X.F., Bao, J.J., Zhang, B., Qian, F.: Data reconciliation and application of NT-MT combined method. J. Chem. Ind. Eng. **58**(11), 2828–2833 (2007)
7. Jin, S., Li, X., Huang, Z.: A new target function for robust data reconciliation. J. Ind. Eng. Chem. Res. **51**(30), 10220–10224 (2012)
8. Wu, S., Ye, Q., Chen, C., Gu, X.: Research on data reconciliation based on generalized T distribution with historical data. Neurocomputing **175**, 808–815 (2016)
9. Llanos, C.E., Sánchez, M.C., Maronna, R.: Robust estimators for data reconciliation. Ind. Eng. Chem. Res. **54**(18), 5096–5105 (2015)
10. Zhou, L., Fu, Y.: Data reconciliation based on robust estimator and MT-NT method. In: 35th Control Conference (CCC), pp. 6426–6430. IEEE Press, Chengdu (2016)

Survey of 3D Map in SLAM: Localization and Navigation

Aolei Yang, Yu Luo, Ling Chen[✉], and Yulin Xu

Shanghai Key Laboratory of Power Station Automation Technology,
School of Mechatronic Engineering and Automation, Shanghai University,
Shanghai 200072, China
aolei_yang@163.com, ly11tea@163.com, {lcheno,xuyulin}@shu.edu.cn

Abstract. 3D mapping is a difficult problem due to real-world places whose appearance and scale can be various. Owing to the rapid development of computer and robot system, remarkable improvements of performance are achieved in 3D map technology, which in turn contribute to the significant advances in SLAM. This paper presents the state-of-the-art 3D map technology and system, which is classified into topological maps, metric maps and semantic maps. Additionally, the advantages and disadvantages of various 3D map technologies are analyzed in different aspects, including navigation performance, localization performance, visual perception, scalability, computation cost and mapping difficulty. In order to better understand them, the key performance parameters of the 3D map technologies are compared in a table. Finally, the paper ends with a discussion on the open problems and future of 3D map technology.

Keywords: 3D map · SLAM · Localization · Navigation

1 Introduction

Robotic map has been a highly active research area in robotics and SLAM for at least two decades. SLAM consists in the pose estimation of the robot and the map of the environment. As the key element of SLAM, the map in nature is the spatial models of physical environments which are acquired through the robot's operation. A good map is the necessary prerequisite to navigation tasks performed by robots. For instance, many applications of robot aim to explore the unknown environment and build a corresponding map of it. On the other way round, through the loop closures detection and other methods which are supported by map, the robot can adjust errors in localization. In recent years, the 2D mapping issue is tackled, where the environment is static, limited scale and closure space. So, IEEE has released a standard for 2D map in robotics [1]. However, 3D cases are not yet to be completely solved due to map capacity, performance and

L. Chen—This work was supported by Natural Science Foundation of China (61403244), Science and Technology Commission of Shanghai Municipality under "Yangfan Program" (14YF1408600, 16YF1403700), Key Project of Science and Technology Commission of Shanghai Municipality (15411953502).

© Springer Nature Singapore Pte Ltd. 2017
M. Fei et al. (Eds.): LSMS/ICSEE 2017, Part I, CCIS 761, pp. 410–420, 2017.
DOI: 10.1007/978-981-10-6370-1_41

robustness, especially in dynamic, large scale and open environment. Meanwhile, applications of 3D map are promising and urgently required, for instance, UAV SLAM etc. So 3D Map still need further exploration.

In this paper, 3D maps are generally divided into three groups: topological maps, metric maps and semantic maps. Topological maps drop scale, distance and directions of objects. They only save linkage information between places and model the environment as a graph [3]. Topological maps can be used to analyze the reachability of some area in the environment. Conversely, metric maps which are more frequently used, represent objects and obstacles with metric data and geometry properties, i.e. point, line, cube, and etc. Therefore, detailed information about things in the environment are saved. Semantic maps take one step further to relate semantic concepts with objects in the map, which allow more cognitive process to be operated in it.

By now, there is no single map can simultaneously meet functional requirements (e.g. localization, navigation and perception) and performance requirements (e.g. precision, compact and speed). So, various types of maps are compared and evaluated in terms of above requirements.

The remainder of the paper is organized as follows. Section 2 introduces the topological map. In Sect. 3, metric maps are presented. In Sect. 4, semantic maps are presented. The paper finally compares the various 3D maps. Meanwhile, it offers some opinions on open problems and future directions.

2   Topological Maps

As mentioned earlier, pure topological maps only contain relative positions of places and do not include any metric information between two places. The topological map can be seen as the graph in data structure. Single node represents certain place and the edges represent paths between places. Robot movement can then be viewed as a visit between nodes and graph will add a new node if robot has reached new area. A common pure topological map is shown in Fig. 1.

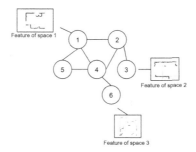

Fig. 1.   General form of pure topological map

Due to their compact structure and high efficiency, pure topological maps were used in early SLAM research [4]. The localization is largely based on finding matches by environment's appearance-based feature. A key drawback of this practice is scalability. The inverted index and bag-of-words method were used to relieve this problem [5].

One limitation of pure topological maps is impossible to navigate a robot due to the lack of metric information. Only the reachability information between two places could be extracted from them. Obviously, abstract topological information and metric information of physical environments should be mixed in a single consistent model. Following this path, one kind of hybrid maps is created, which is called topological-metric maps. As shown in Fig. 2, a global topological map is used to move between places and rely on metric information in local space for precise navigation [8]. So, all topological-metric maps are hierarchical. This hybrid practice is proven to be effective, for instance, topological-metric extension CAT-SLAM [9] and SMART [10] are both proven to outperform their pure topological counterparts FAB-MAP [7] and SeqSLAM [8].

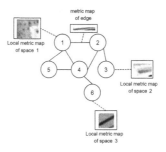

Fig. 2. One kind of topological-metric maps

Generally speaking, due to high-level and abstract representation, topological maps show several advantages in large-scale and dynamic environments. Becasuseplaces are connected with topological relations, topological maps have robustness to changes in the environments. For example, place B could be described as "place C is to the left of place B and to the right of place A". Despite some small changes of size or appearance, the localization which is under topological maps keeps effective. forthese advantages, topological maps are already integrated with long-term SLAM in the research [11].

By now, the biggest challenge to topological maps in SLAM is "perceptual aliasing". This issue is essentially a data association problem where current observation should belong to an already visited place or a completely new one. Moreover, the relationship between abstract topological information and low-level metric information is quite hard to be established in topological-metric maps. So, although there are plenty of corresponding implementations, it will still take some time before a robust and universal solution appears [12].

3   Metric Map

As mentioned in introduction, metric maps capture the geometric properties of the environment and are especially suitable for robot's accurate operation. However, the localization and navigation of most robots are purely dependent on geometry. So, in this context, the more precise metric maps are, the more accurately robots operate. Unfortunately, precise maps also mean high storage requirement and time cost. This general contradiction is quite common and throughout the development of metric maps.

3.1 Landmark-Based Map

In the early SLAM researches, landmark-based maps were introduced to meet the limited storage and computational resources [13]. This kind of maps represents the environment as a set of isolated 3D landmarks which have well-distinguished features and corresponding spatial coordinates (e.g. rocks in the sand). When it comes to the localization, the robot matches the features of observation with the landmarks in maps to get its absolute location. By the discretization of the environment, landmark-based maps dramatically reduce the data volumes, so it is also called as sparse representation. Landmark-based maps are introduced in various SLAM systems, such as ORB-SLAM and PTAM [14, 15]. The PTAM is even transplanted to iPhone.

In fact, when landmark-based maps are built or used, a strong assumption is that there must be distinctive features in the environment. If this assumption is not valid (e.g. a pure white wall), the algorithm will simply fail. However, owing to low storage requirement, landmark-based maps still have advantages in small-scale environment.

3.2 Grid Maps

Grid maps are possibly the best known type of map for robots and their 2D navigation applications were introduced as early as last century. Grid maps were essentially spatial-partitioning representations, which divide the 3D environment into independent and same cubes. As shown in Fig. 3, cubes are arranged in form of grid. Each cell in the maps represents the probability of the corresponding spatial region to be occupied. The occupancy probability is quite useful for navigation.

Fig. 3. 3D grid representation

The most common scheme of 3D grid map is Octomap which is based on Octree [16]. Octree is a derivative data structure of tree, of which each node could be recursively subdivided into eight sub nodes. Since each cube in 3D space could also be divided into eight cubes, Octotree can be used to represent 3D grid map, as shown in Figs. 4 and 5. So, leaf nodes of Octotree store the occupancy probability of spatial space.

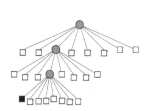

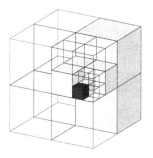

Fig. 4. Octotree data structure   **Fig. 5.** 3D grid map corresponding to Fig. 4.

Since the information of each cube is its occupancy, grid maps are easy to merge information from different sensors, as long as they can provide occupancy information [17]. In mergence, the confidence level can also be modified according to certain sensor's reliability. Because of strong connection with the metrical motion of the robot, grid maps show great advantages on robot navigation, e.g. translation motion of the map to guide the robot movement within the grid is straightforward.

However, data volume of 3D grid maps grows at a much higher rate than the scale of the region. Researchers try to solve this problem by adaptive resolution, i.e. increasing the resolution of interesting places and decreasing the marginal one [18]. What's more, in Octomap, if all sub-cube has the same state (occupied or free), they could be pruned. Experiments prove that this practice effectively relieves the data volume cost [16]. Though the way of representing the environment in cube is easy, detailed information of spatial objects are lost, e.g. smooth changes on surface. As a result, on the occasion which is in no need of objects' details, especially in navigation and path planning, grid maps are most frequently used.

3.3 Point-Based Maps

Point-Based map uses large amounts of unstructured points to represent 3D space and is a relatively primitive map type. Data point sets are also the main output of RGBD camera and 3D radar. In the early SLAM research, this kind of data point is too dense to process by robots. However, along with the quick development of hardware, direct methods in SLAM were introduced, which estimates the trajectory of the robot directly from every data point [19–21]. Because of the high coherence with sensor data and easy mapping process, point-based maps are closely integrated in direct methods. This kind of map provides rich environment information and suits the localization well. If data points are dense enough, they can also meet the need of navigation. However, due to their primitive form, in which data points are just simply cut and mosaicked without changes in form, data is quite redundant, e.g., maps will use thousands of identical white points to represent a white wall. On the other hand, data volume quickly turns unfeasible for large-scale scenarios.

In conclusion, pure point-based maps can be viewed as temporary local maps in SLAM operation. Once there is large-scale mapping or offline usage need, other map types should be taken into consideration.

3.4 SDF-Based Maps

SDF-based maps are essentially one kind of implicit surface representations. They divide the spatial space into identical cubes whose values are defined by SDF (signed distance function). The signed distance function determines the distance of a given cube from the boundary of the nearest object, with the sign determined by whether point is in object (i.e. inside the object is negative and outside is positive), as shown in Fig. 6. The value of the cube in SDF-based maps can be computed in different ways [22]. In the case of ESDF (i.e. Euclidean Signed Distance Function), one cube's value represents the Euclidean distance to the surface of the nearest object. While in the case of projective TSDF (i.e. projective Truncated Signed Distance Function), a cube's value is the distance to the surface which follows the direction from the center of the sensor, and is truncated to one fixed value (e.g. 1 m) when cube is relatively far from the object, as shown in Fig. 7 [23].

(a) Cone is cut by plane.   (b) SDF sign of the corresponding cross-sectional view

Fig. 6. Cone and its corresponding SDF representation

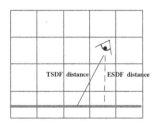

Fig. 7. Comparison of TSDF distance and ESDF distance

SDF-based maps are commonly used in navigation. They can provide not only the occupancy probability of each cube, but also the distances to obstacles, which is necessary for trajectory optimization-based planners (e.g. CHOMP, TrajOpt) [24, 25]. Because the distance to obstacle is directly stored in each cube, ESDF-based maps show greater speed advantage in collision checking [26]. However, ESDF maps are very hard to achieve: they are usually constructed from other map types and sometimes even need human intervention [27]. So, they cannot be achieved simultaneously during localization process, which is defective in SLAM context.

Projective TSDF-based maps come into the frontier of SLAM researchers after Microsoft releases Kinect Fusion, which aims to reconstruct a high-quality, geometrically precise 3D model of the scene by RGBD camera, e.g. Kinect [23]. Projective TSDF is used as the main representation in it. However, Kinect Fusion and its extension are more like a 3D reconstruction technology and can be hardly applied in navigation and localization [28, 29].

4   Semantic Maps

As is evident from the above, current most common map types are either pure geometrical representations which decompose the environment by just metrical data, or a graph which represents reachability among places. Due to the flourish of the artificial intelligence technology, semantic maps become active in recent years, which provide higher levels of abstraction than the ones introduced before and facilitate more complex tasks, e.g. finding a certain person with his feature without knowing his position in the environment.

Semantic maps consist of various semantic elements which are distinguished from each other and can be used explicitly in human-like action, e.g. planning, reasoning, decision making. Semantic elements usually could be either background classes (e.g. ground) or object classes (e.g. desk) [30]. Intuitively, semantic elements are not atomic, that is to say, they could have several attributes and sub elements, e.g. a table may include some cups and is moveable. So, the relations in semantic maps are hierarchical, such as "has–a" relation, as shown in Fig. 8.

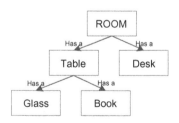

Fig. 8. Semantic maps hierarchical structure example

However, since there are various kinds of things in the world, semantic meanings can be impossibly predefined. So, it is still a difficult problem to model the complicated world. What's more, due to the sophisticated algorithm, semantic maps need large computational resources, which are reluctant for online usage [31]. Above all, though semantic maps are not so mature at present, they will certainly change the SLAM and robot's navigation in future.

5 Conclusion

As mentioned above, there is not yet a 3D map that meets all requirements of SLAM and navigation. Though immature 3D maps do lead to this situation in some ways, varied and even inherently conflicting requirements are main causes. As human beings, people certainly need high resolution and colorful 3D map to provide good visual perception, which is also an intrinsic attribute of the map. However, from the perspective of localization and navigation, a visually pleasing map means great data redundancy which inversely leads to poor real-time performance of robot. Besides, localization and navigation operate in variant environments which have different performance requirements. From the perspective of robot, there is speed, available sensors computational resources, and etc. From the perspective of outside world, there is brightness, space size and etc. From the perspective of tasks, there is measurement accuracy, duration and etc. These environment factors lead to different performance requirements.

According to the study based on the current mainstream 3D maps in the purpose of localization and navigation, Table 1 lists the comparison of various 3D maps.

Table 1. Overview of comparison between 3D maps

MAPS	Navigation performance	Localization performance	Visual perception	Scalability	Computation cost	Mapping difficulty
Topological Map	Hard to accomplish	Medium	Medium	Large space	High	High
Landmark-based Map	Hard to accomplish	Good	Low	2 or 3 rooms	Medium	Medium
Grid Map	Good	Medium	Low	One building	Low	Low
Point-based Map	Low	Good	Medium	2 or 3 rooms	High	Low
ESDF-based Map	Good	Medium	Low	One building	Low	High
TSDF-based Map	Hard to accomplish	Low	High	2 or 3 rooms	High	Medium

Apart from TSDF-based map, nearly all maps in the above table do not have good performance in visual perception, which is exactly a regretful comprise for current level development of robot. Landmark-based maps are the best choice in the pure navigation tasks for its medium computation resource requirement and high navigation performance. Grid maps own best comprehensive performance, which is consistent with its wide application. In conclusion, choice of 3D maps is largely dependent on the task that the robots perform.

There are several open problems in this area:

Maps in the Dynamic Environment: Current mainstream 3D maps rely on a static world assumption, in which the robot is the only dynamic object. However, changes of other agents are very common in the real world, e.g. cars and walking people. Researchers try to increase the robustness of 3D map towards dynamic environment by small-scale 3D reconstruction [32]. However, 3D maps in the large-scale dynamic environment are still unexplored.

Distributed Maps: Modelling the environment individually by multi-robot is an effective way to solve localization and navigation issues in large-scale space. Following this approach, some researches have made several achievements [33]. But how to maintain and transmit the maps between robots in long term and how to use other robots' observation to improve current robot's map still remains unsolved.

Hybrid and adaptive maps: By now, in most circumstances, only one map is available in one task. But requirements on map can be different even in same task. For instance, sometimes we want to model the office by grid map for navigation, while modelling persons in it by TSDF-based map for visual perception. So, a map container which can accommodate maps in different types and resolution, are required on this occasion. This kind of map is still unexplored.

In addition, semantic maps are also a promising research area in 3D map.

Some characteristics of future 3D maps can be guessed for above open problems. Firstly, 3D maps in the future will be smart, which means that high-level information can be extracted from maps and maps can also support more cognitive process. What's more, maps will be robust, which means that maps are able to be used in complicated and dynamic environment. Finally, they should have scalability to operate unlimitedly in large-scale space.

References

1. IEEE Standard for Robot Map Data Representation for Navigation. 1873–2015 IEEE Standard for Robot Map Data Representation for Navigation, pp. 1–54 (2015)
2. Lowry, S., Sünderhauf, N., Newman, P., Leonard, J.J.: Visual place recognition: a survey. IEEE Trans. Robot. **32**(1), 1–19 (2016)
3. Naseer, T., Spinello, L., Burgard, W., Stachniss, C.: Robust visual robot localization across seasons using network flows. In: Twenty-Eighth AAAI Conference on Artificial Intelligence, pp. 2564–2570 (2014)
4. Ulrich, I., Nourbakhsh, I.: Appearance-based place recognition for topological localization. In: IEEE International Conference on Robotics and Automation. Symposia Proceedings 2000 ICRA. Millennium Conference (Cat. No.00CH37065), vol. 2, pp. 1023–1029 (2000)
5. Cummins, M., Newman, P.: Appearance-only SLAM at large scale with FAB-MAP 2.0. Int. J. Robot. Res. **30**(9), 1100–1123 (2011)
6. Blanco, J.L., FernÁndez-Madrigal, J.A., GonzÁlez, J.: Toward a unified bayesian approach to hybrid metric-topological SLAM. IEEE Trans. Robot. **24**(2), 259–270 (2008)
7. Paul, R., Newman, P.: FAB-MAP 3D: topological mapping with spatial and visual appearance. In: 2010 IEEE International Conference on Robotics and Automation, pp. 2649–2656 (2010)
8. Milford, M.J., Wyeth, G.F.: SeqSLAM: visual route-based navigation for sunny summer days and stormy winter nights. In: 2012 IEEE International Conference on Robotics and Automation, pp. 1643–1649 (2012)

9. Maddern, W., Milford, M., Wyeth, G.: Capping computation time and storage requirements for appearance-based localization with CAT-SLAM. In: 2012 IEEE International Conference on Robotics and Automation, pp. 822–827 (2012)

10. Pepperell, E., Corke, P.I., Milford, M.J.: All-environment visual place recognition with SMART. In: IEEE International Conference on Robotics and Automation, pp. 1612–1618 (2014)

11. Krajník, T., Fentanes, J.P., Mozos, O.M., Duckett, T., Ekekrantz, J., Hanheide, M.: Long-term topological localisation for service robots in dynamic environments using spectral maps. In: 2014 IEEE/RSJ International Conference on Intelligent Robots and Systems, pp. 4537–4542 (2014)

12. Boal, J., Sánchez-Miralles, Á., Arranz, Á.: Topological simultaneous localization and mapping: a survey. Robotica 32(5), 803–821 (2014)

13. Montemerlo, M., Thrun, S., Koller, D., Wegbreit, B.: FastSLAM: a factored solution to the simultaneous localization and mapping problem. In: Eighteenth National Conference on Artificial Intelligence, pp. 593–598 (2002)

14. Klein, G., Murray, D.: Parallel tracking and mapping for small AR workspaces. In: 2007 6th IEEE and ACM International Symposium on Mixed and Augmented Reality, pp. 225–234 (2007)

15. Mur-Artal, R., Montiel, J.M.M., Tardós, J.D.: ORB-SLAM: a versatile and accurate monocular SLAM system. IEEE Trans. Robot. 31(5), 1147–1163 (2015)

16. Hornung, A., Wurm, K.M., Bennewitz, M., Stachniss, C., Burgard, W.: OctoMap: an efficient probabilistic 3D mapping framework based on octrees. Auton. Robots 34(3), 189–206 (2013)

17. Castellanos, J.A., Tardos, J.D.: Mobile robot localization and map building: a multisensor fusion approach. Springer, Heidelberg (1999)

18. Ryde, J., Hu, H.: 3D mapping with multi-resolution occupied voxel lists. Auton. Robots 28(2), 169–185 (2010)

19. Labbe, M., Michaud, F.: Online global loop closure detection for large-scale multi-session graph-based SLAM. In: IEEE/RSJ International Conference on Intelligent Robots and Systems, pp. 2661–2666 (2014)

20. Newcombe, R.A., Lovegrove, S.J., Davison, A.J.: DTAM: dense tracking and mapping in real-time. In: IEEE International Conference on Computer Vision, pp. 2320–2327 (2011)

21. Endres, F., Hess, J., Sturm, J., Cremers, D., Burgard, W.: 3-D mapping with an RGB-D camera. IEEE Trans. Robot. 30(1), 177–187 (2014)

22. Oleynikova, H., Millane, A., Taylor, Z., Galceran, E., Nieto, J., Siegwart, R.: Signed distance fields: A natural representation for both mapping and planning. In: RSS Workshop on Geometry and Beyond. (2016)

23. Newcombe, R.A., Izadi, S., Hilliges, O., Molyneaux, D.: KinectFusion: real-time dense surface mapping and tracking. In: IEEE International Symposium on Mixed and Augmented Reality, pp. 127–136 (2011)

24. Zucker, M., Ratliff, N., Dragan, A.D., Pivtoraiko, M., Klingensmith, M., Dellin, C.M., Bagnell, J.A., Srinivasa, S.S.: CHOMP: covariant hamiltonian optimization for motion planning. Int. J. Robot. Res. 32(9–10), 1164–1193 (2013)

25. Schulman, J., Duan, Y., Ho, J., Lee, A., Awwal, I., Bradlow, H., Pan, J., Patil, S., Goldberg, K., Abbeel, P.: Motion planning with sequential convex optimization and convex collision checking. Int. J. Robot. Res. 33(9), 1251–1270 (2014)

26. Kalakrishnan, M., Chitta, S., Theodorou, E., Pastor, P., Schaal, S.: STOMP: stochastic trajectory optimization for motion planning. In: 2011 IEEE International Conference on Robotics and Automation, pp. 4569–4574 (2011)

27. Pan, J., Chitta, S., Manocha, D.: FCL: a general purpose library for collision and proximity queries. In: 2012 IEEE International Conference on Robotics and Automation, pp. 3859–3866 (2012)

28. Whelan, T., Kaess, M., Fallon, M.F., Johannsson, H., Leonard, J.J., McDonald, J.B.: Kintinuous: spatially extended kinectfusion. In: RSS Workshop on RGB-D: Advanced Reasoning with Depth Cameras (2012)

29. Steinbrücker, F., Sturm, J., Cremers, D.: Volumetric 3D mapping in real-time on a CPU. In: IEEE International Conference on Robotics and Automation, pp. 2021–2028 (2014)

30. Cadena, C., Dick, A., Reid, I.D.: A fast, modular scene understanding system using context-aware object detection. In: 2015 IEEE International Conference on Robotics and Automation (ICRA), pp. 4859–4866 (2015)

31. Salasmoreno, R.F., Newcombe, R.A., Strasdat, H., Kelly, P.H.J., Davison, A.J.: SLAM++: simultaneous localisation and mapping at the level of objects. In: Computer Vision and Pattern Recognition, pp. 1352–1359 (2013)

32. Pronobis, A., Jensfelt, P.: Large-scale semantic mapping and reasoning with heterogeneous modalities. In: IEEE International Conference on Robotics and Automation, pp. 3515–3522 (2012)

33. Cunningham, A., Indelman, V., Dellaert, F.: DDF-SAM 2.0: consistent distributed smoothing and mapping. In: IEEE International Conference on Robotics and Automation, pp. 5220–5227 (2013)

Analysis of Cyber Physical Systems Security Issue via Uncertainty Approaches

Hui Ge[1,2(✉)], Dong Yue[1,2,3], Xiang-peng Xie[2], Song Deng[2], and Song-lin Hu[2]

[1] School of Automation, Nanjing University of Posts and Telecommunications,
Nanjing 210023, People's Republic of China
2014050228@njupt.edu.cn
[2] Institute of Advanced Technology, Nanjing University of Posts
and Telecommunications, Nanjing 210023, People's Republic of China
[3] Hubei Province Collaborative Innovation Center for New Energy Microgrid,
China Three Gorges University, Yichang 443002, People's Republic of China

Abstract. From security perspective, cyber physical system (CPS) security issue is investigated in this note. Based on a double-loop security control structure, the typical cyber attack called information disclosure, denial-of-service (DoS), deception attack and stealth attack are analyzed from uncertainty perspective. The performance of these attacks are formulated, meaningful models are proposed meanwhile. According to aforementioned attacks, security control scenarios are obtained via the character of each kind cyber attack. And some novel results are obtained via a well designed double closed loop structure. At last, from traditional standpoint, uncertainties of a separately excited DC motor is taken as an example to demonstrate the problem.

Keywords: Cyber physical systems · Security control · System uncertainty · Separated excited DC motor

1 Introduction

In the past the ICSs are often local control, this model is much more security as its connections to outside is cut off. As the increasing development of computation, communication and control technologies, the study of CPSs security has received growing attention due to its significant functions and widespread applications most recently [1]. All of the critical process and key instruments are tightly connected via integration techniques [2].

To achieve mission and business success, enterprise information systems must be dependable in the face of serious cyber threats. In a word, the security control of CPSs is full of challenges [3,4]. First of all, it is difficult to eliminate the error between the models and the virtual system. In addition, attack process are often

This work is supported by National Natural Science Foundation (NNSF) of China under Grant 61533010, 61374055, 51507084.

M. Fei et al. (Eds.): LSMS/ICSEE 2017, Part I, CCIS 761, pp. 421–431, 2017.
DOI: 10.1007/978-981-10-6370-1_42

stealthy and pretended to be seem normal, and the abnormal information is usually unavailable.

The last but not the least, to identify the security risks from stable systems rather than unstable ones is a greater challenge. If these challenges can be successfully resolved, the control-level and security-level of CPSs will be greatly improved.

Protecting for data and information via the network is the traditional missions of IT security control, however, in control theory field, how to guarantee the stability of system underlying uncertain perturbations or even cyber attacks is the most important issue.

In recent years, a great deal of incidents referring to ICS security have taken place, such as the Stuxnet virus which destroyed the Iranian nuclear program, the breakout in Ukraine and the lose track of light M370 of Malaysia, etc.

All of the accidents aforementioned have caused property losses or even huge casualties. In order to avoid these disadvantages, theoretical and technical scenarios should be operated. During this work, control theory, optimization and game theory approach [5] are mainstream methodologies. Recently, the set-based approach and event-trigger approach are increasing adopted to analyze cyber attacks (see [6] and the reference therein).

Information disclosure is often presented as the first step of the cyber attack process. This step lets the attacker gain the features of the system, consequently, the next step of attack can operate much more easily, such as denial-of-service in [7], stealthy attack models in [8] and synthesis attacks in [9].

The most contributions of this paper are as follows: (1) Fault diagnosis techniques is considered for CPS security issues; (2) Uncertainties are adopted to analyze the character of the ICS based on the cyber attack situation; (3) Theoretical and technical scenarios are proposed to defense the cyber attacks.

Notations: In whole of this paper, $\lambda_{min}(R) = R_{min}$ presents the smallest eigenvalue of matrix R. $E(.)$ and $hash(.)$ are adopted to stand for the encryption function and hash function, respectively.

2   Formulation

2.1   Networked Control Systems (NCSs)

Networked control systems (NCSs) are typical cyber-physical systems (CPSs), which are combined by local-controller, remote-plant, remote-sensors and linking networks to be a entity. The greatest advantage of this structure is that, it can make full use the network to control spatially distributed plants.

A typical model of networked distributed discrete-time system in Fig. 1 is presented as

$$\begin{cases} x(k+1) = (A_p + \Delta A_p)x(k) + B_p\tilde{u}(k) + D_1\omega(k) + Ff(k) \\ y(k) \quad = C_px(k) + D_2\upsilon(k) \end{cases} \tag{1}$$

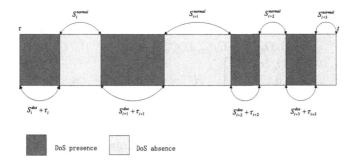

Fig. 1. Flow diagram for security detection for attack with covert agent bedded.

where $x(k) \in R^{n_p}$, $\tilde{u}(k) \in R$ and $y(k) \in R^{n_y}$ denote the state of the plant, control signal input from controller side and measurement output of the plant, respectively. And ΔA_p denotes the uncertainties of the system performance. $w(k) \in R^{n_x}$ and $v(k) \in R^{n_y}$ stand for the input (or model) and the output (or measurement) perturbations, respectively.

Similar to [10], a LTI feedback controller is considered as

$$\begin{cases} z(k+1) = A_c z(k) + B_c \tilde{y}(k) \\ u(k) \quad = C_c z(k) + D_c \tilde{y}_c(k) \end{cases} \tag{2}$$

where $z(k) \in R^{n_z}$, $u(k) \in R^{n_x}$ and $\tilde{y}(k) \in R^{n_y}$ denote controller state, control signal for plant and the feedback measurement of the plants, respectively. In addition, $\tilde{y}_c(k) = \tilde{y}(k) + y_{ref}(k)$, where $y_{ref}(k)$ is the reference input.

Remark 1. In the past, $\tilde{y}(k) = y(k) + \Delta y(k)$ is generally adopted to describe the uncertainties caused by sensor faults or measurement errors. However, in this note, $\Delta y(k)$ is not just for uncertainties, even for cyber attacks in the loop of sensor-to-controller. Particularly, if the cyber attack is absent, $\Delta y(k) = 0$, then $\tilde{y}(k) = y(k)$.

Motivated by the work in [11], an observer-based residual detection structure is formulated as

$$\begin{cases} s(k+1) = A_e s(k) + B_e u_c(k) + E_e \tilde{y}(k) \\ r(k) \quad = C_e s(k) + D_e u_c(k) + F_e \tilde{y}(k) \end{cases} \tag{3}$$

where $s(k) \in R^{n_s}$ and $r(k) \in R^{n_r}$ denote the state of anomaly detector and the residue between estimator and virtual exists for anomalies detection, respectively.

Stacking vectors are defined as $\eta(k) = \left[x(k)^T \; z(k)^T \; s(k)^T \right]^T$, $\xi(k) = \left[w(k)^T \; v(k)^T \right]^T$, $\mu(k) = \left[\Delta u_{att}^T(k) \; \Delta y_{att}^T(k) \right]^T$, and $\mathcal{R}(k)$ represents the residual error of the detector.

$$\begin{cases} \eta(k+1) = \bar{\mathcal{A}}\eta(k) + \bar{\mathcal{B}}\mu(k) + \bar{\mathcal{E}}\xi(k) + \mathcal{H}f(k) + \mathcal{G}_1 y_{ref}(k) \\ \mathcal{R}(k) \quad = \bar{\mathcal{C}}\eta(k) + \bar{\mathcal{D}}\mu(k) + \bar{\mathcal{F}}\xi(k) + \mathcal{G}_2 y_{ref}(k) \end{cases} \tag{4}$$

where
$$\bar{\mathcal{A}} = \begin{bmatrix} A_p + B_p D_c C_p & B_p C_c & 0 \\ B_c C_p & A_c & 0 \\ B_e D_c C_p + E_e C_p & B_e C_c & A_e \end{bmatrix}, \quad \bar{\mathcal{B}} = \begin{bmatrix} B_p & B_p D_c \\ 0 & B_c \\ 0 & B_e D_c + E_e \end{bmatrix},$$

$$\bar{\mathcal{E}} = \begin{bmatrix} D_\omega & B_p D_c D_\nu \\ 0 & 0 \\ 0 & B_e D_c D_\nu + E_e D_\nu \end{bmatrix}, \quad \mathcal{H} = \begin{bmatrix} F \\ 0 \\ 0 \end{bmatrix}, \quad \mathcal{G}_1 = \begin{bmatrix} B_p D_c \\ 0 \\ B_e D_c \end{bmatrix}, \quad \bar{C} =$$

$$\begin{bmatrix} D_e D_c C_p + F_e C_p & D_e C_c & C_e \end{bmatrix}, \quad \bar{\mathcal{D}} = \begin{bmatrix} 0 & D_e D_c + F_e \end{bmatrix}, \quad \bar{\mathcal{F}} = \begin{bmatrix} 0 & D_e D_c D_\nu + F_e D_\nu \end{bmatrix},$$
$$\mathcal{G}_2 = D_e D_c.$$

The reference input $y_{ref}(k)$ is proposed to adjust the outputs of the controller, which can eliminate the fault or attack effectively.

Remark 2. In (4), $y_{ref}(k)$ is special designed to adjust the controller output, which is adopted to fight against the perturbations $\omega(k), \nu(k)$ and attacks $\Delta u_{att}(k), \Delta y_{att}(k)$. Without of losing generality, $y_{ref}(k) = \mathcal{J} x_c(k)$ is employed to evolve the model of (2). And, $D_\omega = I_{n_x}, D_\nu = I_{n_y}$ are defined to for simple.

Then, from Eq. (4), we have

$$\begin{cases} \eta(k+1) = \bar{\mathcal{A}}_c \eta(k) + \bar{\mathcal{B}}_c \mu(k) + \bar{\mathcal{E}}_c \xi(k) + \mathcal{H} f(k) \\ \mathcal{R}(k) \quad = \bar{C}_c \eta(k) + \bar{\mathcal{D}}_c \mu(k) + \bar{\mathcal{F}}_c \xi(k) \end{cases} \tag{5}$$

where
$$\bar{\mathcal{A}}_c = \begin{bmatrix} A_p + B_p D_c C_p & B_p C_c + B_p D_c \mathcal{J} & 0 \\ B_c C_p & A_c & 0 \\ B_e D_c C + E_e C & B_e C_c + B_e D_c \mathcal{J} & A_e \end{bmatrix}, \quad \bar{\mathcal{B}}_c = \begin{bmatrix} B_p & B_p D_c \\ 0 & B_c \\ 0 & B_e D_c + E_e \end{bmatrix}, \quad \bar{\mathcal{E}}_c =$$

$$\begin{bmatrix} I_{n_x} & B_p D_c \\ 0 & 0 \\ 0 & B_e D_c + E_e \end{bmatrix}, \quad \mathcal{H} = \begin{bmatrix} F \\ 0 \\ 0 \end{bmatrix}, \quad \bar{C}_c = \begin{bmatrix} D_e D_c C + F_e C & D_e C_c + D_e D_c \mathcal{J} & C_e \end{bmatrix},$$
$$\bar{\mathcal{D}}_c = \begin{bmatrix} 0 & D_e D_c + F_e \end{bmatrix}, \quad \bar{\mathcal{F}}_c = \begin{bmatrix} 0 & D_e D_c I_{n_y} + F_e I_{n_y} \end{bmatrix}.$$

Assumption 1. We assume $\Delta u_f(k) = \mathcal{K} \Delta x_f(k)$ for simple, then we can easily find $\Delta x_f(k)$ is related with $\Delta u_f(k)$ from state feedback or output feedback control law. According to above work, we assumption $\Delta \tilde{u}(k) = \Delta u_{att}(k) + \Delta u_f(k)$ and $\Delta \tilde{y}(k) = \Delta y_{att}(k) + \Delta y_f(k)$. Then, the definition of $\mu(k)$ can be rewritten as

$$\tilde{\mu}(k) = \begin{bmatrix} \Delta u^T(k) & \Delta y^T(k) \end{bmatrix}^T = \begin{bmatrix} \Delta u_{att}^T(k) + \Delta u_{k_f}^T(k) & \Delta y_{att}^T(k) + \Delta y_{k_f}^T(k) \end{bmatrix}^T$$

In a word, system faults and cyber-attacks, the double closed-loop system model (5) can be further synthesized as

$$\begin{cases} \eta(k+1) = \bar{\mathcal{A}}_z \eta(k) + \bar{\mathcal{B}}_z \tilde{\mu}(k) + \bar{\mathcal{E}} \xi(k) \\ \mathcal{R}(k) \quad = \bar{C}_z \eta(k) + \bar{\mathcal{D}}_z \tilde{\mu}(k) + \bar{\mathcal{F}} \xi(k) \end{cases} \tag{6}$$

where
$$\bar{\mathcal{A}}_z = \begin{bmatrix} A_p + B_p D_c C_p & B_p C_c + B_p D_c \mathcal{J} & 0 \\ B_c C_p & A_c & 0 \\ B_e D_c C + E_e C & B_e C_c + B_e D_c \mathcal{J} & A_e \end{bmatrix}, \quad \bar{\mathcal{B}}_z = \begin{bmatrix} B_p + M & B_p D_c \\ 0 & B_c \\ 0 & B_e D_c + E_e \end{bmatrix},$$

$$\bar{\mathcal{E}} = \begin{bmatrix} I_{n_x} & B_p D_c \\ 0 & 0 \\ 0 & B_e D_c + E_e \end{bmatrix}, \quad \bar{C} = \begin{bmatrix} D_e D_c C + F_e C & D_e C_c + D_e D_c \mathcal{J} & C_e \end{bmatrix}, \quad \bar{\mathcal{D}} = \begin{bmatrix} 0 & D_e D_c + F_e \end{bmatrix}.$$

Remark 3. The coefficient F is changed according to $f(k)$. Since $\Delta u_f(k)$, $\Delta x_f(k)$ and $\Delta y_f(k)$ are related with A_p, B_p and C_p, respectively. Such that, $\begin{bmatrix} B_p & A_p & C_p \end{bmatrix}$ can be used to take place the matrix F.

$$\begin{cases} \eta(k+1) = \bar{\bar{\mathcal{A}}}_z \eta(k) + \bar{\bar{\mathcal{B}}}_z \varpi(k) \\ \mathcal{R}(k) \quad = \bar{\bar{C}}_z \eta(k) + \bar{\bar{\mathcal{D}}}_z \varpi(k) \end{cases} \tag{7}$$

where $\varpi(k) = \begin{bmatrix} \tilde{\mu}^T(k) & \xi^T(k) \end{bmatrix}^T$, which represent all coupled disturbances, and $\mathcal{R}(k)$ represents the residue of the synthesis system.

2.2 Information Disclosure

The attacker capture the information of the system by the approaches monitoring, scanning, enumeration as well as destroy, infection an even advance persistent threat (APT) [12]. These scenarios helps the attacker to find the bypass and back door of the application software.

The transmission sequence via network includes control input (controller-to-actuator) and measurement output (sensor-to-controller). In time k, the adversary's attack scenarios can be modeled as

$$S_{eq}(k) = \begin{bmatrix} \Gamma_u(k) & 0 \\ 0 & \Gamma_y(k) \end{bmatrix} \begin{bmatrix} u(k) \\ y(k) \end{bmatrix}, k \in [0, \infty)$$

Thus, all the sequence of the signals can be formulated as

$$S(k) = \bigcup_{k=0}^{\infty} \left\{ \begin{bmatrix} \Gamma_u(k) & 0 \\ 0 & \Gamma_y(k) \end{bmatrix} \begin{bmatrix} u(k) \\ y(k) \end{bmatrix} \right\}$$

In virtual case, $\{\Gamma_u(k), \Gamma_y(k)\} \in \{0, 1\}$ is generally considered, where "1" represents the both process of the signal's transmission and reception are successful, otherwise "0" means the failure case of transmission and reception.

For further study, we define $\alpha = diag\{\alpha_i, \alpha_j\}$ and $\beta = diag\{\beta_i, \beta_j\}$ as the probability of transmission data arrive rate and intercepted data mission success rate, respectively. Then, the models changed as:

$$S_{eq}(k) = S_{eq}(k-1) \bigcup \left\{ \begin{bmatrix} \alpha_i \Gamma^u & 0 \\ 0 & \beta_i \Gamma^y \end{bmatrix} \begin{bmatrix} u(k) \\ y(k) \end{bmatrix} \right\} \text{ and }$$

$$S_{attack}(k) = S_{attack}(k-1) \bigcup \left\{ \begin{bmatrix} \alpha_j \Gamma_a^u & 0 \\ 0 & \beta_j \Gamma_a^y \end{bmatrix} \begin{bmatrix} u(k) \\ y(k) \end{bmatrix} \right\} \text{ where } \{\alpha_i, \alpha_j, \beta_i, \beta_j\}$$

$\in [0, 1]$, denote the probability of ..., which are often defined as:

$$\begin{cases} P\{\Gamma^u = 1\} = \alpha, \ P\{\Gamma^u = 0\} = 1 - \alpha; \\ P\{\Gamma^y = 1\} = \beta, \ P\{\Gamma^y = 0\} = 1 - \beta; \end{cases} \tag{8}$$

2.3   DoS Attack VS. Packet Lose

In this note, the analysis process of DoS is as follow. For the time interval $[k_0, k_m]$, the sequence transmitted is packaged as $S_{eq}(k) = \{S_{eq}^{k_0}(k), \cdots, S_{eq}^{k_m}(k)\}$, each $S_{eq}^{k_i}(k)$ contains control signal $u(k)$ and sensor measurement $y(k)$, which are considered as key-data of the NCSs in detail form as

$$\chi^u = \bigcup_{k=1}^{N} u(k) \quad and \quad \chi^y = \bigcup_{k=1}^{N} y(k)$$

To facilitate the analysis, the following definitions of DoS are given.

Definition 1. *For given closed-loop system, χ^u and χ^y are control and sensor output sequences, respectively. Stacking them as $\chi = diag\{\chi^u, \chi^y\}$, with $\chi^u \in R^{n_u}, \chi^y \in R^{n_y}$, then models for DoS can be constructed as*

$$\begin{cases} U_{DoS}^k := \chi^u - I_{n_u} \\ Y_{DoS}^k := \chi^y - I_{n_y} \end{cases} \tag{9}$$

By checking the value of U_{DoS}^k and Y_{DoS}^k, it can be determined that when and which physical signal is suffering DoS attack.

In $[k_0, k_m]$, with $0 \le k_0 < k_m$, the DoS presence and absent set can be defined as

$$S_{nor} := \bigcup_{i=1}^{\infty} [s_{att}^i + \tau_i, s_{att}^{i+1}) \tag{10}$$

$$S_{att} := \{s_{att}^i\} \cup [s_{att}^i, s_{att}^i + \tau_i), \quad i \in [0, m) \tag{11}$$

Furthermore, the following two conclusions can be obtained:

1. $[k_0, k_m] = S_{att} \bigcup S_{nor}$;
2. $S_{att} \bigcap S_{nor} = \phi$;

In order to demonstrate it clearly, a diagram is given as Especially, for intermittent DoS, the following diagram can describe the process clearly.

Definition 2. *Effective DoS Duration Time (EDDT):* *Observe the i_{th} DoS interval $[S_{att}^i, S_{att}^i + \tau_i]$, the i_{th} duration time is τ_i, as depicted in Fig. 2. If τ_i can successively caused the system lost stable, we call it as effective DoS duration time (EDDT).*

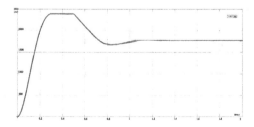

Fig. 2. the speed of the DC motor under the normal case

2.4 Stealth/Covert Attack

Considering the cyber attack, the NCS in (1) can be rewritten as

$$\begin{cases} y_p(k) = C_p x(k) + D_2 \upsilon(k) \\ \tilde{u}(k) = \mathcal{K} x(k) \end{cases} \tag{12}$$

where $\tilde{u}(k)$ denotes the control input from controller transmitted via network, and $\tilde{u}_c(k) = \mathcal{K} x(k)$ is the state feedback control law. For linear case, it is represented as $\tilde{u}_c(k) = u_c(k) + \Delta u_{att}(k)$, where $\Delta u_{att}(k)$ is usually taken as cyber attack in controller-to-sensor channel. Synthesis the two equations in (13), we can get

$$y_p(k) = C_p \mathcal{K}^{-1} \tilde{u}_c(k) + D_2 \upsilon(k) \tag{13}$$

According to Eq. (2),

$$u_c(k) = C_c z(k) + D_c \left[y_p(k) + \Delta y_{att}(k) + y_{ref}(k) \right] \tag{14}$$

Remark 4. On the one hand, for normal case, $\Delta u_{att}(k) = 0, \Delta y_{att}(k) = 0$ and thus $\tilde{u}_c(k) = u_c(k), y_{p-c}(k) = y_p(k)$, then system (12) and (14) reduce to NCS (1) and (2). On the other hand, the system is lost of security, it means $\mu(k) \neq 0, \Delta y_{att}(k) \neq 0$ and $\tilde{u}_c(k) \neq u_c(k), y_{p-c}(k) \neq y_p(k)$

Combining (12)–(14) gives, in the nominal case, the closed loop response as

$$y_p(k) = \Psi \left[C_p \mathcal{K}^{-1} C_c z(k) + C_p \mathcal{K}^{-1} y_{ref}(k) + D_2 \upsilon(k) \right] \tag{15}$$

where $\Psi = \left(I - C_p \mathcal{K}^{-1} D_c \right)^{-1}$.

Since the attacker has been learning and imitating the original system, the model of attacker can be formulated similarly to the original system

$$\begin{cases} \Delta y_{att}(k) = \Pi_\mu \Delta u_{att}(k) \\ \Delta u_{att}(k) = \Theta_\mu \Delta y_{att}(k) + \Theta_{ref} \Delta y_{ref}(k) \end{cases} \tag{16}$$

where Π_μ, Θ_u and Θ_{ref} are the matrices that need to be determined and adjust according to the learning errors. This feedback loop is driven by the $\Delta y_{ref}(k)$ input giving

$$\begin{cases} \Delta u_{att}(k) = (I - \Theta_\mu \Pi_\mu)^{-1} \Theta_{ref} \Delta y_{ref}(k) \\ \Delta y_{att}(k) = \Pi_\mu (I - \Theta_\mu \Pi_\mu)^{-1} \Theta_{ref} \Delta y_{ref}(k) \end{cases} \tag{17}$$

According to (13) and (14), the case $\Pi_\mu = C_p \mathcal{K}^{-1} + D_2 \upsilon(k) u_c^{-1}(k)$ is ideal, which indicates the error between virtual system and covert agent is zero and the original system is well learned and mastered by the attacker.

3   Main Results

3.1   System Design Objections

The information output from controller is packaged as $\{T_{stamp}(k), U_c^w(k),$ $U_c^d(k)\}$, which contains time-stamp, work signals and detection signals.

$T_{stamp}(k)$ represents the time-stamp function at instant k, it is obtained by

$$T_{stamp}(k) = E\left(co\left\{hash_1(k), time(k), date(k)\right\}\right) \qquad (18)$$

Among them, $U_c^w(k) = E(u_c(k))$ denotes the encrypted control signal, and $U_c^d(k) = hash(U_c^w(k))$ is the detection signal.

Based on above works, some significant conclusions can be derived as follows.

Theorem 1. *For given controller output transmission sequence $S_{eq}^c(k) =$ $\{T_{stamp}(k), U_c^w(k), U_c^d(k)\}$ and the actuator received and decrypted sequence is $\tilde{S}_{eq}^c(k) = \{\tilde{T}_{stamp}(k), \tilde{U}_c^w(k), \tilde{U}_c^d(k)\}$, the system (7) is said to be secure with no covert agent, if the following conditions satisfied:*

1. $\tilde{T}_{stamp}(k) > T_{stamp}(k)$;
2. $hash(\tilde{U}_c^w(k)) = \tilde{U}_c^d(k)$;

Proof. As the limitation of pages, the proof is neglected. If the read need it, please contact the author.

Theorem 2. *For given sensor measurement output sequence $\Gamma_p(k) =$ $\{T_{stamp}^p(k), Y_p^w(k), Y_p^{d-all}(k), Y_p^d(k)\}$ and controller side feedback sequences packaged as $\tilde{\Gamma}_p(k) = \{\tilde{T}\ _{stamp}^p(k), \tilde{Y}_p^{w-all}(k), \tilde{Y}_p^{d-all}(k), \tilde{Y}_p^d(k)\}$, synthesis Theorem 1, it is inferred that the system Σ composed of (1)–(3) is security, if the following inequalities holds*

1. $T_{stamp}^p \leq \tilde{T}_{stamp}^p$;
2. $Hash(\tilde{Y}_p^w(k)) = Y_p^d(k)$;
3. $Y_p^{d-all}(k) = Y_p^d(k)$;

4   Examples

The DC motor is widely applied in the industrial systems since its previous performance in large range adjustable speed. Assuming the magnetization curve is taken as linear one. The armature voltage u_a is generally presented as

$$u_a = R_a i_a + L_a \frac{di_a}{dt} + K_m \phi \omega \qquad (19)$$

where ϕ represents the pole flux, which has a hysteretic nonlinearity. In virtual case, it is usually operated in the linear region for simplicity as $\phi = L_f i_f$.

Table 1. Parameter notations of DC motor system

Parameters	Physical concept	Value	Parameters	Physical concept	Value
P	Rated power	3.73 kw	ω_{ref}	Rated speed	183.26 Rad/s
$I_{a(norm)}$	Rated armature current	16.74 A	$I_{f(norm)}$	Rated field current	4 A
T_e	Rated torque	18 Nm	U_f	Field voltage (maximum)	240 V
L_a	Armature inductance	0.01 H	L_f	Field winding inductance	60 H
J_m	Motor inertia	0.208 kgm^2	B_m	Motor damping	0.011 kgm^2
K_m	Motor torque constant	0.3 Nm/A^2	R_a	Armature resistance	1.2 Ω
R_f	Field resistance	60 Ω			

For pages limitation, the notation of the parameters in above equations are give in Table 1 together with the values the examples needed.

According to the equations of AD motor in [8], the DC dynamic model can be given as

$$\dot{x}(t) = (A + \Delta A)x(t) + B\left(u(t) + \Delta u(t)\right) + \tilde{\omega}(t) \qquad (20)$$

where $A = \begin{bmatrix} -\frac{R_a}{L_a} & -\frac{K_m u_f}{L_a R_f} & 0 \\ \frac{K_m u_f}{J R_f} & -\frac{B}{J} & 0 \\ 0 & -1 & 0 \end{bmatrix}$, $B = \begin{bmatrix} \frac{1}{L_a} \\ 0 \\ 0 \end{bmatrix}$, $\tilde{\omega}(t) = \begin{bmatrix} 0 \\ 0 \\ \omega_{ref} \end{bmatrix}$, $u(t) = U_a(t)$.

The obvious relationship of uncertainties and normal parameters are $R_a = R_{a_{normal}} + \Delta R_a$, $R_f = R_{f_{normal}} + \Delta R_f$, and $T_L = T_{L_{normal}} + \Delta T_L$. Furthermore, control input is presented as $u_a(t) = u_{a_{normal}}(t) + \Delta u_a(t)$, which is usually neglected before. Nevertheless, since its important influence to the system, it is reconsider in a great deal of recent publications.

From recent studies, we find that, the uncertainties $\Delta u_a(t)$ is generally caused by cyber attacks. $A = \begin{bmatrix} -\frac{R_{a_{normal}}}{L_a} & -\frac{K_m U_f}{L_a R_{f_{normal}}} & 0 \\ \frac{K_m U_f}{J R_{f_{normal}}} & -\frac{B}{J} & 0 \\ 0 & -1 & 0 \end{bmatrix}$, $\Delta A = $

$\begin{bmatrix} -\frac{\Delta R_a}{L_a} & -\frac{K_m U_f}{L_a \Delta R_f} & 0 \\ \frac{K_m U_f}{J \Delta R_f} & 0 & 0 \\ 0 & 0 & 0 \end{bmatrix} = \begin{bmatrix} \Delta A_{11} & \Delta A_{12} & 0 \\ \Delta A_{21} & 0 & 0 \\ 0 & 0 & 0 \end{bmatrix}$, where the parameter uncertainties

are often decomposed as $\Delta A = HF(t)E$.

In this note, static state feedback controller is designed as $u(t) = -Kx(t) = -\begin{bmatrix} k_1 & k_2 & k_3 \end{bmatrix} x(t)$.

According to the robust controller design theory (see [13] and the references therein), the above controller can be obtained by using the Lemma 2.5 in to get a symmetric definite matrix P. Borrowing from [13], $K = \begin{bmatrix} 0.37265 \ 1.1029 \ -8.0814 \end{bmatrix}$.

If the parameter uncertainties is varying with $\pm 30\%$ of the normal ones. Such that, we have $\Delta A = \begin{bmatrix} -36 & -400 & 0 \\ 192.3 & 0 & 0 \\ 0 & 0 & 0 \end{bmatrix}$ or $\Delta A = \begin{bmatrix} 36 & 400 & 0 \\ -192.3 & 0 & 0 \\ 0 & 0 & 0 \end{bmatrix}$

Applying the uncertainties as effect of attack, then we have

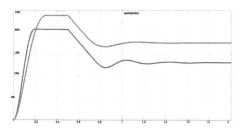

Fig. 3. The speed of the DC motor of normal case and with the uncertainties case

From the Fig. 3, we can find that, the uncertainties caused the decline of the motor speed, the red line represents the normal case, and the blue line denotes the abnormal case. In the sophisticated mentioned in [8], the decline of the speed can not be detected so easily.

5   Conclusion

In this paper, four kinds of cyber attack are considered, the models of these attacks are formulated and the analysis of attack features are given. From the analysis, all of the attack models can formulated to be one unite form. It is the predominant contribution of this paper. Aiming to eliminate these attacks, a double closed-loop structure with defense function is designed, the detailed algorithm is present to demonstrate the anti-attack process as well.

References

1. Sandberg, H., Amin, S., Johansson, K.: Cyberphysical security in networked control systems: an introduction to the issue. IEEE Control Syst. **35**(1), 20–23 (2015)
2. How, J.P.: Cyberphysical security in networked control systems [about this issue]. IEEE Control Syst. **35**(1), 8–12 (2015)
3. Cárdenas, A.A., Amin, S., Sinopoli, B., Perrig, A., Sastry, S.: Challenges for securing cyber physical systems. In: First Workshop on Cyber-Physical Systems Security, pp. 363–369 (2006)

4. Neuman, D.C.: Challenges in security for cyber-physical systems. In: Proceedings of S&T Workshop Future Directions Cyber-Physical System Security, No. 1, pp. 1–2 (2009)
5. Zhu, Q., Basar, T.: Game-theoretic methods for robustness, security, and resilience of cyberphysical control systems: games-in-games principle for optimal cross-layer resilient control systems. IEEE Control Syst. **35**(1), 46–65 (2015)
6. Yue, D., Tian, E., Han, Q.L.: A delay system method for designing event-triggered controllers of networked control systems. IEEE Trans. Autom. Control **58**(2), 475–481 (2013)
7. Householder, A., Manion, A., Pesante, L., Weaver, G.: Managing the threat of denial-of-service attacks. CERT Coord. Center **33**(4), 99–110 (2001)
8. Smith, R.S.: Covert misappropriation of networked control systems: presenting a feedback structure. IEEE Control Syst. **35**(1), 82–92 (2015)
9. Pasqualetti, F., Dorfler, F., Bullo, F.: Attack detection and identification in cyber-physical systems. IEEE Trans. Autom. Control **58**(11), 2715–2729 (2013)
10. Zhou, K., Doyle, J.C., Glover, K.: Robust and Optimal Control. Prentice-Hall, Inc., Upper Saddle River (1996)
11. Ding, S.: Model-Based Fault Diagnosis Techniques: Design Schemes, Algorithms, and Tools. Springer Science & Business Media, Heidelberg (2008)
12. Knapp, E.D.: Industrial Network Security: Securing Critical Infrastructure Networks for Smart Grid, SCADA, and Other Industrial Control Systems. Syngress, Rockland (2011)
13. Zhou, J., Wang, Y., Zhou, R.: Global speed control of separately excited DC motor. In: Power Engineering Society Winter Meeting, vol. 3, pp. 1425–1430 (2001)

Identification Approach
of Hammerstein-Wiener Model Corrupted
by Colored Process Noise

Feng Li, Li Jia$^{(\boxtimes)}$, and Qi Xiong

Department of Automation, College of Mechatronics Engineering
and Automation, Shanghai University, Shanghai 200072, China
jiali@staff.shu.edu.cn

Abstract. For Hammerstein-Wiener model with colored process noise, this paper derives an identification approach. The correlation function between input and output data points is derived by using separable signal to realize that the unmeasurable internal variable is replaced by the correlation function of input, and then correlation analysis method is used to estimate the parameters of the output nonlinear part and linear part. Furthermore, a correction term is added to least square estimation to compensate error caused by process noise, and then to derive an error compensation recursive least square method for the observed data from Hammerstein-Wiener model. Therefore, the parameters of the input nonlinear part can be estimated by error compensation method. Finally, the advantages of proposed algorithm are shown by simulation example.

Keywords: Hammerstein-Wiener model · Separable signal · Process noise

1 Introduction

Hammerstein model and Wiener model are two typical simple nonlinear block oriented models, which consists of the combination of dynamics linear parts and static nonlinear parts. A number of methods have been devoted to identify two types of block oriented models [1–4].

Apart from the above two kinds of models, Hammerstein-Wiener model is a more complex nonlinear model and is composed of a dynamics linear part between two static nonlinear parts. Bai proposed a two-stage identification algorithm [5] and a blind identification approach [6]. In [7], a recursive algorithm is studied for a special form of Hammerstein-Wiener system with dead-zone nonlinearity input block. Recently, many other algorithms have been used to identify Hammerstein-Wiener system with measurement noise or process noise [8, 9]. However, these methods only consider the identification of Hammerstein-Wiener model with white noises. Few papers addressed the Hammerstein-Wiener model in the case of colored noise. Identification method based instrumental variable is proposed for Hammerstein-Wiener system in [10]. In [11], gradient based and least square based iterative learning algorithms are included, which can successfully estimate the matrix of unknown parameter as well as the colored noises. It is worth pointing out that the colored noise mentioned in literatures is a linearly correlated process of white noise.

© Springer Nature Singapore Pte Ltd. 2017
M. Fei et al. (Eds.): LSMS/ICSEE 2017, Part I, CCIS 761, pp. 432–441, 2017.
DOI: 10.1007/978-981-10-6370-1_43

The aim of this paper is to study parameter identification of Hammerstein-Wiener model with colored noise. The identifications of the linear part and input nonlinear part are carried out independently by using special input signals. At first, the parameters of the output nonlinear part and linear part can be identified by correlation analysis method. In addition, a correction term is added to least square estimation to compensate error caused by process noise, and then to derive an error compensation recursive least square method to estimate the parameters of the input nonlinear part.

Briefly, the paper is organized as follows. In Sect. 2, the identification problem of neuro-fuzzy based Hammerstein-Wiener model with colored process noise is briefly described. An identification approach of neuro-fuzzy based Hammerstein-Wiener model with colored process noise is presented in Sect. 3. Section 4 gives simulation example. Finally, the concluding remarks are approached in Sect. 5.

2 Neuro-Fuzzy Based Hammerstein-Wiener Model with Colored Process Noise

Consider the Hammerstein-Wiener model [8] depicted in Fig. 1 that consists of static input nonlinear part $f(\cdot)$, dynamic linear part $H(\cdot)$ and static output nonlinear part $g(\cdot)$ as given by

$$v(k) = f(u(k)), \quad z(k) = \frac{B(z)}{A(z)} v(k) + e(k), \quad y(k) = g(z(k)) \tag{1}$$

Where $v(k)$, $x(k)$ and $z(k)$ are unmeasurable internal variable, $e(k)$ is colored noise, $A(z) = 1 + a_1 z^{-1} + a_2 z^{-2} + \ldots + a_{n_a} z^{-n_a}$ and $B(z) = b_1 z^{-1} + b_2 z^{-2} + \ldots + b_{n_b} z^{-n_b}$ are the parameter vector of dynamic linear part, a_i and b_j are the parameters of the linear part.

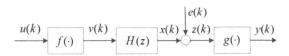

Fig. 1. The structure of Hammerstein-Wiener model

The input nonlinear function $f(\cdot)$ and the output nonlinear function $g(\cdot)$ are approximated by neuro-fuzzy model [12], the linear part is represented by polynomial model. The input nonlinear part and the output nonlinear part can be represented as

$$\hat{v}(k) = \hat{f}(u(k)) = \sum_{l=1}^{L^{input}} \phi_l(u(k)) w_l^{input}, \quad \hat{z}(k) = \hat{g}^{-1}(y(k)) = \sum_{l=1}^{L^{output}} \phi_l(y(k)) w_l^{output} \tag{2}$$

Where $\phi_l(u(k)) = \mu_l(u(k)) / \sum_{l=1}^{L} \mu_l(u(k))$, $\phi_l(y(k)) = \mu_l(y(k)) / \sum_{l=1}^{L} \mu_l(y(k))$,

$\mu_l(u(k)) = \exp\left(-(u(k) - c_l)^2/\sigma_l^2\right)$ and $\mu_l(y(k)) = \exp\left(-(y(k) - c_l)^2/\sigma_l^2\right)$ are Gaussian membership functions, c_l and σ_l is the center and width of membership function, respectively, w_l^{input} and w_l^{output} are weights, and L is total fuzzy rules.

3 Identification Approach of Neuro-Fuzzy Based Hammerstein-Wiener Model with Colored Process Noise

This section presents the identification procedure of Hammerstein-Wiener model with colored process noise. Our previous work [13] pointed out that the cross-correlation function of input and internal variable can be treated as the auto-correlation function of input for Hammerstein model in the case of separable signal.

3.1 Identification of the Output Nonlinear Part and the Linear Part

According to Eqs. (1) and (2), the linear part can be expressed as

$$z(k) = -\sum_{i=1}^{n_a} a_i z(k - i) + \sum_{j=1}^{n_b} b_j v(k - j) + \sum_{i=1}^{n_a} a_i e(k - i) + e(k) \tag{3}$$

Multiplying both sides of Eq. (3) by $u(k - \tau)$, and calculating the mathematical expectations, we obtain

$$R_{zu}(\tau) = -\sum_{i=1}^{n_a} a_i R_{zu}(\tau - i) + \sum_{j=1}^{n_b} b_j R_{vu}(\tau - j) + \sum_{i=1}^{n_a} a_i R_{eu}(k - i) + R_{eu}(\tau) \tag{4}$$

Since $e(k)$ is independent of input $u(k)$, we have $R_{eu}(\tau) = 0$, therefore

$$R_{zu}(\tau) = -\sum_{i=1}^{n_a} a_i R_{zu}(\tau - i) + \sum_{j=1}^{n_b} b_j R_{vu}(\tau - j) \tag{5}$$

According to [13], the cross-correlation function $R_{vu}(\tau)$ can be replaced with auto-correlation function $R_u(\tau)$. We derive

$$R_{zu}(\tau) = -\sum_{i=1}^{n_a} a_i R_{zu}(\tau - i) + \sum_{j=1}^{n_b} \tilde{b}_j R_u(\tau - j) \tag{6}$$

Where $\tilde{b} = b_0 b_j$.
Using Eq. (2) and (5) get

$$E\left(\phi_1^{output}(y(k))u(k-\tau)\right) =$$

$$-\sum_{l=2}^{L^{output}} \tilde{w}_l^{output} E\left(\phi_l^{output}(y(k))u(k-\tau)\right) - \sum_{i=1}^{n_a} a_i E\left(\phi_1^{output}(y(k))u(k-\tau+i)\right) \tag{7}$$

$$-\sum_{i=1}^{n_a} a_i \sum_{l=2}^{L^{output}} \tilde{w}_l^{output} E\left(\phi_l^{output}(y(k))u(k-\tau+i)\right) + \sum_{j=1}^{n_b} \bar{b}_j R_u(\tau-j)$$

Where $\tilde{w}_l^{output} = w_l^{output}/w_1^{output}$ $(l=2,3,\ldots,L^{output})$, $\bar{b}_j = \tilde{b}_j/w_1^{output}$ $(j=1,2,\ldots,n_b)$.

Let $\phi_l(k) = \phi_l^{output}(y(k))$ $(l=1,2,\ldots,L^{output})$, then Eq. (7) can be expressed as

$$R_{\phi_1 u}(\tau) = -\sum_{l=2}^{L^{output}} \tilde{w}_l^{output} R_{\phi_l u}(\tau) - \sum_{i=1}^{n_a} a_i R_{\phi_1 u}(\tau-i) -$$

$$\sum_{i=1}^{n_a} a_i \sum_{l=2}^{L^{output}} \tilde{w}_l^{output} R_{\phi_l u}(\tau-i) + \sum_{j=1}^{n_b} \bar{b}_j R_u(\tau-j) \tag{8}$$

Furthermore, the correlation analysis algorithm [14] is used to estimate the parameters. Suppose $\tau = 1,2,\ldots,P$ $(P \geq n_a L^{output} + n_b + L^{output} - 1)$. Thus

$$\hat{\theta} = R\Phi^T(\Phi\Phi^T)^{-1} \tag{9}$$

Where $\hat{\theta} = \left[\tilde{w}_2^{output},\ldots \tilde{w}_{L^{output}}^{output}, a_1,\ldots.a_{n_a}, a_1\tilde{w}_2^{output},\ldots.a_1\tilde{w}_{L^{output}}^{output},\ldots a_{n_a}\tilde{w}_2^{output},\ldots a_{n_a}\tilde{w}_{L^{output}}^{output}, \bar{b}_1,\ldots\bar{b}_{n_b}\right]$,

$R = \left[R_{\phi_1 u}(1), R_{\phi_1 u}(2),\ldots,R_{\phi_1 u}(P)\right]$, $\Phi =$

$$\begin{bmatrix}
-R_{\phi_2 u}(1) & -R_{\phi_2 u}(2) & -R_{\phi_2 u}(3) & \cdots & -R_{\phi_2 u}(P) \\
\vdots & \vdots & \vdots & & \vdots \\
-R_{\phi_l u}(1) & -R_{\phi_l u}(2) & -R_{\phi_l u}(3) & \cdots & -R_{\phi_l u}(P) \\
-R_{\phi_1 u}(0) & -R_{\phi_1 u}(1) & -R_{\phi_1 u}(2) & \cdots & -R_{\phi_1 u}(P-1) \\
\vdots & \vdots & \vdots & & \vdots \\
0 & 0 & 0 & \cdots & -R_{\phi_1 u}(P-n_a) \\
-R_{\phi_2 u}(0) & -R_{\phi_2 u}(1) & -R_{\phi_2 u}(2) & \cdots & -R_{\phi_2 u}(P-1) \\
\vdots & \vdots & \vdots & & \vdots \\
-R_{\phi_l u}(0) & -R_{\phi_l u}(1) & -R_{\phi_l u}(2) & \cdots & -R_{\phi_l u}(P-1) \\
\vdots & \vdots & \vdots & & \vdots \\
0 & 0 & 0 & \cdots & -R_{\phi_l u}(P-n_a) \\
R_u(0) & R_u(1) & R_u(2) & \cdots & R_u(P-1) \\
\vdots & \vdots & \vdots & & \vdots \\
0 & 0 & 0 & \cdots & R_u(P-n_b)
\end{bmatrix}.$$

3.2 Identification of the Input Nonlinear Part

In the case of random signal with uniform distribution, the procedure of identifying neuro-fuzzy based model is to estimate the parameters c_l, σ_l and w_l^{intput}, the parameters c_l and σ_l are be calculated by cluster method in [15]. However, the key issue is to estimate the parameters w_l^{intput}.

Problem formulation

According to Eq. (3), we can draw the estimation θ_1 by using least square method

$$\hat{\theta}_{LS}(k) = \theta_1 + P(k)\sum_{i=1}^{k}\varphi(i)\left[\psi^T(i)\theta_1 + e(i)\right] \tag{10}$$

Where $P^{-1}(k) = \sum_{i=1}^{k}\varphi(i)\varphi^T(i)$,

$$\theta_1 = [a_1, a_2, \ldots, a_{n_a}, b_1 w_1^{intput}, b_1 w_2^{intput}, \ldots, b_1 w_L^{intput}, \ldots, b_{n_b} w_1^{intput}, b_{n_b} w_2^{intput}, \ldots, b_{n_b} w_L^{intput}]^T$$

$$\varphi(k) = [-z(k-1), \ldots, -z(k-n_a), \phi_1(u(k-1)), \ldots, \phi_L(u(k-1)), \ldots, \phi_1(u(k-n_b)), \ldots, \phi_L(u(k-n_b))]^T$$

$$\psi(k) = [e(k-1), e(k-2), \ldots, e(k-n_a), 0, \ldots, 0]^T$$

Multiplying by $P^{-1}(k)/k$ on both sides of Eq. (10), and taking limit yield get

$$\lim_{k\to\infty}\frac{1}{k}P^{-1}(k)[\hat{\theta}_{LS}(k) - \theta_1] = \lim_{k\to\infty}\left[\frac{1}{k}\sum_{i=1}^{k}\varphi(i)\psi^T(i)\right]\theta_1 + \lim_{k\to\infty}\frac{1}{k}\sum_{i=1}^{k}\varphi(i)e(i) \tag{11}$$

Defining the noise correlation function

$$r_e(j) = \lim_{k\to\infty}\frac{1}{k}\sum_{i=1}^{k}e(i-j)\,e(i),\ j = 0, 1, \cdots, n_a \tag{12}$$

Combing Eqs. (1) and (2) with (10) get

$$\lim_{k\to\infty}\left[\frac{1}{k}\sum_{i=1}^{k}\varphi(i)\psi^T(i)\right]\theta_1 = -R\theta_1,\quad \lim_{k\to\infty}\frac{1}{k}\sum_{i=1}^{k}\varphi(i)e(i) = -p \tag{13}$$

Where

$$R = \begin{bmatrix} \gamma & 0 \\ 0 & 0 \end{bmatrix},\ p = [\rho,\ 0],\ \rho = [r_e(1),\ r_e(2),\ \cdots\ r_e(n_a)],\ \gamma =$$

$$\begin{bmatrix} r_e(0) & r_e(1) & \cdots & r_e(n_a - 1) \\ r_e(1) & r_e(0) & \cdots & r_e(n_a - 2) \\ \vdots & \vdots & \ddots & \vdots \\ r_e(n_a - 1) & r_e(n_a - 2) & \cdots & r_e(0) \end{bmatrix}$$

Therefore, the following equation can be obtained:

$$\lim_{k\to\infty}\frac{1}{k}P^{-1}(k)[\hat{\theta}_{LS}(k) - \theta_1] = -(R\theta_1 + p) \tag{14}$$

We introduce $\Delta\theta(k) = -P(k)k(R\theta_1 + p)$ to Eq. (10) and get

$$\lim_{k\to\infty} \hat{\theta}_{LS}(k) = \theta_1 - \lim_{k\to\infty} P(k)k(R\theta_1 + p) = \theta_1 + \lim_{k\to\infty} \Delta\theta(k) \tag{15}$$

Equation (15) indicates that the least square estimation $\hat{\theta}_{LS}(k)$ is biased. If a compensation term $\Delta\theta(k)$ is introduced in $\hat{\theta}_{LS}(k)$, we can acquire unbiased estimation $\hat{\theta}_B(k)$.

$$\hat{\theta}_B(k) = \hat{\theta}_{LS}(k) + P(k)k[\hat{R}(k)\hat{\theta}_B(k-1) + \hat{p}(k)] \tag{16}$$

Neuro-fuzzy based error compensation recursive least square method
Introduce a stable n_a order filter $1/F(z)$ in the linear part of Hammerstein-Wiener mode with

$$F(z) = (1 - \lambda_1 z^{-1})(1 - \lambda_2 z^{-1}) \cdots (1 - \lambda_{n_a} z^{-1}) = 1 + f_1 z^{-1} + f_2 z^{-2} + \cdots + f_{n_a} z^{-n_a}$$

where its zero meet $0 < \lambda_i < 1$, $i = (1, 2, \cdots, n_a)$.

The extended Hammerstein-Wiener model can be got

$$z(k) = \bar{\varphi}^T(k)\bar{\theta}_1 + \bar{\psi}^T(k)\bar{\theta}_1 + e(k) \tag{17}$$

where

$$\bar{A}(z) = A(z)F(z) = 1 + \bar{a}_1 z^{-1} + \cdots + \bar{a}_{n_1} z^{-n_1},\ \bar{B}(z) = B(z)F(z) = \bar{b}_1 z^{-1} + \cdots + \bar{b}_{n_2} z^{-n_2}$$
$$\bar{\theta}_1 = [\bar{a}_1, \bar{a}_2, \dots, \bar{a}_{n_1}, \bar{b}_1 w_1^{intput}, \bar{b}_1 w_2^{intput}, \dots, \bar{b}_1 w_L^{intput}, \dots, \bar{b}_{n_2} w_1^{intput}, \bar{b}_{n_2} w_2^{intput}, \dots, \bar{b}_{n_2} w_L^{intput}]^T \in \mathbf{R}^{n_1 + L \times n_2}$$
$$\bar{\varphi}(k) = [-z(k-1), \dots, -z(k-n_1), \phi_1(u(k-1)), \dots \phi_L(u(k-1)), \dots, \phi_1(u(k-n_2)), \dots, \phi_L(u(k-n_2)))]^T \in \mathbf{R}^{n_1 + L \times n_2}$$
$$\bar{\psi}(k) = [e(k-1), e(k-2), \dots, e(k-n_a), 0, \dots, 0]^T \in \mathbf{R}^{n_1 + L \times n_2}$$

Similarly, the unbiased estimation $\bar{\theta}_B$ can be easily acquired

$$\hat{\bar{\theta}}_B(k) = \hat{\bar{\theta}}_{LS}(k) + \bar{P}(k)k[\hat{\bar{R}}_1(k)\hat{\bar{\theta}}_B(k-1) + \hat{p}_1(k)] \tag{18}$$

Next, we will discuss how to calculate the estimation $\hat{R}_1(k)$ and $\hat{p}_1(k)$.
Let

$$\bar{A}^*(z) = z^{n_a} F(z) z^{n_a} A(z) = z^{n_1} \bar{A}(z) = z^{n_1} + \bar{a}_1 z^{n_1-1} + \bar{a}_2 z^{n_1-2} + \cdots + \bar{a}_{n_1}$$
$$\bar{B}^*(z) = z^{n_a} F(z) z^{n_b} B(z) = z^{n_2} \bar{B}(z) = \bar{b}_1 z^{n_2-1} + \bar{b}_2 z^{n_2-2} + \cdots + \bar{b}_{n_2}$$

Then

$$\bar{A}^*(\lambda_{a_i}) = \lambda_{a_i}^{n_1} + \bar{a}_1 \lambda_{a_i}^{n_1-1} + \bar{a}_2 \lambda_{a_i}^{n_1-2} + \cdots + \bar{a}_{n_1} = 0 \tag{19}$$

$$\bar{B}^*(\lambda_{b_i}) = \bar{b}_1 \lambda_{b_i}^{n_2-1} + \bar{b}_2 \lambda_{b_i}^{n_2-2} + \cdots + \bar{b}_{n_2} = 0 \tag{20}$$

Introduce a matrix

$$
H = \begin{bmatrix} \lambda_{a_1}^{n_1-1} & \cdots & \lambda_{a_1} & 1 & \lambda_{b_1}^{n_2-1} & \cdots & \lambda_{b_1} & 1 \\ \lambda_{a_2}^{n_1-1} & \cdots & \lambda_{a_2} & 1 & \lambda_{b_2}^{n_2-1} & \cdots & \lambda_{b_2} & 1 \\ \vdots & \ddots & \vdots & \vdots & \vdots & \ddots & \vdots & \vdots \\ \lambda_{a_{n_a}}^{n_1-1} & \cdots & \lambda_{a_{n_a}} & 1 & \lambda_{b_{n_a}}^{n_2-1} & \cdots & \lambda_{b_{n_a}} & 1 \end{bmatrix}^T
$$

Thus, it is easily to get $H^T\bar{\theta} = -\left[\lambda_{a_1}^{n_1}, \ \lambda_{a_2}^{n_1}, \ \cdots, \ \lambda_{an_a}^{n_1}\right]^T$.

Multiplying by matrix H^T on both sides of Eq. (14) get

$$
H^T\hat{\bar{\theta}}_{LS}(k) = -\left[\lambda_{a_1}^{n_1}, \ \lambda_{a_2}^{n_1}, \ \cdots, \ \lambda_{an_a}^{n_1}\right]^T - H^T\bar{P}(k)k(\hat{R}_1(k)\hat{\bar{\theta}}_B(k-1) + \hat{p}_1(k)) \qquad (21)
$$

Defining residual error

$$
\varepsilon_{LS}(k) = z(k) - \varphi^{\mathrm{T}}(k)\hat{\bar{\theta}}_{LS}(k) \qquad (22)
$$

and cost function

$$
J(\hat{\bar{\theta}}_{LS}) = \sum_{k=1}^{N} \left\| z(k) - \bar{\varphi}^{\mathrm{T}}(k)\hat{\bar{\theta}}_{LS}(k) \right\|^2 \qquad (23)
$$

Referring to Eq. (15) derive

$$
\begin{aligned}
\lim_{k\to\infty} \frac{1}{k}J(k) &= \lim_{k\to\infty} \sum_{i=1}^{k} \varepsilon_{LS}(i)[z(i) - \bar{\varphi}^T(i)\hat{\bar{\theta}}_{LS}(i)] \\
&= r_e(0) + p_1^T\bar{\theta}_1 + \bar{\theta}_1^T(p_1 + R_1\bar{\theta}_1) - (p_1 + R_1\bar{\theta}_1)^T\bar{P}(k)k(p_1^T + R_1\bar{\theta}_1)
\end{aligned} \qquad (24)
$$

From the above analysis, we can obtain neuro-fuzzy based error compensation recursive least square form by the following:

$$
\hat{\bar{\theta}}_B(k) = \hat{\bar{\theta}}_{LS}(k) + \bar{P}(k)k[\hat{R}_1(k)\hat{\bar{\theta}}_B(k-1) + \hat{p}_1(k)] \qquad (25)
$$

$$
\hat{\bar{\theta}}_{LS}(k) = \hat{\bar{\theta}}_{LS}(k-1) + L(k)[z(k) - \bar{\varphi}^T(k)\hat{\bar{\theta}}_{LS}(k-1)] \qquad (26)
$$

$$
L(k) = \bar{P}(k-1)\bar{\varphi}(k)[1 + \bar{\varphi}^T(k)\bar{P}(k-1)\bar{\varphi}(k)]^{-1} \qquad (27)
$$

$$
\bar{P}(k) = [I - L(k)\bar{\varphi}^T(k)]\bar{P}(k-1) \qquad (28)
$$

4 Example

Consider the following Hammerstein-Wiener model:

$$v(k) = 2 + 2\tanh(u(k)) - 2\exp(0.1u(k))$$
$$x(k) = 0.5x(k-1) + 0.4v(k-1), \quad e(k) = e_1(k) + 0.6e_1(k-1)$$
$$z(k) = x(k) + e(k), \quad y(k) = 0.9782z(k) + 0.2735z(k)^2$$

Where $e_1(k)$ is white noise with zero mean and variance σ^2.

Define a filter as $F(z) = (1 - 0.5z^{-1})$, noise to signal ratio (NSR) as $\delta_{ns} = \sqrt{\mathrm{var}[e(k)]/\mathrm{var}[z(k) - e(k)]} \times 100\%$.

First, binary signal and corresponding output are used to estimate the output nonlinear part parameters w_l^{output} and linear part parameters a_i, b_j. The approximation of the output nonlinear part and corresponding error are given in Fig. 2. Figure 3 shows the identification results of the linear part.

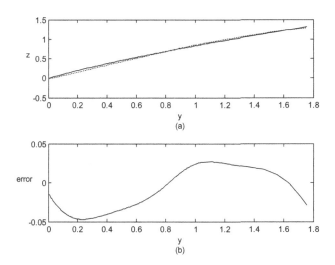

Fig. 2. The approximation of output nonlinearity (a) Actual output (solid line) and estimated-output (dotted line) (b) Estimation error

According to Fig. 2, the proposed algorithm can well approximate the output nonlinear part. As can be easily seen from Fig. 3, the correlation analysis method can well estimate parameters of the linear part.

In the case of random signal, the parameters of the input nonlinear part are determined by using design parameters: $S_0 = 0.98$, $\rho = 1.5$ and $\lambda = 0.01$. The approximation of the input nonlinear part and corresponding error are given in Fig. 4. Figure 3 indicates that the proposed approach can effectively identify the Hammerstein-Wiener model with process noise.

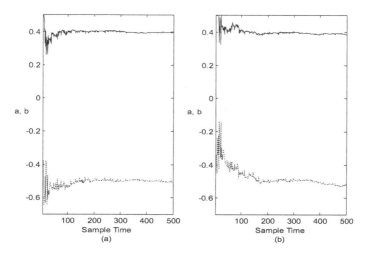

Fig. 3. Identification parameters of the linear part (a)$\delta_{ns} = 14.07\%$ (b)$\delta_{ns} = 37.35\%$

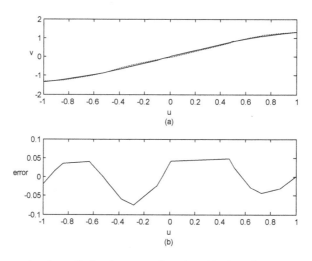

Fig. 4. The approximation of the input nonlinearity (a) Actual output (solid line) and estimatedoutput (dotted line) (b) Estimation error

5   Conclusion

Special input signals that contain separable signal and uniformly random signal are applied to Hammerstein-Wiener model, resulting in the identification issue of the linear part separated from that of the input nonlinear part. Firstly, we make use of the correlation analysis method to estimate the parameters of the linear part and output nonlinear part. In addition, a correction term is added to least square estimation to compensate error caused by process noise, and then to derive an error compensation recursive least square method for the observed data from Hammerstein-Wiener model. In further work, output noise will be considered for Hammerstein-Wiener model.

References

1. Bai, E.W.: Decoupling the linear and nonlinear parts in Hammerstein model identification. Automatica **40**(4), 671–676 (2004)
2. Giri, F., Rochdi, Y., Chaoui, F.Z.: Hammerstein systems identification in presence of hard nonlinearities of preload and dead-zone type. IEEE Trans. Autom. Control **54**(9), 2174–2178 (2009)
3. Hagenblad, A., Ljung, L., Wills, A.: Maximum likelihood identification of Wiener models. Automatica **44**(11), 2697–2705 (2008)
4. Wang, D.Q., Ding, F.: Least squares based and gradient based iterative identification for Wiener nonlinear systems. Signal Process. **91**(5), 1182–1189 (2011)
5. Bai, E.W.: An optimal two-stage identification algorithm for Hammerstein-Wiener nonlinear systems. Automatica **34**(3), 333–338 (1998)
6. Bai, E.W.: A blind approach to the Hammerstein-Wiener model identification. Automatica **38**(6), 967–979 (2002)
7. Yu, F., Mao, Z.Z., Jia, M.X.: Recursive identification for Hammerstein-Wiener systems with dead-zone input nonlinearity. J. Proc. Control **23**(8), 1108–1115 (2013)
8. Wang, D.Q., Ding, F.: Hierarchical least squares estimation algorithm for Hammerstein-Wiener systems. IEEE Signal Proc. Let **19**(2), 825–828 (2012)
9. Zhang, B., Mao, Z.Z.: Adaptive control of stochastic Hammerstein-Wiener nonlinear systems with measurement noise. Int. J. Syst. Sci. **47**(1), 162–178 (2015)
10. Ni, B., Gilson, M., Garnier, H.: Refined instrumental variable method for Hammerstein-Wiener continuous-time model identification. IET Contr. Theor. Appl. **7**(9), 1276–1286 (2013)
11. Salimifard, M., Jafari, M., Dehghani, M.: Identification of nonlinear MIMO block-oriented systems with moving average noises using gradient based and least squares based iterative algorithms. Neurocomputing **94**(3), 22–31 (2012)
12. Li, F., Jia, L., Peng, D.G., Han, C.: Neuro-fuzzy based identification method for Hammerstein output error model with colored noise. Neurocomputing **244**(8), 90–101 (2017)
13. Jia, L., Li, X.L., Chiu, M.S.: The identification of neuro-fuzzy based MIMO Hammerstein model with separable input signals. Neurocomputing **174**, 530–541 (2016)
14. Hu, S.S.: Identification of parameters of MIMO systems by correlation analysis. Acta Aeronaut. et Astronaut. Sin. **11**(7), 400–404 (1990)
15. Jia, L., Chiu, M.S., Ge, S.S.: A noniterative neuro-fuzzy based identification method for Hammerstein processes. J. Proc. Control **15**(7), 749–761 (2005)

Research on Active and Passive Monitoring Fusion for Integrated Lamb Wave Structural Health Monitoring

Qiang Wang[✉], Jie Hua, and Dong-chen Ji

College of Automation, Nanjing University of Posts and Telecommunications,
Nanjing 210046, China
wangqiang@njupt.edu.cn

Abstract. Lamb wave based active and passive monitoring technologies are both hot points in structural health monitoring (SHM). However, active and passive monitoring methods were usually worked independently. The interaction and complementarity between Lamb wave active and passive monitoring techniques was analyzed. According to the advantages and disadvantages of active and passive monitoring methods, the active and passive cooperative working mechanism was proposed which combined the active and passive monitoring approaches. In the new method, active scanning and monitoring was set to be trigged by passive acoustic emission event, and the scanning interval could be greatly extended to save the energy consumption and improving monitoring efficiency as the evolution of damages caused by service and external erosion were usually very long. Meanwhile, the results and diagnosis information of active and passive monitoring method could be fused to improve the monitoring accuracy. In addition, the hardware implementation and software frame of the new integrated system were given. The experimental validation showed that the new approach combined the advantages of active and passive monitoring methods, and improved the damage monitoring efficiency and accuracy.

Keywords: Structural health monitoring (SHM) · Cooperative working mechanism · Passive acoustic emission event · The scanning interval

1 Introduction

The large-scale engineering structures, such as aerospace vehicles, civil engineering, ship railways, oil pipelines etc., are affected from external environmental loads, fatigue, corrosion, material aging which would cause damages. On the surface or inside of the structures [1]. In order to avoid unexpected accidents and even heavy casualties, structural health monitoring technology was proposed in the aviation field and has been widely concerned and developed in the early 1990s [1]. Structural health monitoring (SHM) is one kind of the real-time on-line monitoring technology without destroying

This work is supported by National Natural Science Foundation (NNSF) of China under Grant 61533010.

© Springer Nature Singapore Pte Ltd. 2017
M. Fei et al. (Eds.): LSMS/ICSEE 2017, Part I, CCIS 761, pp. 442–451, 2017.
DOI: 10.1007/978-981-10-6370-1_44

the structure and ensuring the integrity of structural parts. Professional equipment were usually designed and adopted to analyze structural responses continuously to determine the occurrence, location and extension of possible damages [2, 3].

Lamb wave based structural damage monitoring technology is one of the hotspots and frontier technologies in the field of SHM with broad application background. This technology includes two categories, namely active monitoring methods and passive monitoring methods. In the passive monitoring mode, acoustic emission signal would be captured and analyzed to extract the damage related information and locate the damage online [4–6]. In the active monitoring mode, an excitation with certain form which usually was narrowband signal was inspired in the structure firstly through preset actuator, and at the same time the structural responses would be acquired using several distributed sensors around. The received response signals were analyzed by comparing with the ones before damage so that the changes of the responses could be extracted to evaluate the damage features and complete the damage monitoring and diagnosis. No matter active methods or passive methods, most of the existing studies are usually focused on one of the above-mentioned categories; however, less attention was drawn on how to take into account the advantages of these two categories to improve the monitoring effect and accuracy. Based on the existing efforts in Lamb wave based SHM, the cooperative working mechanism of the active and passive categories was proposed which also included the fusion method of active and passive SHM technology. Experimental validation was carried out to verify the improvement of the new method.

2 The Analysis of Lamb Wave Based Active and Passive Monitoring Fusion SHM Mechanism

2.1 The Principle of Mechanism

For the active Lamb wave damage monitoring methods, the certain forms of Lamb waves are firstly excited in the structure, and usually single mode Lamb wave was preferred. The structural responses would be collected and analyzed to extract the information related to the damage and to diagnose the damaged area. So it is more sensitive to the changes of the structural state, but it can't detect the accidental damages timely, such as impact and fracture. Thus, it is necessary to scanning the structure periodically and the interval between two adjacent scanning should be short to avoid missing the possible damages caused by impact, and the shorter the time interval, the greater the power consumption. For the passive monitoring methods, the monitoring system would work only if impact or acoustic emission events occurred. So it is sensitive to the accidental damages, but is not effective for evolutionary damage, such as fatigue, corrosion. At the same time, the signals from the possible damages may be very complicated due to the dispersive nature of Lamb wave and the sensor network should be preset very carefully because the acoustic emission events may occur at any place. Limited sensors would lead to monitoring accuracy, but more sensors would increase equipment costs greatly. Actually, these two categories will effective for most of the damages at different stages. For example, impact may cause the delamination in composite and perforation in metal structure. At the moment of the impact, the passive Lamb wave monitoring methods would be activated to capture the impulses from the damage location and point out the approximate damage position, and then the active Lamb wave monitoring methods can realize the evaluation about the possible damage by using specified excitation. The same sensor network was used in both of the two methods. Thus, in order to combine the advantages of active and passive Lamb wave monitoring mechanisms, a new monitoring methodology was proposed in which the two monitoring methods were functionally combined in time series to realize the full-time monitoring. The principle of the new monitoring methodology can be shown in Fig. 1.

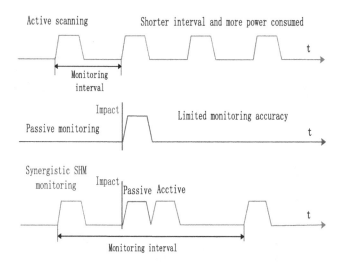

Fig. 1. Principle of the active and passive fusion monitoring methodology

According to Fig. 1, in the new monitoring method the passive monitoring mode would monitor the state of the structures during the active scanning interval. The active monitoring mode would be activated by the impact events or it only worked at the presented time periodically. So the passive monitoring mode was used as the trigger of active monitoring mode. The scanning interval of the active monitoring can be extended greatly to reduce power consumption and improve monitoring efficiency. When the impact or acoustic emission occurred, the passive monitoring mode can sense and locate the damage immediately, after that the active monitoring mode was triggered to start the monitoring process to evaluate the size and extension of damage. Furthermore, the two monitoring information can be fused to achieve more accurate damage diagnosis.

2.2 Lamb Wave Monitoring Information Fusion

The fusion of active and passive monitoring information was based on the probabilistic localization and imaging algorithm. The characteristic parameters of Lamb waves of active and passive monitoring mode were extracted and analyzed respectively. The surface coordinate of the measured structure can be transformed to pixels of an image which can be calculated by the contrast value at each pixel. And two images can be obtained by active and passive monitoring modes. The fusion of damage information was achieved though these two images.

The value assignment of each pixel in passive monitoring mode could be expressed as

$$\alpha_{mn} = \sum_{a=1}^{K-1} \sum_{b}^{K} \beta_{ab} \left| \frac{S_{mn}^a - S_{mn}^b}{(t_a - t_b)v} - 1 \right| \tag{1}$$

Where K was the number of sensors, β_{ab} was the weight coefficients of sensor a and sensor b, t_a and t_b were the impulse response times, S_{mn}^a and S_{mn}^b were the distance of the pixel to sensor a and b. In conjunction with the contrast of each pixel, $M \times N$-order matrix A_{MN} can be obtained.

The value assignment of each pixel in active monitoring mode can be expressed as

$$\beta_{mn} = \frac{1}{(K-1)(K-2)} \sum_{a=1, a \neq z}^{K} \sum_{b \neq a, b \neq z}^{K} \frac{\left| \frac{\Delta t_a}{\Delta t_b} - \frac{S_{mn}^z + S_{mn}^a}{S_{mn}^z + S_{mn}^b} \right|}{\frac{\Delta t_a}{\Delta t_b}} \tag{2}$$

Where K was the number of sensors, Δt_a, Δt_b was the arrival time of the Lamb wave scattering signals received by sensor a and b, and S_{mn}^a, S_{mn}^b, S_{mn}^z were the distance from the pixel (m, n) to the sensor a, b and c. In conjunction with the contrast of each pixel, $M \times N$-order matrix B_{MN} can be obtained.

The elements in the matrixes were mapped to $[0, 1]$ by the min-max normalization. After the normalization processing, we can get the positioning and imaging matrix as A_{MN}^* and B_{MN}^* by the active and passive monitoring mode respectively, and the fused image can be obtained by using the followed formula.

$$C_{MN}^* = \delta_p A_{MN}^* + \delta_a B_{MN}^* = \frac{K}{2(K-1)} [\alpha_{mn}^*] + \frac{K-2}{2(K-1)} [\beta_{mn}^*] \tag{3}$$

δ_a, δ_p were the weights of respectively active and passive monitoring information.

3   SHM Validation System Design Based on New Monitoring Methodology

According to the Lamb wave principle of the active and passive fusion monitoring methodology, a functional verification of the monitoring mechanism was carried out, which used the standard module and was developed to integrate the fusion of the active and passive monitoring methods. The system included man-computer interaction interface and hardware device modules as shown in Fig. 2.

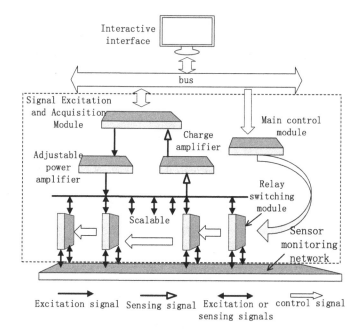

Fig. 2. The integrated system based on new monitoring methodology

3.1 The Hardware Framework of Integrated Systems

The hardware framework of the integrated system was shown in Fig. 2. The main components were the control module, relay switching module and the signal conditioning module. The signal conditioning module included signal excitation and acquisition equipment, power amplifier and charge amplifier. All parts of the hardware were connected through the bus. The bus based interconnection of the integrated system makes it compatible with other systems, and can be extended with more monitoring channels and sensor network. So it was possible to realize the large-area monitoring for the engineering structures.

The control module communicated with the PC through the bus, buffered, processed and transformed the control instruction issued by the man-machine interface. In this way, the system can switch between the active and passive monitoring modes, select the

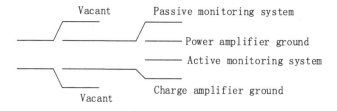

Fig. 3. Topology diagram of channel switching

acquisition channel. The signal conditioning module achieved the Lamb wave excitation signal generation, power amplification, acquisition and storage of sensing signals.

Relay switching module was the core of hardware system design, which solved the switching of piezoelectric array freely in active and passive monitoring modes. The designed relay switching topology (two relays to form a channel) was shown in Fig. 3.

According to Fig. 3, the passive monitoring mode worked during the active scanning interval. The switches of the second level relay were connected with the signal inputs and the ground of the charge amplifiers. When the active scanning or events triggered switch to the active monitoring mode, the switches of the second level relay were connected to the output and ground of the power amplifier. Time-sharing control over time series was used to switch the system working states between the active and passive monitoring modes freely.

3.2 The Software Framework of Integrated Systems

According to the collaborative working mechanism and hardware design of the integrated system, the software framework of the monitoring system was divided into the application and the driver layers which interacted with each other through the man-machine interface. The software design framework was shown in Fig. 4.

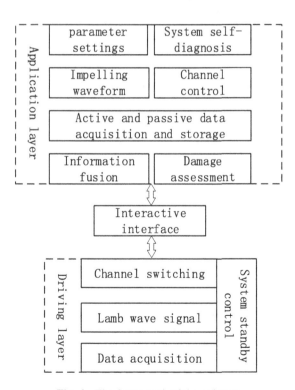

Fig. 4. The framework of the software

The interactive interface implemented functions, such as input of system parameters and user's commands, presentation of damage monitoring results, and so on. The input of the user's instruction was convenient to switch the working state of the system between the active and passive modes, expand the monitoring channels and the sensor network.

The design of application layer included the generation of excitation waveform, the control of monitoring channel, the fusion of active and passive acquisition information, the damage monitoring and evaluation. The monitoring channel control realized the selection of the actuator and the sensor under the active and passive modes, and extension of the channels according to the scope of the monitored structures. The fusion of the active and passive information was based on the extracted characteristic parameters of the damage information from the active and passive Lamb wave responses, and it can be integrated in the imaging algorithm. Driving layer achieved the control of hardware devices, including the channel switching, Lamb wave signal excitation, data acquisition and system standby drive.

4   Experimental Validations

4.1   Object and Platform of Experiment

The experimental specimen and platform were shown in Figs. 5 and 6. The specimen was made of reinforced glass fiber composite material with dimensions of 1000 mm * 500 mm and thickness of 3 mm. PZT sensors were distributed over the structure to compose a sensor array network. The distance between the PZT sensors was 250 mm. The sensor was numbered as shown in Fig. 5. The monitoring area was divided into three parts, named area A, B and C which composed with sensors No. 1, 2, 7, 8, sensors No. 2, 3, 6, 7 sensors and sensors No. 3, 4, 5, 6 respectively.

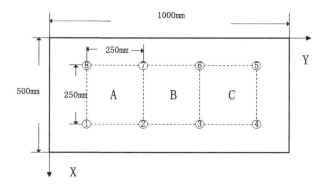

Fig. 5. Schematic diagram of the experimental object

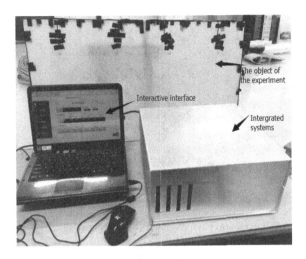

Fig. 6. Experimental platform of integrated system

In the integrated system, the impact damage in the passive monitoring was achieved by the impact hammer knocking the surface of the structure. Cracks and delamination damage were simulated by pasted masses on the structure. The experimental integrated system platform was shown in Fig. 6.

4.2 Localization and Imaging of Integrated System Experiment

A typical 5-wave sinusoidal modulation signal was used as the excitation signal and the central frequency was 60 kHz. Its amplitude was power amplified to100 V. Excited Lamb wave signal was mainly A0 mode [7–9]. In the composite plate, the impact damage was occurred at (125 mm, 187 mm), and the simulated delamination was at the same location. When the impact triggered passive monitoring mode, the acoustic signals were acquired and the damage imaging could be obtained. The coordinates of impact was detect at (124 mm, 182 mm). At the same time active monitoring mode and the data acquisition were triggered after the passive monitoring mode. And the coordinates of the damage were detected at (122 mm, 186 mm) by using the active imaging algorithm. The location coordinates obtained from the active and passive fusion algorithm were (124 mm, 183 mm). The positioning coordinates were shown in Table 1.

Table 1. Comparison of different monitoring methods (mm)

Actual damage location	(125,187)
Passive monitoring	(124,182)
Active monitoring	(122,186)
Collaborative monitoring	(124,183)

The active and passive monitoring results and fused imaging were shown in Fig. 7. The crosses in the figure were the actual approximate impact and damage location.

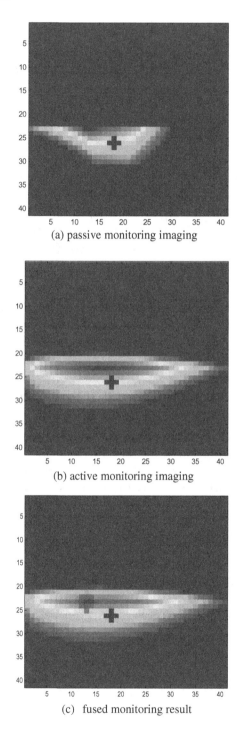

(a) passive monitoring imaging

(b) active monitoring imaging

(c) fused monitoring result

Fig. 7. The single and fusion of information by imaging (Color figure online)

From the three imaging results in Fig. 7, the locations of impact damage and the simulated damage can be seen clearly. The brightened part of the color was the passive or active localization area. After the fusion process, the positioning points were relatively more accurate than that of active and passive monitoring mode. From the trend of the color highlighting, the damage expanded from the left to the right.

5   Conclusions

The cooperative working mechanism of the active and passive Lamb wave damage monitoring methods was studies. Integrated system based on the new method was designed, including the hardware and software. Experimental validation showed that the advantages of both active and passive monitoring modes could be retained in the cooperative working mechanism. The hardware of the system integrated the active and passive monitoring functions. It was realized by the design of the topology of switching control. The software platform provided supports for hardware devices, visualization of human-computer interaction interface, information collection, pre-processing and integration. Finally, the experimental results also showed that the new method and system improved the accuracy of damage location and the evaluation.

References

1. Yuan, S.F.: Structural Health Monitoring. National Defence Industry Press, Beijing (2007)
2. Qiu, L., Yuan, S.F.: On development of a multi-channel PZT array scanning system and its evaluating application on UAV wing box. Sens. Actuators A Phys. **151**(2), 220–230 (2009)
3. Wang, Q., Yuan, S.F., Chen, X.H.: Active Lamb wave synthetic wavefront damage imaging monitoring method. Chin. J. Sci. Instrumen. **32**(11), 2468–2474 (2011)
4. Su, Z., Ye, L.: Identification of Damage Using Lamb Waves. Springer, London (2009)
5. Habib, F., Martinez, M., Artemev, A.: Structural health monitoring of bonded composite repairsa critical comparison between ultrasonic Lamb wave approach and surface mounted crack sensor approach. Compos. Part B Eng. **17**, 26–34 (2013)
6. Dodson, J.C., Inman, D.J.: Thermal sensitivity of Lamb waves for structural health monitoring applications. Ultrasonics **53**(3), 677–685 (2013)
7. Su, Z., Zhou, C., Hong, M.: Acousto-ultrasonics-based fatigue damage characterization Linear versus nonlinear signal features. Mech. Syst. Signal Process. **45**(1), 225–239 (2014)
8. Wang, Q., Li, J.: Active Lamb wave based crack monitoring and evaluation using projection of reflection field. In: 2013 32nd Chinese Control Conference, pp. 6248–6251. IEEE (2013)
9. Wang, Q., Xu, J.: Lamb wave tomography technique for crack damage detection. In: 2014 33rd Chinese Control Conference (CCC), pp. 3094–3099. IEEE (2014)

Image and Video Processing

An Embedded Driver Fatigue Detect System Based on Vision

Huaming Shen[1], Meihua Xu[1(✉)], and Feng Ran[2]

[1] Department of Automation, College of Mechatronics Engineering and Automation, Shanghai University, No. 149, Yanchang Rd., Shanghai 200072, China
mhxu@shu.edu.cn
[2] Microelectronic R&D Center, Shanghai University, No. 149, Yanchang Rd., Shanghai 200072, China

Abstract. The embedded driver fatigue detect system is a real-time system can detect driver fatigue. In order to improve the performance of embedded driver fatigue monitor system, we propose a new system on chip (SOC) structure for accelerating the fatigue estimate. The new SOC consists two parts including the main processor and support vector machine IP core. An embedded Linux was transplanted and run the main algorithm which consists Haar-Adaboost classifier to locate the face and eyes. The SVM IP core accomplished the task of classifying the eyes' statues. At last the system will estimate the state with PERCLOS standard. The results show that the system can content the need of real-world.

Keywords: Embedded system · Driver fatigue · Support vector machine

1 Introduction

Decrease the traffic accidents is one of the main goals of intelligent traffic system (ITS) [1]. The research shows that fatigue is a major cause of traffic accident, accounting for up to 20% of serious accidents on motorways and monotonous roads in the world. The driver will lose the vehicle control when they have a sleepiness episode. Researches [2, 3] show that usually the driver is fatigued after 1 h of driving in the afternoon early hours, after eating lunch and at midnight.

The driver fatigue detection system (DFDS) is a real-time system which can monitors the driver's physical and mental condition base on the processed features from special sensors. The DFDS mostly can be divided into two types: Intrusive and Non-intrusive. The Non-intrusive have been considered as the best solution because of the disturbance from Intrusive system. According to the standard of PERCLOS, the Non-intrusive system are usually implemented by detecting the symptom of driver's eyes. And we call this method as the DFDS based on visions. Some researchers [4] also fused the features such as the driver's head pose, yawning and temperature.

This paper is organized as follows: some previous researches are reviewed in Sect. 2. Section 3 describes the design of driver fatigue detect system. Section 4 presents the experimental results from the proposed structure to detect real-world driver fatigue status. Section 5 summarizes the results and draws a general conclusion.

© Springer Nature Singapore Pte Ltd. 2017
M. Fei et al. (Eds.): LSMS/ICSEE 2017, Part I, CCIS 761, pp. 455–461, 2017.
DOI: 10.1007/978-981-10-6370-1_45

2   Related Work

There are many researches have implemented the DFDS based on vision. The main reason of this large amount is that the main symptoms of fatigue and distraction appear in the driver eyes. These system usually can be divided into 7 parts: (1) capture the image contained the driver's face, (2) image processing, (3) face detection, (4) eyes detection, (5) eye symptom extraction, (6) tracking, (7) estimate the state of driver.

In the most of driver fatigue detection systems, the face detection is a main step for further operations. Color-based face detection is one of the fast and common methods including RGB, YCbCr or HIS [5]. While, in noisy images or in the images with low illuminations, these algorithms have low accuracy. Learning-based face detection uses statistical learning methods and training samples to learn the discriminative features. The benefit of these methods is high accuracy, while these methods usually have more computational complexity. With the adaboost prosed by Viola and Jones [6], mounts of DFDS was adopted this high perform algorithm for face detection [7]. Almost in all driver fatigue detection systems based on vision, on account of symptoms are related to eyes, the eye region is always detected for extracting the symptoms. In most researches, eye detection was realized in the same way as face detection. Some methods like image banalization and projection are also used to detect the eye region [8, 9]. But these solution are in low accuracy. Usually, the entire image will be searched for detecting the face. And the eye will be searched in entire face region. Searching the entire region increases the computational complexity of the system. Therefore, face/eye will be tracked after early detection by Kalman filter [10], extended versions of Kalman Filter [11] or Particle Filter (PF) [12].

After eye detection, the fatigue symptoms will be extracted from the eyes. The symptoms related to eye region include PERCLOS [13], eyelid distance, eye blink speed, eye blink rate, and gaze direction. And the most important symptoms is the PERCLOS.

After symptom extraction, the driver state has to be determined. The simplest method for detecting the driver fatigue is based on applying a threshold on counting the frame number of open eyes. Other methods like knowledge-based approach, fuzzy expert systems and Bayesian network were also used for driver state determination [14]. These approaches are usually more accurate than threshold-based approaches, however, they are more complicated.

3   The Proposed DFDS

In this article, we proposed a real-time embedded DFDS which can setup in cars and get high performance. The flowchart of the proposed system is shown in Fig. 1. After image acquisition, some variables will be initialized including the timing cycle and counters for fatigue estimate. And then the image will be scaled and converted to gray-scale. After that, the driver's face and eyes will be detected by adaboost based on the haar-like feature. And then, the open eyes will be classified by support vector classifier. Finally, we used a threshold to estimate driver fatigue. This system also used Kalman filter to track the face to reduce the detection time. And the MIT face database was used to train the adaboost classifier.

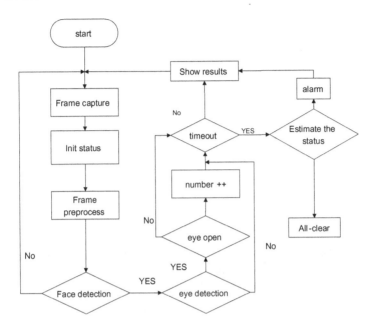

Fig. 1. Flowchart of DFDS

As an embedded system, we adopted the ARM as the main processor because of its low power consumption and simple peripheral. The hardware of the DFDS is shown in Fig. 2. The whole system powered by 12 V, so it can be setup in cars. And a camera was used to acquisition image. All the software are stored in the SD card. Usually, the frames can be displayed in the monitor with HDMI interface.

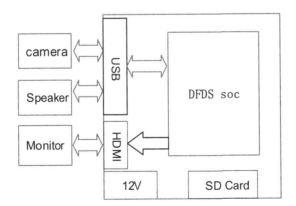

Fig. 2. The hardware system of driver fatigue detection system

The DFDS SOC is designed and implemented on the Zynq platform. And its structure is shown in Fig. 3.

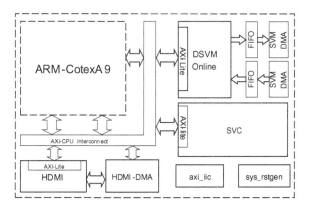

Fig. 3. The DFDS SOC

The support vector classifier (SVC) was designed and implemented in Verilog. And the model was train by the Digital Support Vector Machine (DSVM). To connect with the ARM, we use the AXI bus supported by Xilinx.

3.1 Object Detection

After capturing an image, the image will be reduced and conversed to grayscale. As mentioned in the previous section, the system need to detect human eye before judging the state of eyes. And regularly, face will be detect firstly to reduce the time to search the eye region. Viola and Jones proposed a method of fast target detection framework which is one of the best methods, so this article adopts their framework. In the framework the integral figure should be calculated. Integral figure is in order to accelerate the Haar feature extraction. And the equation is

$$I(x, y) = \sum_{\substack{x' \leq x \\ y' \leq y}} i(x', y')$$
(1)

Then the traversal of integral graph, a rectangular area on the grey value can be calculated by the following formula.

$$I(x, y) = I(x, y) + I(x - 1, y) - I(x, y - 1) - I(x - 1, y - 1)$$
(2)

And the calculation of the integral figure, haar features will be imported to face and eyes Adaboost cascade classifiers. The Adaboost cascade classifier is composed of a series of weak classifiers. And most of the area can be ruled out after previous levels of weak classifiers.

3.2 State Estimate

After detecting the eyes, the features of eyes will be extracted to estimate the state of driver. In this system, a support vector machine IP core was used to distinguish the eyes status with the Local Binary Pattern (LBP) feature. And experiences show that the binary pixels are directly related to the samples' characteristics. The IP core used the liner kernel Due to floating point calculation implementation is complex in FPGA.

And the expression of the liner classifier is as follow.

$$f(x) = sgn\left(\sum_{i=1}^{n} \alpha_i y_i (x_i \cdot x) + b\right) \tag{3}$$

According to the standard of P80 of PERCLOS, the equation to judge the driver's fatigue is

$$PERCLOS = \frac{frames\ of\ closed\ eyes}{all\ frames} \times 100\% \tag{4}$$

Within a certain time T, frames of closed eyes N make up 80% or more of the total number of frames in T will be called fatigue driving which is shown in Fig. 4.

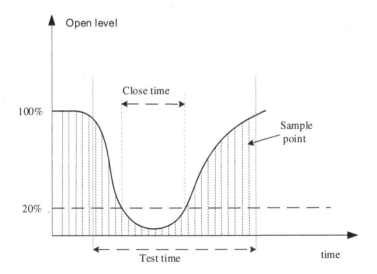

Fig. 4. PERCLOS diagram

The results show that when detection time T is stetted to be one second, the frames of open eyes should be more than 12 when the driver is sober. And the value of PERCLOS will be greater than 68% which can be used to judge whether the driver is fatigue.

4   Experimental Result Discussion

The proposed system was implemented on the ZC702 with a rebuilt Linux kernel and a Linaro file system. The haar-like features extraction and adaboost training was supported by OPENCV. After the system was powered on, the entire system was shown in Fig. 5. The part (a) is the entire system, a zynq board and a monitor. The part (b) is the frame of face detection. And the part (c) is the frame of eye detection.

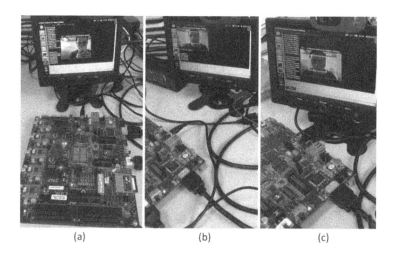

(a)   (b)   (c)

Fig. 5. The entire running system

Experiments have been carried out by 5 people. And each of them pretend to be fatigue 20 times in 5 min. The results was recoded in Table 1.

Table 1. The testing results

Number	Time	Effective alarm	Accuracy	False alarm	Average process time/ frame
1	300 s	18	90%	2	37 ms
2	300 s	17	85%	1	41 ms
3	300 s	19	95%	3	38 ms
4	300 s	19	95%	2	39 ms
5	300 s	18	90%	2	42 ms
Average	300 s	18.2	91%	2	39.4 ms

As seen from the table, the average true alarm accuracy is 91%, and the processing time for each frame is 39.4 ms. There are also some false alarms, which can be caused by the use of liner support vector classifier. Besides this, this driver fatigue detection system get a high performance at the process time and the alarm accuracy.

5 Conclusion

In this paper, we have presented a new DFDS structure implemented in SOC. The results show that this proposed system can achieve a high accuracy in real time. And in the future, we plan to get more eye samples to train the SVC and use a non-linear SVC to detect eyes to get higher accuracy and reduce the process time.

Acknowledgements. This work was financially supported by the national natural science foundation of China under Grant (No. 61376028) and (No. 61674100).

References

1. Fink, W.: Intelligent transportation systems. IEEE Control Syst. Mag. **28**, 28–37 (1995)
2. Lucidi, F., Mallia, L., Violani, C., et al.: The contributions of sleep-related risk factors to diurnal car accidents. Accid. Anal. Prev. **51**, 135–140 (2013)
3. Zwahlen, D., Jackowski, C., Pfaffli, M., et al.: Sleepiness, driving, and motor vehicle accidents: a questionnaire-based survey. J. Forensic Legal Med. **44**, 183–187 (2016)
4. Abtahi, S., Hariri, B., Shirmohammadi, S.: Driver drowsiness monitoring based on yawning detection. In: Proceedings of the Instrumentation and Measurement Technology Conference, Hangzhou, China (2011)
5. Goswami, G., Powell, B.M., Vatsa, M., et al.: FaceDCAPTCHA: face detection based color image CAPTCHA. Future Gener. Comput. Syst. **31**, 59–68 (2014)
6. Viola, P., Jones, M.: Rapid object detection using a boosted cascade of simple features. In: Proceedings of the IEEE Computer Society Conference on Computer Vision and Pattern Recognition, Cambridge, Mass, USA, pp. I511–I518, December 2001
7. Cheng, R., Zhao, Y., et al.: An on-board embedded driver fatigue warning system based on Adaboost method. Acta Scientiarum Naturalism Univ. Pekinensis **5**, 719–726 (2012)
8. Tabrizi, P.R., Zoroofi, R.A.: Drowsiness detection based on brightness and numcral features of eye image. In: Proceedings of the 5th International Conference on Intelligent Information Hiding and Multimedia Signal Processing, Kyoto, Japan, pp. 1310–1313, September 2009
9. Zhang, Y., Hua, C.: Driver fatigue recognition based on facial expression analysis using local binary patterns. Optik **126**, 4501–4505 (2015)
10. Chen, K., Liu, C., Xu, Y., et al.: Face detection and tracking based on Adaboost Camshift and Kalman filter algorithm. In: Fei, M., Peng, C., Su, Z., Song, Y., Han, Q. (eds.) Computational Intelligence, Networked Systems and Their Applications. LSMS/ICSEE 2014. CCIS, vol. 462, pp. 149–158. Springer, Berlin (2014). doi:10.1007/978-3-662-45261-5_16
11. Zhang, Z., Zhang, J.: Driver fatigue detection based intelligent vehicle control. In: International Conference on Pattern Recognition, pp. 1262–1265 (2006)
12. Zheng, W., Bhandarkar, S.M.: Face detection and tracking using a boosted adaptive particle filter. J. Vis. Commun. Image Representation **20**(1), 9–27 (2009)
13. Dinges, D.F., Grace, R.: PERCLOS: a valid psychophysiological measure of alertness as assessed by psychomotor vigilance. National Highway Traffic Safety Administration Final Report. Washington (1998)
14. Sigari, M., Fathy, M., Soryani, M., et al.: A driver face monitoring system for fatigue and distraction detection. Int. J. Veh. Technol. 1–11 (2013)

A Hybrid Generative-Discriminative Learning Algorithm for Image Recognition

Bin Wang[1], Chuanjiang Li[1(✉)], Xiong Li[2], and Hongwei Mao[1]

[1] College of Information, Mechanical and Electrical Engineering,
Shanghai Normal University, Shanghai 200234, China
lichuanjiang2016@hotmail.com
[2] National Computer Network Emergency Response Technical Team,
Beijing 100029, China

Abstract. Feature representation is usually a key point in image recognition. The recognition performance can be potentially improved if the data distribution information is exploited. In this paper, we propose an image recognition approach based on generative score space. Specifically, we first leverage probabilistic latent semantic analysis (pLSA) to model the distribution of images. Then, we derive the mid-level feature from the model in a generative feature learning manner. At last, the derived feature is embedded into a discriminative classifier for image recognition. The advantages of our proposed approach are two folds. First, the probabilistic generative modeling allows us exploiting information hidden in data and has good adaptation to data distribution. Second, the discriminative learning process can utilize the information of label effectively. To confirm the effectiveness of our method, we perform image recognition on three datasets. The results demonstrate its advantages.

Keywords: Feature learning · Image recognition · Probabilistic distribution

1 Introduction

Image recognition has been a hot research subject in computer vision with applications in image retrieval systems, vehicle navigation, and video analysis [1]. Among a few possible taxonomies, the literature present two orthogonal methods: the discriminative and the generative paradigms. Specifically, the discriminative approaches present the boundaries between different classes, instead of modeling the distribution of samples belonging to the same class. Discriminative models directly model the map from images to classes, by capturing the decision bounds among different classes [2, 3]. A variety of discriminative models, such as support vector machines (SVM) [4], discriminative kernel type model [5], and multiple-instance learning [6] have been applied to image recognition. On the other hand, the generative approaches model the distribution of data and tell people prior knowledge by a graph structure, which establishes correspondences between image feature and the model by means of conditional distribution. For instance, in probabilistic latent semantic analysis (pLSA) [7], the representation of latent space can capture high-level relations within and across the class

© Springer Nature Singapore Pte Ltd. 2017
M. Fei et al. (Eds.): LSMS/ICSEE 2017, Part I, CCIS 761, pp. 462–471, 2017.
DOI: 10.1007/978-981-10-6370-1_46

and visual modalities. Generative models therefore can utilize this additional high level information for image recognition.

Different from the above approaches, we here present a hybrid generative-discriminative method aiming to simultaneously benefit from the better of the two paradigms. Specifically, we first derive the feature mapping from generative models, and then the derived features are seamlessly embedded into discriminative models for image recognition. Probabilistic generative models here can discover information hidden in the data via inferring hidden variable, that is of great importance for image recognition. In a word, there are two advantages of our proposed approach: first, the probabilistic generative modeling can exploit information in hidden variables and has good adaptation to image distribution; second, the learning approach in discriminative manner can utilize label information effectively.

The remainder of this paper is organized as follows. Section 2 presents the framework of our proposed approach. Section 3 experimentally evaluates the method for image recognition on three popular datasets. Section 4 draws a conclusion.

2 Our Proposed Hybrid Generative-Discriminative Learning Algorithm

2.1 Generative Feature Learning

Generative feature learning is a branch of probabilistic methods, which derives the explicit feature mapping based on the probabilistic distribution over the data from generative score space [8]. [9] firstly proposed the theory about score space. These methods are composed of two classes [10, 11]. Specifically, let $\mathbf{x} \in \mathbf{R}^D$ and $\theta = \{\theta_1, \cdots, \theta_K\}$ be observed variable and a set of K parameters respectively. Suppose $P(\mathbf{x}|\theta)$ be the marginal distribution of a chosen probabilistic generative model. The first class of methods are represented by fisher score (FS) [9]. FS derives score functions explicitly from an adopted generative model, which measures the way of a sample affecting parameter θ. The advantage of this class of methods is its robustness to hidden variables number [10, 12, 13]. Another class of methods are represented by free energy score space (FESS) [14] and sufficient statistics (SS) [15]. For clarity, we summarize the notations in Table 1.

Table 1. Lists of notation.

Notation	Description
$\mathbf{x} \in \mathbf{R}^D$	Observed data
$S = \{\mathbf{x}^t, y^t\}_{t=1}^N$	Training set of N pairs
θ	Model parameters
Z	A set of hidden variable
$P(\mathbf{x}|\theta)$	Marginal distribution
$Q(Z)$	Approximate posterior
$-F(Z, \theta)$	Variational lower bound
$\Phi(\mathbf{x}^t)$	Score function of sample $\mathbf{x}^t$

2.2 Model Formulation

In this section, we proceed to derive the feature mapping and then embed the learned feature into discriminative classifier for image recognition. We first employ probabilistic latent semantic analysis (pLSA) [7] to model data distribution for its effectiveness in image feature modeling. The effectiveness of pLSA in image representation has been extensively proved before. Then we derive the feature mapping based on pLSA. At last, we perform image recognition using the learned feature mapping.

pLSA is a generative model which is originally designed in text analysis to discover the meaningful topics hidden in a document by means of bag of words representation. pLSA can also be used in image analysis when we have an image as a document and the mission is to discover object classes as topics using bag of words image representation where an image is quantified into visual words.

Let $D = \{d_1, \cdots, d_N\}$ be a set of images with words selecting from a dictionary $W = \{w_1, \cdots, w_V\}$, where N and V are documents number and vocabulary terms number respectively. The data is presented by a $N \times V$ co-occurrence matrix $M_{jk} = m(w_j, d_k)$, where the element $m(w_j, d_k)$ denotes the frequency of w_j taking place in d_k. Suppose $z \in Z = \{z_1, \cdots, z_Z\}$ is the latent variable which associates with a particular observation to be the occurrence of a word. Let $P(d_k)$ be the probability of a particular document; let $P(w_j, z_i)$ denote the conditional probability of a certain word w_j conditioned on the latent topic z_i; let $P(z_i, d_k)$ denote a conditional probability of latent topic variable z_i conditioned on the specific document d_k. Having the above definitions, the pLSA model can be expressed as follows:

1. Choose an image(document) d_k according to the probability $P(d_k)$;
2. Select a latent topic z_i according to the conditional probability $P(z_i|d_k)$;
3. Generate a word w_j according to the conditional probability $P(w_j|z_i)$.

Then we have an observation pair (w_j, d_k). A joint probability $P(w, d, z)$ for pLSA can be expressed as $P(w, d, z) = P(w|z)P(z|d)P(d)$.

We have the marginal distribution by marginalizing out the hidden topic z.

$$
\begin{aligned}
P(w, d) &= \sum_{z \in Z} P(w, d, z) \\
&= P(d) \sum_{z \in Z} P(w|z)P(z|d)
\end{aligned}
\tag{1}
$$

Noting that $P(w, d) = P(w|d)P(d)$, we have the marginal distribution $P(w|d)$ as $P(w|d) = \sum_{z \in Z} P(w|z)P(z|d)$.

Actually, in this model, every document is modeled as mixture of topics, and the histogram for a document comprises of a mixture of the histograms which corresponds to every topic. Particularly, the Z topic combines every document. To learn this model, we need to determine the mixture coefficients that are corresponding to every document. Also, we need to settle the topic parameters for all the documents. Specifically, learning the pLSA model involves the determination of $P(z|d)$ and $P(w|z)$ via maximizing the log likelihood function:

$$L = \log P(D, W)$$
$$= \sum_{d \in D} \sum_{w \in W} m(w, d) \log P(w, d) \tag{2}$$

Maximizing the above log likelihood equals to minimizing the Kullback-Leibler divergence among the measured empirical distribution and the parameterized model. The above model can be efficiently learned by Expectation Maximization (EM) algorithms, which is described in [16].

Although we can obtain the analytic form of $P(w, d; \theta)$, the differential operation over log is very complex since $P(w, d; \theta)$ takes the form of summation. The differential operation on variational lower bound [17] of the log likelihood function is very simple. As a result, we resort to it to solve the problem.

$$\log P(d, w; \theta) \geq KL(Q(Z) \| P(d, w, z; \theta))$$
$$= -F(\theta)$$
$$= -E_Q(Z)[\log Q(Z) - \log P(d, w, z; \theta)]$$
$$= \sum_{d,w} m(d, w) \sum_z Q(z|d, w)[\log Q(z|d, w) - \log P(w|z)P(z|d)] \tag{3}$$
$$= \sum_{d,w} m(d, w) \sum_k g_{kwd}[\log g_{kwd} - \log \beta_{kw}\alpha_{kd}]$$

where $Q(Z)$ is the approximate posterior of z, which takes the same parameterizations with $P(Z)$. It is worth noting that using the lower bound will not loss generality, because the lower bound equals to the real log likelihood and $Q(Z)$ equals to the real posterior when using exact inference. Given the lower bound $F(\theta)$ of $\log P(d, w|\theta)$, the elements of sufficient statistics feature mapping [15] can be written as,

$$\Phi(d) = E_{Q(z)}[\varphi(d, w, z)] \tag{4}$$

where $\varphi(d, w, z) = vec\left(\left\{m(d, w)w_i z_k, m(d, w)z_k, m(d, w)z_k \log g_k^d\right\}_{ik}\right)$.

It is worth noting that the above elements are the expectation over a function of observed variable d, latent variables z and model parameters θ. And the latent variable makes the derived feature mapping exploit hidden information and the model parameter allows the model is adaptive to data distribution.

Then the derived feature is embedded in the chosen classifier to perform image classification. The training procedure of our approach is summarized in Algorithm 1. To confirm the effectiveness of our method, we embed the derived features into two popular discriminative classifiers: localized multiple kernel learning (LMKL) [18] and linear support vector machine (SVM) [19] for their brilliant performance to perform image classification.

Algorithm 1. Train the model

1: input: training set $\chi = (\mathbf{x}^1, \cdots \mathbf{x}^N)$

2: estimate θ and $\{Q^i\}_{i=1}^N$ from training set

3: estimate $Q_{-c}^i \leftarrow Q^i$

4: for $c = 1$ to N do

5: $\theta_{-c} \leftarrow \arg\max_{\theta_{-c}} \sum_{i \neq c} -E_{Q_{-c}^i}[\log Q(Z) - \log P(d, w, z; \theta)]$

6: construct Φ^c using Eq.(4)

7: **end for**

8: output: $\{\theta_c\}_{c=1}^C$ and α

3 Experiments

In this section, we incorporate probabilistic latent semantic analysis (pLSA) with sufficient statistics (SS) [15] for image recognition. We perform image classification on three datasets. We compare our method with some related methods and several state-of-the-art methods.

3.1 Database

We evaluate our approach on three benchmark image databases: Scene-15 dataset [19], OT dataset [20] and UIUC-sports dataset [21].

The Scene-15 dataset [19] is composed of 15 scene categories. Every category comprises 200–400 images. Totally, the number of this dataset is 4485. The OT scene dataset [20] comprises 4 natural scenes and 4 artificial scenes, which has 2688 images totally. The UIUC-Sports dataset [21] comprises 8 categories. Each category comprises 137–250 images. This dataset has 1792 images in total.

3.2 Feature Representation

We leverage SIFT descriptor for image low-level representation. SIFT descriptors are extracted from three scales: 16×16, 24×24 and 32×32 and dense sampling on a grid with the step size of 4 pixels [7, 22]. The number of topics is set to be $K \in [20, 40]$ in accordance with cross-validation. We use SVM and LMKL for classification.

3.3   Experimental Results on Scene-15 Dataset

We randomly choose 100 images from each class to form training set and the rest as test set [19, 23]. The proposed approach will compare with some state-of-the-art methods for image recognition task. Specifically, the compared approaches are listed as follows:

BoW [24] is a baseline method without using feature learning. **DVSA** [23] presents a model to generate natural language descriptions of images and their regions. **CFMD** [21] is an extension of the classical Fourier–Mellin descriptors for image recognition. **AGMM** [24] leverages hard quantization in the feature space results in a loss of discriminative power. **DCN** [25] investigates the effect of the convolutional network depth on its accuracy in the large-scale image recognition setting. **DRL** [26] presents a residual learning framework to ease the training of networks. **DFN** [27] proposes a new version of the Fisher vector image encoding. **FESS** [14] considers the latent structure of the data at various levels when deriving score function from a generative model.

The experimental results on the Scene-15 dataset are presented in Table 2. As shown in Table 2, we find that, compared with the baseline approach BoW, both AGMM and DVSA obtain a significant improvement. CFMD, DCN and DRL show competitive performance and outperform AGMM and CFMD. Meanwhile, DFN and FESS, which are two approaches most close to our proposed method, get better results due to the consideration of probabilistic modeling of data distribution. Our proposed approach, with SVM as the classifier achieves the best performance among the compared methods mentioned in Table 2. The reasons accounting for the improvement are: firstly our proposed approach employs pLSA to model the generative distribution of images; secondly, it derives the feature mapping which encodes information about hidden variables, observed variables and model parameters.

Table 2. The classification accuracy on the scene-15 dataset.

Method	#topic K	Classifier	Accuracy (%)
BoW	–	SVM	79.30 ± 1.64
AGMM	–	SVM	82.40 ± 0.96
DVSA	–	SVM	80.77 ± 1.54
CFMD	–	SVM	79.78 ± 1.33
DCN	–	SVM	81.51 ± 1.32
DRL	–	SVM	80.22 ± 1.21
DFN	20	SVM	81.48 ± 1.55
FESS	20	SVM	81.87 ± 1.21
Ours	40	SVM	**82.97 ± 1.18**
DFN	20	LMKL	82.08 ± 1.03
FESS	20	LMKL	82.21 ± 1.95
Ours	40	LMKL	82.57 ± 1.76

3.4   Experimental Results on OT Dataset

To further validate the effectiveness of our approach, we then perform an experiment on OT dataset. This dataset shares the same experimental setting as scene-15 dataset. It is worth noting that, parameters of the proposed model, except the number of topics K, are learned from the dataset. We compared our proposed approach with DFN [27] and FESS [11] closely related to our method and other state-of-the-art methods, including BoW [24], HGA [24], NPR [29], DVSA [23], CFMD [21] and AGMM [24]. BoW [24] is a baseline approach. HGA [24] uses the posterior distribution of topics as image features. NPR [29] is a nonparametric kernel estimation approach estimating the kernel similarity. CFMD [21] uses color Fourier-Mellin descriptors for image recognition.

The experimental results on the OT dataset are presented in Table 3. Our hybrid generative-discriminative approach shows convincing performance. The results indicate that our method learns features encoding semantic information effectively. When deriving feature mapping, our method benefits from image generative distribution information. Also in the test process, our method benefits from the Bayesian inference of class label. Due to the above reasons, our method is superior to other compared approaches. More specifically, we can see that HGA and CFMD show competitive performance, both achieving a significant improvement over the baseline Euclidean metric BoW. DFN and FESS outperform the other two compared approaches AGMM and DVSA with both linear SVM and LMKL. This is because DFN and FESS take probabilistic modeling of data distribution into account. Meanwhile, NPR is a state-of-art method of image recognition task, sharing a similar methodology with generative feature learning methods. Our method leverages different technique, but it presents brilliant performance with NPR.

Table 3. The recognition accuracy on the OT dataset.

Method	#topic K	Classifier	Accuracy (%)
BoW	–	SVM	83.80 ± 1.69
HGA	–	SVM	83.93 ± 1.47
DVSA	–	SVM	83.79 ± 1.65
CFMD	–	SVM	80.94 ± 1.24
AGMM	–	SVM	84.22 ± 1.56
NPR	–	SVM	88.90 ± 2.01
DFN	20	SVM	85.95 ± 1.92
FESS	20	SVM	85.76 ± 1.15
Ours	40	SVM	87.81 ± 1.34
DFN	20	LMKL	85.88 ± 1.45
FESS	20	LMKL	86.06 ± 1.68
Ours	40	LMKL	88.73 ± 1.78

3.5   Experimental Results on UIUC Dataset

We randomly choose 70 images from each class to form training set and the rest as testing set [28]. As mentioned above, we still choose BoW [24] as the baseline approach. Meanwhile, we evaluate our proposed approach with closely related approaches DFN [27] and FESS [11], and other state-of-the-art methods AGMM [24], DVSA [23], CFMD [21] and LPD [30]. LPD is a new method, which aims to improve the descriptor matching flexibility via leveraging local multi-resolution pyramids in feature space. The experimental results on the UIUC-Sports dataset are presented in Table 4. As shown in Table 4, our approach with LMKL as the classifier shows the best performance among all the compared approaches.

Table 4. The recognition accuracy on the UIUC dataset.

Method	#topic K	Classifier	Accuracy (%)
BOW	–	SVM	78.12 ± 1.23
DVSA	–	SVM	79.25 ± 1.44
CFMD	–	SVM	82.74 ± 1.46
AGMM	–	SVM	82.46 ± 1.89
LPD	–	SVM	82.81 ± 0.68
DFN	20	SVM	80.96 ± 0.96
FESS	20	SVM	80.92 ± 1.55
Ours	40	SVM	83.12 ± 1.08
DFN	20	LMKL	80.88 ± 2.01
FESS	20	LMKL	81.56 ± 1.67
Ours	40	LMKL	**84.35 ± 1.38**

3.6   Discussion on Experimental Results

We perform image recognition on three popular image datasets. As shown in Tables 2, 3, and 4, when comparing with state-of-the-art approaches, our approach presents convincing results. The reasons accounting for this excellent performance can be summarized as follows. First, our method models data distribution via pLSA, which can capture high-level relations within and across the class and visual modalities. Second, the derived feature from generative model is essentially a function over hidden variable, model parameters, and observed data, which encodes high-level information important for image recognition. Third, the discriminative learning process can fully utilize class label information effectively.

4   Conclusions

We here propose a generative feature learning approach to perform image classification. The approach is based on generative score space, in which image feature is derived from the probabilistic distribution of image datasets. The derived feature allows to fully exploit hidden information and is well adapt to data distribution. We conducted extensive experiments on three benchmark datasets. The brilliant experimental results validate the effectiveness of our feature learning approach for the image classification task. Our method can further benefit from the mining of larger dataset. The computational efficiency is a main limitation of the training procedure, when facing to larger dataset. These works will leave in the future.

Acknowledgments. This work was supported by the National Natural Science Foundation of China (No. 61503251), the Science and Technology Commission of Shanghai Municipality (No. 16070502900), the Program of Shanghai Normal University (No. A-7001-15-001005).

References

1. Jiang, Z., Zhang, S., Zeng, J.A.: Hybrid generative/discriminative method for semi-supervised classification. Knowl.-Based Syst. **37**(2), 137–145 (2013)
2. Wang, B., Liu, Y.: Collaborative similarity metric learning for semantic image annotation and retrieval. KSII Trans. Internet Inf. Syst. **7**(5), 1252–1271 (2013)
3. Wang, C., Wang, B., Zheng, L.: Learning free energy kernel for image retrieval. KSII Trans. Internet Inf. Syst. **8**(8), 2895–2912 (2014)
4. Zhou, X., Jiang, P., Wang, X.: Recognition of control chart patterns using fuzzy SVM with a hybrid kernel function. J. Intell. Manuf., 1–17 (2015)
5. Moran, S., Lavrenko, V.: A sparse kernel relevance model for automatic image annotation. Int. J. Multimedia Inf. Retriev. **3**(4), 209–229 (2014)
6. Sadeghi, M.A., Farhadi, A.: Recognition using visual phrases. In: Proceedings of IEEE Conference on Computer Vision and Pattern Recognition, pp. 1745–1752 (2011)
7. Bouguila, N.: Hybrid generative/discriminative approaches for proportional data modeling and classification. IEEE Trans. Knowl. Data Eng. **24**(12), 2184–2202 (2012)
8. Wang, B., Li, X., Liu, Y.: Learning discriminative fisher kernel for image retrieval. KSII Trans. Internet Inf. Syst. **7**(3), 532–548 (2013)
9. Jaakkola, T.S., Haussler, D.: Exploiting generative models in discriminative classifiers. Adv. Neural. Inf. Process. Syst. **11**(11), 487–493 (1998)
10. Wang, B., Wang, C., Liu, Y.: Exploiting class label in generative score spaces. Neurocomputing **145**(18), 495–504 (2014)
11. Perina, A., Cristani, M., Castellani, U., Murino, V., Jojic, N.: Free energy score spaces: Using generative information in discriminative classifiers. IEEE Trans. Softw. Eng. **34**(7), 1249–1262 (2012)
12. Chatfield, K., Lempitsky, V., Vedaldi, A., Zisserman, A.: The devil is in the details: an evaluation of recent feature encoding methods. In: Proceedings of British Machine Vision Conference, pp. 76.1–76.12 (2011)
13. Carreira, J., Caseiro, R., Batista, J., Sminchisescu, C.: Free-form region description with second-order pooling. IEEE Trans. Pattern Anal. Mach. Intell. **37**(6), 1177–1189 (2015)
14. Cinbis, R.G., Verbeek, J., Schmid, C.: Approximate fisher kernels of non-iid image models for image categorization. IEEE Trans. Pattern Anal. Mach. Intell., 1 (2015)
15. Li, X., Wang, B., Liu, Y., Tai, S.L.: Stochastic feature mapping for pac-bayes classification. Mach. Learn. **101**, 5–33 (2015)
16. Jordan, M.I., Ghahramani, Z., Jaakkola, T.S., Saul, L.K.: An introduction to variational methods for graphical models. Mach. Learn. **37**(2), 183–233 (2012)
17. Nen, M., Alpaydin, E.: Localized multiple kernel learning. In: Proceedings of International Conference on Machine Learning, pp. 1531–1565 (2008)
18. Feng, J., Ni, B., Tian, Q., Yan, S.: Geometric lp-norm feature pooling for image classification. In: Proceedings of IEEE Conference on Computer Vision and Pattern Recognition, pp. 2697–2704 (2011)

19. Girshick, R., Donahue, J., Darrell, T., Malik, J.: Rich feature hierarchies for accurate object detection and semantic segmentation. In: Computer Science, pp. 580–587 (2014)
20. Mennesson, J., Saint-Jean, C., Mascarilla, L.: Color fourier-mellin descriptors for image recognition. Pattern Recogn. Lett. **40**(1), 27–35 (2014)
21. Chen, J., Li, Q., Peng, Q., Wong, K.H.: Csift based locality-constrained linear coding for image classification. Formal Pattern Anal. Appl. **18**(2), 441–450 (2015)
22. Karpathy, A., Fei-Fei, L.: Deep visual-semantic alignments for generating image descriptions. In: Computer Vision and Pattern Recognition, pp. 3128–3137. IEEE (2015)
23. Dixit, M., Rasiwasia, N., Vasconcelos, N.: Adapted gaussian models for image classification. In: Proceedings of IEEE Conference on Computer Vision and Pattern Recognition, pp. 937–943 (2011)
24. Simonyan, K., Zisserman, A.: Very deep convolutional networks for large-scale image recognition. Comput. Sci. **11**(11), 487–493 (2014)
25. He, K., Zhang, X., Ren, S., Sun, J.: Deep residual learning for image recognition. Comput. Sci. (2015)
26. Simonyan, K., Vedaldi, A., Zisserman, A.: Deep fisher networks for large-scale image classification. In: Advances in Neural Information Processing Systems, pp. 163–171 (2013)
27. Zhang, C., Liu, J., Tian, Q., Xu, C., Lu, H., Ma, S.: Image classification by non-negative sparse coding, low-rank and sparse decomposition. In: Proceedings of IEEE Conference on Computer Vision and Pattern Recognition, pp. 1673–1680 (2011)
28. Poczos, B., Xiong, L., Sutherland, D.J., Schneider, J.: Nonparametric kernel estimators for image classification. In: Proceedings of IEEE Conference on Computer Vision and Pattern Recognition, pp. 2989–2996 (2012)
29. Seidenari, L., Serra, G., Bagdanov, A.D., Del Bimbo, A.: Local pyramidal descriptors for image recognition. IEEE Trans. Pattern Anal. Mach. Intell. **36**(5), 1 (2013)
30. Vapnik, V.: The Nature of Statistical Learning Theory. Springer, New York (2000)

Multi-channel Feature for Pedestrian Detection

Zhixiang He, Meihua Xu$^{(\boxtimes)}$, and Aiying Guo

School of Mechatronics Engineering and Automation, Shanghai University,
No. 149, Yanchang Rd, Shanghai 200072, China
mhxu@shu.edu.cn

Abstract. Multi-channel feature for pedestrian detection is proposed to solve problems of real-time and accuracy of pedestrian detection in this paper. Different from traditional low level feature extraction algorithm, channels such as colours, gradient magnitude and gradient histogram are combined to extract multi-channel feature for describing pedestrian. Then classifier is trained by AdaBoost algorithm. Finally the performance of the algorithm is tested in MATLAB. The result demonstrates that the algorithm has an excellent performance on both detection precision and speed.

Keywords: Pedestrian detection · Image channels · Multi-channel feature · AdaBoost

1 Introduction

Advanced Driver Assistance Systems (ADAS) [1] has already been widely used in automobiles. As a part of ADAS, pedestrian detection plays a key role in safe driving. Pedestrian is one of the main objectives in drivers' view. Drivers need to analyze the behaviors of pedestrians and make the right decisions so as to avoid traffic accidents. Therefore pedestrian detection is a direct factor influencing safe driving.

Pedestrian detection algorithms are generally divided into two types: background modeling [2, 3] and statistical learning [3]. Background modeling firstly obtains the region of interest (ROI) which contains pedestrians. Then pedestrian features are extracted from ROI and classified. It mainly includes template matching [4], frame differential [5], optical flow [6], etc. Statistical learning requires features like color, gray, gradient and texture, which is extracted from a lot of pedestrian samples. Then it trains pedestrian detection classifiers by AdaBoost [7, 8], Support Vector Machine (SVM) [9], neural network [10], etc.

A novel pedestrian detection method based on statistical learning is proposed in this paper. As the main contribution of this paper, we extract features from multi- channel like Luv color space, gradient magnitude and gradient histogram. Integral image is introduced into the process of extracting features, which reduces a lot of computation. Features from various channels are utilized to train pedestrian classifier by AdaBoost.

The rest of the paper is organized as follows. We review related work in Sect. 2. In Sect. 3 we discuss implementation details. We perform an experimental evaluation in detail in Sect. 4 and make a conclusion in Sect. 5.

© Springer Nature Singapore Pte Ltd. 2017
M. Fei et al. (Eds.): LSMS/ICSEE 2017, Part I, CCIS 761, pp. 472–480, 2017.
DOI: 10.1007/978-981-10-6370-1_47

2 Related Work

Due to the influence of complex road circumstance, a real-time and accurate pedestrian detection in front of vehicles is very difficult. Pedestrian occlusion, variable light and weather factors will make a difference to the performance of pedestrian detection. A lot of researches have been down as follows.

Dalal et al. [11] extracted HOG from pedestrian dataset and trained them with SVM. Zhu et al. [12] calculated the HOG features with integral image to reduce the operation time. Two variants of Local Binary Patterns (LBP): S - LBP and F - LBP were applied to the pedestrian detection by Mu et al. [13]. Wang et al. [14] combined the advantages of HOG and LBP to detect pedestrian when sheltered.

Inspired from the researches above, we try to describe pedestrian with features extracted from different channels. The corresponding channel of a given image is an output response. Channels can be computed by a variety of linear or nonlinear transformations. There are many available image channels such as gradient, color, texture and so on. Below we present the details of the channels used in this paper.

Luv color space: the CIE 1976 (L*, u*, v*) color space [15], which is also known as CIELUV, is a color space adopted by the International Commission on Illumination in 1976. A Luv image can be obtained by simple coordinate transformation from RGB image, which requires two steps.

(1) RGB to CIE XYZ

$$
\begin{bmatrix} X \\ Y \\ Z \end{bmatrix} = \frac{1}{0.17697} \begin{bmatrix} 0.49 & 0.31 & 0.20 \\ 0.17697 & 0.81240 & 0.01063 \\ 0.00 & 0.01 & 0.99 \end{bmatrix} \begin{bmatrix} R \\ G \\ B \end{bmatrix} \tag{1}
$$

(2) CIE XYZ to CIE Luv

$$
L^* = \begin{cases} 116 \left(\frac{Y}{Y_n} \right)^{\frac{1}{3}}, \frac{Y}{Y_n} > \left(\frac{6}{29} \right)^3 \\ \left(\frac{29}{3} \right)^3 \frac{Y}{Y_n}, \frac{Y}{Y_n} \le \left(\frac{6}{29} \right)^3 \end{cases} \tag{2}
$$

$$
u^* = 13L^* \cdot \left(u' - u'_n \right) \tag{3}
$$

$$
v^* = 13L^* \cdot \left(v' - v'_n \right) \tag{4}
$$

While $u' = \frac{4X}{X + 15Y + 3Z}$ and $v' = \frac{9Y}{X + 15Y + 3Z}$.

Gradient magnitude: The gradients of each pixel in horizontal and vertical have to be calculated before the calculation of gradient magnitude. Then gradient magnitude could be obtained by formula (5).

$$M(x,y) = \sqrt{(I(x+1,y) - I(x-1,y))^2 + (I(x,y+1) - I(x,y-1))^2} \qquad (5)$$

Gradient Histograms: The gradient angle is calculated by the tangent function of the vertical gradient and the horizontal gradient, which is shown as formula (6).

$$\theta(x,y) = \arctan\left(\frac{I(x,y+1) - I(x,y-1)}{I(x+1,y) - I(x-1,y)}\right) \qquad (6)$$

Six gradient histograms are calculated in this paper. We choose every 30° of gradient angle as a gradients histogram channel. Six new images with the same size of the given image are created. Then we fill the images with gradient magnitude according to its gradient angle so as to get six gradient histograms.

3 Implementation Details

3.1 Extraction of Features

Multi-channel features are extracted as shown in Fig. 1. We acquire the output response from original image in the channels of Luv color space, gradient magnitude and gradient histogram through a variety of image transformations. Then rectangular regions with random channels, locations and sizes are selected. We define a feature as a sum of pixels in this rectangular region. The process is repeated for several times. Finally we obtain a feature pool within features from different channels.

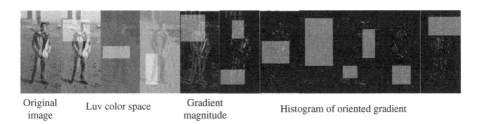

| Original image | Luv color space | Gradient magnitude | Histogram of oriented gradient |

Fig. 1. Multi-channel features

We introduce the integral image to reduce the amount of computation in the process of extracting features. The integral image is utilized to compute rectangle features. Example of integral image is shown in Fig. 2. Firstly integral image is calculated by

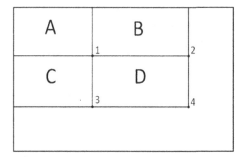

Fig. 2. Example of integral image

formula (7). Then we get the sum of all pixels in rectangular region D according to formula (8).

$$ii(x, y) = \sum_{x' < x, y' < y} i(x', y') \tag{7}$$

$$S = ii(x_1, y_1) + ii(x_4, y_4) - ii(x_2, y_2) - ii(x_3, y_3) \tag{8}$$

3.2 AdaBoost Algorithm

As an iterative algorithm, the key idea of AdaBoost is training different weak classifiers based on the same training dataset. The classification ability of a single weak classifier is low, but several weak classifiers could be gathered as a strong classifier in high classifying ability. Then, a series of strong classifiers are concatenated to achieve an excellent performance in classification.

AdaBoost classifier is trained by changing data distribution. It modifies the weights of each sample according to previous classifying accuracy in each training dataset. The next data with new weights are trained in next level of the classifier. All classifier are combined to be the final decision classifier.

The weak classifier has the advantages of simple structure and fast classifying speed while its accuracy is low. The classification of samples is shown in formula (9).

$$h_j(x) = \begin{cases} 1, p_j f_j(x) < p_j \theta_j \\ 0, others \end{cases} \tag{9}$$

While: x represents the type of rectangular feature and $f_j(x)$ represents value of feature and θ_j is the threshold. p_j equals to 1 when inputting positive samples.

The classification error of weak classifier could be calculated according to formula (10):

$$\varepsilon_j = \sum_{i=1}^{n} w_i |h_j(x_i) - y_i| \tag{10}$$

While, w_i is the sample weight, n is the number of samples and y_i is the sample type (positive sample equals to 1 and negative sample equals to 0).

The training process of strong classifier is shown as following steps:

(1) normalizing sample weights:

$$w_{1,j} = \frac{1}{n}, i = 1, 2, 3 \ldots, n \tag{11}$$

(2) Training T round, t = 1,2,3,..., T:
   (1) Normalizing weights of samples;

$$w_{t,i} = \frac{w_{t,i}}{\sum\limits_{k=1}^{n} w_{t,k}}, i = 1, 2, 3, \ldots, n \tag{12}$$

   (2) Choosing the weak classifier which has a best classification of x;
   (3) Updating sample weight;

$$w_{t+1,i} = \begin{cases} w_{t,i}\beta_t, x_i \text{ was classified correctly} \\ w_{t,i}, others \end{cases} \tag{13}$$

   While x_i represents sample I and $\beta_t = \frac{\varepsilon_t}{1-\varepsilon_t}$.

(3) Strong classifier trained:

$$h(x) = \begin{cases} 1, \sum\limits_{t=1}^{T} \alpha_t h_t(x) \geq \frac{1}{2}\sum\limits_{t=1}^{T} \alpha_t \\ 0, others \end{cases} \tag{14}$$

While $\alpha_t = - \log \beta_i$.

3.3 Architecture of Pedestrian Detection

The architecture of pedestrian detection system is shown in Fig. 3. In offline training section, we extract multi-channel features samples whose size is 64 * 128. Then we train classifier with features by AdaBoost.

In the process of real-time pedestrian detection, we preprocess traffic images and extract features as the input of classifier. While pre-processing images we compute multi-channel and extract features. In online detection, we input the features to the AdaBoost classifier trained offline. Finally we mark pedestrian out with rectangle boxes.

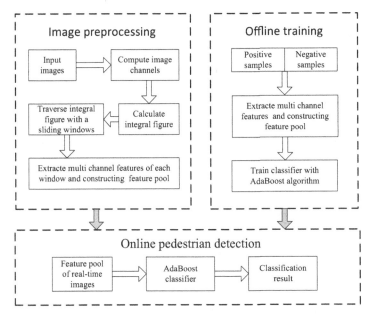

Fig. 3. Architecture of pedestrian detection

4 Experiments

4.1 Multi-channel

There are many kinds of multi-channel as Luv colour space, gradient magnitude and six gradient histograms could be combined in different ways. We extracted features of these multi-channel combinations from INRIA dataset and trained classifiers. We use same test samples to test each classifier and the DET curves are shown in Fig. 4.

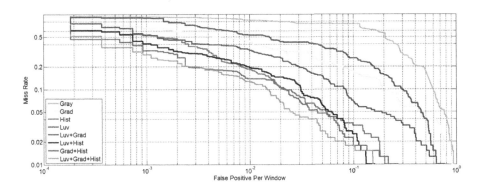

Fig. 4. DET curves of multi-channel feature classifiers

The test result shows that the classifier, which is trained with the features extracted from Luv, gradient magnitude and six gradient histograms, behaves the best in detection rate. Therefore, we choose these 10 channels as an optimal multi-channel compounding.

4.2 Experimental Data Analysis

We implement the pedestrian detection function with MATLAB on PC. Main configuration parameters of our PC are Intel 3.3 GHz processor and 8.0 GB memory. We compare the performance between HOG-SVM and this paper, which is shown in Table 1.

Table 1. Performance comparison between HOG-SVM and this paper.

Algorithm	Detection rate	Detection time
HOG-SVM	89%	7.5 s
This paper	93.75%	1.97 s

It could be found from the data in Table 1 that the algorithm has an obvious improvement on detection rate while compared with the classical pedestrian detection algorithm HOG-SVM. Furthermore the detection speed is 4 times more than HOG-SVM. The improvement of speed and accuracy in this paper mainly comes from feature extraction. Although HOG is an important feature describing pedestrian, a single feature can hardly describe pedestrian completely. Multi-channel feature used in the paper covered multiple aspects of pedestrian, such as like color, gradient and so on, which has a better performance on detection rate. On the other hand, HOG needs to traverse the entire detection window for exhaustive extraction. Large amount of computation causes a long detection time. But in this paper we randomly select a rectangular area in detection window and calculate features with integral image, which greatly reduces the amount of calculation. So it achieves a remarkable performance on detection time.

We design a pedestrian detection classifier on MATLAB and input traffic images. The detection results are shown in Fig. 5.

As it shows in Fig. 5, the algorithm has a nice detection effect on the main pedestrian targets in front of the vehicle. But it is difficult to detect the small pedestrian targets in the distance. As those pedestrians are far away from the vehicle, they make little influence on the safe driving. So the algorithm is feasible in practical application.

Fig. 5. Detection results

5   Conclusion

We optimize feature extraction of pedestrian and study a pedestrian detection method based on multi-channel feature. The pedestrian detection function is implemented in MATLAB. The multi-channel features of real time traffic image are extracted with integral image. Features are input to AdaBoost classifier for fast classification. The experimental result shows that the algorithm used in the paper has improved significantly in both detection speed and detection accuracy when compared with the classical algorithm HOG+SVM. It contains practical application value in pedestrian detection.

Acknowledgement. This work was financially supported by the national natural science foundation of China under Grant (No. 61376028) and (No. 61674100).

References

1. Gietelink, O.J., Ploeg, J., De Schutter, B., et al.: Development of advanced driver assistance systems with vehicle hardware-in-the-loop simulations. Veh. Syst. Dyn. **44**(7), 569–590 (2006)
2. Zhen, L.I., Wei, Z.Q., Xiao-Peng, J.I., et al.: Pedestrian detection based on adaptive background modeling. J. Syst. Simul. (2009)
3. Yand, T., Li, J., Pan, Q., et al.: Scene modeling and statistical learning based robust pedestrian detection algorithm. Acta Autom. Sin. **36**(4), 499–508 (2010)

4. Hao, Z., Wang, B., Teng, J.: Fast pedestrian detection Based on Adaboost and probability template matching. In: International Conference on Advanced Computer Control, pp. 390–394. IEEE (2010)

5. Meng, X.Z., Xing, J.P., Wang, Y.Z., et al.: Pedestrian detection using frame differential method and improved HOG feature. Adv. Mater. Res. **461**, 7–12 (2012)

6. Hariyono, J., Hoang, V., Jo, K.: Moving object localization using optical flow for pedestrian detection from a moving vehicle. Sci. World J. Article id. 8196415 (2014)

7. Viola, P., Jones, M.: Rapid object detection using a boosted cascade of simple features. In: Computer Vision Pattern Recognition, pp. 511–518 (2001)

8. Viola, P., Jones, M.: Robust real-time face detection. Int. J. Comput. Vis. **57**(2), 137–154 (2004)

9. Adankon, M.M., Cheriet, M.: Support Vector Machine. Comput. Sci. **1**(4), 1–28 (2002)

10. Zhao, L., Thorpe, C.E.: Stereo- and neural network-based pedestrian detection. IEEE Trans. Intell. Transp. Syst. **1**(3), 148–154 (2000)

11. Dalal, N., Triggs, B.: Histograms of oriented gradients for human detection. In: Computer Vision and Pattern Recognition, pp. 886–893 (2005)

12. Zhu, Q., Yeh, M., Cheng, K., et al.: Fast human detection using a cascade of histograms of oriented gradients. In: Computer Vision and Pattern Recognition, pp. 1491–1498 (2006)

13. Mu, Y., Yan, S., Liu, Y., et al.: Discriminative local binary patterns for human detection in personal album. In: Computer Vision and Pattern Recognition, pp. 1–8 (2008)

14. Wang, X., Han, T.X., Yan, S., et al.: An HOG-LBP human detector with partial occlusion handling. In: International Conference on Computer Vision, pp. 32–39 (2009)

15. CIELUV. https://en.wikipedia.org/wiki/CIELUV Accesed 30 March 2017

16. Wang, Z., Yoon, S., Xie, S.J.: A high accuracy pedestrian detection system combining a cascade adaboost detector and random vector functional-link net. Sci. World J. Article id. 7105089 (2014)

17. Zhang, S., Benenson, R., Schiele, B., et al.: Filtered feature channels for pedestrian detection. In: Computer Vision and Pattern Recognition, pp. 1751–1760 (2015)

Detection Method of Laser Level Line Based on Machine Vision

Xiaozhen Wang[1]($\boxtimes$), Haikuan Wang[1], Aolei Yang[1], Minrui Fei[1], and Chunfeng Shen[2]

[1] Shanghai Key Laboratory of Power Station Automation Technology,
School of Mechatronic Engineering and Automation, Shanghai University,
Shanghai 200072, China
wangxzh92@gmail.com
[2] Shanghai Baosight Software Co., Ltd., Shanghai 201203, China

Abstract. Laser lines emitted by the laser level are mostly detected manually and laser particle and optical effects also bring difficulties on measurement. In this paper, we design a detection system for the five-line laser level and propose a laser line measurement method based on ma- chine vision. Image processing is divided into two stages: in the first stage, we use random sample consensus (RANSAC) algorithm combined with Hough transform to fit the laser axis, which can get its position information. In the second stage, a laser edge extraction method based on conditional random fields (CRFs) is proposed, and the sub-pixel width of laser line is obtained by spline interpolation algorithm. The results confirm that the laser level detection method proposed in this paper can realize the corresponding detection precision and requirement.

Keywords: Laser lines measurement · RANSAC · Laser level · CRFs

1 Introduction

Nowadays, the laser level, instead of the plumb line and rulers, has been used in a large number of fields such as interior design, building planning, tool measurement, geospatial information and so on. But during its production process, each laser level is required to undergo quality inspection. The traditional method of detection is through the human eye observation, which is inefficient and without unified testing standards. However, with the continuous development of machine vision technology, more and more indus- trial production processes use machine vision instead of the human eye for on-line detection.

In recent years, the laser line width detection is based on the optical theory. The diameter of the laser spot is obtained by optical parameters. The general methods include the two-beam heterodyne method, the delay from the heterodyne method and the indirect measurement [1–4]. These laser line detection methods are based on physical optics

This work is supported by Key Project of Science and Technology Commission of Shanghai Municipality under Grant No. 14JC1402200

© Springer Nature Singapore Pte Ltd. 2017
M. Fei et al. (Eds.): LSMS/ICSEE 2017, Part I, CCIS 761, pp. 481–490, 2017.
DOI: 10.1007/978-981-10-6370-1_48

which require complex equipments. Thus, it is difficult to promote a large-scale use in the industrial production. Another way to detect laser points is to combine the CCD camera with the image processing algorithm to measure the laser diameter [5–8]. Therefore, this paper took the actual production demand into account, and proposed a method using image processing algorithms to obtain laser line position and width information based on machine vision. The position information is mainly related to the straight line detection, the general straight line detection method involving RANSAC algorithm, Hough transform and others. A number of researchers focus on road straight line extraction using RANSAC [9, 10], radar linear information detection [11], contour fitting of the mass production food packaging [12] and so on. In this paper, we presented a method to detect the axis using RANSAC combined with Hough transform based on the characteristics of laser line to achieve the laser line position information acquisition. We also analyzed image segmentation in the research of laser beam. The traditional segmentation is not ideal for laser line image. More and more scholars have applied Lafferty's concept of CRFs [13] to the image segmentation to extract features and continue to implement it in different cases [14–16]. In this paper, we refer to the thought of CRFs and put forward a set of rules to obtain the laser line features and the line-width information.

The rest of the paper is organized as follows. Section 2 designs the laser level detection system. Section 3 introduces the laser line image detection and proposes the line-width measurement algorithm. Section 4 illustrates the laser line detection experiments and the analysis of results. Finally, a conclusion is given in Sect. 5.

2   Design of Laser Level Detection System

2.1   Laser Features Under the Industrial Camera

The traditional light source is emitted in all directions, and cannot be focused at one point through the lens. On the contrary, due to the single frequency, the light from the laser can be spread in the same direction in a consistent manner. Correspondingly, the single frequency also brings the problems of coherence and strong diffraction. Under shooting of the industrial camera, some unique phenomena occur in laser image processing. Thus, in this section, we will discuss the laser line images taken by industrial cameras and introduce a system design for laser level detection.

As shown in Fig. 1, due to the optical effects of laser emission and particle effects, there are three aspects in the picture taken which cause great interference to the

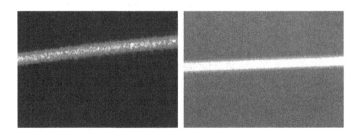

Fig. 1. Laser image under different light intensities

extraction of laser line and measurement of width. The first one is that the laser line appears spot stacking. The second one is that the internal chromaticity distribution of the laser line is uneven and random. The third one is that there are scattered noises at the edge of the laser line.

2.2 Design of Laser Level Detection System

For the five-line laser level, like Fig. 2, a system is designed to detect the position, inclination and width of the laser line in the image. To accurately judge whether the laser line meets the standard, a laser level detection structure is proposed, as shown in Fig. 3.

Fig. 2. Laser actual object

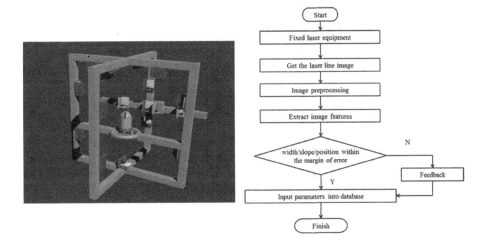

Fig. 3. Laser level detection platform design, system detection flow chart

Firstly, we collect images of the laser line from the laser level in real time by using the industrial camera. Then, we process these images by calculating the position and width of the laser line and comparing them with the measurement standard to judge whether the laser level conforms to the standard. After the laser level is started, it will project a plurality of laser lines. Considering the view size of the industrial camera and the speed of image processing, images should not be too large, therefore only three to five points on a laser line should be measured, and the measurement position is 5 meters outside the laser level. Through the image acquisition and processing of all the lines by machine vision, the purpose of the laser level detection could be achieved. The whole system detection flow chart is shown in Fig. 3.

3   Laser Line Image Detection and Measurement

This paper aimed at the red laser to do image preprocessing, including grayscale images through the red single channel and laser line regions by threshold segmentation. The contents of this section will concentrate on these pixels.

3.1   Central Axis of Laser Line Fitting

Due to the laser spot stacking and grating effect, the edge of the laser line is rough and will interfere with the linear position detection of the laser. Therefore, this paper uses RANSAC algorithm to determine the range of dense laser spots, remove noise and the interference pixel points of laser edge, and then determine the laser line central axis geometrically through the Hough transform or least squares method. The random sample consensus is an algorithm that can be used to estimate the mathematical model from a set of observational datasets. In this paper, we use the two-dimensional array of binary image pixels as the observation data to test by repeating the selection from a subset of random laser points. First, select two reference points to make a straight line, and establish the laser line axis model based on this line. Then, we set the region of interest (ROI) around the line which width is d. The probability of selecting a trusted point from the ROI at a time can be described by:

$$w = \frac{\text{the number of pixels in the ROI}}{\text{the number of pixels in the full image}} \tag{1}$$

Assuming that there are n laser line pixels in the ROI, the probability that the current n points are the confidence points is w^n; if at least one of n points is a non-trusted point, then the probability can be expressed as $1 - w^n$. The probability of failure to make the n pixels to meet the point on the central axis after k iterations is $(1 - w^n)^k$.

$$1 - p = (1 - w^n)^k \tag{2}$$

In (2), take logarithm on both sides at the same time,

$$k = \log(1 - p) / \log(1 - w_n) \tag{3}$$

Where p is used to denote the probability that the points randomly selected from the pixels of the laser line which are all trusted during the iteration. However, in the actual processing, proper parameter w cannot be given accurately, so we choose parameter t, the threshold of minimum pixel points in the model, instead of parameter w.

It can be seen from the formula (3) that the more times of iterations, the more accurate the central axis model is. But after a certain parameter k, the subsequent iteration of the model is not much better than the former, so parameter k selection also rely on the points selection. In this paper, we use parameter d as a key variable to deal with. Figure 4 illustrates different parameters d, t and k in the calculation process. On the one hand, the larger d and k values are selected, the greater axis alignment, which can be found in Fig. 4(a–c). On the other hand, Fig. 4(b–d) show that the effect of increasing the minimum t value in the current picture is not significant. Therefore, the choice of $d =$ 4.0, $t = 500$, $k = 20$ is the most reasonable.

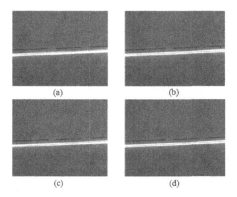

(a)   (b)

(c)   (d)

Fig. 4. d, w, k values are expressed as follows: (a) (4.0,500,10); (b) (4.0,500,20); (c) (10.0,500,10); (d) (10.0,800,20)

The pixel points obtained by RANSAC are the new data sources, and the feature information in the image space is transformed into the parameter space by Hough transform to detect the maximum point in the parameter space. The least squares method is used as the contrast algorithm, and the central axis of the laser line is shown in Fig. 5.

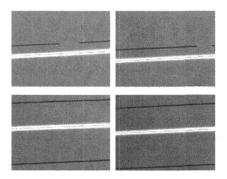

Fig. 5. The position of the laser axis

3.2   CRFs of Laser Line Modeling

The conditional random fields (CRFs) is a conditional probability model proposed by Lafferty, which contains two sequences: the observed data sequence and the tag sequence. Conditional random fields modeling thought is very similar to laser images with spots stacking, noise and other forms by optical effects. The laser is formed by the excitation energy transition, which can be regarded as a discrete random point. However, laser spots within a certain range is not relatively independent, featuring consistency which is in line with optical principles like grating effect. In this paper, we set pixel points of the laser line as the observation sequence, and the confidence pixel points under the width measurement condition are modeled as the label sequence. During the laser line processing, we use CRFs modeling to remove points out of the confidence interval, to avoid the interference of optical effect like grating on machine vision, so that the results are closer to those of human observation.

The definition of a conditional random field [13] is a representation of a graph: $G = (V, E)$, where $Y = (y_i)i{\in}V$ is the set of nodes for the graph. Then (Y, X) called CRFs only if following conditions are satisfied: random variable node y_i on the graph satisfies the Markov property when given X:

$$p(y_i|X, y_{V-1}) = p(y_i|X, y_{N_i}) \tag{4}$$

In the above equation, y_{V-1} denote other nodes in the figure except for y_i, and y_{N_i} denote neighborhood points of y_i. Considering the observed sequence X has given, the conditional probability distribution of the label sequence Y is as follows:

$$P(Y|X) = \frac{1}{Z} \exp\left(\sum_{i \in V} A_i(y_i, X)\right) + \sum_{i \in V} \sum_{i \in N_i, j \neq i} I_{ij}(y_i, y_j, X) \tag{5}$$

Where Z is a normalized constant, A_i and I_{ij} denote the first and second order potential energy functions. In consideration of the laser line characteristics, the first-order potential function with the central axis to the laser line decline on both sides:

$$A_i(y_i, F) = \sum_{i=1}^{N} \delta(y_i)p(y_i|f_i) \tag{6}$$

Here $\delta(y_i)$ denote characteristic function, when y_i is a point of the laser line, $\delta(y_i)$ value is 1; when y_i is not the background point, $\delta(y_i)$ value is 0. $p(y_i|f_i)$ is the current observation probability function in ROI, which defined by the probability distance to the laser line axis? Because the second-order potential energy function mainly for region processing, so in this paper, we proposed the function based on 8 neighborhood points of the image.

$$\sum_{j \in N_i} I_{ij}(y_i, y_j, X) = \frac{1}{C} \sum_{y_i \in N(y_i)} p(y_i|f_i, y_j) \tag{7}$$

Where C is the normalized constant, $N(y_i)$ contains 8 neighborhood points, through $p(y_i | f_i, y_j)$ to determine the 8 pixels weight value, showing whether the pixel belongs to the edge of the laser line.

As shown in the above figure, the 8 neighborhood is divided into two parts of the function, $g1(y_i)$ denote weight that whether the central point is at upper edge of the laser line or lower; $g2(y_j)$ denote weight by the slope of the straight line parameter. The expression is as follows (Fig. 6).

$$g1(y_i) = \begin{cases} \alpha i > j, upperedge; \ i < j, loweredge \\ \beta i < j, upperedge; \ i > j, loweredge \end{cases}$$

$$g2(y_j) = \begin{cases} \alpha i > j, positiveslope; \ i < j, negetiveslope \\ \beta i < j, positiveslope; \ i > j, negetiveslope \\ 0.5 \quad slope \ is \ zero \end{cases} \tag{8}$$

$$\alpha + \beta = 1$$

$g1(y)$	$g1(y)$	$g1(y)$
$g2(y)$		$g2(y)$
$g1(y)$	$g1(y)$	$g1(y)$

Fig. 6. Strategy of weight for 8 neighborhood

Figure 7 shows that CRFs can effectively remove the noise and irrelevant points outside the laser line, and have a good effect on the roughness laser edge, but it cannot completely smooth the edge of the laser line. On the one hand, the first-order and second-order potential function selection strategies and methods have affected the smoothing result. On the other hand, there are rough edges details of the laser line originally under observation of human eye, and the purpose of machine vision processing algorithm is to replace the human eye work. Hence, rough edges in the processed image seemed

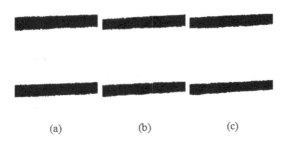

(a)   (b)   (c)

Fig. 7. Edge of line with different algorithms: (a) original image; (b) classic filtering; (c) CRFs

reasonable. Although classical filtering also has some effect of filtering out noise and smoothing edges, the method makes the edges fuzzy, and the resulting fuzzy boundaries are difficult for subsequent measurements.

4   Experiments and Analysis

The algorithm has been tested on the normal PC, using C# combined with Visual Studio 2010. We have collected a number of images with the resolution of 2048 × 1536. In order to make sure that one laser line can be measured with accuracy and stability, a series of images are taken 20 times on one laser line. The results are summarized in Table 1. On the laser level nameplate, it indicates that the laser wavelength is 650 nm, laser line width is 2 mm at 5 m with the horizontal and vertical accuracy is plus or minus 1 mm.

Table 1. The comparison between the measured pixels number and the mean pixel number in 20 images.

Laser line width in images						
	No.	Pixels number	Offset	No.	Pixels number	Offset
Images	1	45.2274	0.1144	11	44.2295	−0.8835
	2	45.3061	0.1931	12	44.4554	−0.6576
	3	44.8382	−0.2748	13	45.4897	0.3767
	4	44.9149	−0.1981	14	44.8349	−0.2781
	5	44.8271	−0.2859	15	45.0321	−0.0809
	6	45.0126	−0.1004	16	44.8572	−0.2558
	7	44.2408	−0.8722	17	45.2025	0.0895
	8	45.4375	0.3245	18	44.3119	−0.8011
	9	45.5962	0.4832	19	44.9226	−0.1904
	10	45.0056	−0.1074	20	44.7661	−0.3469

Even though we can get the width range of the laser line from nameplate of products, it is difficult to compare the measured value with the exact value. Therefore, the

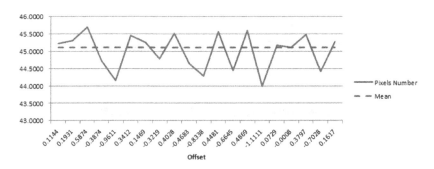

Fig. 8. Offset distribution of 20 images

experiment mainly discusses the stability and pixel-width ratio accuracy of multiple measurements. Figure 8 shows the number of pixels under multiple measurements relative to the mean value fluctuation, but the offset is all in one-pixel scale.

Considering the view angle of lens in the x-axis direction is 14, the length of lens is 46 mm, the distance from the lens to the laser image is 400 mm. The field of view in the x direction is

$$L = 2h = u * \tan(\theta/2) \tag{9}$$

According to the formula of camera resolution

$$field\ of\ view\ (X)/resolution\ (X) = resolution\ of\ camera\ (X) \tag{10}$$

Here, we get the pixel-width ratio is 23 px/mm. Combined with the offset of one pixel obtained above, the true distance error is less than 0.0435 mm, which proves that the method presented in this paper can measure the width of the laser line accurately and stably.

5 Conclusion

We have proposed an efficient and practical method for detecting laser lines based on machine vision. Initially, after the basic image preprocessing, the method of RANSAC combined with Hough transform is applied to obtain the line feature points. Next, CRFs is used to build the model of the laser image, which solves the noise removal problem at the edge of the laser line. Finally, the cubic spline interpolation is used to get sub-pixel width of the laser line. The experimental results show that it works accurately in various images within the allowable range of error. However, this paper is mainly based on the static situation, and the image just need to be taken only one at any time. But in the actual detection environment, the laser level placed or moved will cause the attenuate jitter of the laser line. On the one hand, waiting for its static condition consumes detection time. On the other hand, taking dynamic images may cause offset with its real position, resulting in false detection. Therefore, further research will be focused on laser line jitter detection to meet higher standards.

References

1. Xie, D.: Line-width measurement method of narrow line width laser. Laser Optoelectron. Progress **50**(1), 61–66 (2013)
2. Yue, Y.L., Qin, B., Hong-Wei, L.V., et al.: Comparison of several ultra-narrow laser linewidth measurement. Opt. Commun. Technol. **37**(3), 19–21 (2013)
3. An, P., Zhao, R., Zheng, Y., et al.: Linewidth rapid measurement of narrow fiber laser by spectrum analyzer. Infrared Laser Eng. **44**(3), 897–900 (2015)
4. Sutili, T., Figueiredo, R.C., Conforti, E.: Laser linewidth and phase noise evaluation using heterodyne offline signal processing. J. Lightwave Technol. **34**(21), 4933–4940 (2016)
5. Vázquez-Otero, A., Khikhlukha, D., Solano-Altamirano, J.M., et al.: Laser spot detection based on reaction diffusion. Sensors **16**(3), 315–326 (2016)

6. Zhang, H.Z., Yao, M., Lei, P., et al.: Research of image processing method of far-field laser spots. Laser Technol. **4**(37), 460–463 (2013)
7. Xiang, X.Y., Chen, H.Q., Peng, W.U., et al.: Adaptive measuring algorithm of laser beam width based on CCD. Laser Technol. **30**(5), 552–554 (2006)
8. Wang, Y., Wang, Q., Ma, C., et al.: Factors affecting the accurate measurement of laser beam width with CCD camera. Chin. J. Lasers **41**(2), 303–308 (2014)
9. Guo, J., Wei, Z., Miao, D.: Lane detection method based on improved RANSAC algorithm. In: 12th IEEE International Symposium on Autonomous Decentralized Systems (ISADS), pp. 285–288. IEEE Press, Taichung (2015)
10. Pollard, E., Gruyer, D., Tarel, J.P., et al.: Lane marking extraction with combination strategy and comparative evaluation on synthetic and camera images. In: 14th International IEEE Conference on Intelligent Transportation Systems Council (ITSC), pp. 1741–1746. IEEE Press, Washington (2011)
11. Jacobs, L., Weiss, J., Dolan, D.: Object tracking in noisy radar data: comparison of Hough transform and RANSAC. In: 2013 IEEE International Conference on Electro/Information Technology (EIT), pp. 1–6. IEEE Press, Rapid City (2013)
12. Kröger, M., Sauer-Greff, W., Urbansky, R., et al.: Performance evaluation on contour extraction using Hough transform and RANSAC for multi-sensor data fusion applications in industrial food inspection. In: 2016 Signal Processing: Algorithms, Architectures, Arrangements, and Applications (SPA), pp. 234–237. IEEE Press, Poznan (2016)
13. Lafferty, J., McCallum, A., Pereira, F.: Conditional random fields: probabilistic models for segmenting and labeling sequence data. In: Proceedings of the 18th International Conference on Machine Learning (ICML), vol. 1, pp. 282–289. ACM, Williamstown (2001)
14. Tran, L.C., Pal, C.J., Nguyen, T.Q.: View synthesis based on conditional random fields and graph cuts. In: 17th IEEE International Conference on Image Processing (ICIP), pp. 433–436. IEEE Press, Hong Kong (2010)
15. Hur, J., Kang, S.N., Seo, S.W.: Multi-lane detection in urban driving environments using conditional random fields. In: 2013 IEEE Intelligent Vehicles Symposium (IV), pp. 1297–1302. IEEE Press, Gold Coast City (2013)
16. Qian, S., Chen, Z.H., Lin, M.Q., et al.: Saliency detection based on conditional random field and image segmentation. Acta Automatica Sinica **41**(4), 711–724 (2015)

An Accurate Calibration Method of a Multi Camera System

Song Han, Xiaojing Gu, and Xingsheng Gu[✉]

Key Laboratory of Advanced Control and Optimization for Chemical Process,
Ministry of Education, East China University of Science and Technology, Shanghai, China
{xjing.gu,xsgu}@ecust.edu.cn

Abstract. In this paper, we proposed a novel method of geometric calibration and synchronization for a multi camera system. Traditional calibration methods of visible cameras can't be applied to thermal cameras. According to the imaging characteristics of thermal cameras, we designed a new calibration board using materials with different emissivity for calibration. Our calibration board can accurately calibrate RGBD cameras and thermal cameras. In general, thermal cameras have regular non-uniformity corrections, which will result in camera interruption about 1.5 to 2 s and can impact synchronization. In this respect, we adopted the method named nearest adjacent time using timestamp to solve the problem of non-uniformity corrections and synchronization. We evaluated our methods and the experiments showed that our methods had an ideal result for camera calibration and synchronization.

Keywords: Multi camera system · Thermal cameras · RGBD cameras · Calibration · Synchronization

1 Introduction

Consumer level RGBD cameras, such as Kinect [1], are popular in the field of computer vision in recent years. A large number of Kinect performance tests have been completed by [2–4]. Kinect is applicable to many fields, such as robotic field, 3D reconstruction field and so on [5]. Recently, with the decrease of thermal camera price, infrared imaging has become a helpful and useful tool in the field of computer vision. Thus, the idea of combining a RGBD camera with a thermal camera has started to become more popular. More and more people design multi camera systems including Kinect and a thermal camera to perform a 3D reconstruction of a thermal scene [6]. The 3D reconstruction of thermal scene can be applied to building inspection [7] and fire rescue [8].

Geometric calibration and temporal calibration are the precondition of a multi camera system. Without above calibrations, the multiple image information from a multi camera system can't be integrated together. The infrared image represents the temperature information of the object in the scene. Therefore, traditional calibration boards can't calibrate thermal cameras. In this paper, we designed a novel calibration board to calibrate RGBD and thermal cameras. The temporal calibration means synchronizing scene images from different cameras. The scene information will be confusing without

© Springer Nature Singapore Pte Ltd. 2017
M. Fei et al. (Eds.): LSMS/ICSEE 2017, Part I, CCIS 761, pp. 491–501, 2017.
DOI: 10.1007/978-981-10-6370-1_49

temporal calibration. We adopted the method of nearest adjacent time to synchronize RGBD and thermal cameras. The contribution of this paper is as follows:

1) We designed a novel calibration board which can calibrate both RGBD and thermal cameras. The advantages and disadvantages between the circular calibration plate and the checkerboard calibration plate are compared.
2) We considered the effect of the interruption from non-uniformity corrections of the thermal camera on synchronization. And we adopted the method of nearest adjacent time by using timestamps to solve camera synchronization.

The rest of the paper is organized as follows: In Sect. 2 we provide an overview of prior works in camera calibration and synchronization of multi camera systems. And then, we introduce our methods of camera calibration and synchronization in detail in Sect. 3 and 4. Next, we carry out some experiments and evaluation in Sect. 5 and then provide conclusion and future work in Sect. 6.

2   Related Work

Camera calibration is a core problem in the field of 3D reconstruction and robot navigation [9]. The most widely accepted calibration method is the strategy from Zhang [10]. The traditional calibration boards usually use printed chessboards and the calibration points are easily located in a visible image, but they cannot be accurately located in an infrared image. This is because the thermal camera mainly acquires the temperature information of objects and the printed chessboard generally maintains near-uniform temperature with low contrast. In recent years, many researches have designed different calibration boards in order to calibrate thermal cameras. In [11–13], authors cut rectangular holes through a board and put it in front of different temperature objects. The method can calibrate both RGBD and thermal cameras. However, the thickness of holes will result in inaccurate calibration. A thermal calibration rig with 42 small LED lights located on the intersections of the conventional checkerboard was designed. Which can calibrate thermal cameras when the LED lights were on [14]. However the light will affect the precise positioning of feature points in infrared images. Kim [15] presented a line-based grid pattern board, which included a line based grid of regularly sized squares pattern. Michael [16] adopted a black and white checkerboard with one resistor mounted in the center of each checkerboard square. In [17], the authors used rubber heater to warm up a plastic mask with a grid of holes for thermal camera calibration. It worked well but the price of the calibration board is a little high. We proposed a novel calibration board made of aluminum foil and cardboards, which is easy to make and low cost. Besides, the calibration board only need to be heat one minute by a hair drier before calibrating cameras. We designed square and circle pattern calibration boards and carried out experiments to study which one is more accurate.

The process that all images from all sensors of a multi camera system have corresponding timestamps is called camera temporal calibration or synchronization. Many previous studies adopted hardware to solve synchronization. Soonmin [18] used the master-slave synchronization technique to synchronize visible and thermal cameras. The visible and thermal cameras are synchronized by sending trigger signals from master to slaves. In [19], the author adopted beam splitter to synchronize cameras. The synchronization result is ideal with the method of hardware, but most cameras don't have trigger generators or beam splitters. Therefore, the cost of synchronization will increase. Some other works such as [20] dealt with extrinsic calibration and synchronization jointly.

The previous works have taken little account of the impact of non-uniformity corrections from thermal cameras on synchronization. Most thermal cameras have regular non-uniformity corrections, which will lead to camera interruption about 1.5 to 2 s. To overcome this problem, we adopted the method of nearest adjacent time, which can quickly synchronize cameras and can solve the impact of non-uniformity corrections.

3   Calibration

3.1   Intrinsic Calibration

Pinhole model is one of the simplest types of camera models, which can mathematically describe the projection of points in 3D space onto an image plane. P is the point in 3D space, p is the perspective projection point from P to a 2D image plane. In homogeneous coordinates, the 3D world point P and perspective projection point p can be respected as:

$$P = \begin{bmatrix} X & Y & Z & 1 \end{bmatrix} \tag{1}$$

$$p = \begin{bmatrix} u & v & 1 \end{bmatrix} \tag{2}$$

The geometric relation between P and p can be written as:

$$z \begin{bmatrix} u \\ v \\ 1 \end{bmatrix} = \begin{bmatrix} f_x & 0 & u_0 & 0 \\ 0 & f_y & v_0 & 0 \\ 0 & 0 & 1 & 0 \end{bmatrix} \begin{bmatrix} R & T \\ 0^T & 1 \end{bmatrix} \begin{bmatrix} X \\ Y \\ Z \\ 1 \end{bmatrix} \tag{3}$$

Where z is the distance from the point P to the camera, R and T are the spatial relation matrixes between the world coordinate system and the camera coordinate system. The intrinsic matrix consists of focal length f_x, f_y and the principal point (u_0, v_0).

The traditional calibration boards can't calibrate the thermal camera. The thermal camera acquires the information of the scene through the reflected and radiate infrared rays of the objects. Materials with high emissivity reflect most of the infrared rays back because they don't absorb much infrared energy. Therefore we can calibrate the thermal camera with the calibration board made of different materials. We use the second generation Kinect as the RGBD camera. The second generation Kinect contains both a color camera and a thermal camera. And the spatial depth information is detected by the thermal camera. This means that the depth camera and the thermal camera are the same camera. Thus the calibration of the depth camera can be converted to calibrated the thermal camera.

In this paper, we proposed two different calibration pattern boards which can calibrate both thermal cameras and RGBD cameras. One is the square planar pattern calibration board. The other is the circle pattern calibration board. The square planar pattern calibration board was consisted of white cardboard squares and aluminum foil squares. The aluminum squares were made by sticking aluminum foil on the white cardboard squares. The circle pattern calibration board was made of circle aluminum foil and a thick white cardboard. Both square planar pattern and circle pattern calibration boards are suitable, convenient and low cost. In order to know which kind of calibration boards has higher calibration accuracy, some tests was carried out by using OpenCV. Figure 1 shows the two types of calibration boards and the results of calibration. The accuracy of calibration can be analyzed by calculating the mean reprojection error (MRE), which indicates the distance between a detected point and a reprojection point in a photo. The unit of MRE is pixel.

Fig. 1. Calibration boards and the results of calibration. First column: square and circle planar pattern calibration boards. Second column: RGB camera. Third column: near-infrared (depth) camera. Last column: thermal camera.

Table 1 shows a comparison of the proposed calibration patterns and the pattern presented by Skala et al. [13]. From Table 1 we can see that our method is more accurate than the method from Skala et al. Besides, the MRE of circle pattern is less than square pattern. It is suggested that the circle pattern is more accurate than the square pattern.

Table 1. The MRE of different calibration patterns.

Cameras Types	MRE of Circle	MRE of Square	MRE of Square with holes [13]
Thermal Camera	0.0183	0.0230	0.1312
Depth Camera	0.0272	0.0281	0.1861
Color Camera	0.0294	0.0334	0.1464

The main reason of affecting calibration precision is the precise localization of feature points. The localization of feature points is very difficult when the calibration board is non-front parallel to the camera imaging plane [21]. The calibration board is usually not parallel to the camera imaging plane. As a result, the square in the photo becomes a diamond and the circle becomes an ellipse. The algorithm detecting the center of a circle is divided into the three steps. Firstly, an input image is converted into a set of binary images by a series of continuous threshold values. Secondly, the connected region of each binary image is extracted by detecting their boundaries. The connected region is the blob of corresponding a binary image. Finally, the blob is fitted to a circle and the center of the circle can be calculated. And it can also detect the center of an ellipse by fitting the ellipse edge. The method is very simple and efficient in detecting the center of a circle or an ellipse. For square pattern, the algorithm is detecting the corners of a chessboard. When the square in the photo becomes a diamond, the localization error of feature points becomes larger. According to Table 1, the calibration accuracy of circle pattern is higher than square's. So we adopted the circle pattern calibration board to calibrate cameras.

3.2 Stereo Calibration of Thermal and RGBD Cameras

In order to obtain fusion information from Kinect and the thermal camera, it is require to make them in a common geometric coordinate system. The process of acquiring the rotation and translation relationships between different cameras is known as stereo calibration. The new calibration board that is presented in this paper can calibrate thermal camera, RGB camera and depth camera. Thus we are able to use calibration-based to carry out stereo calibration.

According to pinhole projection model, a spatial point in scene can be project onto the camera image plane. And there is an infinite number of spatial points can be project onto the same point on the camera image plane. All these spatial points lie on a straight line that goes through the camera center. The points on the camera image plane can't be reprojected into spatial scene because they lost the distance information. Nevertheless, the pixels of a depth image represent the distance between the spatial object and the depth camera. So the depth information is able to be used to reproject points from the camera image plane to the spatial scene.

$$z_d p_{depth} = K_{depth} \begin{bmatrix} R_{depth} & t_{depth} \end{bmatrix} P \tag{4}$$

$$z_t p_{thermal} = K_{thermal} \begin{bmatrix} R_{thermal} & t_{thermal} \end{bmatrix} P \tag{5}$$

Where P is the point in a 3D space scene, p_{depth} and $p_{thermal}$ are the projection points from P to the image plane of the depth camera and the thermal camera. K_{depth} and $K_{thermal}$ are the intrinsic matrixes of the depth camera and the thermal camera. $\begin{bmatrix} R_{depth} & t_{depth} \end{bmatrix}$ and $\begin{bmatrix} R_{thermal} & t_{thermal} \end{bmatrix}$ are the extrinsic matrixes of the depth camera and the thermal camera, including the rotation matrix and the translation vector. z represents the distance from P

to camera in the camera coordinate system. The projection point P_{depth} from P in the camera coordinate system can be written as:

$$P_{depth} = z_d K_{depth}^{-1} p_{depth} = R_{depth} P + t_{depth} \qquad (6)$$

The spatial position relationship between the thermal camera and the depth camera can be written as:

$$P_{thermal} = R P_{depth} + T \qquad (7)$$

Where R is the rotation relationship between two cameras, T is the translation relationship between two cameras. The following equation can be acquired by combining Eqs. (4), (5) and (6).

$$P_{thermal} = R_{thermal} R_{depth}^{-1} P_{depth} + t_{thermal} - R_{thermal} R_{depth}^{-1} t_{depth} \qquad (8)$$

Comparing Eqs. (7) and (8), R and T are able to be calculated.

$$R = R_{thermal} R_{depth}^{-1} \qquad (9)$$

$$T = t_{thermal} - R t_{depth} \qquad (10)$$

From Eqs. (9) and (10), we can come to a conclusion that we only need to obtain the extrinsic matrixes of the thermal camera and the depth camera in the same scene, and then the transformation matrixes of two cameras can be calculated. Similarly, the relationship between visible and depth cameras can obtain using the same method.

4   Synchronization

It is very important for synchronization of a multi camera system to integrate various information. The authors in [12] adopted motion statistic alignment to synchronize cameras. They used the TV-L1 optical flow to calculate the average flow vector magnitude of RGBD and thermal images streams. However, it cannot be used in low-frame cameras and the thermal camera with non-uniformity corrections (NUCs). In general, most thermal cameras need regular intensity correction in order to increase the accuracy of receiving the infrared radiation signal, which is known as NUCs. When the thermal camera performs NUCs, a baffle of uniform temperature will temporarily block the imaging sensor. Therefore, this leads to a data stream interruption about 1.5 to 2 s. NUCs are significant for thermal images. If a thermal camera doesn't have NUCs, the pixel values of thermal images can't represent an accurate estimate of infrared radiance in the scene. On the other hand, the sampling frequency of Kinect and thermal camera is different, which maybe cause an infrared photo can't match a corresponding depth photo.

In order to solve above problems, we improved the method from [22]. We used the nearest adjacent time between an infrared image and a depth image to quickly find the corresponding frames between the Kinect and the thermal camera. The RGB and depth image frames from the Kinect are already synchronized. It only need to synchronize the depth and thermal cameras. We first obtain the timestamps of depth and thermal frame streams $\{t_n^d, n = 1, 2, 3 \cdots\}$ and $\{t_m^{th}, m = 1, 2, 3 \cdots\}$. After that, we calculate the time difference between depth frames and corresponding thermal frames:

$$\Delta t = \min\left\{ \left|t_{n-1}^d - t_m^{th}\right|, \left|t_n^d - t_m^{th}\right|, \left|t_{n+1}^d - t_m^{th}\right| \right\} \tag{11}$$

The frame rate of the Kinect is almost 30 frames per second and frame rate of the thermal camera is almost 20 frames per second. So every infrared frame maybe correspond more than one depth frames. Considering that, we select the three depth frames adjacent to the infrared frame to calculate and keep the minimum time difference as Δt. The threshold σ represents half of the inter-frame interval of the camera with the slowest frame rate. So the value of σ is set to 25 ms. If Δt is less than the threshold σ, the depth frame and corresponding thermal frame are the synchronized frames and we will keep two fames. Otherwise, we will abandon them temporarily. Considering NUCs of thermal camera, many depth frames don't have corresponding thermal frames during NUCs. When Δt exceeds 1.5 s, we can consider that the thermal camera is carrying out NUCs. Thus, the depth frame should be abandoned and the corresponding infrared frame is kept to search next depth frames. If Δt is greater than 25 ms but no more than 1.5 s, relative infrared and depth frames are both abandoned.

5 Experiment and Evaluation

In our experiment, we adopted the second generation Kinect as RGBD camera and FLIR T420 as thermal camera. We first carried out the experiment of camera calibration including single camera calibration and stereo camera calibration. For the visible camera, the calibration needs to be carried out in a weak light environment. And a light source needs to be arranged on one side of the calibration board. After we obtained the intrinsic parameters and extrinsic parameters of cameras, we registered depth images with infrared images and RGB images. Figure 2 shows the results of registration. We can find that results are particularly desirable and indicate that we can obtain an accurate camera calibration using our calibration board.

Fig. 2. The results of registration. First row: left column is original infrared images, middle column is original depth images, right column is registered images. Second row: left column is original color images, middle column is original depth images, right column is registered images.

And then we evaluated the efficacy of synchronization using infrared and depth image frames. Figure 3 shows the offset between thermal camera and the depth camera. Figure 4 shows the results of synchronization. We selected a key frame as a start frame. From Fig. 3, as time goes by, the offset is increasing because of different frame rates between two cameras. We can clearly see two curve pulses, which are NUCs operation of the thermal camera. In Fig. 4, the depth frames those in NUCs time are abandoned. Comparing with pictures from Figs. 3 and 4, the nearest adjacent time has an ideal effect on synchronization.

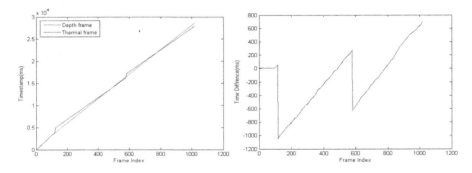

Fig. 3. The offset between the thermal camera and the depth camera. Left image shows the timestamps of two cameras. Right image shows the time difference of thermal frames and depth frames.

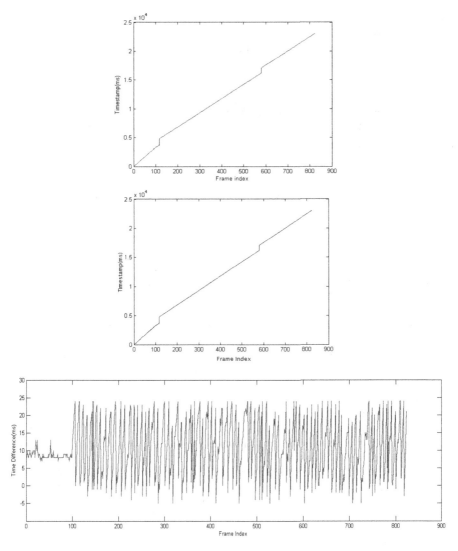

Fig. 4. The results of synchronization using our method. First row: left image shows the result of depth frames after synchronization; right image shows the result of thermal frames after synchronization. Second row shows the time difference between thermal frames and depth frames after synchronization.

6 Conclusion

In this paper we presented novel methods for geometric calibration and temporal calibration of the Kinect and the thermal camera. We designed a new calibration board for camera calibration, which can be used to calibrate thermal camera, depth camera and RGB camera at the same time. Comparing with existing methods that require high cost and complex operation, our calibration board is easy to product and use. The calibration board only consists of aluminum foil and a cardboard, so the manufacturing cost is lower than other kinds of calibration boards. In addition, the method of the nearest adjacent time we adopted can synchronize the image streams from a multi cameras system. By calculating the time difference and eliminating invalid frames, the method of the nearest adjacent time can solve the problem of NUCs quickly.

In the future, we will focus on the fusion of depth datas, thermal datas and RGB datas from the Kinect and the thermal camera. Besides, we will study indoor three dimensional temperature field reconstruction. In addition, we will explore a real time reconstruction system using the Kinect and the thermal camera.

Acknowledgments. This work is supported by National Nature Science Foundation under Grant No. 61573144, 61502293 and 61205017.

References

1. Zennaro, S., Munaro, M., Milani, S., Zanuttigh, P., Bernardi, A., Ghidoni, S., et al.: Performance evaluation of the 1st and 2nd generation Kinect for multimedia applications. In: International Conference on Multimedia and Expo, pp. 1–6. IEEE, Turin (2015)
2. Butkiewicz, T.: Low-cost coastal mapping using Kinect v2 time-of-flight cameras. In: Oceans, pp. 1–9. IEEE, St. John's (2015)
3. Gonzalez-Jorge, H., Rodríguez-González, P., Martínez-Sánchez, J., González-Aguilera, D., Arias, P., Gesto, M., et al.: Metrological comparison between kinect I and kinect II sensors. Measurement **70**, 21–26 (2015)
4. Corti, A., Giancola, S., Mainetti, G., Sala, R.: A metrological characterization of the Kinect V2 time-of-flight camera. Robot. Auton. Syst. **75**(8), 584–594 (2015)
5. Lachat, E., Macher, H., Landes, T., Grussenmeyer, P.: Assessment and calibration of a RGB-D camera (Kinect v2 Sensor) towards a potential use for close-range 3D modeling. Remote Sens. **7**(10), 13070–13097 (2015)
6. Moghadam, P., Vidas, S., Lam, O.: Spectra: 3D multispectral fusion and visualization toolkit. In: Australasian Conference on Robotics and Automation. ACRA, Melbourne (2014)
7. Borrmann, D., Nüchter, A., Djakulovi'C, M., Maurovi'C, I., Petrovi'C, I., Osmankovi'C, D., et al.: The Project thermalmapper-thermal 3D mapping of indoor environments for saving energy. Ifac Symp. Robot. Control **45**(22), 31–38 (2012)
8. Nauer, C., Vonach, E., Gerstweiler, G., Kaufmann, H.: 3D building reconstruction and thermal mapping in fire brigade operations. In: Augmented Human International Conference, pp. 1–2. IEEE, Lake Buena Vista (2013)
9. Borrmanna, D., Nüchtera, A., Đakulovic, M., Maurovic, I., Petrovic, I., Osmankovic, O., et al.: A mobile robot based system for fully automated thermal 3D mapping. Adv. Eng. Inform. **28**(4), 425–440 (2014)
10. Zhang, Z.: A flexible new technique for camera calibration. IEEE Trans. Pattern Anal. Mach. Intell. **22**(11), 1330–1334 (2000)
11. Vidas, S., Moghadam, P., and Bosse, M. (2013).: 3D thermal mapping of building interiors using an RGB-D and thermal camera. In: International Conference on Robotics and Automation, pp. 2311–2318. IEEE, Karlsruhe (2013)

12. Lussier, J. T., Thrun, S.: Automatic calibration of RGBD and thermal cameras. In: International Conference on Intelligent Robots and Systems, pp. 451–458. IEEE, Chicago (2014)
13. Skala, K., Lipić, T., Sović, I., Gjenero, L., Grubišić, I.: 4D thermal imaging system for medical applications. Periodicum Biologorum **113**(4), 407–416 (2011)
14. Ham, Y., Golparvar-Fard, M.: An automated vision-based method for rapid 3D energy performance modeling of existing buildings using thermal and digital imagery. Adv. Eng. Inform. **27**(3), 395–409 (2013)
15. Kim, N., Choi, Y., Hwang, S., Park, K.: Geometrical calibration of multispectral calibration. In: International Conference on Ubiquitous Robots and Ambient Intelligence, pp. 384–385. IEEE, Goyang (2015)
16. Gschwandtner, M., Kwitt, R., Uhl, A., Pree, W.: Infrared camera calibration for dense depth map construction. In: Intelligent Vehicles Symposium, pp. 857–862. IEEE, Baden-Baden (2011)
17. St-Laurent, L., Prévost, D., Maldague, X.: Fast and accurate calibration-based thermal/color sensors registration. In: 10th International Conference on Quantitative Infrared Thermography, QIRT, Québec, Canada, pp. 27–30 (2010)
18. Hwang, S., Choi, Y., Kim, N., Park, K.: Low-cost synchronization for multispectral cameras. In: International Conference on Ubiquitous Robots and Ambient Intelligence, pp. 435–436. IEEE, Goyang (2015)
19. Baar, J.V., Beardsley, P., Pollefeys, M., Gross, M.: Sensor fusion for depth estimation, including TOF and thermal sensors. In: 2nd International Conference on 3d Imaging, Modeling, Processing, Visualization and Transmission, pp. 472–478. IEEE, Zurich (2012)
20. Miller, S., Teichman, A., Thrun, S.: Unsupervised extrinsic calibration of depth sensors in dynamic scenes. In: Intelligent Robots and Systems, pp. 2695–2702. IEEE, Tokyo (2013)
21. Datta, A., Kim, J.S., Kanade, T.: Accurate camera calibration using iterative refinement of control points. In: 12th Computer Vision Workshops, pp. 1201–1208. IEEE, Kyoto (2009)
22. Bahnsen, C.: Thermal-visible-depth image registration (2017). http://projekter.aau.dk/projekter/en/studentthesis/termiskvisueldybde-billedregistering(8564c183-bdc1-4a9c-8bda-c71878cb0814).html

A Novel Memory Gradient Based for Efficient Image Segmentation

Kun Zhang[1,2], Jianguo Wu[1(✉)], and Peijian Zhang[1]

[1] School of Electrical Engineering, Nantong University, Nantong, China
zhangkun_nt@163.com, {wu.jg, zhang.pj}@ntu.edu.cn
[2] Nantong Research Institute for Advanced Communication Technologies,
Nantong, China

Abstract. Image segmentation is a very important phase in automatic image analysis. Of the developed techniques for image segmentation, iterative methods have been proven to be one of the most effective algorithms in the literature. Mean shift algorithms is one of the iterative approaches which have been successfully deployed to many applications. However, despite its promising performance, mean shift has shown its weaknesses in convergence in some of the application areas. In this paper, an improved version of the standard mean-shift algorithm using a memory gradient method is proposed and implemented in order to achieve fast convergence rates by integrating mean shift and memory gradient. Experimental results on real images demonstrate that our proposed algorithm not only improves the efficiency of the classical mean shift algorithm, but also achieves better segmentation results.

Keywords: Memory gradient · Mean shift · Convergence property · Image segmentation

1 Introduction

Mean shift algorithm is a non-parametric iterative clustering method [1–3] which has been deployed successfully to many applications such as image segmentation [4–6], image smoothing [7–9], tracking [10, 11] and clustering [12–14]. In recent years, the theoretical basis of mean shift has been widely studied, e.g. [15, 16]. Mean shift essentially is a gradient decent algorithm of gradual convergence. It is well known that steepest descent method is fast with a simple stucture but turns to be inefficient at seeking global optimization [17, 18], especially in the presence of complex and irregular density functions. However, there is not any research work to discuss the use of steepest descent method in mean shift.

Yang et al. [19] investigated a Quasi-Newton method combined with the classical mean shift algorithm in order to improve the convergence rate of the latter. However, searching an approximation to the Hessian Matrix is still time consuming during each

© Springer Nature Singapore Pte Ltd. 2017
M. Fei et al. (Eds.): LSMS/ICSEE 2017, Part I, CCIS 761, pp. 502–512, 2017.
DOI: 10.1007/978-981-10-6370-1_50

iteration. Li et al. [20] introduced a conjugate gradient approach to mean shift so as to improve the convergence rate. However, to a generalised continuously differentiable objective function, conjugate gradient is difficult to guarantee convergence in a finite number of steps. Based on Local Sensitive Hashing (LSH), Collins [21] has proposed a new way to reduce the computational complexity, which has the advantages in high dimensional data clustering but does not suit the applications of low dimensionality such as segmentation and vision tracking. On the other hand, many researchers have tried to improve the efficiency of boundary optimization algorithms [22–24]. For example, Salakhutdinov [24] provided theoretical convergence analysis of edge optimization to handle the efficiency problem.

Memory gradient method is a powerful line search method like conjugate gradient method for solving large scale problems [25–27], as they avoid the computation and storage of matrices associated with Newton type methods. In this paper, a new memory gradient method is proposed. The search direction generated by the new method satisfies the requested descent conditions and the angle property. We give an appropriate initial step value at each iteration of the searching, which can reduce the computational effort of grandient evaluations so as to improve the performance of the new method.

The rest of this paper is organized as follows. The new memory gradient method is introduced in Sect. 2. Experimental results are presented and disscussed in Sect. 3. Finally, we give conclusions and future work in Sect. 4.

2   New Memory Gradient Based Mean Shift Algorithm

The unconstrained minimization problem which is studied in this paper is as follows

$$minf(x), x \in \Re^n \tag{1}$$

where $\Re^n$ denotes an n dimensional Euclidean space, and g denotes the gradient of f. A line search method can be used here. The line search process is given by

$$X_{K+1} = X_K + \alpha_K d_K \tag{2}$$

First of all, we introduce the assumptions made to f as follows.

(H_1): The function f has a lower bound of the level- set $Ls_0 = \{x \in \Re^n | f(x) \leq f(x_1).\}$, where x_1 is an initial point.

(H_2): The gradient g is *Lipschitz* continuous in an open convex set B which includes Ls_0, i.e., there exists $L > 0$ such that $\|g(x) - g(y)\| \leq L_s\|x - y\|, \forall x, y \in B$

Now, we begin to describe the proposed memory gradient method.

Algorithm 1(proposed memory gradient algorithm)

Step 0 Given an initial point $x_1 \in \Re^n$, $0 < \rho < \dfrac{2}{3}$, $\sigma \in (0, 0.5)$ and set $d_1 = g_1$,

$k := 1$.

Step 1 if $\| g_k \| = 0$ then stop. Else, go to step 2;

Step 2 $x_{k+1} = x_k + \alpha_k d_k$ where

$$d_k = \begin{cases} -g_k, & if\ k = 1 \\ -(1 - \alpha_k)g_k + \alpha_k g_{k-1}, & if\ k \geq 2 \end{cases} \tag{3}$$

Set

$$s_k = \begin{cases} \rho, & if\ k = 1 \\ \dfrac{\rho \| g_k \|^2}{1 \| g_k \|^2 + | g_k^T g_{k-1} |}, & if\ k \geq 2 \end{cases} \tag{4}$$

α_k is chosen by Armijo rule: Let α_k be the largest α in $\{s_k, \rho s_k, \rho^2 s_k, \ldots\}$ such that

$$f_k - f(x_k + \alpha d_k) \geq -\sigma \alpha [g_k^T d_k + \tfrac{1}{2} \alpha \| g_k \|^2] \tag{5}$$

Step 3 $k := k + 1$, go to step 1

It is easy to find out that the proposed algorithm has an important charactor: the searching direction and the stepset are made at the same time so that it is less difficult to search for the best searching direction and the stepset. Based on this proposed memory gradient algorithm, we have a new mean shift algorithm.

Algorithm 2 (our proposed mean shift algorithm)
Step 0 Initialization.
Step 1 Repeat.

Step 2 Execute mean shift algorithm, calculate entropy E : $E = -\hat{f}(x)log\,\hat{f}(x)$.

Step 3 If $E \geq \tau$, and then turn to step 4.

Step 4 Execute the proposed memory gradient algorithm, and calculate entropy E.

Step 5 If $E < \tau$, and then turn to step 2, otherwise finish.

In order to analyse the global convergence properties, we firstly introduce necessary lemmas.

In this section, it is assumed that $g_k \neq 0$, for $k > 0$. The lemmas introduced below imply that d_k is used to implement a satifactory descent direction of f.

Lemma 2.1. For all $k \geq 1$, we have

$$g_k^T d_k \leq -(1 - \rho)\|g_k\|^2 \tag{6}$$

Proof:

$$g_k^T d_k = -(2-\alpha)\|g_k\|^2 + \alpha g_k^T d_{k-1}$$
$$\leq -(2-\alpha)\|g_k\|^2 + \alpha|g_k^T d_{k-1}|$$
$$= -(2-\alpha)s_k\|g_k\|^2 + (\rho - 2s_k)\|g_k\|^2/s_k$$
$$\leq -(1-\rho)\|g_k\|^2$$

This completes the proof.

Lemma 2.2. For all $k \geq 0$, $\alpha \in (0, s_k)$, we have

$$g_k^T d_k + \frac{1}{2}\alpha\|g_k\|^2 \leq 0 \qquad (7)$$

Proof:
Using lemma 2.1, we have

$$g_k^T d_k \leq -(1-\rho)\|g_k\|^2$$

and

$$s_k = \frac{\rho\|g_k\|^2}{2\|g_k\|^2 + |g_k^T d_{k-1}|}$$

if $k \geq 2$ we have $s_k \leq \rho$, and denote $\alpha \in (0, s_k)$
$g_k^T d_k + \frac{1}{2}\alpha\|g_k\|^2$ can be changed to $(\rho - 1 + \frac{1}{2}\rho)\|g_k\|^2$
For $0 < \rho < \frac{2}{3}$, $(\rho - 1 + \frac{1}{2}\rho) < 0$
We can get $g_k^T d_k + \frac{1}{2}\alpha\|g_k\|^2 \leq 0$
This completes the proof.

Lemma 2.3. For all $k \geq 1$, $\alpha \in (0, s_k)$, we get $\|d_k\| \leq \max_{1 \leq i \leq k}\{\|g_i\|\}$

Proof:
When $k = 1$, we have $\|d_1\| \leq \|g_1\|$
From Eq. (2)

$$\|d_k\| = \|(\alpha_k - 1)g_k + \alpha_k g_{k-1}\|$$
$$\leq \|(\alpha_k - 1)g_k\| + \|\alpha_k g_{k-1}\|$$
$$= (\alpha_k - 1)\|g_k\| + \alpha_k\|d_{k-1}\|$$
$$\leq (\alpha_k - 1)\|g_k\| + \alpha_k\|d_{k-1}\|$$
$$(1 - \alpha_k)\|d_k\| \leq (\alpha_k - 1)\|g_k\|$$

For $\alpha_k \in (0, s_k)$ and $0 < \rho < \frac{2}{3}$
So $\alpha_k - 1 < 0$, $\|d_k\| \leq \|g_k\|$
This completes the proof.

3 Experiment and Discussion

We have evaluated the proposed method for the segmentation of a set of color images. The experimental computer is of Intel core2 E7500 3.0 GHZ processor and 4G memory. In this section, we present the experimental results by comparing the proposed method with *EDSION* [28] and *Li's* approach [20].

3.1 Convergence Rate Analysis

First of all, we evaluate the proposed algorithm using extensive simulations. In the experiment, we consider a popular Rosenbrock benchmark function and the function is described as follows:

$$f_0(\vec{x}) = \sum_{i=1}^{n-1} \left(100(x_{i+1} - x_i^2)^2 + (x_i - 1)^2\right) \tag{8}$$

In Fig. 1, the tradeoff between the convergence rate and the state performance is illustrated. The curves in the figure were generated recursively using a sphere function. The number of runs in the simulation is 20. The graph of Fig. 1 was obtained as follows: The gradient and conjugate gradient algorithms cannot reach a steady state within 20 runs, while memory gradient can enjoy a satisfied convergence rate and reach to a steady state within 12 runs. In the first 7 runs, the memory gradient descent algorithm is significantly better than the others. The proposed mean shift algorithm has the best execution efficiency because it well handled computational complexity, while the *EDSION* and *Li's* systems require significantly higher computational complexity.

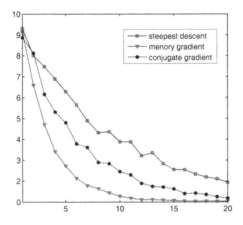

Fig. 1. Convergence comparsion among three gradient based algorithms

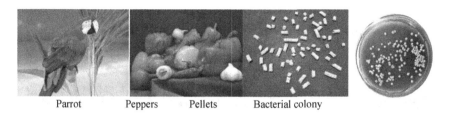

| Parrot | Peppers | Pellets | Bacterial colony |

Fig. 2. The original colour images

Table 1. Region number and executing time

	Parrot $(hs, hr) = (20, 10)$		Peppers $(hs, hr) = (20, 10)$		Pellets $(hs, hr) = (20, 10)$		Bacterial colony $(hs, hr) = (20, 10)$	
	Num	Time	Num	Time	Num	Time	Num	Time
EDISON	689	4.3	952	5.2	158	3.1	195	5.9
YANG	596	4.1	759	4.9	43	2.7	161	5.4
Proposed method	509	2.6	296	2.1	32	1.5	108	3.8

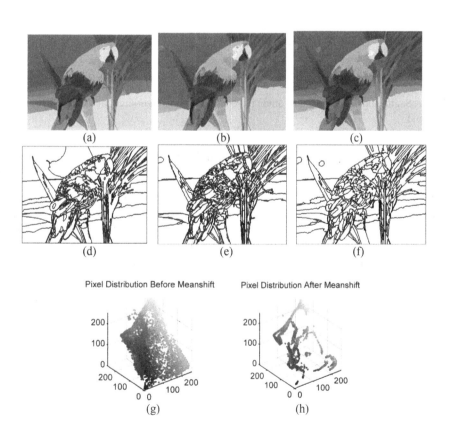

Pixel Distribution Before Meanshift   Pixel Distribution After Meanshift

Fig. 3. Performance comparsion of three mean shift algorithms and pixel distribution(parrot)

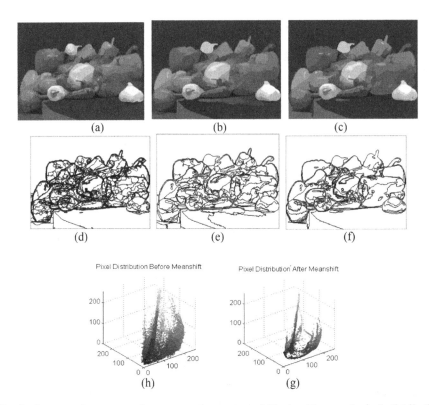

Fig. 4. Segmentation comparsion among there mean shift algorithms and pixel distribution (peppers)

3.2   Comparison of Different Algorithm in Terms of Segmentation Performance

In this section, we compare the segmentation performance of the proposed method against the *EDSION* system and *Li* systems. Figure 2 includes the original color images"parrot", "peppers", "pellets", and "Bacterial colony", whose sizes are respectively 400×300, 512×384, 608×486, 1900×1000. In the paper, all the segmentation experiments are carried out by using Gaussian kernel. to conduct fair comparisons, we choose the same parameters (*hs* and *hr*) and the same iteration trials for all the involved algorithms.

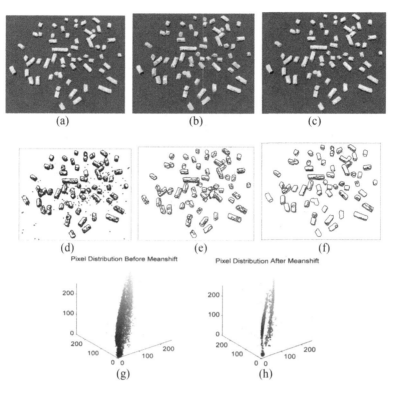

Fig. 5. Segmentation comparsion among there mean shift algorithms and pixel distribution (pellets)

The value of hs is related to the spatial resolution while the value hr defines the range resolution. It is necessary to note that the spatial resolution hs has a different effect on the output image, compared to the image resolution. Only features with large spatial support are represented in the segmented image with our algorithm when hs is increased. On the other hand, only features with high contrast survive when hr is large. Therefore, the quality of segmentation is controlled by the spatial value hs and the range hr, with the resolution parameters defining the radii of the windows in the respective domains.

The number of regions generated by each method is showed in Table 1. Figures 3, 4, 5 and 6 illustrate the mean shift segmentation results based on the three mean shift algorithms. We set 20 iteration trial to be the pixels distribution after applying memory mean shift.stopping criterion due to the balance of the algorithm efficiency and stability. (a) and (d) refer to the results of *EDISON*. (b) and (e) on the results of *Li*. (c) and (f) on the results of our proposed method. (g) is color image pixels distribution before mean shift is applied, and (h) is color image pixels distribution after applying memory mean shift..

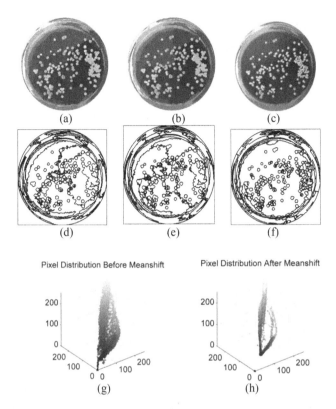

Fig. 6. Segmentation comparsion among there mean shift algorithms and pixel distribution (Bacterial colony)

4 Conclusion

In this paper, a new memory gradient based fast mean shift algorithm for color image segmentation is proposed. The proposed algorithm firstly uses a memory gradient algorithm to optimize the mean shift vector. The memory gradient algorithm possesses sufficient descent conditions without carrying out any line search and satisfies the angle property, which is independent of the convexity of the objective function. Evidence shows that the performance of our method is comparable to the conjugate gradient method and the steep decent gradient method. The proposed mean shift algorithm is carried out to update the output of the memory gradient algorithm. Experiment results show that the proposed algorithm is an effective method and outperforms the *EDISON* system and *Li* system.

Acknowledgements. This work was financially supported by the Natural Science Foundation of Jiangsu Province, China. Nantong University-Nantong Joint Research Center for Intelligent Information Technology under Grant No. KFKT2016A08 and by Nantong Research Program of Application Foundation under Grant No. GY12016022.

References

1. Aiazzi, B., Alparone, L., Baronti, S., Garzelli, A., Zoppetti, C.: Nonparametric change detection in multitemporal sar images based on mean-shift clustering. IEEE Trans. Geosci. Remote Sens. **51**(4), 2022–2031 (2013)
2. Yuan, X.T., Hu, B.G., He, R.: Agglomerative mean-shift clustering. IEEE Trans. Knowl. Data Eng. **24**(2), 209–232 (2012)
3. Varga, B., Karacs, K.: High-resolution image segmentation using fully parallel mean shift. J. Adv. Sig. Process. **1**, 1–17 (2011)
4. Freedman, D., Kisilev, P.: KDE paring and a faster mean shift algorithm. SIAM J. Imaging Sci. **3**(4), 878–903 (2010)
5. Zhou, H., Schaefer, G., Emre, C.M.: Gradient vector flow with mean shift for skin lesion segmentation. Comput. Med. Imaging Graphics **35**(2), 121–127 (2011)
6. Tsai, D.-M., Luo, J.-Y.: Mean shift-based defect detection in multicrystalline solar wafer surfaces. IEEE Trans. Indus. Inf. **7**(1), 125–135 (2011)
7. Rodriguez, R., Suarez, A.G., Sossa, J.H.: A segmentation algorithm based on an iterative computation of the mean shift filtering. J. Intell. Robot. Syst. **63**(3–4), 447–463 (2011)
8. Lin, L., Garcia-Lorenzo, D., Li, C., Jiang, T., Barillot, C.: Adaptive pixon represented segmentation (APRS) for 3D MR brain images based on mean shift and Markov random fields. Pattern Recogn. Lett. **32**(7), 1036–1043 (2011)
9. Shim, S.-O., Malik, A.S., Choi, T.-S.: Noise reduction using mean shift algorithm for estimating 3D shape. Imaging Sci. J. **59**(5), 267–273 (2011)
10. Khan, Z.H., Gu, I.Y.H., Backhouse, A.G.: Robust visual object tracking using multi-mode anisotropic mean shift and particle filters. IEEE Trans. Circuits Syst. Video Technol. **21**(1), 74–87 (2011)
11. Leichter, I.: Mean shift trackers with cross-bin metrics. IEEE Trans. Pattern Anal. Mach. Intell. **34**(4), 695–706 (2012)
12. Comaniciu, D., Meer, P.: Mean shift: A robust approach toward feature space analysis. IEEE Trans. Pattern Anal. Intell. **24**(5), 603–619 (2002)
13. Chang-Chien, S.-J., Hung, W.-L., Yang, M.-S.: On mean shift-based clustering for circular data. Soft Comput. **16**(6), 1043–1062 (2012)
14. Craciun, S., Wang, G., George, A.D.: A Scalable RC Architecture for mean-shift clustering. In: IEEE 24th International Conference on Application-Specific Systems, Architectures and Processors, pp. 370–374 (2013)
15. Ghassabeh, Y.A., Linder, T., Takahara, G.: On some convergence properties of the subspace constrained mean shift. Pattern Recogn. **46**(11), 3141–3147 (2013)
16. Carreira-Perpinan, M.A.: Gaussian mean-shift is an EM algorithm. IEEE Trans. Pattern Anal. Mach. Intell. **29**(5), 767–776 (2007)
17. Bento, G.C., Ferreira, O.P., Oliveira, P.R.: Unconstrained steepest descent method for multicriteria optimization on riemannian manifolds. J. Optim. Theor. Appl. **154**(1), 88–107 (2012)
18. Li, C., Shi, G.: Weights optimization for multi-instance multi-label RBF neural networks using steepest descent method. Neural Comput. Appl. **22**(7–8), 1562–1569 (2013)
19. Yang, C., Duraiswami, R., DeMenthon, D., Davis, L.: Mean-shift analysis using quasi-newton methods. In: IEEE International Conference on Image Processing, vol. 2, pp. 447–450 (2003)
20. Li, Y.L., Shen, Y.: Fast mean shift for image segmentation based on conjugate gradient. Opto-Electron. Eng. **36**(8), 94–99 (2009)

21. Georgescu, B.,Shimshoni, I.,Meer, P.: Mean shift based clustering in high dimensions: a texture classification example. In: Proceedings of IEEE International Conference on Computer Vision. vol. 1, pp. 456–463 (2003)
22. Cao, K., Huang, B., Wang, S.: Sustainable land use optimization using boundary-based fast genetic algorithm. Comput. Environ. Urban Syst. **36**(3), 257–269 (2012)
23. Kong, L., Wang, Z., Wang, W.: Global boundary optimization for automobile engine based on genetic algorithm. In: International Conference on Artificial Intelligence and Computational Intelligence, vol. 315, pp. 342–349 (2012)
24. Salakhutdinov, R.,Roweis, S.: On the convergence of bound optimization algorithms. In: Proceedings of the Nineteenth conference on Uncertainty in Artificial Intelligence, pp. 509–516 (2002)
25. Hager, W.W., Zhang, H.: The limited memory conjugate gradient method. SIAM J. Optim. **23**(4), 2150–2168 (2013)
26. Zheng, Y., Wan, Z.: A new variant of the memory gradient method for unconstrained optimization. Optim. Lett. **61**(12), 1491–1509 (2012)
27. Sun, M., Bai, Q.: A new descent memory gradient method and its global convergence. J. Syst. Sci. Complex. **24**(4), 784–794 (2011)
28. Christoudias, C.M., Georgescu, B., Meer, P.: Synergism in low level vision. In: 16th International Conference on Pattern Recognition. vol. 4, pp. 150–155 (2002)

Research on Cigarette Filter Rod Counting System Based on Machine Vision

Hongjun Qu[1], Peijian Zhang[1(✉)], Kun Zhang[1,2], and Jianguo Wu[1]

[1] School of Electrical Engineering, Nantong University, Nantong, China
737043942@qq.com, {zhang.pj,wu.jg}@ntu.edu.cn,
zhangkun_nt@163.com
[2] Nantong Research Institute for Advanced Communication Technologies, Nantong, China

Abstract. The traditional method for on-line detection of cigarette filter stick packing is manual sampling, which has low efficiency, high labor cost and can't detect all products. Therefore, it is necessary to establish a set of image processing system [10] based on machine vision. This system uses CCD image sensor to get the filter rod section image, through the image smoothing, edge detection, Binarization and feature extraction, the region of interest is analyzed, and finally get the number of filter rods. The filter rod arrangement on the production line is not completely flat. Using the dynamic area threshold method to calculate the number of filter rod in the flat area. 3D reconstruction of a single image is used to calculate the number of filter rod in the uneven area. The accuracy and practicability of the system are verified by theoretical analysis and experimental comparison.

Keywords: Machine vision · Binarization · Feature extraction · Dynamic threshold · Single image 3D reconstruction

1 Detection System Design

1.1 Imaging Principle

The system is mainly aimed at the existing filter rod production equipment. First of all, because unable to install the line scan camera in the narrow space of equipment, and line scanning equipment is expensive, so can't obtain the ideal image; Secondly, in the process of cigarette filter rod packing arrangement is not completely flat, there will be a small part of the filter rod slightly convex; Furthermore, the size of the packing box is 800 mm * 600 mm, which belongs to a large field of vision. Based on the limitations of the above conditions, we use CCD plane array industrial camera, the system schematic shown in Fig. 1.

Firstly, the packaging box is under the light field composed of striped light source. The side of the filter rod is imaged in the imaging area of CCD after irradiation of the scattered light. In addition, CCD industrial camera using a fisheye lens so as to adapt to the wide field and filter rod boxes each dock position uncertainty leads to the problem of object distance is changed. Fisheye lens provides a great perspective (generally can reach 220°) and large depth of field (generally can reach several millimeters to several

© Springer Nature Singapore Pte Ltd. 2017
M. Fei et al. (Eds.): LSMS/ICSEE 2017, Part I, CCIS 761, pp. 513–523, 2017.
DOI: 10.1007/978-981-10-6370-1_51

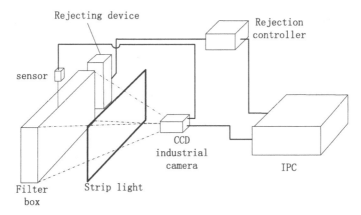

Fig. 1. Schematic diagram of detection system [15]

centimeters), made in the case of the changes in object distance can also get a clear image. CCD in the driving of the driving circuit, the output pulse signal containing information of filter rod. The signal after the processing of the signal processing circuit is converted into digital image (see Fig. 2). In the digital image transmission to the computer after the software processing can get the appropriate parameters and make the appropriate action.

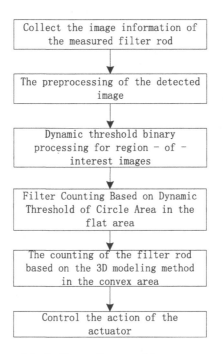

Fig. 2. Frame diagram of the system

1.2 System Structure

The ultimate goal of the system is in order to realize the on-line counting of filter rod [17], and according to the count value action on current box (excluding or shipped to the next station). For this purpose, design system should fully meet the design requirements, to ensure that the system has good effectiveness, stability, ease of use, and extensibility. System is mainly composed of image acquisition module, image processing module, an actuator [5]. Image acquisition module is composed of CCD industrial camera and image acquisition card, the function of measured object is the output of digital image; Industrial computer to perform the work for the image processing module, including image preprocessing, image binarization, filter rod counting, and communication of the actuator; Actuators according to the results given by the industrial computer to manipulate: eliminate or will it into the next station. System framework is shown in Fig. 2.

As we can see from the above figure, two methods are used to count the number of filter rod. The first method is to use a circular area based on the dynamic threshold of the filter bar. This method is applied in the area where the filter rods are arranged in a relatively flat area. This method runs speed blocks and has high stability. This method is not applicable for cases where the filter rod arrangement protrudes more seriously. For this reason, the filter rods are positioned and counted based on the 3D reconstruction of the uneven areas, but the algorithm is complex and time-consuming, so it is only applied to uneven areas. System synthesis of these two methods to detect, that is to ensure the accuracy of the test, but also to ensure the speed of the system.

2 Image Processing

2.1 The Preprocessing of the Detected Image

The image preprocessing [2] is to extract the region of interest in the image and remove the invalid feature. The image acquisition process was conducted in production equipment inside, so the acquisition process is affected by the surrounding environment and photographic equipment, the image will inevitably exist in the noise. In order to get high quality image in order to improve the detection accuracy, need to filter the image.

There are two main types of filter rod packing image noise: The first is electromagnetic interference from surrounding equipment, the noise is the pulse noise; The second is due to the electronic circuit and low intensity of illumination, it causes Gaussian noise. Using median filter for impulse noise, median filtering is to deal with the current center pixel domain, it is a kind of nonlinear filtering. Filtering results of the current pixel value is to use all the grey value of pixels in the filtering window median instead. Formula is as follows:

$$f(x, y) = \underset{(x,y) \subseteq S_{xy}}{median} \{g(x, y)\} \tag{1}$$

Among them, f(x, y) is the filter output, S_{xy} is the coordinates of all points of filtering window, g(x, y) is the grey value. Median filtering can keep the line of output image clear, and the image is relatively uniform. F We use a Gaussian filter to handle Gaussian noise. Gaussian filter is thought to pixel in filter window after the weighted average instead of the current pixel values.

2.2 Dynamic Threshold Binary Processing

The environment when shooting images is complicated, we can't decorate a uniform light source, and the ambient light to imaging effect also has great influence, so can't use a single threshold for binarization image. Dynamic threshold binary processing of the basic idea is: Dynamic threshold binary processing of the basic idea is: according to the pixel's neighborhood block pixels to determine the distribution of the current pixel binarization threshold. The threshold in the area of higher brightness is higher. On the contrary, in the area of low brightness threshold is low. By using dynamic threshold binarization after rendering as shown in Fig. 3.

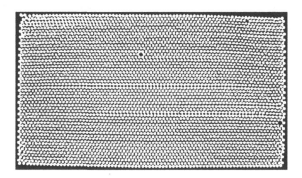

Fig. 3. Dynamic threshold binarization figure

2.3 Filter Rods Counting Method for Flat Area

In the area where the filter rods arranged in order. After the dynamic threshold binarization processing can get as shown in Fig. 4 binary figure with better effect, local amplification figure (as shown in Fig. 4) clearly show filter rod appearance pattern [3].

In this case, with the method of based on dynamic threshold area [7] statistics the number of filter rod. For any one of the flat areas, there are at least two filter rods that are tangent to it [6]. Using this principle, we can find with the current filter rod circumscribed filter rods. The statistical method is: First locate the position of a filter rod in the image, and then spread out in a spider web way. In the process of diffusion by judging the current target area and fitting circle roundness to determine whether the effective target. We use the upper and lower thresholds to limit the area of the statistical process. The specific formula is as follows:

$$\begin{cases} D_S \leq S \leq D_B \\ D_S = \xi_S \times \eta(x,y) \times S_S \\ D_B = \xi_B \times \eta(x,y) \times S_S \end{cases} \tag{2}$$

Among them, S is the area of the current target; D_S is the area down-threshold; D_B is the area up-threshold. Because the object to be detected is a large field of view and a fisheye lens is used, the distortion of the image increases as the distance from the center of the image becomes far from the center axis, resulting in a larger area of the filter rod near the edge of the binarized image. Therefore, the area threshold in the formula is a dynamic threshold, whose value is related to the position of the current point, and the closer it is to the edge, the greater the value. S_S is the standard cross-sectional area of filter rod; $\eta(x,y)$ is the coordinates (x, y) threshold amplification coefficient; ξ_S is on the down-threshold coefficient; ξ_B is the up-threshold coefficient. Figure 5 shows the effect of the first diffusion of this method.

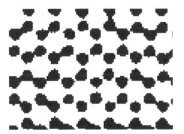

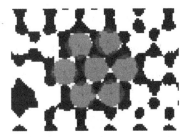

Fig. 4. Local amplification binarization image   **Fig. 5.** The first diffusion diagram

The operation of the above is in the case of filter stick flat pattern, Fig. 6 shows the encounter in the process of diffusion of the two cases. The effect of binarization in the area in the box in Fig. 6(a) is not very good, due to the shadow of the bulge. We can see from the graph, the algorithm in the process of diffusion successfully avoided the area, the area of detection method is specified in 2.4. The Fig. 6(b) shows the success of avoiding a void, which is caused by the lack of packing in the boxing process.

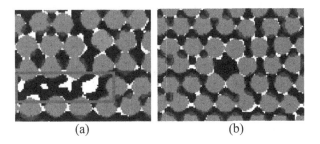

(a)   (b)

Fig. 6. Two cases in the process of diffusion

For the entire picture, if only one starting point is required to perform 40–50 times of diffusion to find all the points in the entire graph, there is a great risk of doing so. First, the accumulation of 40–50 times of error may present a greater error; secondly, if the point stops unexpectedly during the diffusion process, the remaining target can't be found. In order to solve these two problems, we set six starting points for different locations, each point only need to spread 10–20 times to complete, but also to avoid the risk of accidental termination. Figure 7 shows the effect of different diffusion times in the case of flattening.

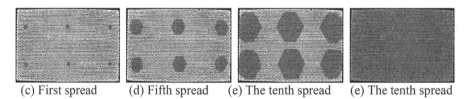

(c) First spread   (d) Fifth spread   (e) The tenth spread   (e) The tenth spread

Fig. 7. Schematic diagram of flat area

2.4 Detection of Filter Rods in Uneven Areas

In the actual production, due to machinery and packing technology and other reasons, so FCL box filter can not be completely flat, there will be a small amount of filter rods protruding. The bulge have less effect on the image center area. If the edge of the image has a bulge of filter rod, the method based on dynamic threshold area can't find out the filter rod in the shaded area of filter rod, and the bulging filter rod statistical quantity more than real. Figure 8 shows the detection results of image containing bulging filter rods.

We can see from the figure that there are two regions with poor binarization, which can't be quickly detected by the dynamic area threshold detection method. For these areas, we use slower method of 3D reconstruction based on single image [4, 12], establish a cylinder model of bulge of filter rod, and then according to the cylindrical form judge the number of filter rod in these areas.

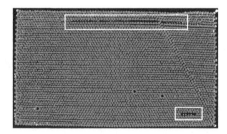

Fig. 8. The detection results of image containing bulging filter rods

Due to the loss of depth information in 2D images and the influence of environmental factors, want from 2D image reconstruction of 3D model is difficult. The traditional 3D reconstruction [16] is to calibrate the camera coordinate relationship with reality, and then according to several 2D image to reconstruct 3D model. Based on a single 2D image 3D reconstruction is a fuzzy reconstruction process, need to calculate the characteristic value of 2D images in various information, then according to the geometry knowledge and experience knowledge to deal with these eigenvalues, restore the 3D feature information. Due to the increasing ability of the modern computer calculation, the calculation has been able to finish quickly.

The filter rod protruding part of the image is shown in Fig. 9. For 3D reconstruction of cigarette filter rod, feature extraction [13] is the key step in the process of reconstruction. The eigenvalues we need are the circular information at the top of the cylinder and the linear information of the cylindrical edge. These two pieces of information can be reconstructed by fuzzy reconstruction of the 3D structure of the rod protrusion, which can be counted out of the number of rods in the protruding area. Straight lines and circles are the basic features of the image, and we can find the edge of the filter rod by straight fitting and circular fit [8]. Hough transform [9] is to analyze the global characteristics of a specific shape of the edge of the connection, Hough transform can also effectively remove the image texture and background noise [19].

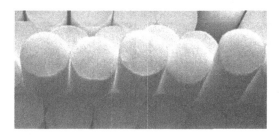

Fig. 9. Protruding local image

The idea of hough transform [14] is:

$$y_0 = kx_0 + b \tag{3}$$

convert to polar coordinate system is:

$$\rho = x \cos \theta + y \sin \theta \tag{4}$$

ρ is the distance from the origin to the straight line, the value of θ is $-\pi \sim \pi$. By plotting the ρ values to find the local maximum (maximum) point, a straight line can be obtained by setting a threshold.

Hough circle transformation [18]: In the Cartesian coordinate system, the circle equation is:

$$(x - x_0)^2 + (y - y_0)^2 = r^2 \tag{5}$$

converted to polar coordinate system is:

$$\begin{cases} a = x - r \cos \theta \\ b = y - r \sin \theta \end{cases} \tag{6}$$

For any one of the edge points (x_0, y_0), a and r respectively traverse $(0, 2\pi)$ and $[R_1, R_2]$ with the steps of Δa and Δr, to obtain the subspace of the parameter space (a, b, r), and add 1 to the corresponding accumulator unit $A(a, b, r)$. For all the points of the binarized image, the accumulator array A is obtained, where any of the array elements $A(a, b, r)$ represents the number of edge pixels on the circle (radius r, center is (a, b)). The larger the value of $A(a, b, r)$, the greater the likelihood that the circle is (a, b) and the radius r is, so the maximum $A(a, b, r)$ needs to be found in the program.

We can see from Fig. 10 found by hough transform [11] circle is relatively clear, but the straight line part is disorder. The ends of the line had been marked by a small circle, you can see there are some short invalid texture information, 3D reconstruction needs according to the geometrical characteristics of the cylindrical and the location of the current point, and it can get a cylindrical oblique angle. Based on each Hough circle, the protruding part of the cylinder is reconstructed by searching the nearby straight line to meet the requirements [20]. As shown in Fig. 11.

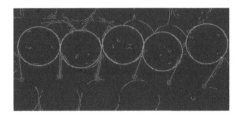   

Fig. 10. Hough transform image   **Fig. 11.** 3D reconstruction of the image

Through the analysis of the height and tilt Angle of each cylinder can deduce the number of filter rod in the shadow, and can remove protruding filter stick more lateral root number, improve the accuracy of the statistics.

Table 1. Test comparison table

(a) Machine vision statistics

Serial number	Accurate value	Machine vision statistics	Statistical time	Absolute error
1	3998	3998	723 ms	0
2	4000	4003	698 ms	3
3	4000	4000	753 ms	0
4	4001	4003	780 ms	2
5	4002	4003	812 ms	1
6	4001	4003	708 ms	2
7	4003	4002	892 ms	−1
8	3999	3999	768 ms	0
9	3975	3977	744 ms	2
10	3972	3974	815 ms	2

(b) Artificial statistics

Serial number	Accurate value	Artificial statistics	Statistical time	Absolute error
1	3998	3995	152 s	−3
2	4000	4002	248 s	2
3	4000	3999	184 s	−1
4	4001	4001	195 s	0
5	4002	4000	178 s	−2
6	4001	3999	235 s	−2
7	4003	4004	205 s	1
8	3999	3999	187 s	0
9	3975	3978	332 s	3
10	3972	3973	229 s	1

2.5 The Experimental Contrast

In order to verify the design of the filter plate online counting system of practicality, we use the equipment with a rating of 4000 to verify, and manual statistics with the comparative test. In the previous production, the count of detection of filter rod is artificial sampling. This approach is characterized by: low efficiency, slow speed, and can't realize online full inspection. Table 1 shows the test results.

As can be seen from the test results, the precision of the machine vision detection and artificial detection difference is very small, and the errors are within the error of the standard (±10错误!未找到引用源。) required; In the test time has the very big difference, machine vision detection is much faster than the artificial detection, machine vision test meet the requirements of the rapidity of on-line detection.

3   Conclusion

The online detection system of Cigarette filter rod counting based on machine vision designed in this paper completes the counting of cigarettes on-line production, by using the combination of 2D and 3D detection, the accuracy of the count is guaranteed. This system is stable, fast and robust. The experimental results show that the error of detection is within ±5, and meets the required error standard of ±10. It replaces the traditional inefficiency manual statistics mode, effectively improving the quality of the cigarette filter packing products. In future studies, it is necessary to improve the adaptability of the system so that it can count different shapes of filter rods.

Acknowledgements. This work was financially supported by Nantong University-Nantong Joint Research Center for Intelligent Information Technology under Grant No. KFKT2016A08 and by Nantong Research Program of Application Foundation under Grant No. GY12016022.

References

1. Labudzki, R., Legutko, S.: The essence and application of machine vision. J. Tehnicki Vjesnik **21**, 903–909 (2014)
2. Du, H., Zhao, Z., Wang, S., Hu, Q.: Two-dimensional discriminant analysis based on Schatten p-norm for image feature extraction original research. J. Vis. Commun. Image Represent. **45**, 87–94 (2017)
3. Ding, W., Gu, J., Shang, Z., Tang, S., Wu, Q., Duodu, E.A., Yang, Z.: Semantic recognition of workpiece using computer vision for shape feature extraction and classification based on learning databases. J. Optik Int. J. Light Electron Optics **130**, 1426–1437 (2017)
4. Hamledari, H., McCabe, B., Davari, S.: Automated computer vision-based detection of components of under-construction indoor partitions. Autom. Constr. **74**, 78–94 (2017)
5. Ying-Jie, L., Fu-Cheng, Y.: Postmark date recognition based on machine vision. Phys. Procedia **33**, 819–826 (2012)
6. Igathinathane, C., Ulusoy, U.: Machine vision methods based particle size distribution of ball- and gyro-milled lignite and hard coal. Powder Technol. **74**, 78–94 (2017)
7. Židek, K., Hošovský, A.: Image thresholding and contour detection with dynamic background selection for inspection tasks in machine vision. Int. J. Circ. **8**, 545–554 (2014)
8. Han, J.H., Poston, T.: Chord-to-point distance accumulation and planar curvature: a new approach to discrete curvature. Pattern Recogn. Lett. **22**, 1133–1144 (2001)
9. Kaehler, A., Bradski, G.: Learning OpenCV 3. O'Reilly Media, Sebastopol (2016)
10. Beyerer, J., León, F.P., Frese, C.: Machine Vision: Automated Visual Inspection: Theory Practice and Applications. Springer, New York (2015)
11. Gonzalez, R.C., Woods, R.E., Eddins, S.L.: Digital Image Processing Using MATLAB, 2nd edn. Gatesmark Publishing, Knoxville (2009)
12. Cho, J., Lee, M., Oh, S.: Single image 3D human pose estimation using a procrustean normal distribution mixture model and model transformation. Comput. Vis. Image Underst. **155**, 150–161 (2017)
13. Kang, X., Yau, W.-P., Taylor, R.H.: Simultaneous pose estimation and patient-specific model reconstruction from single image using maximum penalized likelihood estimation (MPLE). Pattern Recogn. **57**, 61–69 (2016)
14. China Software Developer Network. http://www.csdn.net

15. Industrial Forum. http://bbs.gongkong.com
16. Márquez, G., Pinto, A., Alamo, L., Baumann, B., Ye, F., Winkler, H., Taylor, K., Padrón, R.: A method for 3D-reconstruction of a muscle thick filament using the tilt series images of a single filament electron tomogram. J. Struct. Biol. **186**, 265–272 (2014)
17. Ramos, P.J., Prieto, F.A., Montoya, E.C., Oliveros, C.E.: Automatic fruit count on coffee branches using computer vision. Comput. Electron. Agric. **137**, 9–22 (2017)
18. Ayub, M.A., Mohamed, A.B., Esa, A.H.: In-line inspection of roundness using machine vision. Procedia Technol. **15**, 807–816 (2014)
19. Wang, Y., Cheng, G.: Application of gradient-based Hough transform to the detection of corrosion pits in optical images. Appl. Surf. Sci. **366**, 9–18 (2016)
20. Schneider, M., Hirsch, S., Weber, B., Székely, G., Menze, B.H.: Joint 3-D vessel segmentation and centerline extraction using oblique Hough forests with steerable filters. Med. Image Anal. **19**, 220–249 (2015)

Circular Mask and Harris Corner Detection on Rotated Images

Le Wang, Minrui Fei[✉], and Taicheng Yang

Shanghai Key Laboratory of Power Station Automation Technology,
School of Mechatronic Engineering and Automation, Shanghai University,
Shanghai 200072, China
wangle_shu@shu.edu.cn, mrfei@staff.shu.edu.cn,
t.c.yang@sussex.ac.uk

Abstract. Corners are the key feature of image. Stable corners are particularly important in the industrial pipelining of beer cap surface defects detection, greatly affecting the efficiency of image matching and detection precision. To find a stable algorithm for the cap surface defects detection, Stable Corner and Stable Ration are proposed to evaluate the stability of corner detectors, which are able to give an intuitive and unified stability description of various corner detection algorithm. After comparing the stability with Difference of Gaussian (DOG) and Features from Accelerated Segment Test (FAST), Harris is selected as the detector of cap surface images due to its high stability. To eliminate the redundant corners detected by Harris, Circular Mask and Harris (CMH) corner detection is proposed. In CMH, a circular mask with an adaptive threshold is adopted to remove the redundant corners, whereby comparing the intensity between the center pixel and others on the mask in a rapid way, more stable corners are obtained eventually. The effectiveness and robustness of CMH are verified in this paper, and the Stable Ratio increased by 16.7% relatively.

Keywords: Image processing · Corner detection · Harris algorithm · Rotated image · Stability Ratio

1 Introduction

Cap surface defects detection is important in the whole production process of beer cap. As the first step of cap surface defects detection, corner detection plays a crucial role in cap surface image matching and recognition. Due to the noise and vibration in industrial environment, it is easy for cap to exist rotation and offsets in the production line. Meanwhile, the image acquisition system is unable to get cap surface image at a standardized position, which adds difficulty to cap surface defects detection, so an effective corner detection is needed on rotated cap surface images. The classification of corner detection usually can be divided into three groups: contour-based detection [1, 2], intensity-based detection [3] and model-based detection [4].

Model-based detection detects corners by predefined models. However, it is difficult for it to cover all corners, which limits the application in corner detection. Thus, contour-based detection and intensity-based detection are more applied. Contour-based

© Springer Nature Singapore Pte Ltd. 2017
M. Fei et al. (Eds.): LSMS/ICSEE 2017, Part I, CCIS 761, pp. 524–535, 2017.
DOI: 10.1007/978-981-10-6370-1_52

corner detection requires high precision in edge detection and contour extraction [5], which is mainly based on curvature scale space (CCS) [6]. Because of complex method and compute costs in the image segment, the application range of this detector is narrow. Also this detector may not operate well especially when the images carry many noises. Intensity-based detection operates on gray-scale pixels without requiring edge detection [7]. Compared to Contour-based corner detection, this detector has a high precision, strong applicability and extensive use in corner detection.

Moravec corner detection considers corners with low self-similarity [8]. On the basis of Moravec [9], Harris corner detection calculates local variation by shifting in various orientations [10]; SUSAN defines corners as smallest univalue segment assimilating nucleus (USAN) points [11], only using the features of the corner itself, FAST algorithm could extract corners in a fast way [12], a circular template region is adopted to select the candidates whose intensity is below or above a certain threshold after comparing enough contiguous pixels on the template. However, it loses information about other angles and direction of corners, which could increase difficulties in the follow-up image matching.

Though Harris detector is stable and has a good performance in detecting corners, there are false corners detected easily. In cap surface defects detection system, a stable and fast detecting algorithm is needed. For a good performance of detector on rotated images, the Stable Corners and Stability Ratio are proposed to evaluate the stability of different corner detectors. According to the experimental results, Harris is chosen for its higher stability compare to that of FAST and DOG detectors. In order to prune the redundant corners detected by Harris in an effective way, the Circular Mask and Harris corner detection (CMH) is proposed. Same as the FAST, a circular mask is employed to pick out the ideal corners. Besides, CMH detector does not adopt the same corner response value related to the circular mask, so the gradient information will be saved. Experimental results demonstrate that CMH is superior to Harris on the problems of eliminating redundant corners with high stability in cap surface image corner detection.

2 Image Acquisition Platform and Harris Algorithm

In order to ensure the validity of experimental results, an image acquisition platform is built. It is convenient for the platform to acquire cap surface images that rotated at arbitrary angle under the same experimental environment, images are also free from the disturbance of uninterested factors. Meanwhile, this image acquisition platform could be directly transplanted to the industrial field of cap surface defects detection as well.

2.1 Image Acquisition Platform

The image acquisition platform is mainly composed of lighting system, CCD Camera and photoelectric sensor. Cap is placed on a disc whose speed is under control of the motor, and the Lighting system is settled over the disc. CCD Camera is triggered by a photoelectric sensor to capture images, communicating with computer via Gigabit Network so as to finish images transmission. Delay time is decided by the speed of disc and the distance between photoelectric sensor and CCD Camera (Fig. 1).

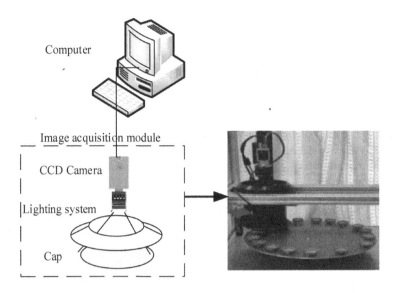

Fig. 1. Image acquisition platform

2.2   Harris Algorithm

Harris algorithm utilizes a window of N*N size in the image $I(x, y)$ to calculate the grayscale variation after shifting the window over the area (u, v) in an arbitrary direction. The center point of the window is (x, y), $E(u, v)$ represents the change of grayscale intensity after the center point (x, y) is move to $(x + u, y + v)$, the analytical equation of $E(u, v)$ can be generated as:

$$E(u, v) = \sum_{x,y} \omega(x, y)[I(x + u, y + v) - I(x, y)]^2 \tag{1}$$

According to Taylor expansion, $E(u, v)$ can be approximated as:

$$E(u, v) \approx \sum_{x,y} \omega(x, y)[I_x \times u + I_y \times v]^2 = [u \quad v]M\begin{bmatrix} u \\ v \end{bmatrix} \tag{2}$$

Where I_x and I_y are the first order partial derivative of the area (u, v) in horizontal and vertical directions. $\omega(x, y)$ is a weighting function which can be Gaussian weighting function or a constant.

To improve the anti-noise ability of image, the Gaussian weighting function $\omega(x, y)$ is utilized to apply convolution operation to I_x^2, I_y^2 and $I_x I_y$, then could get a self-correlation matrix M shown in which

$$M = \sum_{x,y} \omega(x, y)\begin{bmatrix} I_x^2 & I_x I_y \\ I_x I_y & I_y^2 \end{bmatrix} = \begin{bmatrix} a & c \\ c & b \end{bmatrix} \tag{3}$$

The two eigenvalues λ_1 and λ_2 of the symmetrical matrix M, after doing diagonalization to the matrix M, the analytical equation $E(u, v)$ can be given by:

$$E(u, v) = K^{-1} \begin{bmatrix} \lambda_1 & 0 \\ 0 & \lambda_2 \end{bmatrix} K \tag{4}$$

Where K is the twiddle factor. The flat rate of the area (u, v) and size of the function $E(u, v)$ are determined by eigenvalues λ_1 and λ_2, they have corresponding relationship with corners. If the center point (x, y) is a corner, eigenvalues λ_1 and λ_2 will be large, and the gray level intensity will change violently when it is shifted by (x, y) in any direction. So a large corner response value R of the point (x, y) could be get as follows:

$$R = \det(M) - k \times \mathrm{trace}(M)$$
$$\det(M) = \lambda_1 \times \lambda_2 \qquad \mathrm{trace}(M) = \lambda_1 + \lambda_2 \tag{5}$$

Where $\det(M)$ is the determinant of the matrix M, and k is an empirical constant, ranging from 0.04 to 0.07 typically. It's not necessary to calculate the value of λ_1 and λ_2, they have be involved in $\det(M)$ and $\mathrm{trace}(M)$.

$$\det(M) = \lambda_1 \times \lambda_2 = ab - c^2 \qquad \mathrm{trace}(M) = \lambda_1 + \lambda_2 = a + b \tag{6}$$

The center point (x, y) is consider to be a corner when its response value is greater than a given threshold. To avoid assign empirical value manually, this paper takes a new response function as a substitution as:

$$R = \det(M)/(\mathrm{trace}(M) + \varepsilon) \tag{7}$$

The cap surface image obtained by the Image Acquisition Platform shows as the left one in Fig. 2, and the right one is the result of Harris corner detection, the red points are detected corners.

Fig. 2. Cup surface image and the results of Harris

3   CMH Corner Detection

3.1   The Problem of Harris

The results of Harris detector is shown in Fig. 3, the red points are detected corners. For the distribution of the red points, the most obvious corners in the middle of cap surface are detected. Because Harris detects corners only in a single scale, it's easy to detect some redundant corners which will low down the efficiency of the follow-up image matching. The redundant corners are marked by ellipse at the angle of 45°, 135°, 225° and 315° in Fig. 3, all of these corners need a further operation for a high precise detection result. Although many corners are detected in the middle caps, they may be unstable and could not be selected as final corners to be utilized in matching section, we'd better find a good way to filter the redundant corners and pick out the stable corners, that's why CMH is proposed.

Fig. 3. Corner detection results of Harris

Whether the corner is stable or not, it has a close relation to the detector itself. The definition of Stable Corner and Stability Ratio is proposed to value the stability of corner detectors, so the best algorithm is chosen according to the valuation.

If a corner could be detected from multiple views, it could be defined as a Stable Corner. That means the Stable Corner could be extracted at the same relative location in the image when the image rotated. The Stability Ratio is the proportion of Stable Corners to all corners detected in one image, the formula is defined as:

$$Stability\,Ratio = \frac{The\,Number\,of\,Stable\,Corners}{The\,Number\,of\,all\,Corners} \times 100\% \tag{8}$$

In cap surface defects detection experiment, 6 cap surface images which are counterclockwise rotated at a different angle are selected to test the stability of DOG, Harris and FAST. All the 6 images are detected by DOG, Harris and FAST detectors. For a clear observation, the same local region of each image is utilized as the calculating data samples. If a corner is detected no less than 4 times in the local region of the 6 images, indicating it is not sensitive to rotation, therefore it is considered as a Stable

Table 1. The results of Harris in local region image

Number	1	2	3
Local Region Image			
Corners	16	17	17
Stable Corners	8	9	9
Number	4	5	6
Local Region Image			
Corners	17	21	18
Stable Corners	7	10	8

Corner. The detailed data of Harris detector are shown as Table 1. The red points are corners detected by Harris, results of local region is shown in Table 1. The Stable Corners are marked by white circle. It's clear for us to see that how many corners and Stable Corners there are in each image, the Stable Ration of Harris is:

$$Stable\ Ration\ of\ Harris = \frac{8 + 9 + 9 + 7 + 10 + 8}{16 + 17 + 17 + 17 + 21 + 18} \times 100\% \approx 48\%$$

If the data sample expanded beyond 6 images, the definition of Stable Corners should be changed according to the amount of images. No matter how, the times of Stable Corners be detected should more than half of the amount. Experiment of comparison has been done by using DOG and FAST detector, the results are shown in Table 2.

From the Table 2, it's clear for us to see the corners distribution and the stability of different detectors. The Stable Corner that detected not less than 4 times in 6 images is marked by a white circle. The Stable Ration of DOG is 24%, Harris is 45% and FAST is 35%. Most corners detected by DOG distribute on the edge of cap surface, which indicates that DOG is suitable for edge detection. There is only one Stable Corner in the local region image of cap surface, so its stability is the lowest among the three detectors. The Stable Ratio of Harris is the highest, thus Harris is the right corner

Table 2. Stable ratio comparison

Detector	DOG	Harris	Fast
Image			
Local image			
Stable Ratio	24%	48%	35%

detector in the cap surface image. Nevertheless, there are more redundant corners detected by Harris in the cap surface compared to FAST, particularly the corners at the angle of 45°, 135°, 225° and 315°. FAST has less execute time due to utilize corner information only, which also result in a lower Stable Ration than Harris. So it is unable to detect reliable corner from multiple views.

3.2 Circular Mask and Harris Corner Detection (CMH)

As shown in Table 2, Harris has the highest Stable Ratio, FAST detected less redundant corners than Harris. The stability of corner detection algorithm is so important that Harris is selected as the appropriate detector. The redundant corners detected by Harris should be discarded for a good performance detector in cap surface defects detection, therefore CMH detector is proposed. In FAST detector, a circular mask shown in Fig. 4 is adopted to determine whether the pixel is a corner or not without calculating the gradient of local region. So it is faster than Harris by almost 10 times. Thus this paper also applies the circular mask to eliminate the redundant corners detected by Harris for a better detection results, that's the main thoughts of CMH.

The circular mask with the radius of 3.4 pixels has 16 pixels in all. The number 1 to 16 represent 16 pixels in the circular mask. It selects the pixel p as a corner when there are enough contiguous pixels are all above or below the intensity of the center pixel p.

The steps of CMH is summarized as follows.

Step 1. Utilize a window of 3 * 3 size and shift it over the cap surface image, calculating the first order partial derivative in horizontal and vertical directions and

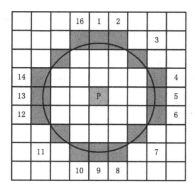

Fig. 4. Circular mask

Fig. 5. Corner detection results of CMH (Color figure online)

getting a self-correlation matrix M. Then get the corner response value of each pixel. The corner response formula is as:

$$R = \det(M)/(\text{trace}(M) + \varepsilon) \tag{9}$$

Step 2. Utilize Non-maxima suppress to eliminate the redundant corners and false corers at the first time.

Step 3. Apply a circular mask shown as Fig. 4 to filter the redundant corners and select the final corners.

The procedure of applying the circular mask of 16 pixels can be divide into Step 3.1 and Step 3.2 as follows:

Step 3.1: Examine the intensity of pixel 1, pixel 3, pixel 5 and pixel 9 whether they are all below or above the center pixel p to an adaptive threshold t. The intensity of pixels denoted as $I_{p \to x}$ and the intensity of p is I_p. If the difference between $I_{p \to x}$ and I_p denoted as s_{x1} is lower than the threshold t, set the value of s_{x1} to 0. If the difference s_{x1} is greater than the threshold t, set the value of s_{x1} to 1. When the total difference denoted as n_1 is no smaller than 3, the center pixel p is taken as a candidate. The difference s_{x1} and the total difference n_1 could be described as follows.

$$s_{x1} = \begin{cases} 0, & |I_p - I_{p \to x}| < t \\ 1, & |I_p - I_{p \to x}| \geq t \end{cases} \quad x = 1, 5, 9, 13 \tag{10}$$

$$n_1 = \sum s_{x1} \tag{11}$$

Step 3.2: Select final corners from the candidates. For each pixel of the candidates, set it as the center pixel p of the circular mask. If the difference between I_p and $I_{p \to x}$ denoted as s_{x2} is greater than the threshold t, set s_{x2} to 1, otherwise, set s_{x2} to 0. Examine whether there are 12 contiguous pixels on the circular mask that whose intensity $I_{p \to x}$ are all above or below the center pixel p's intensity I_p to an adaptive threshold t. If there are, the total difference denoted as n_2 will be greater than 12, then the center pixel p could be regarded as a final corner meeting our requirements.

If the pixel is not come from the candidates, it will be discarded and could not participate in following procedure. This strategy conduce to decrease the amount of computation and, accordingly, the circular mask is applied to eliminate the redundant corners of step 2 but no much time consumption.

The difference s_{x2} and the total difference n_2 could be describe as follows.

$$s_{x2} = \begin{cases} 0, & |I_p - I_{p \to x}| < t \\ 1, & |I_p - I_{p \to x}| \geq t \end{cases} \quad x = 1, 2, 3 \ldots 16 \tag{12}$$

$$n_2 = \sum s_{x2} \tag{13}$$

The threshold of the circular mask is usually set by experience, here an improved adaptive threshold is put forward to replace the traditional one. With the adaptive threshold, t could vary according to different image region, avoiding assign threshold value manually. It is defined as follows:

$$t = \frac{1}{2} \sum_{1}^{4} (I_{p \to x} - I_{avg})^2 \quad I_{avg} = \frac{1}{4} \sum I_{p \to x} \quad x = 1, 5, 9, 13 \tag{14}$$

The algorithm thoughts of CMH are based on Harris and FAST detectors. As the improved algorithm of Harris, the Accelerated segment test of FAST is employed for reference in CMH, which contribute to extract high quality corners in a rapid and simple way without taking up too much time. While a circular mask is used as a filter in CMH, the difference of intensity between the center pixel and other pixels on the mask will not change when the cap surface image rotated at arbitrary angle, thus a more stable corner be obtained. The circular mask also makes up for the disadvantage of the square neighborhood of Harris and just care about corner information without any gradient calculation, all of these help to improve the precision of corner detection and produce strong corners.

3.3   The Experimental Result of CMH

The experimental result of CMH is shown as Fig. 5. The red points represent the final corners detected by CHM detector, the blue points represent the corners detected by Harris but rejected by CHM detector. Points locate at four directions (top, bottom, left and right) of the corner are all colored by red. The experimental results show that CHM detector does good performance in eliminating redundant corners, particularly the corners at the angle of 45°, 135°, 225°, and 315°, all of them are rejected. Besides, other useless points in the cap surface image are also eliminated such as the points at cap margin and the points of background. The red points are almost centered on the cap surface image could symbolizing all contents of the cap surface image, and the amount of corners are reasonable, therefore the detection results will conduct to image matching and defects detecting of cap surface.

To verify the stability of CMH, the same experiment as before has been done. 6 cap surface image rotated are utilized as sample data to detect corners, then 6 detected corners distribution images are obtained. For each of the 6 detected corners distribution images, the local region image of Chinese character "珠" is cut out and list as Table 3.

Table 3. The results of CMH in local region image

Number	1	2	3
Local Region Image			
corners	10	10	12
Stable Corners	7	6	7
Number	4	5	6
Local Region Image			
corners	15	15	13
Stable Corners	7	8	7

The Stable Corners detected not less than 4 times in 6 cap surface images are marked by white circle. Table 3 shows the detail data of Stable Corners and corners of each images. So the Stable Ration of CMH could be get:

$$Stable\ Ration\ of\ CMH = \frac{7 + 6 + 7 + 7 + 8 + 7}{10 + 10 + 12 + 15 + 15 + 13} \times 100\% \approx 56\%$$

The Stable Ration of CMH is 56% and Harris is 48%. The Relative Increase of the Ration between CMH detector and Harris detector shown as follows.

Relative Increase of the Ration

$$= \frac{Stable\ Ration\ of\ CMH - Stable\ Ration\ of\ Harris}{Stable\ Ration\ of\ Harris} = \frac{0.56 - 0.48}{0.48} \times 100\% \approx 16.7\%$$

The calculation results show the progress of CMH detector, the Stable Ration improved from 48% to 56%. Although there will lose some information about other corners in the cap surface image when rejecting redundant corners, the final corners obtained have a higher stability than any other corners. Even the cap surface image rotated at any angle, the circular mask will retain the Stable Corners in a rapid way without much complicate calculation. The algorithm speed is 41 ms, which means that 1463 cap images could be detected per minutes. The speed can satisfy the demand of industrial pipelining of beer cap surface defects detection. The Relative Increase of the Ration between CMH detector and Harris detector is 16.7% meaning there are more redundant corners rejected. Then can conclude that not only the CMH detector eliminates redundant corners well but also has fine stability.

4   Conclusion

A stable corner will make positive effects on cap surface defects detection. In order to choose a stable and reliable corner detection algorithm for cap surface image, this paper presented Stable Corner and Stable Ratio to evaluate the stability of different algorithm. Harris corner detection was finally chosen because of higher Stable Ratio comparing to that of DOG and FAST. For the sake of an accurate way in corner detection of cap surface image, Circular Mask-Harris corner detection (CMH) was proposed in this paper, which is an improved algorithm of Harris using a circular mask with an adaptive threshold. With the use of the circular mask, even when the cap surface image rotated at arbitrary angle, the intensity difference between the center pixel and other pixels on the circular mask will not change, so more stable corners could be obtained. Meanwhile, the redundant corners appeared in Harris detector were pruned to a great extent. Experimental data showed that the stability ration of CMH increased by 16.7% relatively, and the algorithm speed of CMH is 41 ms, demonstrating that the proposed methodology has a better performance.

Acknowledgements. This work is supported by Key Project of Science and Technology Commission of Shanghai Municipality under Grant No. 14JC1402200.

References

1. Awrangjeb, M., Lu, G., Fraser, C.S.: Performance comparisons of contour-based corner detectors. IEEE Trans. Image Process. **21**, 4167–4179 (2012)
2. Shui, P.L., Zhang, W.C.: Corner detection and classification using anisotropic directional derivative representations. IEEE Trans. Image Process. **22**, 3204–3218 (2013)
3. Gao, X., Sattar, F., Venkateswarlu, R.: Multiscale corner detection of gray level images based on log-gabor wavelet transform. IEEE Trans. Circ. Syst. Video Technol. **17**, 868–875 (2007)
4. Florentz, G., Aldea, E.: SuperFAST: model-based adaptive corner detection for scalable robotic vision. In: 2014 IEEE/RSJ International Conference on Intelligent Robots and Systems, pp. 1003–1010 (2014)

5. Ma, X., Wang, H., Xue, B., Zhou, M., Ji, B., Li, Y.: Depth-based human fall detection via shape features and improved extreme learning machine. IEEE J. Biomed. Health Inform. **18**, 1915–1922 (2014)
6. Topal, C., Özkan, K., Benligiray, B., Akinlar, C.: A robust CSS corner detector based on the turning angle curvature of image gradients. In: 2013 IEEE International Conference on Acoustics, Speech and Signal Processing, pp. 1444–1448 (2013)
7. Dutta, A., Mandal, A., Chatterji, B.N., Kar, A.: Bit-plane extension to a class of intensity-based corner detection algorithms. In: 2007 15th European Signal Processing Conference, pp. 267–271 (2007)
8. Song, H.S., Lu, S.N., Ma, X., Yang, Y., Liu, X.Q., Zhang, P.: Vehicle behavior analysis using target motion trajectories. IEEE Trans. Veh. Technol. **63**, 3580–3591 (2014)
9. Liu, D., Wang, X., Song, J.: A robust pedestrian detection based on corner tracking. In: 2015 5th International Conference on Information Science and Technology (ICIST), pp. 207–211 (2015)
10. Harris, C.: A combined corner and edge detector. In: Proceedings of Alvey Vision Conference, pp. 147–151 (1988)
11. Smith, S.M., Brady, J.M.: SUSAN—a new approach to low level image processing. Int. J. Comput. Vis. **23**(1), 45–78 (1997)
12. Rosten, E., Porter, R., Drummond, T.: Faster and better: a machine learning approach to corner detection. IEEE Trans. Pattern Anal. Mach. Intell. **32**, 105–119 (2010)

MEG Source Imaging Algorithm for Finding Deeper Epileptogenic Zone

Yegang Hu[1,2], Yicong Lin[3], Baoshan Yang[1,2], Guangrui Tang[1,2], Yuping Wang[3], and Jicong Zhang[1,2(✉)]

[1] School of Biological Science and Medical Engineering, Beihang University,
Beijing 100191, China
{huyegang,bshyang,tangguangrui,jicongzhang}@buaa.edu.cn
[2] Beijing Advanced Innovation Center for Big Data-Based Precision Medicine,
Beihang University, Beijing 100191, China
[3] Department of Neurology, Xuanwu Hospital, Capital Medical University,
Beijing 100053, China
Linyc_1@163.com, wangyuping01@sina.cn

Abstract. In recent years, magnetoencephalography (MEG) has played a prominent role on neocortical epilepsy preoperative evaluation. However, its clinical utility with locating deeper sources may be more challenging such as the mesial temporal structures. We proposed a new source imaging algorithm for finding the epileptogenic zone in mesial temporal lobe epilepsy (mTLE). Since the localization results using the Elekta MEG method are very sensitive to some MEG noises, the source modeling was modified by spatial filtering in wavelet domain and cortex constraint. Two surgical patients randomly selected with medically refractory mTLE, which were diagnosed based on a comprehensive preoperative evaluation, had been studied in this manuscript. The localization results using proposed method on individual MRI showed that the deeper regions had been exactly found in the mesial temporal lobe. Yet, the results using the Elekta Neuromag Software only appeared in the lateral temporal lobe. Thus, the proposed algorithm maybe become an effective method in detecting deeper epileptogenic zone.

Keywords: Epileptogenic zone · Magnetoencephalography · Magnetic source imaging · Mesial temporal lobe epilepsy

1 Introduction

MEG is a neuroimaging modality that captures neural activity with high spatiotemporal resolution and minor signal deterioration from the skull and scalp [1–3]. In addition, MEG is a noninvasive and worthful tool for improving patient management in the evaluation of epilepsy and pre-surgical spike mapping for epilepsy surgery, which can help delineate the epileptogenic zone in three dimensions using magnetic source imaging (MSI) techniques [4–6].

MTLE plays an important role in pharmacoresistant or refractory epilepsy, and represents one of the most common forms of focal epilepsy [7]. Thus, mTLE is

© Springer Nature Singapore Pte Ltd. 2017
M. Fei et al. (Eds.): LSMS/ICSEE 2017, Part I, CCIS 761, pp. 536–544, 2017.
DOI: 10.1007/978-981-10-6370-1_53

considered to be one of the most important categories of the symptomatic focal or localization-related epilepsies in epileptology, and its pathophysiological substrate is usually hippocampal sclerosis (HS) [8], which is the most common structural abnormality in human epilepsy [9]. Around two thirds of patients with temporal lobe epilepsy could achieve seizure freedom via resective epilepsy surgery, and MEG examination results become a key role in preoperative evaluation with epileptogenic zone and in guiding surgical placement of intracranial electrodes [10, 11].

Some researchers have given the view that MSI methods seem to be effective and helpful for finding the epileptic foci in neocortical epilepsy [12, 13]. Nevertheless, its ability about detecting deep sources such as those in mesial temporal lobe remains in question [14–16]. Although the magnetometer has been considered for spike detection in patients with mesial temporal epileptic focus [17], the clinical utility of MEG has not been completely accepted by consensus in preoperative evaluation with epileptogenic foci in deep regions [18]. Thus, a technical challenge is whether signals from deep sources could be well detected and correctly located using MEG device and MSI techniques.

In this manuscript, two patients randomly selected with focal epilepsy had been diagnosed as medically refractory mTLE according to a comprehensive preoperative evaluation, and almost all of the MEG localization results using the Elekta Neuromag Software method [19] appeared in the lateral temporal lobe. Of note, the Elekta Neuromag method for MEG localization is more stable than the other magnetic source imaging methods in clinical epilepsy preoperative evaluation [20–22]. Since the Elekta Neuromag method applied a snapshot of the MEG recordings under the time domain, the localization results generated by this method can be sensitive to some MEG noises. Herein we proposed a new method and applied it to the MEG recordings, which is characterized by the spatial filtering in wavelet domain and the cortex constraint.

2   Deeper Epileptogenic Zone Localization Algorithm

2.1   Data Preprocessing and Artifact Rejection

In this manuscript, three primary toolboxes including Matlab R2014a (The MathWorks Inc), SPM8 [23], and FieldTrip [24] were used jointly for MEG data analysis. The MEG signal with continuous data, which would not be divided into many segments, was filtered by a band-pass filter of 0.5–60 Hz, notch-filtered at 50 Hz, and detrended via removing the linear trend from the data. Then, the MEG channels including noises were detected by the module of manual artifact rejection, and the bad channels were repaired automatically using spline interpolation algorithm. To improve the quality of MEG recordings signals, the independent component analysis (ICA) and principal component analysis (PCA) were used for removing the artifacts related to heartbeats, muscles and eye blinks. We hoped that the interested signal containing spike waves was very clean.

2.2   Source Localization Algorithm

Two experienced clinical epileptologists visually marked epileptic spikes respectively in MEG signals after the previous step and ruled out drowsiness. Then, the individual anatomical magnetic resonance images (MRI; T1-weighted) and the digitised head shapes were co-registered to the MEG coordinate system using anatomical landmarks via an iterative closest point (ICP) algorithm [25]. Then, the detailed procedure of inverse solution would be addressed as follows.

For the preprocessed signal matrix $\mathbf{X} = [X_1, X_2, \ldots, X_q]$, which is acquired from MEG sensors, the inverse solution model is given as

$$\mathbf{X} = \mathbf{GW} + \boldsymbol{\varepsilon} \tag{1}$$

where X_i is an $N \times 1$ vector of the MEG measurements at i-th time point, N is the number of MEG sensors, $\mathbf{G}$ is the $N \times m$ (lead-field) gain matrix, m denotes the number of unknown dipole moment parameters, $\mathbf{W}$ is an $m \times q$ dipole moment matrix for given time series, and $\boldsymbol{\varepsilon}$ denotes the $N \times q$ noise matrix. Then the original observed signal would be transformed into a novel space for representing the intrinsic features from time domain to transform domain. The transformed procedure is expressed as

$$\Phi: \mathbf{X} \mapsto \mathbf{D} \tag{2}$$

where Φ is a Haar Wavelet transform function, $\mathbf{D}$ denotes the transformed data in novel space. Then, we employ an optimization algorithm of regularization minimizing the interference (rMinInf) [26] for passing the activity at position $r = r_0$ with unit gain, while inhibiting contributions from all other sources. The mathematical formulation is

$$\min_{\mathbf{A}} [E(\|\mathbf{AD}\|^2) + \alpha\|\mathbf{A}\|^2] \quad with \quad \mathbf{AG}(r_0) = 1 \tag{3}$$

where $E(\cdot)$ denotes the expectation value function and α is the regularization coefficient. The optimal solution can be computed by minimizing the corresponding Lagrange function as

$$\mathbf{A}(\mathbf{r}) = \left(\mathbf{G}^T(\mathbf{r})(\mathbf{C} + \alpha\mathbf{I})^{-1}\mathbf{G}(\mathbf{r})\right)^{-1}(\mathbf{C} + \alpha\mathbf{I})^{-1}\mathbf{G}(\mathbf{r}) \tag{4}$$

where $\mathbf{A}(\mathbf{r})$ indicates the projection matrix at the grid location $\mathbf{r}$, T denotes matrix or vector transpose and $\mathbf{C}$ is the covariance matrix of random variables based on row vectors of the transformed domain matrix $\mathbf{D}$. Besides, the power value at each grid point, which can also be seen as the dipole moment, would be expressed as

$$\text{pow}(\mathbf{r}) = \lambda_1\left(\mathbf{A}(\mathbf{r})\mathbf{CA}(\mathbf{r})^{*T}\right) \tag{5}$$

where $\text{pow}(\mathbf{r})$ indicates the power value at the grid location $\mathbf{r}$ and $\lambda_1(\cdot)$ denotes the maximum eigenvalue of the expression in braces. In fact, a similar approach has been attempted to find the epileptogenic zone based on MEG, and has achieved some preliminary results in an earlier study [27]. However, it was designed only for finding epileptogenic zone in neocortical epilepsy, and has not drawn much attention in clinical application.

Thus, the new procedure is summarized into an algorithm as follows. It is assumed that data preprocessing and artifact rejection have been accomplished for the MEG data and individual MRI of all the patients.

Proposed algorithm: Deeper epileptogenic zone localization procedure.

Forward solution:

1. Construct the volume conduction model $\mathbf{V}$ based on a single shell approximation under the cortex constraint.

2. Calculate the lead field matrices $\mathbf{G}$ under the cortex constraint.

3. Denote the lead field matrix $\mathbf{G}_i$ corresponding to the i-th grid point.

MEG data space transformation:

1. Pick out the k-th data segment matrix $\mathbf{X}^k$ containing spike.

2. Transform the data matrix $\mathbf{X}^k$ into wavelet domain $\mathbf{D}^k$ using the equation (2).

Inverse solution:

1. Compute the covariance matrix $\mathbf{C}^k$ of the matrix $\mathbf{D}^k$.

2. Obtain the optimized projection matrix $\mathbf{A}^k(\mathbf{r})$ at the grid location $\mathbf{r}$ based on the equation (4).

3. Calculate the power value $\mathrm{pow}^k(\mathbf{r})$ at the grid location $\mathbf{r}$ via the equation (5).

Localization results display:

1. Select those grid points corresponding to the larger power values.

2. Visualize the result on the individual MRI using FieldTrip toolbox.

Feasibility of the above algorithm will be demonstrated by experimental results in the next section.

3   Experimental Results

3.1   Patients and Data Description

For describing the preliminary effectiveness of proposed algorithm, we adopted two patients randomly with medically refractory unilateral mTLE retrospectively. Mesial TLE was diagnosed based on a comprehensive preoperative evaluation, including seizure history and semiology, neurologic examination, 3 T MR imaging, scalp electroencephalography, invasive electroencephalography and pathology. Two patients had already undergone a standard clinical presurgical evaluation including clinical seizure semiology, long-term video-EEG monitoring, high-resolution MRI, MEG as well as neuropsychological testing. According to all of these examination results, the preoperative assessment conclusion was given by Beijing epilepsy center expert group members. And then, two patients had also accomplished anterior temporal lobectomy [28] for focal epilepsy at Xuanwu Hospital Capital Medical University (XWHCMU) on June 2013, which included three years postoperative follow-up. The results showed that two patients were free of disabling seizures (Engel class IA) after surgery. The study was performed under a protocol approved by the medical ethics committee of the XWHCMU Committee.

MEG Acquisition. The MEG recordings were acquired inside a magnetically shielded room by the 306 channels in total with a helmet-shaped whole-head system (Vector-View, Elekta Neuromag Oy, Finland), comprising 102 locations at triplets including one magnetometer and two orthogonal planar gradiometers. For mTLE patients, continuous data were recorded at a sampling rate of 1000 Hz for the MEG signal, and the electro-cardiography data was recorded simultaneously. Each recording consisted of six 10-min epochs while they were lying in a supine position with eyes closed and resting-state. A three-dimensional digitizer, which was the Polhemus^{TM} system (Colchester, NH, USA), was used to determine the location based on anatomical fiducial points (nasion, bilateral preauricular points) for the following MRI-MEG co-registration. Through checking the uniform distribution of points as far as possible covering the whole scalp, the head shape of each patient was ascertained ensuring that the head of patient cannot be moved in the whole procedure.

3.2   Deep Source Localization Results

Clinical characteristics of the two patients were listed in Table 1, and the two epilepsy patients were judged as mesial Temporal Lobe (TL) origination based on preoperative evaluation and postoperative follow-up. From the clinical experience, the localization results could be found in the mesial TL region, and may also be found in lateral or extra TL region, because the discharge sources may spread from one location to another during a short time. Then, MEG source imaging results would be shown from coronal and sagittal views as follows.

Table 1. Clinical characteristics of the two patients and localization results.

Clinical index	Patient 1 (Results)	Patient 2 (Results)
MRI	Hyper T2 in right hippocampal	Normal
Preoperative evaluation	**mesial temporal lobe origination**	**mesial temporal lobe origination**
Surgical procedure	right anterior temporal lobectomy (include hippocampal)	right anterior temporal lobectomy (include hippocampal)
Postoperative follow-up	Seizure Free (Engel class IA)	Seizure Free (Engel class IA)
MEG (Elekta Neuromag)	Lateral temporal lobe (right)	Lateral temporal lobe (right)
MEG (Proposed method)	**Mesial temporal lobe(right)**	**Mesial temporal lobe(right)**

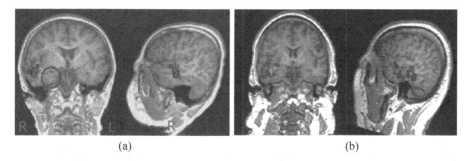

(a)   (b)

Fig. 1. Magnetic source imaging results based on Elekta Neuromag Software. (a) The localization results with the 1st patient displayed on individual MRI from coronal and sagittal views, wherein the hippocampal region was denoted by blue circle on coronal view. (b) The localization results with the 2nd patient displayed on individual MRI from coronal and sagittal views. (Color figure online)

First, Fig. 1 gave the localization results using Elekta Neuromag Software, which were reviewed by clinical MEG neurologists, and these results obviously showed that almost all of the dipoles were displayed in lateral TL region. Second, the localization results using proposed scheme were displayed on individual MRI of epilepsy patient. For clearly observing the deeper regions around hippocampal, we exhibited a figure

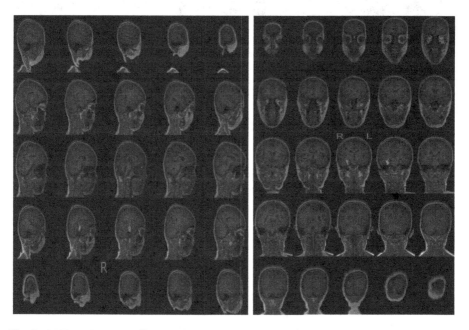

Fig. 2. MSI results of the 1st patient displayed on individual MRI using this paper method from coronal and sagittal views respectively.

including a total of 25 slices, 5 rows by 5 columns, from coronal and sagittal views respectively. Figure 2 showed that the localization results had appeared in the right hippocampal region, thus this patient could be judged as mesial TL origination. Analogously, the second patient could also be judged as mesial TL origination from Fig. 3. Of note, MRI examination results showed that the signal of hippocampus region was normal for the second patient, and it perhaps illustrate that MRI examination often generates false negative results. In addition, we observed that the source activity regions were also located in anterior TL based on MEG using the two methods.

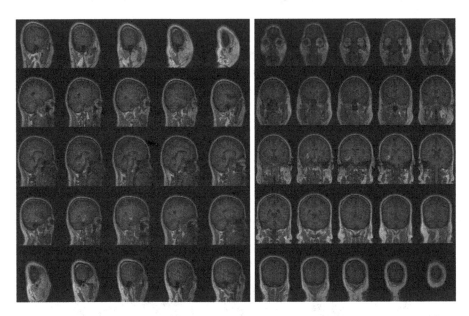

Fig. 3. MSI results of the 2nd patient displayed on individual MRI using this paper method from coronal and sagittal views respectively.

In a word, the proposed method results were almost consistent with the clinical comprehensive preoperative evaluation, surgical outcome and postoperative follow-ups (See Table 1). Compared with the Elekta Neuromag method, the proposed algorithm has obviously prominent for detecting the discharge sources in mesial TL region with the two epilepsy patients. Thus, MEG source imaging plays an important role for epileptogenic zone localization in preoperative evaluation of epilepsy.

4   Conclusion and Discussion

In conclusion, we designed a new MSI algorithm by combining a kind of spatial filtering in the wavelet domain and the cortex constraint to identify the deep epileptogenic zone in mTLE patients detected from preoperative MEG recordings. The proposed method results seem to be more consistent with the multi-modality neuroimages, clinical characteristics, and postoperative follow-ups of those patients. Compared with the Elekta

Neuromag method, this paper method is capable of detecting deep sources in the brain. Thus, it may help increase the clinical utility of MEG in preoperative evaluation with epileptogenic foci in deep brain regions. Yet, the proposed method only provided preliminary framework on finding deeper epileptogenic zone, because of the limited number of cases. In future, to generalize the proposed method in clinical application, more number of patients will be adopted for verifying this method.

Acknowledgments. This work was supported by National Key Research and Development program of the Ministry of Science and Technology (grant number 2016YFF0201002), the Natural Science Foundation of China (grant numbers 61301005 and 61572055), the Natural Science Project of National Statistical Bureau (2014LY088), the project of Brain Functional Disease and Neuromodulation of Beijing Key Laboratory, 'Thousands of People Plan' Workstation between Beihang University and Jiangsu Yuwell Medical Equipment & Supply Co. Ltd.

References

1. Nissen, I.A., Stam, C.J., Citroen, J., Reijneveldb, J.C., Hillebranda, A.: Preoperative evaluation using magnetoencephalography: experience in 382 epilepsy patients. Epilepsy Res. **124**, 23–33 (2016)
2. Barnes, G.R., Hillebrand, A.: Statistical flattening of MEG beamformer images. Hum. Brain Mapp. **18**, 1–12 (2003)
3. Zumer, J.M., Attias, H.T., Sekihara, K., Nagarajan, S.S.: A probabilistic algorithm integrating source localization and noise suppression for MEG and EEG data. Neuroimage **37**, 102–115 (2007)
4. Wu, J.Y., et al.: Magnetic source imaging localizes epileptogenic zone in children with tuberous sclerosis complex. Neurology **66**, 1270–1272 (2006)
5. Nissen, I.A., et al.: Identifying the epileptogenic zone in interictal resting-state MEG source-space networks. Epilepsia **58**, 137–148 (2017)
6. Hillebrand, A., Singh, K.D., Holliday, I.E., Furlong, P.L., Barnes, G.R.: A new approach to neuroimaging with magnetoencephalography. Hum. Brain Mapp. **25**, 199–211 (2005)
7. Engel, J.: Introduction to temporal lobe epilepsy. Epilepsy Res. **26**, 141–150 (1996)
8. Engel, J.: Mesial temporal lobe epilepsy: what have we learned? Neuroscientist **7**, 340–352 (2001)
9. Engel, J.: Recent advances in surgical treatment of temporal lobe epilepsy. Acta Neurol. Scand. **86**(S140), 71–80 (1992)
10. Sutherling, W.W., et al.: Influence of magnetic source imaging for planning intracranial EEG in epilepsy. Neurology **71**, 990–996 (2008)
11. Murakami, H., et al.: Correlating magnetoencephalography to stereo-electroencephalography in patients undergoing epilepsy surgery. Brain **139**, 2935–2947 (2016)
12. Bast, T., et al.: EEG and MEG source analysis of single and averaged interictal spikes reveals intrinsic epileptogenicity in focal cortical dysplasia. Epilepsia **45**, 621–631 (2004)
13. Wennberg, R., Cheyne, D.: Reliability of MEG source imaging of anterior temporal spikes: analysis of an intracranially characterized spike focus. Clin. Neurophysiol. **125**, 903–918 (2014)
14. Leijten, F.S.S., et al.: High-resolution source imaging in mesiotemporal lobe epilepsy: a comparison between MEG and simultaneous EEG. J. Clin. Neurophysiol. **20**, 227–238 (2003)

15. Shigeto, H., et al.: Feasibility and limitations of magnetoencephalographic detection of epileptic discharges: simultaneous recording of magnetic fields and electrocorticography. Neurol. Res. **24**, 531–536 (2002)
16. Wennberg, R., Valianteb, T., Cheynec, D.: EEG and MEG in mesial temporal lobe epilepsy: where do the spikes really come from? Clin. Neurophysiol. **122**, 1295–1313 (2011)
17. Enatsu, R., et al.: Usefulness of MEG magnetometer for spike detection in patients with mesial temporal epileptic focus. Neuroimage **41**, 1206–1219 (2008)
18. Bagić, A., Ebersole, J.S.: Does MEG/MSI dipole variability mean unreliability? Clin. Neurophysiol. **126**, 209–211 (2015)
19. Elekta Neuromag Oy: The MEG signal processor (graph) users guide and reference manual. http://www.martinos.org/meg/Neuromag-manuals.php. Accessed 13 June 2017
20. Oishi, M., et al.: Single and multiple clusters of magnetoencephalographic dipoles in neocortical epilepsy: significance in characterizing the epileptogenic zone. Epilepsia **47**, 355–364 (2006)
21. Knake, S., et al.: The value of multichannel MEG and EEG in the presurgical evaluation of 70 epilepsy patients. Epilepsy Res. **69**, 80–86 (2006)
22. Medvedovsky, M., et al.: Sensitivity and specificity of seizure-onset zone estimation by ictal magnetoencephalography. Epilepsia **53**, 1649–1657 (2012)
23. Litvak, V., et al.: EEG and MEG data analysis in SPM8. Comput. Intell. Neurosci. **2011**, 852961 (2011)
24. Oostenveld, R., Fries, P., Maris, E., Schoffelen, J.M.: FieldTrip: open source software for advanced analysis of MEG, EEG, and invasive electrophysiological data. Comput. Intell. Neurosci. **2011**, 156869 (2011)
25. Besl, P.J., McKay, N.D.: A method for registration of 3D shapes. IEEE T Pattern Anal. **14**, 239–254 (1992)
26. Gross, J., Ioannides, A.A.: Linear transformations of data space in MEG. Phys. Med. Biol. **44**, 2081–2097 (1999)
27. Heers, M., et al.: Frequency domain beamforming of magnetoencephalographic beta band activity in epilepsy patients with focal cortical dysplasia. Epilepsy Res. **108**, 1195–1203 (2014)
28. Schaller, K., Cabrilo, I.: Anterior temporal lobectomy. Acta Neurochir. **158**, 161–166 (2016)

A New Meanshift Target Tracking Algorithm by Combining Feature Points from Gray and Depth Images

Lu Lu[1], Minrui Fei[1,2], Haikuan Wang[1(✉)], and Huosheng Hu[3]

[1] School of Mechatronics Engineering and Automation,
Shanghai University, Shanghai, China
HKWang@shu.edu.cn
[2] Shanghai Key Laboratory of Power Station Automation Technology,
Shanghai University, Shanghai, China
[3] School of Computer Science and Electronic Engineering,
University of Essex, Colchester, UK

Abstract. The traditional MeanShift algorithm cannot obtain accurate tracking results in some complex situations where tracking targets have scale changes or similar color with background. In this paper, a new MeanShift target tracking algorithm, namely DEPTH & SIFT-MeanShift algorithm, is proposed by using a depth camera and SIFT (Scale Invariant Feature Transform) feature metric. The algorithm firstly combines feature points extracted from gray and depth images respectively, and then represents tracked objects with Modulus, i.e. Direction distribution histogram of feature points in the tracking object field, so that targets can be effectively tracked. Experimental results show that the proposed algorithm can achieve good tracking performance when the tracking target changes its scale, and have the strong adaptability to occlusion. Moreover, it is very robust to illumination changes, and able to discriminate targets from background very well.

Keywords: Depth camera · Depth image · SIFT · MeanShift · Target tracking

1 Introduction

MeanShift [1] is an adaptive step-size iterative algorithm and has been successfully applied to image segmentation, nonparametric density estimation, video tracking [2] and many other areas. It uses the spatial kernel weighted color histogram as a template to search for the maximum Bhattacharya coefficient to achieve the target tracking and positioning. However, its tracking results are unstable and the scope of its practical application is limited since the color information is subject to lighting changes or color characteristic instability. By weakening the background noise in the target color

This work is supported by Key Project of Science and Technology Commission of Shanghai Municipality under Grant No. 14JC1402200.

M. Fei et al. (Eds.): LSMS/ICSEE 2017, Part I, CCIS 761, pp. 545–555, 2017.
DOI: 10.1007/978-981-10-6370-1_54

histogram [3], the effect of the MeanShift algorithm on the target location is reduced, but the problem of the fixed bandwidth is not solved.

As the SIFT feature has the advantages of rotation and scale invariance, strong adaptability to illumination changes, Cheng *et al.* [4] solved the tracking unstable problem associated with tracking direction and scale change by updating the MeanShift kernel function bandwidth and direction. Such an updating was conducted through a similarity transformation combined with the SIFT features, but not its tracking features. Since the traditional SIFT method cannot describe the global information of objects or scenes in image, it is necessary to improve the traditional SIFT feature point extraction method [5].

In recent years, the depth data extraction and depth camera [6] technology continue to be mature. Depth images are widely used in the field of measurement and monitoring since they have the advantages of good stability, resistance to light changes and 3D spatial information that can accurately reflect the scene. Instead of the traditional MeanShift tracking algorithm, this paper presents a new MeanShift tracking algorithm that is integrated with the SIFT feature metric of feature points in depth images and gray images. It has good tracking performance with short-term partial shelter, scale invariance and the target color similar to the background because the depth images can accurately reflect the global information.

The rest of this paper is organized as follows. Section 2 introduces the use of Modulus- Direction Distribution Histogram of feature points in the tracking object field for effective target tracking. Section 3 describes how feature points are extracted by using depth camera. In Sect. 4, we propose a new Depth and SIFT Meanshift tracking algorithm. Experimental results and analysis are given in Sect. 5 to show performance of the proposed algorithm. Finally, a brief conclusion and future work are presented in Sect. 6

2   Modulus - Direction Distribution Histogram

The SIFT operator is a traditional image feature description operator based on scale space with the invariant image scaling and rotation [7]. Based on the idea of image feature scale selection, the SIFT algorithm is divided into four steps: (1) find key points in the scale space; (2) determine the final key points which are called feature points; (3) determine the main direction of feature points; (4) generate SIFT feature points descriptors.

As shown in Fig. 1(a), the traditional SIFT operator takes an 8×8 window near the feature point and calculates the direction and modulus of the 64 pixels in the window. In Fig. 1(b), the gradient direction histogram of 8 directions on every 4×4 small pieces is calculated, and the cumulative value of each gradient direction to form a seed point is plotted. A feature point can form four seed points.

Different from the traditional SIFT operator, we don't need to get the SIFT feature seed node in this paper, but calculate the Modulus -Direction vector of each pixel on a 9×9 window centered on the feature point which are obtained in the tracking area. The tracking target can then be represented by the Modulus -Direction distribution histogram.

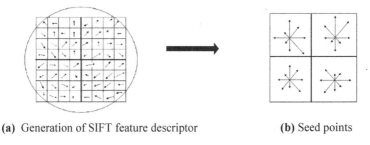

(a) Generation of SIFT feature descriptor   **(b)** Seed points

Fig. 1. (a) Generation of SIFT feature descriptor (b) Seed points

Firstly, we rotate the axis to the direction of the feature point to ensure the rotation invariance, and then take a 9×9 window with the feature point as the center. Obtain the direction $\theta(x,y)$ and the modulus m (x, y) of each pixel in the window and divide the range of the direction into n aliquots to quantize the direction. Assume that $n = 8$ (the bigger the n, the more accurate the tracking and the higher the time complexity [10]), that is, each value interval is $\left[\frac{k}{4}\pi, \frac{k+1}{4}\pi\right]$, $k = 0,1,\ldots,7$. We quantify each $\theta(x, y)$, if $\theta(x,y) \in \left[\frac{k}{4}\pi, \frac{k+1}{4}\pi\right]$, and $\theta(x,y)$ corresponds to a quantized value of $k + 1$. $\theta(x,y)$ is used to represent the direction value after pixel quantization.

As shown in Fig. 2, the black arrow represents the modulus direction of the pixel and the length of the arrow represents the modulo value of the pixel. A feature point can produce 81 Modulus - Direction feature vectors. If there are N feature points in a tracking object area, a moving object can produce less than or equal to $N \times 81$ eigenvectors. Each eigenvector can be expressed as $<x,y,\theta,m>$, where (x, y) represents the position of the feature point to which the eigenvector belongs, θ represents the direction, and m is the normalized modulus. At this point the Modulus - Direction feature vector has removed the influence of geometric deformation factors such as scale change, rotation and so on.

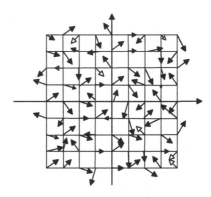

Fig. 2. Feature vector distributions in the feature point field

Here, we generate a histogram h = {H_t | t = 1, 2,... T} by the abscissa of θ, where T denotes the histogram is equal to T and H_t denotes the number of feature points in the t-th interval of the Modulus - Direction feature as follows.

$$H_t = \frac{1}{C} \sum_{i=1}^{n} \sum_{j=1}^{k} \delta[\theta(i,j) - t] \tag{1}$$

where $\delta[x]$ is the discrete impact function:

$$\delta[\theta(i,j) - t] = \begin{cases} 1 & \theta(i,j) = t \\ 0 & \theta(i,j) \neq t \end{cases} \tag{2}$$

where $\theta(x,y)$ represents the direction of the j-th pixel near the i-th feature point on the tracking target, and C is the normalization coefficient.

3 Feature Points Extraction Using Depth Camera

Although the SIFT method can contain the entire selected area information, it also introduces redundant information that affects the accuracy of tracking recognition and the global information of objects or scenes in the image cannot be described. Therefore, it is necessary to use the feature extracted in the depth image to compensate for the deficiency of the SIFT method, that is, the global information of the object or scene in the image needs to be taken into account.

Since the depth values of the pixels in the depth image are very sensitive to the viewing angle and different from the pixel values of the RGB images, the depth values of the depth images are either strongly changing in the edge region or almost unchanging in the background area, the gradient-based feature descriptor is not suitable for the direct description of the depth image feature. So in this paper, we will transform the depth image feature and then use the SIFT feature.

First, according to the paper [11], we use the depth comparison feature. For each given pixel x in the depth image I, we calculate its depth response value, as shown in the following equation:

$$MGoD_{(I,x)} = d_I\left(x + \frac{u}{d_I(x)}\right) - d_I\left(x + \frac{v}{d_I(x)}\right) \tag{3}$$

where $d_I(x)$ represents the depth value of pixel x in the depth image. The parameter (u, v) describes the offset, which is normalized by multiplying $1/d_I$.

Normalization ensures that the feature is not affected by the change in depth values. The example effect of the feature points extracted from the depth image by the above-described depth comparison feature is shown in Fig. 3. We mark the feature points on the original depth image by using the purple points as shown in Fig. 4.

Although the depth comparison feature method above is simple and efficient, it only reflects the relevant information of the depth image. To fully represent the image information, we will combine the gray image features with depth image features.

Fig. 3. The extracted feature points   **Fig. 4.** The marked feature points (Color figure online)

Assuming that the position of the feature point with intense response is equal to the same position in the gray image, we can transfer the depth feature to the gray image. At this time, the extracted SIFT feature points in the gray image and the extracted feature points in the depth image can be combined together, and the DEPTH & SIFT feature points are formed which will be described by gradient histogram in the gray image.

4 DEPTH and SIFT Meanshift Tracking

4.1 MeanShift Tracking Algorithm

Given n sample points x_i in d-dimensional space R_d, i = 1,2,...., N, the basic form of the MeanShift vector [2] at the x point is defined as $M_h = \frac{1}{k} \sum\limits_{x_i \in S_k} (x_i - x)$. Here S_h is a high-dimensional sphere with a radius h and a set of y points that satisfy the following relationship $S_k = \{y | (y - x)^T (y - x) \leq h^2\}$. K is the number of sample points falling into the S_h area.

The MeanShift vector [1] represents the average offset of the sample points, pointing to the region where the sample is most distributed, that is, the gradient direction of the probability density function. In general, in the high dimension sphere region represented by S_h, the closer the sample point is to x, the more effective the statistical feature around x is. So we introduce the concept of kernel function and give it different weight according to the importance of each sample point. In the beginning of the algorithm, the target model is initialized into a weighted histogram form $\hat{q}_u = \{\hat{q}_u\}_{u=1,...m}$ and:

$$\hat{q}_u = C_h \sum_{i=1}^{n} k(||\frac{x_0 - x_i}{h}||^2)\delta[b(x_i) - u] \quad u = 1,...m \quad (4)$$

where C_h is the normalization coefficient, h is the kernel bandwidth, x_0 is the center pixel coordinate and x_i is the i-th pixel in tracking window with a total of n pixels. K(x) is a non-increasing and continuous kernel function; b (x_i) maps the color feature to the corresponding histogram space.

When y is the center, the candidate target model $\hat{p}_u = \{\hat{p}_u\}_{u=1,\ldots m}$ can be established as follows:

$$\hat{p}_u = C_2 \sum_{i=1}^{n} k(\|\frac{y - x_i}{h}\|^2)\delta[b(x_i) - u] \quad u = 1,\ldots m \tag{5}$$

Similarly, C_2 is the normalization coefficient. Using Bhattacharyya coefficient:

$$\rho(y) = \rho[\hat{p}(y), \hat{q}] = \sum_{i=1}^{m} \sqrt{\hat{p}_u(y), \hat{q}_u} \tag{6}$$

As the candidate model and the reference model are similar, they can be obtained by MeanShift algorithm iteration with the maximum of the Bhattacharyya coefficient.

$$y_1 = \frac{\sum_{i=1}^{n} y_i w_i g\left(\left\|\frac{\hat{y}_0 - y_i}{h}\right\|^2\right)}{\sum_{i=1}^{n} m_i g\left(\left\|\frac{\hat{y}_0 - y_i}{h}\right\|^2\right)} \tag{7}$$

$$w_i = \sum_{u=1}^{m} \sqrt{\frac{\hat{q}_u}{\hat{p}_u(y_0)}}\delta[b(y_i) - u] \tag{8}$$

where y_1 is the new target center coordinate and y_0 is the center coordinate of the current candidate target area.

The target tracking method based on the MeanShift algorithm integrates Eq. (7) repeatedly so that the target is continuously moved from the current y_0 to the new position y_1 until the Bhattacharyya coefficient is the largest or reaches the specified number of iterations, and the result y_1 is the new center coordinates of the MeanShift offset tracking window.

4.2   Meanshift Tracking Based on DEPTH & SIFT Feature

Assuming that the pixel corresponding to the Modulus -Direction vector of the template is $\{x_{i,j} \mid i = 1, 2, \ldots, n; j = 1, 2, \ldots, m\}$, which is the coordinate position of the j-th pixel near the i-th feature point in the target object. The radius of the template is R, and the normalized distance of each pixel to the template center point y is $\left\|\frac{y - x_{i,j}}{R}\right\|^2$. K $(x_{i,j})$ is a kernel function and C is a normalized coefficient. If the center of a tracked object is x_0, the object can be described by using the eigenvector distribution near the critical point in the tracking object region:

$$\hat{H}_t = C \sum_{i=1}^{n} \sum_{j=1}^{m} k(\|\frac{x_{i,j} - x_0}{h}\|^2)\delta[\theta(i,j) - t] \tag{9}$$

$$C = \cfrac{1}{\displaystyle\sum_{i=1}^{n}\sum_{j=1}^{m} k(||\frac{x_{i,j}-x_0}{h}||^2)} \qquad (10)$$

The candidate object located at y can be described as:

$$H_t(y) = C\sum_{i=1}^{n}\sum_{j=1}^{m} k(||\frac{x_{i,j}-x_0}{h}||^2)\delta[\theta(i,j)-t] \qquad (11)$$

Therefore, the tracking problem of the object can be simplified to find the optimal y to make $\hat{H}_t$ and $H_t(y)$ the most similar. Define the Bhattacharrya coefficient $\rho(y)$ between the two distributions as the measure of similarity, that is:

$$\rho(y) = \rho[h(y),\hat{h}] = \sum_{t=1}^{T}\sqrt{H_t(y),\hat{H}_t} \qquad (12)$$

Try Taylor unfolds in the H_t (y_0) point on the above formula:

$$\rho(y) \approx \frac{1}{2}\sum_{t=1}^{T}\sqrt{H_t(y_0)\hat{H}_t} + \frac{C}{2}\sum_{i=1}^{n}\sum_{j=1}^{m} w(i,j)k(||\frac{x_{i,j}-y}{h}||^2) \qquad (13)$$

$$w(i,j) = \sum_{t=1}^{T}\sqrt{\frac{\hat{H}_t}{H_t(y_0)}}\delta[\theta_{(i,j)}-u] \qquad (14)$$

The larger the Bhattacharrya coefficient $\rho(y)$ is, the more similar the two objects are. So we can use the MeanShift algorithm to optimize the second term of the formula (13) to determine the next position of the tracking target.

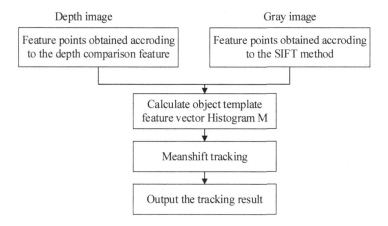

Fig. 5. Flowchart of DEPTH & SIFT-MeanShift tracking algorithm

According to the analysis above, this paper proposes a new MeanShift tracking algorithm based on DEPTH & SIFT feature. Figure 5 is a flow chart of this algorithm.

5   Experimental Results and Analysis

To test the robustness of the proposed algorithm when the target scale changes, short-term partial shelter and the target and the background have similar color, three videos are taked in the room including 320 * 240 resolution, 400 frames, and tracking targets are pedestrian. In the video sequence 1, a pedestrian wearing the same color clothes with the background and moves from left to right. In the video sequence 2, two pedestrians do relative movement and shelter in frame 215 to frame 286. In the video sequence 3, a pedestrian moves from far and near relative to the camera.

5.1   The Target Color is Similar to the Background

As seen in Fig. 6(a), tracking error occurs when using traditional MeanShift tracking method with the tracking target color similar to the background. This is because the spatial color histogram can not significantly distinguish targets from backgrounds with the same color. Figure 6(b) shows the accurate tracking result of the DEPTH & SIFT-MeanShift tracking method. That is because the DEPTH & SIFT-MeanShift tracking method is based on the depth comparison feature of the target and it can still match the target when the tracking target and background have the same color, so that the target can always be tracked throughout the whole process.

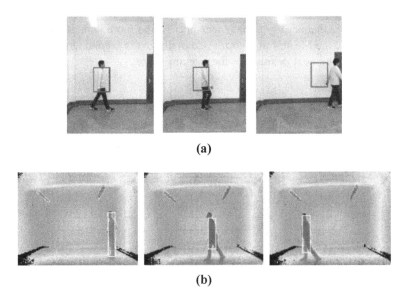

(a)

(b)

Fig. 6. (a) Traditional MeanShift tracking algorithm results, (b) DEPTH & SIFT-MeanShift tracking algorithm results (Color figure online)

5.2   Short-Term Partial Shelter

Figure 7 shows the tracking renderings of MeanShift method based on the spatial color histogram and DEPTH & SIFT feature when the object is occluded. From Fig. 7(a) we can see that, similar to the case 5.1, due to the tracked objects have similar color, so there is a tracking bias when using MeanShift tracking method based on the spatial color histogram. Figure 7(b) shows the tracking rendering of the tracking method proposed in this paper, which has good effect and robustness.

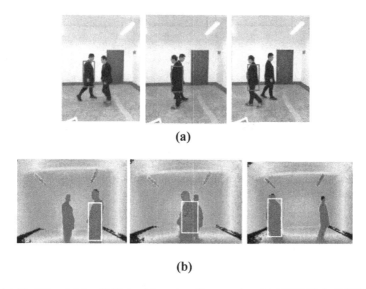

(a)

(b)

Fig. 7. (a) Traditional MeanShift tracking algorithm results, (b) DEPTH & SIFT-MeanShift tracking algorithm results

5.3   Scale Invariance

Figure 8 shows the tracking renderings of MeanShift method based on the spatial color histogram and DEPTH & SIFT feature. As can be seen from the second and third images of Fig. 8(a), the tracking effect has been deviated, but in Fig. 8(b), there is no error inrenderings based on the DEPTH & SIFT-MeanShift tracking method. This is because the method proposed in this paper uses the target local SIFT feature. Then the scale of the target is obtained by using the scale information of the feature, and the scale is consistent with the true value. The accuracy of the obtained scale parameter is high, so it can adjust according to the size of the moving target.

In order to quantitatively test the time of single frame calculation in each step of the DEPTH & SIFT-MeanShift tracking algorithm and its real-time feature compared with the original MeanShift algorithm, the above three sequences are selected to measure the processing speed of the algorithm. Using these two algorithms, run the test sequence multiple times (20 times), and take the average of the running time. The results are shown in Table 1.

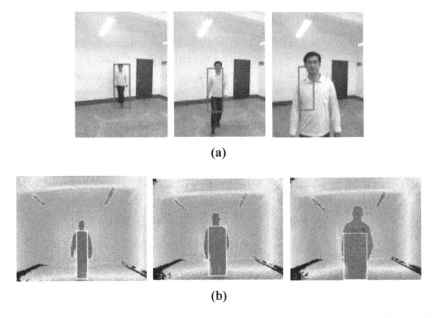

Fig. 8. (a) Traditional MeanShift tracking algorithm results, (b) DEPTH & SIFT-MeanShift tracking algorithm results

Table 1. The number of frames algorithm processed per second

	Traditional Meanshift	DEPTH & SIFT-MeanShift
Sequence one	24	20.4
Sequence two	20	18.5
Sequence three	22.1	18.3

It can be seen from Table 1 that when tracking objects in the 3-segment video sequence, the tracking method using the spatial color histogram can process 22 frames per second and the DEPTH & SIFT-MeanShift histogram can process 19 frames. So the DEPTH & SIFT-MeanShift algorithm is worse than the original MeanShift algorithm in real-time performance. This is because the algorithm adds the scale updating process based on the SIFT feature point. Because of the time complexity of the SIFT operator, the overall real-time performance of the algorithm is inferior to that based on the spatial color histogram.

6   Conclusions

In order to solve the limitation of traditional MeanShift algorithm based on spatial color histogram, this paper proposes a new tracking method, i.e. DEPTH & SIFT-MeanShift algorithm, based on depth and gray image sift feature histogram. Experimental results show that the proposed DEPTH & SIFT-MeanShift algorithm can achieve good tracking performance when the tracking target scale changes, and has strong adaptabilities and robustness to occlusion, illumination change, and the discrimination of target from background.

Based on this paper, our future work will concentrate on optimizing the SIFT operator and improving the real-time performance of the DEPTH & SIFT-MeanShift algorithm.

References

1. Palafox, G.D.L., Ortíz, A.L.S., Melendez, O.M., et al.: Hippocampal segmentation using mean shift algorithm. In: 12th International Symposium on Medical Information Processing and Analysis, pp. 101600D-1–101600D-7. International Society for Optics and Photonics (2017)
2. Jeong, J., Yoon, T.S., Park, J.B.: Mean shift tracker combined with online learning-based detector and Kalman filtering for real-time tracking. Expert Syst. Appl. **79**, 194–206 (2017)
3. Ke, Y., Sukthankar, R.: PCA-SIFT: a more distinctive representation for local image descriptors. In: Proceedings of the 2004 IEEE Computer Society Conference on Computer Vision and Pattern Recognition, CVPR 2004, vol. 2, p. II. IEEE (2004)
4. Dong, R., Li, B., Chen, Q.M.: Multi-degree-of-freedom mean-shift tracking algorithm based on SIFT feature. Control Decis. **27**(3), 399–407 (2012)
5. Castillo-Carrión, S., Guerrero-Ginel, J.E.: SIFT optimization and automation for matching images from multiple temporal sources. Int. J. Appl. Earth Obs. Geoinf. **57**, 113–122 (2017)
6. Luna, C.A., Losada-Gutierrez, C., Fuentes-Jimenez, D., et al.: Robust people detection using depth information from an overhead Time-of-Flight camera. Expert Syst. Appl. **71**, 240–256 (2017)
7. Wu, Y., Ming, Y.: A multi-sensor remote sensing image matching method based on SIFT operator and CRA similarity measure. In: IEEE International Conference on Intelligence Science and Information Engineering (ISIE), pp. 115–118 (2011)
8. Cai, K., Yang, R., Chen, H., et al.: A framework combining window width-level adjustment and Gaussian filter-based multi-resolution for automatic whole heart segmentation. Neurocomputing **220**, 138–150 (2017)
9. Yu, W., Tian, X., Hou, Z., et al.: Region edge histogram: a new feature for region-based tracking. In: 12th International Conference on Signal Processing (ICSP), pp. 1180–1185 (2014)
10. Jiang, J., Li, X., Zhang, G.: SIFT hardware implementation for real-time image feature extraction. IEEE Trans. Circ. Syst. Video Technol. **24**(7), 1209–1220 (2014)
11. Ye, M., Wang, X., Yang, R., et al.: Accurate 3D pose estimation from a single depth image. In: IEEE International Conference on Computer Vision (ICCV), pp. 731–738 (2011)

A Novel 3D Expansion and Corrosion Method for Human Detection Based on Depth Information

Xiexin Qi[1], Minrui Fei[1,2], Huosheng Hu[3], and Haikuan Wang[1(✉)]

[1] School of Mechatronics Engineering and Automation, Shanghai University, Shanghai, China
HKWang@shu.edu.cn
[2] Shanghai Key Laboratory of Power Station Automation Technology, Shanghai University, Shanghai, China
[3] School of Computer Science and Electronic Engineering, University of Essex, Colchester, UK

Abstract. The existing body detection methods based on depth images mostly depend on the extraction of image gradient features, which is the evolution of the traditional 2D plane image processing method for human body detection. Although their detection accuracy is high, the algorithms consume a large amount of computing and storage resources. Aiming at the real-time demand of safe-driving of forklift trucks in industry, this paper presents a novel 3D expansion and corrosion method for human detection by using depth information. A depth image of human body is detected based on the characteristics of human Head-Shoulder-Body Density (HSBD), which can reduce the error and loss of the depth information caused by changing light conditions, complex background scenes and various distances from objects. Experimental results show that the recognition rate of the proposed method is over 96%, and the recognition speed is over 15 frames per second. This can satisfy the safe-driving demands of forklift truckers in factory.

Keywords: Human detection · Depth image · TOF camera · 3D expansion and corrosion · HSBD

1 Introduction

Human detection and recognition have become one of the key technologies in a wide range of real-world applications such as intelligent robots, self-driving cars and industry processes. Traditional human body detection is mostly processed by grayscale images or colour images, which is not robust and unable to handle lighting changes, environment complexity, different human positions, and various materials of clothes. Moreover, it is very time-consuming and resources-consuming to calculate the gradient characteristics of individual pixels in the scene and obtain the scale-invariant body characteristics by constructing the image pyramid. Therefore, this method cannot be used for detecting

This work is supported by Key Project of Science and Technology Commission of Shanghai Municipality under Grant No. 14JC1402200.

M. Fei et al. (Eds.): LSMS/ICSEE 2017, Part I, CCIS 761, pp. 556–565, 2017.
DOI: 10.1007/978-981-10-6370-1_55

humans in front of industrial forklift trucks that normally have limited computing power onboard.

In recent years, with the continuous development of depth cameras, more and more researchers try to use them to obtain the depth images. The use of its depth information on the scene for human body detection and identification has achieved good experimental results. Inspired by the HOG (Histogram of Oriented Gradients) feature descriptor, Wu, et al. proposed an HDD (Histogram of Depth Difference) feature descriptor for detecting the human body based on the depth images taken by a TOF camera [5]. In this method, a unit is composed of 8 × 8 pixels, 4 units form a block. The depth gradient of each pixel is calculated and the direction of the gradient is divided into 9 parts from 0° to 360° to form the HDD character descriptor. The detection rate is over 99% and the FPPW is about 10-4 after using SVM. However, the speed of depth image processing is about 10 frames per second, which cannot meet the actual needs of the safety detection of engineering vehicles.

Xia, et al. proposed a method based on 2D head contours and 3D head surface models [7] for human detection in deep images. A 2D head contour model was firstly used to target the screening, and 3D head surface model was then deployed to identify the human body. However, this method cannot effectively obtain the surface model of human body due to the long distance between the depth camera and the target object and the light condition changes. The SLTP (Simplified Local Ternary Patterns) features proposed in [6] were used to classify the contours of the detected targets according to the threshold of the depth gradient by extracting the body edge contour of the depth transformation characteristics. This method had good detection results and was less time-consuming with a detection rate at 86 depth images per second.

However, this method is only suitable for simple indoor detection, where light source is relatively stable and the contour of targets is relatively simple. However, it cannot be used in an industrial field where the light source is changeable, the background is complex, and the detection distance is far. The depth information captured by a depth camera may have depth drift, distortion, and even the information loss, which can heavily affect the extraction accuracy of targets contour. So this method cannot be directly deployed in industrial applications.

In this paper, a 3D expansion and corrosion method based on probabilistic statistics is proposed to pre-process the depth images taken in the industrial field. It aims to effectively suppress the drift and error of depth information caused by various kinds of lighting conditions, object reflections and detection distances. Also, HSBD (Head-Shoulder-Body Density) based features are used to detect and identify the human body in depth images. The recognition rate of this method is over 96% and the algorithm time cost is within 60 ms. The proposed method has been applied to logistics forklifts with stable operation for several months in a domestic automobile parts and components factory.

The remainder of this paper is organized as follows. Section 2 describes the entire process of using HSBD feature to detect human in depth images. The experimental results and analysis are presented in Sect. 3 to show the feasibility and performance of the proposed approach. Finally, a brief conclusion is given in Sect. 4.

2 Proposed Human Detection Method

Figure 1 shows the framework of the proposed human detection method, and Fig. 2 shows the process of human detection with HSBD. At the pre-processing stage, the original image is processed to reduce the noise and locate the depth range of suspicious targets. Subsequently, the proposed 3D expansion and corrosion is employed to fill the holes of depth image and erase the error data based on probabilities statistics. Finally, HSBD feature is used to recognize the human in the image. The details of each step are described in the following subsections.

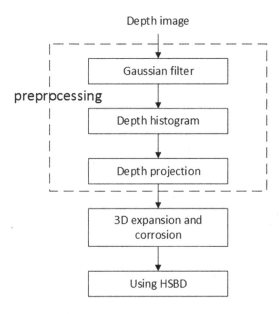

Fig. 1. Flow chart of human detection with HSBD

2.1 Preprocessing of Depth Image

The TOF camera can catch all the depth information of every pixel directly. Due to the change of light refraction and reflection and various kinds of lighting conditions in factory, errors may occur during the acquisition process. It is well known that the Gaussian filter has a good performance in noise reduction. Then the histogram of depth information is counted to roughly locate the depth range where the suspect targets are. As the depth information of the same object is relatively centralized, the abnormal depth interval can be quickly detected by analysing the depth histogram in this case. Then the whole image is projected into different depth intervals to get images with the same resolution and different depth ranges. The result of depth projection is shown in Fig. 3.

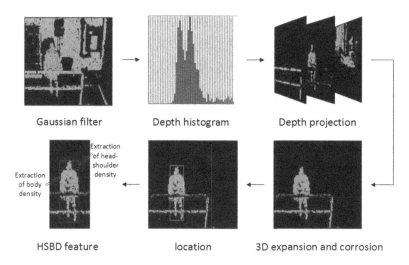

Fig. 2. Flow chart of human detection with HSBD

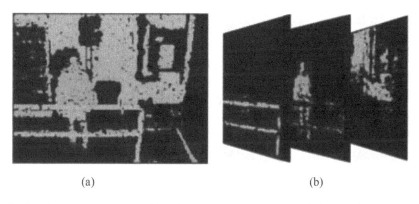

Fig. 3. Depth projection diagram (a) is the original depth image and (b) are images after depth projection

2.2   3D Expansion and Corrosion Based on Probabilities Statistics

A novel expansion and corrosion method based on depth information and probabilities statistics is proposed. The depth image is projected into different depth interval in the preprocessing. To avoid dividing the same object into different depth levels, each interval $[D_{near}, D_{far}]$ needs to be extended and the extension range is half of the step size, as shown in Eq. 1.

$$\Delta d = (D_{far} - D_{near})/2 \tag{1}$$

Formula (2) is used to decide whether the depth information of point is in the activated range or not.

$$\begin{cases} D(x,y) \in [D_{near}, D_{far}], D(x,y) \in DR \\ (D(x,y) + \Delta d) \in [D_{near}, D_{far}], D(x,y) \in DR \\ (D(x,y) - \Delta d) \in [D_{near}, D_{far}], D(x,y) \in DR \end{cases} \tag{2}$$

where $D(x, y)$ is the depth information of $P(x, y)$ and DR is the selected depth range. The criterion of the $P(x, y)$ is as follow

$$P(x,y) = \begin{cases} 1, D(x,y) \in DR \\ 0, D(x,y) \notin DR \end{cases} \tag{3}$$

The ratio of positive points around the negative point $P_{pos}(x, y)$ and the ratio of negative points around the positive point $P_{neg}(x, y)$ in an $m \times n$ window are calculated with Formula (4) and (5).

$$P_{pos}(x,y) = \frac{\sum_{i=1}^{m} \sum_{j=1}^{n} P(i,j)}{m \times n - 1} \tag{4}$$

$$P_{neg}(x,y) = 1 - P_{pos}(x,y) \tag{5}$$

If the point $P(x, y)$ is negative and $P_{pos}(x, y) \geq T_{epd}$, the depth information of this point will be set with Formula (6).

$$D(x,y) = \frac{\sum_{i=1}^{m} \sum_{j=1}^{n} [P(i,j) \cdot D(i,j)]}{\sum_{i=1}^{m} \sum_{j=1}^{n} P(i,j)} \tag{6}$$

When the point $P(x, y)$ is negative and $P_{neg}(x, y) \geq T_{crs}$, the depth value will be replaced with 0. T_{epd} is the threshold of expansion and T_{crs} is the threshold of corrosion.

Compared with the traditional method of expansion and corrosion with a fixed template, the proposed 3D expansion and corrosion method can erase the abnormal data and fill the black hole caused by missing information effectively.

2.3   Human Detection with HSBD

After the process of 3D expansion and corrosion, the target object can be located with horizontal and vertical histograms. The center of big target object can be quickly located by using the Formula (7) and (8), where the X is the amount of pixels in horizontal direction and the Y is the amount of pixels in vertical direction.

$$C_x = \frac{\sum\limits_{i=1}^{X} \sum\limits_{j=1}^{Y} i \cdot D(i,j)}{\sum\limits_{i=1}^{X} \sum\limits_{j=1}^{Y} D(i,j)} \qquad (7)$$

$$C_y = \frac{\sum\limits_{i=1}^{X} \sum\limits_{j=1}^{Y} j \cdot D(i,j)}{\sum\limits_{i=1}^{X} \sum\limits_{j=1}^{Y} D(i,j)} \qquad (8)$$

Then use the region growing algorithm to get the contours of the objects.

To realize safe driving of a forklift in factory, an HSBD feature is deployed here to quickly detect workers in front of vehicles. There is a fixed ratio between the width of head and shoulder, which will not change by height and weight of different people. The upper trunk of people can stay stable no matter they are standing or walking, and the positive points of the upper trunk occupies a stable density of the contour.

The TOF camera can catch depth information of every pixel in the scene. The depth information describes the scale information of the objects. Both the height of the head zone and the width of body zone can be determined by the depth interval where the object is. In this case, the image pyramid is not necessary to be built to form the multi-scale space. The head-shoulder density D_{head} is calculated by the Formula (9),

$$D_{head} = \frac{\sum\limits_{(i,j) \in hz} P(i,j)}{Harea} \qquad (9)$$

where hz is the zone of head and the $Harea$ is the size of hz.

At the meantime, the body density D_{body} is calculated with Formula (10).

$$D_{body} = \frac{\sum\limits_{(i,j) \in bz} P(i,j)}{Barea} \qquad (10)$$

The bz is the zone of the upper trunk and the $Barea$ is the area of this zone.

Using these two density value as feature can detect and recognize the human quickly and effectively.

3   Experiments and Analysis

The 3D expansion and corrosion needs to calculate the ratio of every pixel in a window. The performance of this method is getting better along with the window size growing and the repeat times increasing. However the bigger of window and the more repeat time is, the more time is consuming. Its relationship is shown in the Fig. 4.

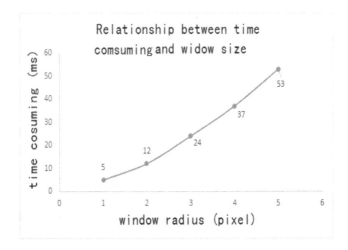

.**Fig. 4.** Time cost chart of 3D expansion and corrosion algorithm

Through several experiments, a good balance between high performance and low time cost can be reached when repeating the operation with a 3×3 window.

The performance of 3D expansion and corrosion is shown in Fig. 5, where $T_{crs} = 0.55$ and $T_{epd} = 0.35$.

(a)   (b)   (c)

Fig. 5. The performance of 3D expansion and corrosion (a) is image after depth projection (b) is expansion and corrosion with morphology (c) is 3D expansion and corrosion with probabilities statistics

Results of human detection and recognition by using HSBD feature are as follows (Fig. 6).

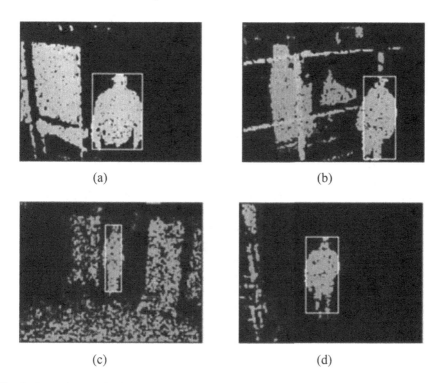

(a)   (b)

(c)   (d)

Fig. 6. Detection result with HSBD feature (a) is human standing upright (b) is human standing sideways (c) is human standing in complex outdoor environment (d) is walking human

243 depth photos have been taken in the factory, for the use of the experiments and tests. There are 176 pictures with human in it. The algorithm is written in C# and run in Visual Studio 2013. The CPU of the test computer is 3.2 GHz inter i5 processor. The results of the experiments are shown in Table 1.

Table 1. The result of detection with HSBD feature

Test images	Amount	Detection rate	False Positive rate	False Negative rate	Time cost
With human	176	96.6%		3.4%	31 ms
Without human	67		7.5%		54 ms

In experiments, the human detection rate reaches 96%. Using HSBD feature can detect the people in the complex environment no matter they are standing or walking. Firstly, preprocessing and depth projection are used to simplify the depth image. The proposed 3D expansion and corrosion based on probabilities statistics can reduce the interference resulted from various light sources and clothes material. According to the stable rate between head and shoulder and the stable trunk density, people with different postures, heights and weights can be well detected by using the HSBD feature. The time cost of whole process of the algorithm is within 60 ms, which means that 16 depth images can be handled in 1 s.

Meanwhile, a 3D computer vision system based on the proposed method has been used on the forklift in an auto parts production plant. The system can alert drivers and workers when detecting human in the alarm zone. The alarm zone is defined 2 m in front of the forklift and the actual distance is between 4 m and 6 m because of the TOF camera placement.

4   Conclusion

In order to meet the real-time requirements of the safe driving of forklift trucks in factories, depth projection, this paper presents a novel 3D expansion and corrosion method that are used in the entire process of human detection, by combining with HSBD features. The experiment results show that 3D expansion and corrosion based on probabilities statistics can have a better suppression on error data caused by different light sources, background and clothes materials. Using HSBD features can detect and recognize people in the image with different postures and body shapes. The detection rate and time cost reach the standards of factory.

In the future, different depth intervals can be dealt with in the same time by parallel computing. As a result of that, the time cost of the algorithm could be effectively reduced.

References

1. Dollar, P., Wojek, C., Schiele, B.: Pedestrian detection: a benchmark. In: CVPR, pp. 304–311 (2009)
2. Enzweiler, M., Gavrila, D.M.: Monocular pedestrian detection: survey and experiments. IEEE Trans. Pattern Anal. Mach. Intell. 31(12), 2179 (2009)
3. Bajracharya, M., Moghaddam, B., Howard, A., Brennan, S., Matthies, L.H.: Results from a real-time stereo-based pedestrian detection system on a moving vehicle. In: Workshop on People Detection & Tracking IEEE ICRA (2009)
4. Navarro-Serment, L.E., Mertz, C., Hebert, M.: Pedestrian detection and tracking using three-dimensional ladar data. Int. J. Robot. Res. 29, 1516–1528 (2010)
5. Wu, S., Yu, S., Chen, W.: An attempt to pedestrian detection in depth images. In: Intelligent Visual Surveillance, pp. 97–100. IEEE (2011)
6. Yu, S., Wu, S., Wang, L.: SLTP: a fast descriptor for people detection in depth images. In: IEEE Ninth International Conference on Advanced Video and Signal-Based Surveillance, pp. 43–47. IEEE (2012)
7. Lu, X., Chen, C.C., Aggarwal, J.K.: Human detection using depth information by kinect. Appl. Phys. Lett., 15–22 (2011)
8. Ikemura, S., Fujiyoshi, H.: Real-time human detection using relational depth similarity features. In: Kimmel, R., Klette, R., Sugimoto, A. (eds.) ACCV 2010. LNCS, vol. 6495, pp. 25–38. Springer, Heidelberg (2011). doi:10.1007/978-3-642-19282-1_3
9. Wang, N., Gong, X., Liu, J.: A new depth descriptor for pedestrian detection in RGB-D images, pp. 3688–3691 (2012)

10. Li, R., Liu, Q., Gui, J., Gu, D., Hu, H.: Night-time indoor relocalization using depth image with Convolutional Neural Networks. In: International Conference on Automation and Computing, pp. 261–266 (2016)
11. Song, L., Hu, R., Zhong, R.: Depth similarity enhanced image summarization algorithm for hole-filling in depth image-based rendering. Wirel. Commun. Over Zigbee Automot. Inclinat. Measur. China Commun. **11**(11), 60–68 (2014)
12. Yu, H.W., Jeon, J.D., Lee, B.H.: Surface normal smoothing for superpixels in noisy depth images. Electron. Lett. **52**(5), 359–361 (2016)
13. Lasang, P., Shen, S.M., Kumwilaisak, W.: Novel edge preserve and depth image recovery method in RGB-D camera systems. In: 2014 IEEE Fourth International Conference on Consumer Electronics – Berlin (ICCE-Berlin), pp. 346–349 (2014)
14. Hernández-Aceituno, J., Arnay, R., Toledo, J., Acosta, L.: Using kinect on an autonomous vehicle for outdoors obstacle detection. IEEE Sens. J. **16**(10), 3603–3610 (2016)

An Adaptive Edge Detection Algorithm Based on Improved Canny

Aolei Yang, Weiwei Jiang, and Ling Chen[✉]

Shanghai Key Laboratory of Power Station Automation Technology,
School of Mechatronic Engineering and Automation,
Shanghai University, Shanghai 200072, China
aolei_yang@163.com, jww704285932@163.com,
lcheno@shu.edu.cn

Abstract. Edge detection is the key to image processing and has a significant impact on the high level of description, classification and matching of subsequent images. The traditional Canny algorithm requires human intervention in the selection of Gaussian function and its fixed parameters. To solve these problems, an improved algorithm based on Canny algorithm is proposed in this paper. The approach introduces the edge preserving filter to replace the original Gaussian filter, and calculates the magnitude and direction of image gradient with a new designed templates from x direction, y direction, and two oblique directions (45°, 135°). Meanwhile, the Otsu algorithm is used to calculate the thresholds, which avoids the problem that the thresholds need to be set repeatedly. The proposed method is successfully applied to the metal plate detection system. Experimental results show that the algorithm has good performance in bright and dark domains.

Keywords: Edge preserving filtering · Canny algorithm · Otsu method · Machine vision · Defect detection

1 Introduction

Image is an important source of human cognitive world. The edge exists in the discontinuous and irregular structure of the image, which can describe the target outline and contain information about the relative and contour positions within the target area. The edge detection could be then considered as the basis of shape detection, and its result will directly impact the image analysis.

With the rapid development of digital image processing technology, edge detection [1, 2] has been paid more and more attention. In 1986, Canny [3] proposed an edge detection operator based on the optimization. Compared with the traditional edge

This work was supported by Natural Science Foundation of China (61403244), Science and Technology Commission of Shanghai Municipality under "Yangfan Program" (14YF1408600, 16YF1403700), Key Project of Science and Technology Commission of Shanghai Municipality (15411953502).

© Springer Nature Singapore Pte Ltd. 2017
M. Fei et al. (Eds.): LSMS/ICSEE 2017, Part I, CCIS 761, pp. 566–575, 2017.
DOI: 10.1007/978-981-10-6370-1_56

detection operator (Such as Sobel operator, Laplace operator), the Canny operator is widely used because of good signal to noise ratio and detection accuracy. However, since there exists the problems of the choices of Gaussian filter parameter and threshold value, various improvements had been carried out. Jinmin Zhong [4] presented the integration of the wavelet transform and the Canny algorithm. Wavelet transform is used to remove the approximation sub image in low frequency, and the edge detail information is reconstructed by the detail sub image in high-frequency. The traditional Canny algorithm is then adopted to extract edge in the high frequency image. Although the wavelet transform suppresses the influence of noise, the improvement of image quality cannot totally solve the defect of traditional Canny algorithm. Qang Sheng [5] introduced the concept of "breakpoint" to the Canny algorithm. Low threshold is used to select the point to be connected at the breakpoint. This method splits the broken edges as much as possible, and it could detect more virtual edges for the picture in noisy background, which is not helpful to the subsequent processing. Wang X [6] depicted the iterative algorithm to calculate the optimal thresholds, but the number of iterations need to be set manually, and the time consuming will increase along with the number. Additionally, Tan Lei [7] used the Canny algorithm to detect the edge of the outdoor obstacle and introduced the Otsu method to calculate the low threshold in Canny algorithm, but its high threshold was obtained by taking the low threshold twice and did not achieve the best segmentation effect.

Canny algorithm requires human intervention in the selection of Gaussian function and the parameters of the traditional Canny edge detection algorithm is not adaptive. In response to these problems, this paper introduced an edge preserving filter to replace the traditional Gaussian filter, and calculate the magnitude of image gradient with a new template at x direction, y direction and two oblique directions (45°, 135°). At the same time, the Otsu algorithm is used to calculate the thresholds, which avoids the problem that the thresholds need to be set repeatedly. Finally, the designed algorithm is applied to the defect detection system in industrial environment to verify its effectiveness and robustness.

2   Traditional Canny Algorithm and Analysis

2.1   The Traditional Canny Algorithm

Canny algorithm is an edge detection algorithm to find the best edge, and its detection complies with the three major criteria [8].

1. Signal-noise ratio
2. Localization precision
3. Unilateral response

The specific steps of algorithm are as follows [9]:

1. The image is smoothed by Gaussian filtering function shown in (1), and the standard deviation σ controls the smooth degree and set artificially.

$$G(x,y) = \frac{1}{2\pi\sigma^2} \exp\left(-\frac{x^2+y^2}{2\sigma^2}\right) \tag{1}$$

2. Calculate the gradient amplitude and direction of image.
3. In order to get pixel edge in image, it is needed to perform non-maximum suppression, and the non-maximum points are removed.
4. Apply the double thresholds to select the potential boundary, in which the choice of high and low thresholds will directly affect the quality of edge detection.
5. Track the edge by hysteresis: finalize the detection of edges by suppressing all the other edges that are weak and not connected to strong edges.

2.2   The Disadvantages of Traditional Canny Algorithm

The step of Canny algorithm is clear, but it still has the following disadvantages:

1. The standard deviation σ played a very important role in smooth. When σ becomes larger, the positioning accuracy of the edge will become lower, and the edge of the image becomes excessive blurring [10].
2. The traditional Canny adopts finite difference of 2×2 neighboring area to calculate the value and direction of image gradient. Because the points are relatively few, so it's sensitive to noise and easy to detect false edge.
3. The high and low thresholds is depended on manual settings, and this requires much prior experience and with non-adaptability. Using two thresholds is more flexible than using one threshold, but in order to find an appropriate threshold, we need to test several times repeatedly. Thresholds setting too high may miss important edge information, and too low thresholds could lead to many pseudo edges in an image. It is difficult to search a generic threshold for all images.

3   The Improved Adaptive Canny Algorithm

3.1   Edge Preserve Filtering

For the Gaussian filtering, the standard deviation should be selected manually and the edge of the image are blurred after filtering. In this paper, a new edge holder is used to process the image, which the denoising ability is strong while the detail of the image edge can be kept. The specific algorithm steps are as follows:

1. Select an appropriate neighborhood size for each pixel [i, j]. The scale of neighborhood used in this paper is 5×5. (Figure 1 show the neighborhood of point)
2. The following formula define the gray scale uniformity:

$$V = \sum g^2[i,j] - \left(\sum g[i,j]\right)^2 / N \tag{2}$$

$$\begin{pmatrix} I_{11} & I_{12} & I_{13} & I_{14} & I_{15} \\ I_{21} & I_{22} & I_{23} & I_{24} & I_{25} \\ I_{31} & I_{32} & I_{33} & I_{34} & I_{35} \\ I_{41} & I_{42} & I_{43} & I_{44} & I_{45} \\ I_{51} & I_{52} & I_{53} & I_{54} & I_{55} \end{pmatrix}$$

Fig. 1. Diagram of edge-preserving filtering algorithm

where represents the gray value of the current point and N is the size of neighborhood.

3. Calculate four regions of the gray scale uniformity V in the top left (points I_{11}, I_{12}, I_{21}, I_{22}), the top right (points I_{14}, I_{15}, I_{24}, I_{25}), lower left (points I_{41}, I_{42}, I_{51}, I_{52}) and lower right (points I_{44}, I_{45}, I_{45}, I_{45}) of [i, j] separately.
4. The gray value of the region corresponding to the minimum gray scale uniformity is selected as the new pixel value of the point [i, j].

3.2 Image Gradient Calculation

The traditional Canny uses a finite difference between regions around the center point to calculate the gradient. The approach is sensitive to noise and easy to lose some important edge information, especially some information in the bevel edge. For this shortcomings, the magnitude of the gradient is achieved by computing the finite difference of the first derivative in x direction, y direction and two oblique directions in [11]. But this is only calculated based on the original Canny algorithm, and it is with low ability to restrict noises. Hence, referring to the first-order gradient template in Sobel operator, and the first-order gradient template in the four directions is proposed as follows (Fig. 2):

$$\begin{pmatrix} -1 & -2 & -1 \\ 0 & 0 & 0 \\ 1 & 2 & 1 \end{pmatrix}$$
(a) x direction

$$\begin{pmatrix} -1 & 0 & 1 \\ -2 & 0 & 2 \\ -1 & 0 & 1 \end{pmatrix}$$
(b) y direction

$$\begin{pmatrix} -2 & -1 & 0 \\ -1 & 0 & 1 \\ 0 & 1 & 2 \end{pmatrix}$$
(c) 45° direction

$$\begin{pmatrix} 0 & 1 & 2 \\ -1 & 0 & 1 \\ -2 & -1 & 0 \end{pmatrix}$$
(d) 135° direction

Fig. 2. First-order gradient template

The first-order gradient components in the four directions can be obtained from the four first-order gradient templates described above, and the Eqs. (3) and (4) can be applied to calculate the gradient magnitude and direction.

$$M(x,y) = sqrt\left(G_x^2 + G_y^2 + G_{45}^2 + G_{135}^2\right) \tag{3}$$

$$\theta(x,y) = \arctan\left(\frac{G_y(x,y)}{G_x(x,y)}\right) \tag{4}$$

3.3 Adaptive Threshold Selection

After calculating the gradient magnitude and direction, the edge can be obtained by non-maximum suppression. The gradient is divided into L level, and the L level can be then divided into three categories C_0, C_1, C_2, where C_0 represents non-edge pixel, C_1 is a pixel that may be edge points or non-edge points, and C_2 is a pixel of edge points.

In order to achieve the better performance in segmentation, the maximum interclass variance method [12] is used to automatically select the thresholds The idea is to make the image into two types. One is the background and the other is the objective. By searching and computing the maximum between class variance, the optimal threshold value could be achieved.

Assume that the higher and lower thresholds for edge detection is t_k and t_m, then C_0 contains the pixels gradient magnitude $\{t_1, t_2, \ldots, t_k\}$, C_1 being with $\{t_k, t_{k+1}, \ldots, t_m\}$, and C_2 being with $\{t_m, t_{m+1}, \ldots, t_L\}$. Suppose N is the total number of gradient histograms and n_j is the number of points corresponding to the gradient magnitude t_j, its probability is given in (5).

$$p_j = n_j/N \tag{5}$$

The probability of C_0, C_1, C_2 is:

$$p(0,k) = \sum_{j=1}^{k} p_j \tag{6}$$

$$p(k,m) = \sum_{j=k}^{m} p_j \tag{7}$$

$$p(m,L) = \sum_{j=m}^{L} p_j \tag{8}$$

The mean value of the whole gradient amplitude is:

$$\mu = \sum_{j=1}^{L} t_j \cdot p_j \tag{9}$$

The mean value of the gradient magnitude in three different categories is:

$$\mu(0,k) = \frac{\sum_{j=1}^{k} t_j \cdot p_j}{\sum_{j=1}^{k} p_j} \tag{10}$$

$$\mu(k,m) = \frac{\sum_{j=k+1}^{m} t_j \cdot p_j}{\sum_{j=k+1}^{m} p_j} \tag{11}$$

$$\mu(m,L) = \frac{\sum_{j=m+1}^{L} t_j \cdot p_j}{\sum_{j=m+1}^{L} p_j} \tag{12}$$

The evaluation function is defined as:

$$\tau = [\mu(0,k) - \mu]^2 \cdot p(0,k) + [\mu(k,m) - \mu]^2 \cdot p(k,m) + [\mu(m,L) - \mu]^2 \cdot p(m,L) \tag{13}$$

The best threshold values of Otsu algorithm are the t_k and t_m value, which makes interclass variance value maximum. The larger the variance value is, the greater the probability that the target is separated from the background in gradient histogram is, better the classification effect is.

After obtained the two thresholds, the points above the high threshold are retained as the edge, but points between the two thresholds is not sure. Typically, the weak edge pixels caused by the real edges will be connected to the strong edge pixels while the noise response is not connected. To track the edge connection, blob analysis is applied to search weak edge pixels and its 8-connected neighborhood pixels. The weak edge points can be identified as ones that should preserved when there is one strong edge pixel involving in the blob.

The Otsu method is introduced into the Canny algorithm, so that the improved Canny algorithm can be used to adaptively select the high and low thresholds, which effectively solves the problem of inaccurate threshold setting.

4 Engineering Verification and Result Analysis

4.1 Engineering Applications

The improved algorithm proposed in this paper has been applied to the metal plate experimental system for the detection of surface defects. Its overall block diagram is shown in Fig. 3. The specifications are as follows:

1. The detection speed is 3 m/s.
2. The corresponding horizontal and vertical resolution is 0.5 mm.
3. The detection rate of common defects on the production line is over 85%.

The reflections of light on the metal plate is mainly specular and diffuse reflection, and one lighting method does not detect all defects well. For this reason, the combination of bright and darkness is adopts in the structure of the light source [13]. The bright domain configuration is mainly used to detect the defects of reflection and

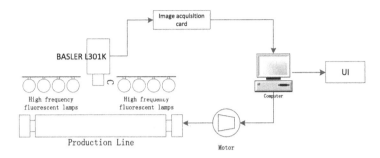

Fig. 3. System block diagram

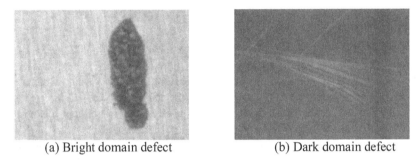

(a) Bright domain defect   (b) Dark domain defect

Fig. 4. Example of bright and dark domain defects

absorption of light. In most bright domain images, the background is bright and defects are dark, such as spots, rust, foreign objects and so on. Figure 4 shows the different types of defects.

In the detection system, the computer adopts Intel i5 6500 processor, 8G memory. The software is realized by VS2013 and OpenCV3, which functions include parameter setting, surface defect detecting, defect location marking and display of all defect details. The entire software screenshot is shown in Fig. 5:

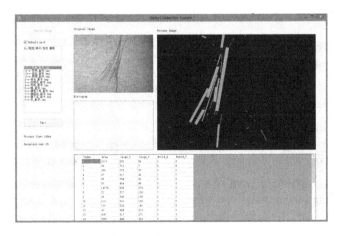

Fig. 5. Software screenshot

4.2   Result Analysis

In this paper, the images selected for experiments include bright domain defect or dark domain defect. The comparative experiments are carried out based on the traditional Canny and the proposed algorithm respectively under the premise of edge preserve filtering and Gaussian filtering. The experimental results are shown from Figs. 6, 7, 8 and 9.

(a) Bright domain defect image (b) Effect of traditional Canny algorithm (c) Effect of improved Canny algorithm

Fig. 6. Image effect comparison of bright domain using Gaussian filter

(a) Bright domain defect image (b) Effect of traditional Canny algorithm (c) Effect of improved Canny algorithm

Fig. 7. Image effect comparison of bright domain using edge preserve filter

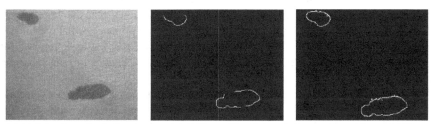

(a) Dark domain defect image (b) Effect of traditional Canny algorithm (c) Effect of improved Canny algorithm

Fig. 8. Image effect comparison of dark domain using Gaussian filter

(a) Dark domain defect image (b) Effect of traditional Canny algorithm (c) Effect of improved Canny algorithm

Fig. 9. Image effect comparison of dark domain using edge preserve filter

Figures 6 and 7 are the comparison results of the Gaussian filter and the edge preserve filter in the bright domain, Figs. 8 and 9 being related to the comparison results in dark domain. The standard deviation of the Gaussian filter is 0.65 (corresponding to the filter kernel size of 3), and the traditional Canny algorithm has a high threshold of 30 and a low threshold of 15. It can be seen from Figs. 6(b), (c) and 8(b), (c) that the improved Canny algorithm is able to detect more edge details in both, while the edge detected using the traditional Canny algorithm are not continuous and cannot form valid regions. Compared Fig. 7 with Fig. 9, it is shown that the edge preserve filter could be used to greatly improve the detection performance, whatever in traditional and improved algorithms. This is mainly because the edge preserve filter effectively removes noise and makes edges clear, which is beneficial to the subsequent algorithm detection.

In the paper, 4 kinds of defects: rust, white spots, pits and scratches, are selected to verify the effectiveness of the improved algorithm, and each defect is with 30 images. The average time and detection rate of the improved algorithm and the traditional Canny algorithm are shown in Table 1.

Table 1. Comparison of time consuming and detection rate in two algorithm

Defect category	Rust spot	White spot	Pit	Scratch
Traditional Canny time consuming/ms	17.8	21.7	25.5	20.5
Improved Canny time consuming/ms	19.6	22.3	27.7	22.3
Traditional Canny detection rate (%)	73.3	76.7	53.3	63.3
Improved Canny detection rate (%)	90	93.3	86.7	86.7

It can be seen that the time consuming of the improved algorithm is more than traditional Canny algorithm but the detection rate is higher. Because the improved Canny algorithm needs to use the Otsu method to calculate the thresholds, the average time consuming is 5%–10% higher than the traditional Canny algorithm, but the corresponding detection rate for each defect is greatly improved. The detection rate of the improved algorithm can be maintained at 85% in the dark domain, where it is more difficult to detect in the bright domain. The results show that the improved Canny algorithm is more suitable for the experimental environment.

5   Conclusion

Edge detection is an important research subject in image processing and machine vision, and Canny operator is a kind of more effective method. Aiming at the drawbacks of the traditional Canny edge detection algorithm, this paper proposed the corresponding improvement method, which is successfully applied to the online detection system. Experimental results show that the improved algorithm can detect more edge details in both bright domain and dark domain, and the performance of rejecting noise and preserving edge is better than that in the traditional Canny algorithm. The detection system combined with improved algorithm is more suitable for industrial application, and could quickly complete the processing to defective pictures.

References

1. Medina-Carnicer, R., Muñoz-Salinas, R., Yeguas-Bolivar, E., Diaz-Mas, L.: A novel method to look for the hysteresis thresholds for the Canny edge detector. Pattern Recogn. **44**, 1201–1211 (2011)
2. Wu, Z., Sun, C., Liu, J.: Oil-seal surface defect automatic detection and recognition method based on image processing. Yi Qi Yi Biao Xue Bao/Chin. J. Sci. Instrument. **34**, 1093–1099 (2013)
3. Xu, Q., Varadarajan, S., Chakrabarti, C., Karam, L.J.: A Distributed Canny Edge Detector: Algorithm and FPGA Implementation. IEEE Trans. Image Process. **23**, 2944–2960 (2014)
4. Zhong, J., Han, Y., Shi, P.: Fish-bone detection based on multiresolution wavelet. Yi Qi Yi Biao Xue Bao/Chin. J. Sci. Instrum. **27**, 2198–2199 (2006)
5. Niu, S., Wang, S., Yang, J., Chen, G.: A fast image segmentation algorithm fully based on edge information. Jisuanji Fuzhu Sheji Yu Tuxingxue Xuebao/J. Comput. Aided Design Comput. Graph. **24**, 1410–1419 (2012)
6. Wang, X., Jin, J.Q.: An edge detection algorithm based on improved CANNY operator. In: Seventh International Conference on Intelligent Systems Design and Applications (ISDA 2007), pp. 623–628 (2007)
7. Tan, L., Wang, Y., Shen, C.: Vision based obstacle detection and recognition algorithm for transmission line deicing robot. Yi Qi Yi Biao Xue Bao/Chin. J. Sci. Instrum. **32**, 2564–2571 (2011)
8. Bao, P., Zhang, L., Wu, X.: Canny edge detection enhancement by scale multiplication. IEEE Trans. Pattern Anal. Mach. Intell. **27**, 1485–1490 (2005)
9. Rong, W., Li, Z., Zhang, W., Sun, L.: An improved Canny edge detection algorithm. In: 2014 IEEE International Conference on Mechatronics and Automation, pp. 577–582 (2014)
10. Zhou, W., Fei, M., Zhou, H., Li, K.: A sparse representation based fast detection method for surface defect detection of bottle caps. Neurocomputing **123**, 406–414 (2014)
11. Wang, Z., He, S.: An adaptive edge-detection method based on Canny algorithm. Yi Qi Yi Biao Xue Bao/Chin. J. Sci. Instrum. **9**, 957–962 (2004)
12. Zhang, J., Hu, J.: Image segmentation based on 2D Otsu method with histogram analysis. In: 2008 International Conference on Computer Science and Software Engineering, pp. 105–108 (2008)
13. Fernandez, C., Platero, C., Campoy, P., Aracil, R.: Vision system for on-line surface inspection in aluminum casting process. In: International Conference on Industrial Electronics, Control, and Instrumentation, Proceedings of the IECON 1993, vol. 3, pp. 1854–1859 (1993)

Design of the Traffic Sign Recognition System
Based on Android Platform

Jie Qiang, Shujing Wang[✉], and Zhenhua Shan

Shanghai University, Shanghai 200072, China
wshujing@shu.edu.cn

Abstract. An algorithm based on HOG (Histograms of Oriented Gradients) and SVM (Support Vector Machine) is developed for traffic sign recognition on Android platform, and the dynamics link library is used as the native layer of Android end by employing Android NDK (Native Development Kit) technology. The test results show that the algorithm can be successfully applied to the Android platform, and Android NDK technology can implement the cross-platform and portability of the programs, while improving the detection and recognition speed.

Keywords: Traffic sign · Android NDK · Mobile devices · Cross-platform

1 Introduction

The interest in intelligent transportation mainly arises from the increasingly prominent contradiction between the existing road traffic facilities and the increasing traffic demand. And road traffic sign recognition, as an important part of intelligent transportation, has been widely studied [1]. However, the existing research on the algorithm and application of road traffic signs recognition is mostly based on PC platform [2], and the research based on mobile device has higher practical application value.

In addition, traffic sign recognition and detection in complex environment still needs further development and perfection. And HOG (Histograms of Oriented Gradients) has been validated as a highly effective feature extraction in the field of target recognition [3].

In this paper, in view of the lack of the research on road traffic sign recognition based on the mobile device, an algorithm based on HOG and SVM (Support Vector Machine) [4] was developed for road traffic sign recognition, and then innovatively applied to the android platform.

2 System Software Design

2.1 System Software Frame

The System software, based on modularized design pattern, consists mainly of image acquisition module, image preprocessing module, HOG feature extraction module, SVM feature recognition module and result display module. The system software architecture is shown in Fig. 1.

© Springer Nature Singapore Pte Ltd. 2017
M. Fei et al. (Eds.): LSMS/ICSEE 2017, Part I, CCIS 761, pp. 576–584, 2017.
DOI: 10.1007/978-981-10-6370-1_57

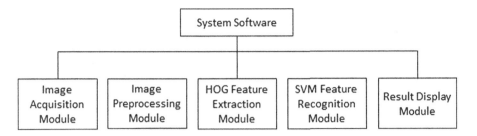

Fig. 1. Software architecture diagram

2.2 System Software Design Scheme

There are mainly two design schemes for the traffic signs recognition based on the Android platform:

(1) Call the OpenCV for Android interface for image processing and recognition.
(2) Based on PC, call the OpenCV interface for image processing and recognition, and then use the resultant C++ code to generate dynamic link library (i.e., 'so' library), finally employ the NDK Android technology to call dynamic link library for image processing and recognition.

In this paper, the second scheme was chosen because of its cross-platform performance and portability. However, there is a serious problem during the use of OpenCV vision library for traffic signs, i.e., training detector takes too long time [5]. In order to solve this problem, we employed the OpenCV combined with Dlib for traffic sign recognition. As an open source library for the target identification which is frequently used abroad but rarely used in China, Dlib has huge advantage in HOG feature extraction and SVM image recognition [6].

3   Traffic Signs Recognition Based on the Android Platform

3.1   Working Process of the Traffic Sign Recognition Based on the Android Platform

The system work process of the traffic sign recognition based on the Android platform is shown in Fig. 2. First of all, select China road traffic sign sample library as the training sample library, extract the features of all the traffic signs in samples library, and train the traffic sign detector for future image recognition. Then, on the basis of getting the traffic sign detector, select any one of images in the mobile phone system, complete the image preprocessing, extract the image features, and finally employ the trained detector to identify whether the selected image has the feature region satisfying the criteria. If the selected image has the feature region satisfying the criteria, its SVM category can be determined and further the result can be obtained. Otherwise, stop the traffic sign recognition.

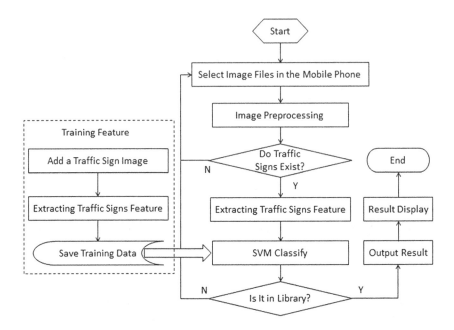

Fig. 2. System operation flow chart

3.2   Image Preprocessing

Appropriate image preprocessing is an important premise for accurate traffic sign recognition.

As the preparation for traffic sign recognition, the image preprocessing of the selected image was carried out, including grey-scale processing, Gamma correction, de-noise processing, illumination compensation, and so on [7].

(1) Grey-scale Processing: Zoom the image and convert it into gray image in the purpose of better improving computational efficiency.
(2) Gamma Correction: In the case that the gray image is too dark, Gamma correction can be used to brighten the whole image, and vice versa [8].
(3) De-noise processing: Noise is a great interference factor during obtaining traffic sign image. In this paper, the median filtering method was employed to eliminate the influence of random noise on image [9].
(4) Illumination Compensation: After histogram equalization processing, the illumination distribution of the traffic sign image is more even, compensating the illumination quality. The rendering of using Gamma correction for lighting compensation is shown in Fig. 3.

Fig. 3. Comparison diagram before and after gamma correction

3.3 Traffic Sign Feature Extraction

Selection of Feature Extraction Algorithm. The algorithm for describing characteristics is divided into three categories: the algorithm describing the color characteristics, the algorithm describing the texture characteristics and the algorithm describing the shape characteristics. This paper compares the four common feature description operators: SIFT feature, haar-like feature, LBP feature and HOG feature.

The SIFT operator requires strong hardware support; The haar-like operator is susceptible to light and target movement, it is applied to static target detection; The rotation invariance of the LBP operator causes it to lose its orientation information, so the LBP operator can't be used on many occasions; Compared with other operators, the HOG operator maintains a good invariance for image geometry and optical deformation, and also has good characteristics fault tolerance. So the HOG operator is chosen for feature extraction.

HOG Feature Extraction Algorithm. HOG feature is a feature descriptor used for object detection in computer vision and image processing. It is obtained through calculating and counting the gradient orientation histogram of the image local area [10].

In the 20×20 window, for instance, the steps of extracting HOG feature are shown below:

(1) The window is divided into small squares of 5×5 size which is called 'cell'. And four adjacent cells constitute a large square which is called 'block'.
(2) Calculate the horizontal gradient H(x, y) and vertical gradient V(x, y) of each pixel by gradient operator.
(3) Calculate the gradient direction $\theta(x, y)$ and the amplitude m(x, y) of each pixel by:

$$\theta(x,y) = \tan^{-1}\left[V(x,y)/H(x,y)\right] \tag{1}$$

$$m(x,y) = \left[H(x,y)^2 + V(x,y)^2\right]^{1/2} \tag{2}$$

(4) The value range of $\theta(x, y)$ is from $-90°$ to $90°$. We divide it into 9 equal portions. Then count each cell in the 9 gradient directions, and the weight is m(x, y). Thus

each cell has a 9-dimensional vector. And, a block consisted of 4 adjacent cells has a 36-dimensional vector. Further, the HOG feature vector of the window can be obtained by combining all the vectors.

Classifier Selection. The common classification algorithms are k-Means, Adaboost, and Support Vector Machines.

The k-Means algorithm needs to select the appropriate K values and determine the initial partition, and it takes a lot of time to calculate the large amount of data; Adaboost algorithm can generate strong classifier through continually integrating the training set of sample data, iterating unceasingly, optimizing itself unceasingly, it has no obvious advantages and disadvantages; SVM presents unique advantages in high dimensional pattern recognition, solving problems on small sample and nonlinear aspects. So I'm going to use SVM here.

Train the Detector. The steps of feature extraction by Dlib library are shown below: Provide a few training samples with similar or proportional traffic signs; Prepare some test samples with or without traffic signs Do calibration to obtain the rectangular box with the sign in it for every sample.; Put the created xml file into the object tracking trainer of Dlib; Set the object tracking parameters, including the paths of the training samples and testing samples); Start training; Zoom the image by image pyramid, then add the trainer as the training sample (In general, the scale of image pyramid will be set as 1–3 times). Test samples, pass the training when the pass rate reaches the set value; Then, the detector can be used after the training. The flow chart of training detector by Dlib is shown in Fig. 4.

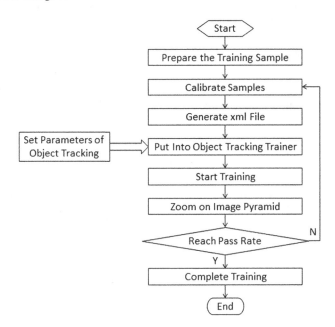

Fig. 4. Flow chart of training detector by Dlib

3.4 Traffic Sign Recognition

The findSign () function should be called when using Dlib library for image recognition. It is a self-defining function and its core method is provided by Dlib library. The function is as follows:

```
Void evaluate_detectors (
  const std::vector<object_detector<scan_fhog_pyramid<
  pyramid_type>>>& detectors,
  constimage_type&img,
  std::vector<rect_detection>&dets,
  const double adjust_threshold = 0)
```

The parameters of the above function include trained detector, image for identifying, location of traffic signs area.

This function can identify the type of the inputted traffic sign by employing trained detector, and then determine the final recognition result.

3.5 Script File Writing and Link Library Generation

The Native code in Android programs must be used by the link library, rather than the Java code. And the link library has two kinds: One is dynamic link library, its name is generally in the form of 'lib**.so'; The other one is static link library, its name is generally in the form of 'lib**.a'. The file compiled by static link library is commonly big, because all the data of the whole function library will be integrated into the target code during the process of compiling through static link library, while that will not be integrated into the target code for the dynamic link library.

```
static{System.loadLibrary("libdlib_android")}
```

This is the standard program usage of loading the dynamic link library. The library's name does not end with '.so', and the program will automatically search for corresponding library in the directory of 'lib' folder.

4 Experimental Verification

Before this section, the practical application of traffic sign recognition system based on the Android platform has been described in detail. In order to verify the validity and reliability of the algorithm, experimental tests are taken based on images under different situations in this section.

The tests in this section are based on MI - 3 mobile phone, the phone's specific parameters are as follows:

```
Version: Android OS 4.4.2
Processor: NVIDIA Tegra4
Resolution: 1920×1080
RAM: 3GB
```

The tests are taken for investigating the influence of the image size, the types of traffic signs, and shooting Angle on recognition results, and providing basis for further research.

Firstly, we tested the images of different sizes in the same scene.

From Table 1, the recognition results are affected hardly by image size, and the recognition time increases almost linearly with the image size.

Table 1. Influence of image size on recognition result (ms)

Test result	1536 kb	880 kb	544 kb	80 kb
Recognition result	Correct	Correct	Correct	Correct
Recognition time	1511	805	453	221

Secondly, we tested the images with different kinds of traffic signs.

From Table 2, the recognition results are affected hardly by the type of traffic signs, it is only affected by the reliability of the detector. More detector training samples are collected, more accuracy the recognition result is.

Table 2. Influence of the type of traffic sign on recognition result (ms)

Test result	No left turn	Danger!	Go straight	Go slow
Recognition results	Correct	Correct	Correct	Correct
Recognition time	484	431	410	399

Thirdly, we tested the images taken from different shooting angles.

From Table 3, the system can't recognize the traffic signs when the shooting angle is greater than a certain angle. The main reason is that the angle of the training sample is chosen within 60°, and the image whose angle is greater than 60 is not recognized because of the severe deformation of the traffic sign.

Table 3. Influence of shooting angle on recognition result (ms)

Test result	0°	15°	60°	79°
Recognition results	Correct	Correct	Correct	Incorrect
Recognition time	471	462	493	327

Finally, we tested 327 pictures containing the traffic signs in natural scene, and all the traffic signs in the images are trained in training detector. The test results show that recognition error occurred only for three of the images because of too fuzzy shooting. Therefore, the recognition accuracy of traffic signs is calculated as 324/327 = 99.083%, and the expectation of the system design is achieved.

The designed interface is shown in Fig. 5, which select image in the system gallery, and display the recognition result. When image with traffic sign is detected, this design will display the recognition result and recognition time.

Fig. 5. Human-computer interaction interface

5   Conclusion

In view of the lack of the research on road traffic sign recognition based on the mobile devices, a traffic sign recognition was innovatively designed based on android platform. And an algorithm based on HOG and SVM was developed for traffic signal feature extraction and recognition, obtaining accurate recognition results. In actual industrial validation process, huge image size, great shooting angle, and overlay of multiple images bring challenges to the road traffic sign recognition, which is the research focus of next step.

References

1. Zhu, S.D., Lu, X.F.: A survey of the research on traffic sign recognition. Comput. Eng. Sci. **28**(12), 50–52 (2006)
2. Tang, H.H.: Research on Traffic Sign Recognition Algorithm. Beijing Jiaotong University, Beijing (2014)
3. Mu, C.L.: The Research of Face Recognition System Based on Hog Feature. University of Electronic Science and Technology of China, Chengdu (2013)

4. Xu, C., Gao, M.Z., Cha, Y.F., Cao, L.M.: Bus passenger flow calculation algorithm based on HOG and SVM. Chin. J. Sci. Instr. **36**(2), 446–452 (2015)
5. Liu, Z.Y., Jiang, C.Z.: People number detection of image in fixed place with OpenCV and Haar-like classifier. J. Univ. Sci. Technol. Liaoning **34**(4), 384–388 (2011)
6. Xu, Y., Xu, X.L., Li, C.H., Jiang, M., Zhang, J.G.: Pedestrian detection combining with SVM classifier and HOG feature extraction. Comput. Eng. **42**(10), 56–65 (2016)
7. Lu, Y.J.: The Research on Traffic Sign Automatic Detection and Recognition Algorithm. Wuhan University of Science and Technology, Wuhan (2015)
8. Shao, G.Z.: Study on the Algorithm of Road Traffic Sign Recognition. Jilin University, Changchun (2008)
9. Gao, Z.X.: The Research on Detection of Motion Blurred Traffic Sign. Zhejiang University, Hangzhou (2012)
10. Guo, J.X., Chen, W.: Face recognition based on HOG multi-feature fusion and random forest. Comput. Sci. **40**(10), 279–283 (2013)

Apical Growing Points Segmentation by Using RGB-D Data

Pengwei Liu[1], Xin Li[1,2(✉)], and Qiang Zhou[3]

[1] School of Mechatronics Engineering and Automation, Shanghai University,
Shanghai 200072, China
su_xinli@aliyun.com

[2] Shanghai Science and Technology Commission Project: Development of Intelligent Inspection
Robot and Expert System for Greenhouse Crop Growth (14DZ1206302), Shanghai, China

[3] Shanghai Dushi Green Co., Ltd., Shanghai, China

Abstract. Generally, plant grows slowly and is difficult to be observed, the apical growing points can reflect the changes of plant, such that the extraction of apical growing points is helpful for the analysis of plant growth. In this paper, a new digital visual-based method of tomato apical growing points segmentation is proposed, which is depended on depth segmentation, color segmentation and position histogram statistic. First of all, use the depth image captured by KinectV2 to remove complex background through depth segmentation. Then, position histogram of the two value image after depth segmentation has been obtained to get the column position of the apical growing points. Using the KinectV2 coordinate mapping mechanism to restore the color information of the two value image, and then the RBG-D image can be color segmented. Finally, the region of the apical growing points is segmented by coordinate mapping, and the apical growing point is extracted by the contour detection. The experimental results show that the method to segment the growth environment is effective.

Keywords: Apical growing points · Image segmentation · Position histogram statistic · Deep segmentation

1 Introduction

In recent years, image segmentation has been widely used in agriculture. Slaughter et al. considered image processing techniques for detection and discrimination of plants and weeds in some detail [1]. Besides identification of weeds to permit precision weeding, plant segmentation is also useful for other proposes, and applied in several applications such as plant species recognition [2], growing phase determination [3], and plant disease detection [4]. Bai et al. introduced a new method for crop segmentation based on the CIE (Commission Internationale de L'Eclairage) L*a*b* color space, using morphological modeling. The study was conducted in a rice paddy field in China, and the images were taken under various conditions for different growth status of rice plant [5].

Crop growth refers to the development of crops and its change trend. The real-time dynamic monitoring of crop condition can understand the distribution and the growth

© Springer Nature Singapore Pte Ltd. 2017
M. Fei et al. (Eds.): LSMS/ICSEE 2017, Part I, CCIS 761, pp. 585–596, 2017.
DOI: 10.1007/978-981-10-6370-1_58

tendency of crops. And then develop reasonable crop cultivation measures, co-ordinate the operation of water and fertilizer, optimize the breeding program, establish a high-yield, efficient crop production system. Generally, plant grows slowly and is different to be observed, the apical growing points can reflect the changes of plant, such that the extraction of apical growing points is helpful for the analysis of plant growth. In the intelligent agricultural greenhouses, advanced temperature control strategy can bring desired tomato outputs, and the apical growing points reflect the influence of strategy. Traditional method for measuring the characteristics of apical growing points is the artificial way. However, the artificial way has destructiveness, hysteresis, weak effectiveness and high cost. Besides, it requires much manpower and material resources. Compared to it, for the machine vision method with real-time monitoring and high-precision, much manpower and material resources are saved, simultaneously, the state of plant growth can be accurately obtained for signal plant and crop colony within an effective measurement range. However, crop image segmentation is complicated by the fact that crop color varies with the illumination condition in the outdoor field, the sharp of the crop and the selected imaging device [6–8]. Kataoka et al. proposed the CIVE segmentation method which takes values of R, G and B channels of each pixel in the crop image into consideration. CIVE method also adopts the Ostu algorithm to distinguish the crop and the background. However, OTSU or manually appointed threshold based segmentation method cannot bring stable segmentation results under different illumination conditions [9]. George et al. proposed the ExG-ExR (Super Green Super red) color indicators, which can extract the green plant from the growth background, but it is easy to be affected by light [10]. An adaptive threshold algorithm is proposed by Wang et al. The algorithm can be used to separate the single blade from the randomly selected image of the blade group. But the method is complex and slow [11]. Xiang et al. used binocular stereo vision technology to identify tomatoes. But the accuracy is not high. And tomatoes can't be blocked by leaves [12]. Dionisio et al. used the depth image collected by KinectV2 to separate the corn from the land and weeds, and then use the RGB color to distinguish the land from the weeds. However, the plant populations were separated, and the plant was not separated [13]. Efi et al. used the adaptive threshold method to segment the red sweet pepper based on the fusion of depth image and color image. The correct recognition rate reaches 90.9% under natural light. [14] RGBD-based segmentation has attracted much interest because of the wide availability of affordable RGBD sensors [15–17]. Mengyao Zhao et al. proposed a novel temporal-coherent approach for real-time foreground object extraction from RGBD videos [18]. Huazhu Fu et al. present an RGB and Depth (RGBD) video segmentation method that takes advantage of depth data and can extract multiple foregrounds in the scene [19]. Xiang Rong et al. used the depth image based on binocular stereo vision to improve the applicability of the recognition method for clustered tomatoes, and the recognition accuracy rate of clustered tomatoes was 87.9% when the leaf or branch occlusion rate was less than 25% [12].

Due to the greenhouse tomato grow in complex environment, and picture taken by KinectV2 are affected by natural light, only through the color information is not easy to extract the apical growing points of tomato. In this paper, we use the threshold segmentation in depth image to get the image of the front tomato, and then use the color information to remove the influence of the hanging line. Because we need to segment the apical growing points of Tomato, using histogram statistics to find the apical growing points column position of the tomato in the image, which after the depth segmentation and color segmentation. The apical growing point is at the highest point of the plant. Use this location information to find the row position of the apical growing points. Then the whole apical growing points region is segmented. After finding the apical growing points, the parameters of the apical growing points can be calculated.

The paper is organized as follows: Sect. 2 talks about the detail section of the segmentation algorithm discussed in this paper. Section 3 describes the results of the experiments done. Section 4 provides a conclusion and outlook on future works.

2 Materials and Methods

2.1 Data Collection System

Picture were acquired from the Kinectv2 sensor running Software Development Kit (SDK) on an IPC (Industrial Personal Computer) with Windows 8 (Microsoft, Redmond, WA, USA). The data acquisition software was developed starting from the official Microsoft SDK 2.0 (Microsoft, Redmond, WA, USA)), which includes the drivers for the Kinect sensor. It allows the necessary drivers for the sensor to be obtained, as well as the use of a set of customizable sample functions that were implemented for the measurements, combined with some OpenCV functions. The aim was to store both depth

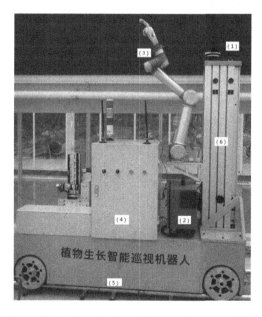

Fig. 1. Practicality picture of agricultural robot. (1) Kinectv2 sensor; (2) IPC (Industrial Personal Computer); (3) robotic arm system; (4) the central controller; (5) drive mechanism; (6) the lifting and lowering device

and RGB information at a rate of at least 25 frames per second. The experimental platform used in this paper is the visual system of agricultural robot (Fig. 1). Agricultural inspection robot was developed by Shanghai University, which is running in agricultural greenhouses located at National Engineering Research Center for Facility Agriculture, Chongming island, Shanghai.

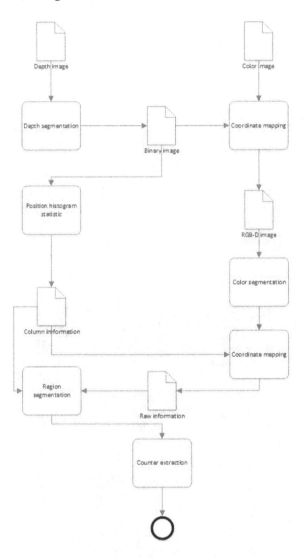

Fig. 2. The flowchart of the picture segmentation method for apical growing points

2.2 Picture Segmentation Method for Apical Growing Points

The flowchart of the picture segmentation method for apical growing points is shown in Fig. 2.

Depth Segmentation
The purpose of deep segmentation is to eliminate the background using depth information. The depth segmentation can preserve the target plant information and eliminate the influence of complex background on segmentation. A detailed description of the depth segmentation is shown in Eq. (1).

$$P_d = \begin{cases} 0, & d > d_{\max} \ or \ d < d_{\min} \\ 255, & d_{\min} < d < d_{\max} \end{cases} \tag{1}$$

Where P_d is a pixel value in two value image; $[d_{\min}, d_{\max}]$ is the depth range of the target tomato plant.

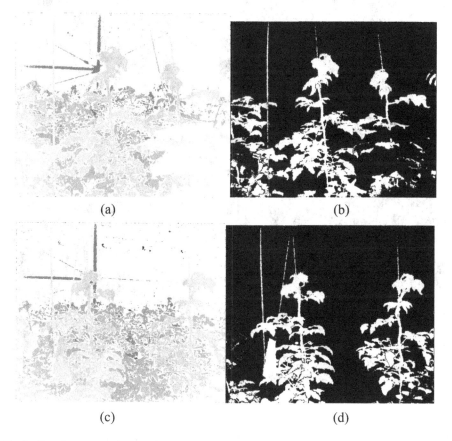

(a)

(b)

(c)

(d)

Fig. 3. Depth segmentation results. (a) Depth image of sample 1; (b) Depth segmentation result for example 1; (c) Depth image of sample 2; (d) Depth segmentation result for example 2

The vision system used in this paper is included in the intelligent agricultural inspection robot. Intelligent agricultural inspection robot is running in track between the rows of Tomato in the greenhouse. The distance of KinectV2 from the target tomato plant is within the range of 0.5–1.5 m. After depth segmentation, the target tomato can be segmented from complex background. Figure 3 shows the depth segmentation results produced from our method.

Position Histogram Statistic
Based on the idea of histogram statistics and combining the characteristics of apical growing points, this paper presents a new method of position histogram segmentation. There are two main characteristics of the apical growing points, the apical growing point is the highest point in the image, and the apical growing point is arranged in the image. The position histograms take the abscissa of the image as abscissa, the number of

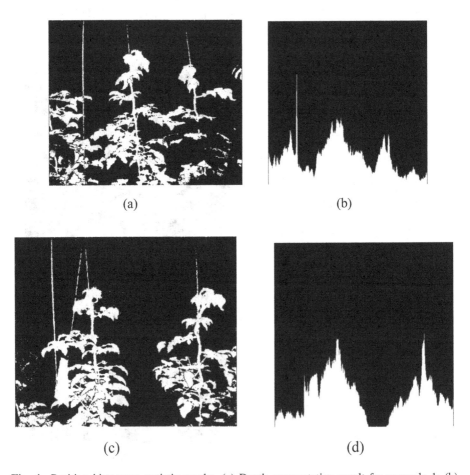

(a)   (b)

(c)   (d)

Fig. 4. Position histogram statistic results. (a) Depth segmentation result for example 1; (b) Position histogram statistic result for example 1; (c) Depth segmentation result for example 2; (d) Position histogram statistic result for example 2

nonzero pixel values on the abscissa position of the image as ordinate. Vector form is shown in Eq. (2).

$$V = \{(h[c_1], h[c_2], \cdots, h[c_W]) | 0 \le h[c_k] \le H, k = 0, 1, \cdots, W\} \tag{2}$$

Where W is the width of the image; H is the height of the image; $h[c_k]$ is the number of nonzero pixels on the image k column. The detailed description is shown in Eqs. (3) and (4).

$$h[c_k] = \sum_{i=0}^{H} \delta(I(i, k)) \tag{3}$$

$$\delta(x) = \begin{cases} 1, & \text{if } x > 0 \\ 0, & \text{if } x = 0 \end{cases} \tag{4}$$

Where $I(i, j)$ is pixel value at (i, j) position in the image.

The result of the position histogram of the two value image after depth segmentation is shown in the Fig. 4.

The purpose of position histogram statistics is to find the location of the column where the apical growing point is located. The results of position histogram statistical has the following characteristics: the column of the apical growing points is the maximum and the second largest value of the statistical results, and all nonzero pixels in the region of the apical growing points exceed a certain threshold value. Therefore, we can have enough information to construct the apical growing points. The detailed description is shown in Eqs. (5) and (6).

$$h[c_m] = \max(h[c_1], h[c_2], \cdots, h[c_W]) \tag{5}$$

$$N = \sum_{i=m-d}^{m+d} h[c_i] \tag{6}$$

Where m is the column where the apical growing points is located; d is the half span of apical growing points; N is the information threshold of apical growing point.

The algorithm for finding the location of tomato apical growing points by the method of position histogram statistic is shown in the Fig. 5.

The experimental result of Fig. 4(b): $m = 234$; the experimental result of Fig. 4(d): $m = 427$. The result value is the column position of the apical growing points after check the value of the result in the two value image. The best results were obtained at d = 37, N = 10000 through many tests.

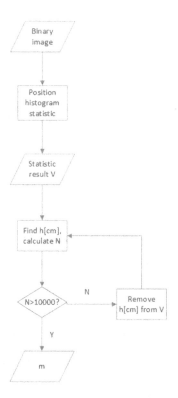

Fig. 5. The method of position histogram statistic

(a)   (b)

Fig. 6. Results of color segmentation. (a) Color segmentation result for example 1; (b) Color segmentation result for example 2 (Color figure online)

Color Segmentation

Through the coordinate mapping mechanism of KinectV2 SDK, the color information is imported into the depth image to create the RGBD image. There is a relationship between the pixels in the green object surface as shown in Eq. (7).

$$\left(g_{value} > b_{value}\right) \cap \left(g_{value} > r_{value}\right) \tag{7}$$

Where r_{value}, g_{value}, b_{value} is the gray value of RGB three channel.

Traversing the whole RGBD image, use the Eq. (6) to complete the color segmentation. Specific Results of color segmentation are shown in Fig. 6.

Find the Row Position of Apical Growing Points

It is easy to obtain the top row position of the apical growing points region by using the apical growing points position information obtained from the Sect. 2.2. Going through column position of the apical growing points from the top to the bottom in the RGBD image, the point where top row position of apical growing points region located is the

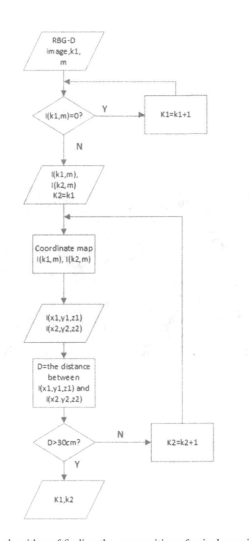

Fig. 7. The algorithm of finding the row position of apical growing points

first not zero pixel value. The useful area of the apical growing points is the area between the top of the apical growing points and 30 cm from the top of the apical growing points. Therefore, it is necessary to find the location that far from the top of apical growing points 30 cm. The coordinates of each point in the depth image in the camera coordinate system can be obtained by the coordinate mapping mechanism of KinectV2 SDK. Traversing the apical growing points from top to bottom, the location which is 30 cm from the top of apical growing points can be found by using Euclidean distance. This position is the bottom of the apical growing points. The region between the top row of apical growing points and the bottom row of apical growing points is where the apical growing points located. The algorithm of finding the row position of apical growing points is shown in Fig. 7.

3   Result and Discussion

Using the row position of apical growing points r_{top} and r_{down} obtained from the Sect. 2.2, RBGD images can be segmentation again. All parts of the image that are not in this area are set to the background color. Then extract all the contours in the image. Finally, find the minimum external rectangle for each contour. The rectangular region where the apical growing point is located contains the column m obtained in Sect. 2.2. According to this information, the final segmentation result is obtained. The segmentation results are shown in Fig. 8.

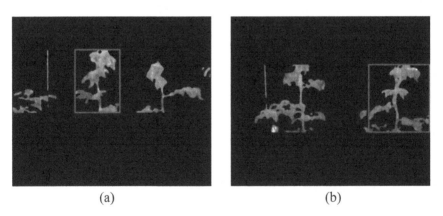

(a)   (b)

Fig. 8. Segmentation results. (a) Segmentation result for example 1; (b) Segmentation result for example 2.

It can be seen from the results that the complex background can be removed from the picture. And then the area of apical growing points can be reserved after segmentation. The key point of the algorithm is that the depth information can be used to remove the complex background behind the target plant. Then a single target plant is segmented by the position histogram.

4   Conclusions

In this paper, we make full use of the depth information, location information and color information of tomato plant, which can realize the segmentation of apical growing points. The experimental results show that the proposed algorithm is reliable and robust. This algorithm is easy to implement and runs fast on the computer. In this paper, using the idea of histogram statistics combined with the location characteristics of apical growing points, it is easy to separate the single apical growing points of Tomato. The extraction of single apical growing points can be used to measure the characteristics of apical growing points. In this paper, we use the original data collected by KinectV2 for depth segmentation. However, the noise in the original depth data, will have some impact on the depth of segmentation. Therefore, the problem that needs to be solved is to filter the depth image collected by KinectV2.

References

1. Slaughter, D.C., Giles, D.K., Downey, D.: Autonomous robotic weed control systems: a review. Comput. Electron. Agric. **61**, 63–78 (2008)
2. Zhang, L., Kong, J., Zeng, X., Ren, J.: Plant species identification based on neural network. In: IEEE Fourth International Conference on Natural Computation, vol. 5, pp. 90–94 (2008)
3. Kataoka, T., Kaneko, T., Okamoto, H., Hata, S.: Crop growth estimation system using machine vision. In: Proceedings of the IEEE/ASME International Conference on Advanced Intelligent Mechatronics (AIM 2003), vol. 2, pp. 1079–1083. (2003)
4. Camargo, A., Smith, J.S.: An image-processing based algorithm to automatically identify plant disease visual symptoms. Autom. Emerg. Technol. **102**, 9–21 (2009)
5. Bai, X.D., Cao, Z.G., Wang, Y., Yu, Z.H., Zhang, X.F., Li, C.N.: Crop segmentation from images by morphology modeling in the CIE L*a*b* color space. Comput. Electron. Agric. **99**(12), 21–34 (2013)
6. Jones, M.J., Rehg, J.M.: Statistical color models with application to skin detection. Comput. Vis. **46**, 81–96 (2002)
7. Kim, C., You, B.-J., Jeong, M.-H., Kim, H.: Color segmentation robust to brightness variations by using B-spline curve modeling. Pattern Recogn. **41**, 22–37 (2008)
8. Onyango, C.M., Marchant, J.A.: Physics-based color image segmentation for scenes containing vegetation and soil. Image Vis. Comput. **19**, 523–538 (2001)
9. Kataoka, T., Kaneko, T., Okamoto, H., Hata, S.: Crop growth estimation system using machine vision. In: Proceedings of the IEEE/ASME International Conference on Advanced Intelligent Mechatronics (AIM 2003), vol. 2, pp. b1079–b1083 (2003)
10. Mayer, G.E., Neto, G.C.: Verification of color vegetation indices for automated crop imaging applications. Comput. Electron. Agric. **63**(2), 282–293 (2008)
11. Wang, J., He, J., Han, J., et al.: An adaptive thresholding algorithm of field leaf image. Comput. Electron. Agric. **96**, 23–29 (2013)
12. Xiang, R., Jiang, H., Ying, Y.: Recognition of clustered tomatoes based on binocular stereo vision. Comput. Electron. Agric. **106**, 75–90 (2014)
13. Dionisio, A., José, D., César, F.Q., Angela, R.: An approach to the use of depth cameras for weed volume estimation. Sensors **16**(7) (2016)
14. Vitzrabin, E., Edan, Y.: Adaptive thresholding with fusion using a RGBD sensor for red sweet-pepper detection. Biosys. Eng. **146**, 45–56 (2016)

15. Silberman, N., Hoiem, D., Kohli, P., Fergus, R.: Indoor segmentation and support inference from RGBD images. In: Proceedings of the European Conference on Computer Vision, pp. 746–760, October 2012

16. Gupta, S., Girshick, R., Arbeláez, P., Malik, J.: Learning rich features from RGB-D images for object detection and segmentation. In: Proceedings of the European Conference on Computer Vision, pp. 345–360, September 2014

17. Banica, D., Sminchisescu, C.: Second-order constrained parametric proposals and sequential search-based structured prediction for semantic segmentation in RGB-D images. In: Proceedings of the IEEE Conference on Computer Vision and Pattern Recognition (CVPR), pp. 3517–3526, June 2015

18. Zhao, M., Fu, C.-W., Cai, J., Cham, T.-J.: Real-Time and temporal-coherent foreground extraction with commodity RGBD camera. IEEE J. Sel. Top. Sig. Process. 9(4), 449–461 (2015)

19. Huazhu, F., Dong, X.: Object-Based multiple foreground segmentation in RGBD video. IEEE Trans. Image Process. 26(3), 1418–1427 (2017)

Towards Visual Human Tracking of Quadcopter: A Survey

Ling Chen, Xinxing Pan, Aolei Yang[(✉)], and Yulin Xu

Shanghai Key Laboratory of Power Station Automation Technology, School of Mechatronic Engineering and Automation, Shanghai University, Shanghai 200072, China
{lcheno,xuyulin}@shu.edu.cn, pxxjack@qq.com, aolei_yang@163.com

Abstract. In recent years, visual human tracking of quadcopter has become a topic of interest to many research institutions. To overview the recent research status of visual human tracking based on quadcopter, firstly, the problem of human tracking is divided into quadcopter control and vision based human tracking which are discussed separately. The present controlling means and the latest applications of quadcopter are summarized systematically. The advantages and disadvantages of each human tracking method are compared and the tracking strategies are summarized. Then, the difficult issues on visual human tracking are discussed specifically. Finally, the future research directions of visual human tracking based on quadcopter are prospected by summarizing related literatures.

Keywords: Quadcopter · Computer vision · Human tracking

1 Introduction

An unmanned aerial vehicle (UAV) is an aircraft without a human pilot abroad using radio remote control or built-in control programs. There are several common UAV such as fixed-wing UAV [1], vertical take-off and landing UAV [2], unmanned helicopter [3] and multi-rotor UAV [4]. In the past decade a four-rotor UAV, also called quadcopter, is especially popular [5]. Meanwhile, with the development of computer vision and the progress of sensor technology, UAV human tracking has become a hot research topic.

Four-rotor unmanned aerial vehicles have been widely concerned, mainly because of the following advantages:

- Excellent maneuverability [6]. Quadcopter is suitable for performing specific tasks in complex environments. Compared with fixed-wing UAV, four-rotor UAV can travel freely in the narrow space, meanwhile track and monitor the observation target.

This work was supported by Natural Science Foundation of China (61403244), Science and Technology Commission of Shanghai Municipality under "Shanghai Sailing Pro-gram" (14YF1408600, 16YF1403700), Key Project of Science and Technology Com-mission of Shanghai Municipality (15411953502), Shanghai University Youth Teacher Training Assistance Scheme (ZZSD15088).

© Springer Nature Singapore Pte Ltd. 2017
M. Fei et al. (Eds.): LSMS/ICSEE 2017, Part I, CCIS 761, pp. 597–606, 2017.
DOI: 10.1007/978-981-10-6370-1_59

– Compact structure and simple control. Because the use of four compact motor and fan, quadcopter mechanical structure is relative simple. In addition, compared with single-rotor UAV, the lift control of four rotors can better realize the precise position function.

– Low noise and good safety. The rotors employed by a quadcopter are smaller than a single-rotor UAV, which makes the flight safer and reduces the noise.

With the improvement of people's quality of life, the research topic about human motion object tracking [7–11] is popular in recent years. At present, the mature tracking algorithm mainly includes computer vision tracking strategies [12] and laser sensor based tracking methods [13].

Vision based tracking refers to obtain motion parameters of human moving objects, such as position, velocity, acceleration and trajectory, by the detection, extraction, identification and tracking of human motion objects in image sequence, to carry out the processing and analysis for the next step, achieving the comprehension of human moving objects, to finally complete the higher-level detection task.

The laser sensor based tracking is to scan a human motion object, and the distance and orientation information of the object can be obtained according to the laser signal reflected. Compared with visual tracking, laser sensor based tracking method is almost unaffected by ambient illumination, but its accuracy is relatively low.

The purpose of this paper is to introduce a human tracking method based on the quadcopter operation platform. As laser sensors or depth cameras are so heavy that a four-rotor UAV is not able to withstand the weight of them, only the monocular camera based visual tracking algorithm is discussed.

The remainder of the paper is organized as follows. Section 2 describes the UAV operation platform based on a Parrot AR. drone 2.0. Section 3 introduces two common methods of motion object detection. Several representation approaches of tracking object are listed in Sect. 4. Human tracking algorithms, technical difficulties and develop direction are discussed in Sects. 5 and 6 respectively.

2   UAV Operation Platform

Unlike ground robots, the navigation of a flying robot is more challenging, as a flight robot needs feedback control to keep itself stable. The ability to automatically track the human motion object is very meaningful because this ability can be applied directly to people's daily life. For example, there have been some commercial UAVs used to aerial photography for individual users, such as DJI-Phantom 4 Pro, Parrot BEBOP 2 FPV. So far, many viable projects have validated that visual servo controls can operate well in some relatively spacious unknown outdoors environments.

Recent studies show that AR. drone is a reliable vertical take-off and landing (VTOL) UAV operation platform based on visual navigation algorithm prototypes [14, 15]. For instance, AR. drone has been applied to the following researches: autonomous navigation in indoor environments [16], a monocular SLAM system [17] and autonomous navigation in natural forest environments [18].

As shown in Fig. 1, Parrot AR. drone 2.0 and its body reference frame (left). X_m and Y_m points towards the front and the right of the drone respectively, Z_m points towards ground, satisfying the orthonormal right-handed reference frame. The pose of drone is defined by Euler angles, $\{\phi, \text{roll}\}$, $\{\theta, \text{pitch}\}$, $\{\psi, \text{yaw}\}$. While performing a visual tracking task, the drone always tracks the object from a constant distance (right). The relative position of the object can be estimated by comparing image feedback with an expected size value of the object. $\psi_{telemref}$ specifies the preferred relative tracking direction [19].

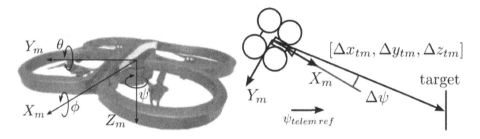

Fig. 1. (left) Parrot AR. drone 2.0 and its body reference frame. (right) a diagram of drone tracking [19].

3 Motion Object Detection

Human detection is the first step to realize automatic human tracking, the purpose of which is to extract the moving human region from the background in the image sequence. The common motion object detection methods are also used to detect human motion objects.

3.1 Background Difference Method

Background difference is a method of detecting object region by the "difference" between current image and background image, which is often used in video surveillance system [20]. The specific operation process is: the difference value can be calculated by subtracting the current image captured from the background image saved before, if the calculated value exceeds a certain threshold, then it can be determined that there is an object. Meanwhile the difference value also gives the basic information about the location, shape and size of the object. In addition, if multiple images are taken in a certain period, the motion state of the object can also be obtained. However, the background difference method is especially sensitive to the drastic change of external illumination, the object motion and the change of shadow position due to the wind direction and other external factors.

3.2   Frame Differential Method

Frame difference method is a method to obtain the moving object region by the difference operation of two adjacent frames in the video image sequence [21]. Because the time interval between two continuous frames is shorter, the method is not sensitive to the change of light and other scenes, and can adapt to a variety of dynamic environments with better stability. However, the objects separated by the frame difference method are often incomplete, and only the boundary can be extracted, which is apt to produce unfavorable model matching phenomenon such as internal voids and ghosting. Moreover, the selection of interval between frames is very important. For fast moving objects, a smaller interval needs to be chosen, otherwise, when the object does not overlap between the two frames, it will be recognized as two independent objects. But for slow moving objects, it is necessary to select a large time lag, otherwise the system is not able to identify the object, because the object is almost completely coincident in the two frames.

4   Representations of Tracking Object

In the tracking scene, to facilitate further analysis and improve tracking accuracy, some characteristics are often used to represent the tracking object [22].

Some common approaches for tracking object representations are shown in Fig. 2, whose details are presented as follows.

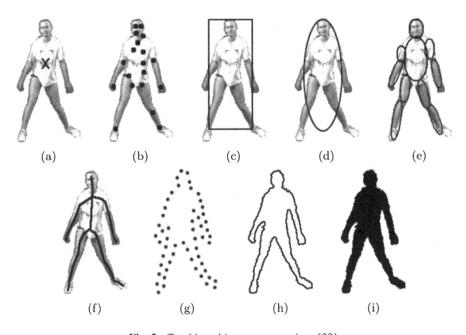

Fig. 2. Tracking object representations [22].

Points: The object is represented by a point, which can be a central point (a), or a set of points (b). In general, the point representation is suitable for tracking objects occupying smaller regions in an image.

Primitive geometric shapes: The object shape is replaced by a rectangle (c), an ellipse (d), and so on. Although this method is usually suitable for representing a simple rigid object, it can also be used for tracking non-rigid objects.

Object silhouette and contour: The contour represents the boundary of the object (g, h), and the inner area of the contour is called the silhouette of the object (i). This method is suitable for tracking the objects with complex non-rigid shapes.

Articulated shape models: An articulated object is made up of the body parts connected by the joint. For instance, the human head, torso, arms and legs are connected by joints. To denote an object with joints, an ellipse or a cylinder can be used to represent the component between joints (e).

Skeletal models: The "skeleton" model of the objects can be extracted by the medium axis transformation of the object silhouette, which is usually used to identify the shape description of the objects and can be used to model both articulated and rigid objects (f).

5   Human Tracking Algorithms

Due to the potential application foreground and its inherent complexity, automatic capture and analysis of human motion has been a hot research field which contains many difficult problems such as fast motion and occlusions [23–28]. This complexity makes it challenging to study from a purely academic standpoint. In addition, from an application perspective, the non-invasive solution provided by the computer vision approach is very attractive [29].

5.1   Model Based Human Tracking

The basic idea of model based human tracking is to build human model based on prior knowledge. At the same time, the pose of the model in the next frame is predicted, then the result is analyzed and integrated, where some potential results are obtained. Finally, the results are compared with the actual image, to find the model in the image that best matches these results and keep the model up-to-date. In practice, this method can combine the body characteristics of the human motion and remove the model which does not conform to features, thus making the selection of tracking object more reliable. However, it is difficult to construct a precise human geometric model, which hinders the further development of the model based human tracking method.

Pose estimation is the process of estimating the basic kinematics or skeleton structure of human, which can be observed directly from each frame. The pose estimation algorithm is divided into three categories, namely, model-free, indirect model and direct model in [30]. There is no explicit a priori model when the model-free method is utilized. The indirect model uses a priori model as a reference or lookup table to interpret the measured data, and the direct model uses an explicit geometric human model to represent the human shape and motion posture.

Here the human model mainly divides into the 2D model and the 3D model, due to the limited UAV load capacity, the general machine body is equipped with only a monocular camera, therefore the human model should choose the 2D model when this tracking method is employed.

5.2 Region Based Human Tracking

This method obtains motion regions in the image by using the frame difference method, which represents the whole human or a component of the human body, and the tracking process is to locate all these regions in the image, meanwhile establishing the correspondence among them [31]. If the tracking object is the whole human, then the object region can be directly added to the tracking object after the geometric structure and area of these detected object regions are constrained. When the tracking object is a component of the human body, more accurate image information is required. Firstly, the color, shape and other features of the object are extracted and analyzed, then the next state of the object is estimated, and finally the correction is made to achieve the goal of stable tracking. The difficulty of this method is how to deal with problems about overlap errors [32] and occlusions.

5.3 Feature Based Human Tracking

The feature based human tracking method usually ignores the whole human, but chooses some outstanding characteristics of human body as tracking object, such as color [33] based human tracking. However, to ensure the reliability of tracking, the tracking features often use composite information consisting of multiple human characteristics. Furthermore, because the tracking object is the local feature of a human body, the high-resolution image is required.

5.4 Contour Based Human Tracking

This method gives a precise description of the shape of the object human, which aims to find the human region in the current frame through the human model generated by the previous frame image [34]. The form of the human model can be the color histogram, the human boundary or the human contour. The contour tracking is usually divided into two categories, namely, shape matching and contour tracking. The most important advantage of the contour based human tracking method is that it is flexible to deal with a variety of object shapes, and this method is a good choice for the irregular shape of human bodies.

5.5 Convolution Neural Network

Convolution Neural Network (CNN) is a feedforward neural network, and the biggest difference between CNNs and other neural networks is that convolution neural network adds the convolution layer as the front layer of the input layer of neural network, which makes the convolution layer become the actual data input layer.

Unlike conventional learning methods, CNNs learn both temporal and spatial characteristics of the object from image pairs of two adjacent frames [35], thus reducing misjudgment and avoiding object loss. A CNN based method of effectively detecting 2D pose in the image is proposed in [36], where the method combining multiple feature points is used to improve the accuracy of the human detection, but in most cases the accuracy is less than 80%. Moreover, the consumption of computing resources is huge and the real-time optimization is not good enough, so there is still much room for improvement.

For more intuitive presentation, Table 1 gives the performance comparison of five human tracking algorithms.

Table 1. Overview of human tracking algorithm.

Human tracking algorithms	Accuracy	Robustness	Real-time	Computation cost	Implementation difficulty
Model based human tracking	High	Good	Low	High	High
Region based human tracking	Medium	Low	Medium	Medium	Medium
Feature based human tracking	Medium	Medium	Good	Low	Low
Contour based human tracking	Medium	Medium	Medium	Medium	Medium
Convolution Neural Network	High	Good	Low	High	High

6 Technical Difficulties and Development Direction

The realization of visual human tracking based on monocular camera is not easy, and the reasons are as follows: the use of 2D images to reflect the 3D world causing some information loss, noise in the images, complex human motion, non-rigid and articulated human bodies, partial occlusion or entire occlusion, complex human shapes, scene illumination changes, the real-time requirement and so on.

In recent years, the research status of human tracking algorithms shows that a good algorithm often merges a variety of factors, such as the combination of model and features. The algorithm based on a single element has been difficult to meet the real-time and robustness of human tracking simultaneously, therefore, the future research direction of human tracking is bound to develop in the direction of multi-factor fusion.

7   Conclusion

In this paper, the problem of vision based human tracking of quadcopter is divided into two components, namely, quadcopter control and visual human tracking. A quadcopter based UAV operation platform is described and several detection and tracking methods of human motion objects are surveyed.

Due to the influence of complexity of human motion, dynamic changes of a background image and partial occlusion or entire occlusion, it becomes very difficult for conventional methods to obtain a visual human tracking algorithm with real-time, accuracy and robustness. With the appearance of the concept of Deep Learning, a new method of pose estimation based on CNNs is proposed, but this new method still has a long way to go because of its huge calculated amount and real-time problems.

Based on previous research and analysis, this paper emphatically introduces the current research status of visual human tracking in recent years, and summarizes the technical difficulties and development directions of the research for vison based human tracking, hoping to be helpful for the related scholars.

References

1. Beard, R.W., Kingston, D., Quigley, M., Snyder, D., Christiansen, R., Johnson, W., Goodrich, M.: Autonomous vehicle technologies for small fixed-wing UAVs. J. Aerosp. Comput. Inf. Commun. 2(1), 92–108 (2005)
2. Goossen, E.: U.S. Patent No. 8,328,130. U.S. Patent and Trademark Office, Washington, DC (2012)
3. Johnson, E., Kannan, S.: Adaptive flight control for an autonomous unmanned helicopter. In: AIAA Guidance, Navigation, and Control Conference and Exhibit, p. 4439, August 2002
4. Sámano, A., Castro, R., Lozano, R., Salazar, S.: Modeling and stabilization of a multi-rotor helicopter. J. Intell. Robot. Syst. 69, 161–169 (2013)
5. Luukkonen, T.: Modelling and control of quadcopter. Independent Research Project in Applied Mathematics, Espoo (2011)
6. Mellinger, D., Michael, N., Kumar, V.: Trajectory generation and control for precise aggressive maneuvers with quadrotors. Int. J. Robot. Res. 31(5), 664–674 (2012)
7. Sminchisescu, C., Triggs, B.: Kinematic jump processes for monocular 3D human tracking. In: Proceedings of the IEEE Computer Society Conference on Computer Vision and Pattern Recognition, vol. 1, p. I. IEEE, June 2003
8. Zhou, J., Hoang, J.: Real time robust human detection and tracking system. In: IEEE Computer Society Conference on Computer Vision and Pattern Recognition-Workshops, CVPR Workshops, p. 149. IEEE, June 2005
9. Khan, S., Javed, O., Rasheed, Z., Shah, M: Human tracking in multiple cameras. In: Proceedings of the Eighth IEEE International Conference on Computer Vision, ICCV, vol. 1, pp. 331–336. IEEE (2001)
10. Zhang, D., Xia, F., Yang, Z., Yao, L., Zhao, W.: Localization technologies for indoor human tracking. In: 5th International Conference on Future Information Technology (FutureTech), pp. 1–6. IEEE, May 2010
11. Ji, S., Xu, W., Yang, M., Yu, K.: 3D convolutional neural networks for human action recognition. IEEE Trans. Pattern Anal. Mach. Intell. 35(1), 221–231 (2013)

12. Wu, Y., Lim, J., Yang, M.H.: Online object tracking: a benchmark. In: Proceedings of the IEEE Conference on Computer Vision and Pattern Recognition, pp. 2411–2418 (2013)
13. Glas, D.F., Miyashita, T., Ishiguro, H., Hagita, N.: Laser tracking of human body motion using adaptive shape modeling. In: IEEE/RSJ International Conference on Intelligent Robots and Systems, IROS, pp. 602–608. IEEE, October 2007
14. Bristeau, P.J., Callou, F., Vissiere, D., Petit, N.: The navigation and control technology inside the AR. drone micro UAV. IFAC Proc. Vol. **44**(1), 1477–1484 (2011)
15. Krajník, T., Vonásek, V., Fišer, D., Faigl, J.: AR-drone as a platform for robotic research and education. In: Obdržálek, D., Gottscheber, A. (eds.) EUROBOT 2011. CCIS, vol. 161, pp. 172–186. Springer, Heidelberg (2011). doi:10.1007/978-3-642-21975-7_16
16. Bills, C., Chen, J., Saxena, A.: Autonomous MAV flight in indoor environments using single image perspective cues. In: IEEE International Conference on Robotics and Automation (ICRA), pp. 5776–5783. IEEE, May 2011
17. Engel, J., Sturm, J., Cremers, D.: Camera-based navigation of a low-cost quadrocopter. In: IEEE/RSJ International Conference on Intelligent Robots and Systems (IROS), pp. 2815–2821. IEEE, October 2012
18. Ross, S., Melik-Barkhudarov, N., Shankar, K.S., Wendel, A., Dey, D., Bagnell, J.A., Hebert, M.: Learning monocular reactive uav control in cluttered natural environments. In: IEEE International Conference on Robotics and Automation (ICRA), pp. 1765–1772. IEEE, May 2013
19. Pestana, J., Sanchez-Lopez, J.L., Saripalli, S., Campoy, P.: Computer vision based general object following for gps-denied multirotor unmanned vehicles. In: American Control Conference (ACC), pp. 1886–1891. IEEE, June 2014
20. Horie, D.: U.S. Patent Application No. 10/413,662. (2003)
21. Zhan, C., Duan, X., Xu, S., Song, Z., Luo, M.: An improved moving object detection algorithm based on frame difference and edge detection. In: Fourth International Conference on Image and Graphics, ICIG 2007, pp. 519–523. IEEE, August 2007
22. Yilmaz, A., Javed, O., Shah, M.: Object tracking: a survey. ACM Comput. Surv. (CSUR) **38**(4), 13 (2006)
23. Chen, L., Wei, H., Ferryman, J.: A survey of human motion analysis using depth imagery. Pattern Recogn. Lett. **34**(15), 1995–2006 (2013)
24. Lao, W., Han, J., De With, P.H.: Automatic video-based human motion analyzer for consumer surveillance system. IEEE Trans. Consum. Electron. **55**(2) (2009)
25. Ganapathi, V., Plagemann, C., Koller, D., Thrun, S.: Real time motion capture using a single time-of-flight camera. In: IEEE Conference on Computer Vision and Pattern Recognition (CVPR), pp. 755–762. IEEE, June 2010
26. Ji, X., Liu, H.: Advances in view-invariant human motion analysis: a review. IEEE Trans. Syst. Man Cybern. Part C (Appl. Rev.) **40**(1), 13–24 (2010)
27. Hasler, N., Rosenhahn, B., Thormahlen, T., Wand, M., Gall, J., Seidel, H.P.: Markerless motion capture with unsynchronized moving cameras. In: IEEE Conference on Computer Vision and Pattern Recognition, CVPR, pp. 224–231. IEEE, June 2009
28. Junior, J.C.S.J., Musse, S.R., Jung, C.R.: Crowd analysis using computer vision techniques. IEEE Sig. Process. Mag. **27**(5), 66–77 (2010)
29. Moeslund, T.B., Hilton, A., Krüger, V.: A survey of advances in vision-based human motion capture and analysis. Comput. Vis. Image Underst. **104**(2), 90–126 (2006)
30. Moeslund, T.B., Granum, E.: A survey of computer vision-based human motion capture. Comput. Vis. Image Underst. **81**(3), 231–268 (2001)
31. Hager, G.D., Belhumeur, P.N.: Efficient region tracking with parametric models of geometry and illumination. IEEE Trans. Pattern Anal. Mach. Intell. **20**(10), 1025–1039 (1998)

32. Brox, T., Rosenhahn, B., Gall, J., Cremers, D.: Combined region and motion-based 3D tracking of rigid and articulated objects. IEEE Trans. Pattern Anal. Mach. Intell. **32**(3), 402–415 (2010)
33. Theobalt, C., Magnor, M.A., Schüler, P., Seidel, H.P.: Combining 2D feature tracking and volume reconstruction for online video-based human motion capture. Int. J. Image Graph. **4**(04), 563–583 (2004)
34. Yilmaz, A., Li, X., Shah, M.: Contour-based object tracking with occlusion handling in video acquired using mobile cameras. IEEE Trans. Pattern Anal. Mach. Intell. **26**(11), 1531–1536 (2004)
35. Fan, J., Xu, W., Wu, Y., Gong, Y.: Human tracking using convolutional neural networks. IEEE Trans. Neural Networks **21**(10), 1610–1623 (2010)
36. Cao, Z., Simon, T., Wei, S.E., Sheikh, Y.: Realtime multi-person 2D pose estimation using part affinity fields. arXiv preprint arXiv:1611.08050 (2016)

Author Index

Printed in the United States
By Bookmasters